Applications of Seasonal Climate Forecasting in Agricultural and Natural Ecosystems
By G.L. Hammer, N. Nicholls and C. Mitchell
ISBN 0-7923-6270-5

Corrigendum *figure 3. p.174*

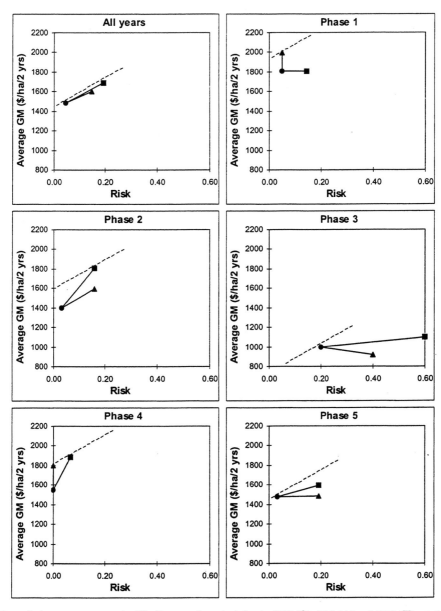

Figure 3. Average gross margin ($/ha/2 years of rotation) for the SFC (●), SSC (▲) and SCC (■) rotations plotted against a measure of risk (defined as the proportion of years where the accumulated 2-year gross margin was less than $500/ha) for all years and for subsets of years associated with each SOI phase. The dashed line represents the nominal farmer utility function adopted for this analysis.

APPLICATIONS OF SEASONAL CLIMATE FORECASTING IN AGRICULTURAL AND NATURAL ECOSYSTEMS

ATMOSPHERIC AND OCEANOGRAPHIC SCIENCES LIBRARY

VOLUME 21

The titles published in this series are listed at the end of this volume.

Contents

Section IV - Regional and National Scale Agricultural Decisions

Section V - Natural Systems

Preface

By now, nearly everyone has heard of El Niño and has some notion that the El Niño and Southern Oscillation (ENSO) system is a major effector of world climate. We have been made increasingly aware of the impacts of ENSO-related climate anomalies around the world. Developments in understanding of ENSO have introduced some skill in seasonal to inter-annual climate forecasting. Is this skill at all useful? Or do we just continue to observe the impacts, albeit with some notice of their likelihood? Is it possible to modify decisions in response to a skilful, but imprecise, seasonal climate forecast in a manner that leads to improved outcomes? This book is about the research and development struggle with this question and is why the word "applications" leads the title. This is what applications of seasonal climate forecasts are about.

Climate and its variability are major factors driving the dynamics of agricultural and natural ecosystems. There is a rich history and body of experience in research and development on these systems in Australia. Over the past two decades there has been significant and increasing attention given to applying seasonal climate forecasts in these systems. This attention has lead to identification of key issues and considerable learning about process. In this book, the focus is thus on agricultural and natural ecosystems in Australia but with the clear intent of seeking general insight about the processes involved in applying seasonal climate forecasts.

The motivation to produce this book was generated by a need to better document and communicate the ideas and understanding emerging from the body of Australian research and development on applications of climate forecasting. The combination of three critical factors –
- target systems that are influenced greatly by ENSO-related climate variability
- scientific disciplinary strength in agriculture, natural systems, and climate, and
- a research environment focussed on outcomes

has generated the requisite interdisciplinary systems approach in Australia. There are people in many countries in the world struggling with the same challenging issue of applying seasonal climate forecasts. It is hoped that the Australian experience documented here will provide valuable insight to inform those efforts.

The book is based on papers derived from a symposium conducted in Brisbane in November 1997. The symposium was held in response to the motivation to draw the Australian experience together in a forum that could summarise progress and discuss lessons and ideas. The Queensland Departments of Primary Industries and Natural Resources funded the symposium. While most of the chapters in the book are derived from the symposium, for completeness, a number have been invited subsequently.

The book synthesises the content of the symposium and is organised in six sections, each containing a number of chapters: (I) Introduction, (II) Seasonal climate forecasting, (III) Farm scale agricultural decisions, (IV) Regional and national scale agricultural decisions, (V) Natural systems, and (VI) Synthesis. The sections are designed to lead the reader through a logically structured sequence. Section I provides an awareness of relevance of climate variability and forecasting and reviews current understanding of world climate and the general concepts for applying seasonal climate forecasts in target systems. This introductory section sets the scene for the review of seasonal climate forecasting and the decision-making focus throughout the subsequent sections. Section II reviews current and potential approaches to seasonal climate forecasting and considers both statistical and dynamic modelling methods. The key issue of connecting dynamic climate models with dynamic agricultural models is also considered. Section III addresses agricultural applications at farm scale with two series of detailed examples on cropping and animal production systems. These papers highlight the key role of systems analysis, modelling, decision analysis, and decision-maker participation in the approach to applications of seasonal climate forecasts that has developed in Australia. The examples highlight the need for a participatory and interdisciplinary approach among decision-makers and scientists/analysts to bring about effective applications. The section ends with a consideration of the educative issues involved in building relevant skills among decision-makers. Section IV examines broader scale issues at regional and national level that involve agribusiness and government policy. While tools developed for this scale differ, the general approach to applying seasonal climate forecasts is consistent with the systems concepts detailed for the farm-scale decisions. Section V explores possibilities for applications in natural systems, such as invertebrate pest population systems, wildlife population systems, water resource systems, and vector-borne viral disease systems. The lesser knowledge of the underpinning dynamics of these systems provides a major challenge to effective applications of climate forecasts. Section VI seeks to synthesise lessons learnt from the body of work and, while addressing key issues of interaction among disciplines and between researchers and practitioners, promotes the concept of a culture of connectivity as a means to advance the effective application of seasonal climate forecasts.

It is hoped that *Applications of Seasonal Climate Forecasting in Agricultural and Natural Ecosystems - The Australian Experience* will serve as a valuable reference for practitioners, professionals and students around the world as they tackle the interdisciplinary challenge this topic presents. While the contents provide good lessons on how to, and how not to, proceed, it is clear that the journey has only just begun.

I wish to thank my co-editors Neville Nicholls and Chris Mitchell for their assistance in pursuing authors and reviewers and Mackey Vogler who provided invaluable support during the editorial process. These contributions have ensured that a quality product has resulted.

Graeme Hammer

List of Contributors

ALLAN, R., *CSIRO Division of Atmospheric Research, PMB1 Aspendale, Vic 3195, Australia*

ASH, A.J., *CSIRO Tropical Agriculture, Private Mail Bag, Aitkenvale, Qld 4814, Australia*

BANGE, M., *Cotton Research Unit, CSIRO Plant Industry, Narrabri, NSW 2390, Australia*

BATES, B.C., *CSIRO Land and Water, Private Bag PO, Wembley, WA 6014, Australia*

BROOK, K.D., *Queensland Department of Natural Resources, PO Box 631, Indooroopilly, Qld 4068, Australia*

BUTLER, D.G., *Queensland Department of Primary Industries, PO Box 102, Toowoomba, Qld 4350, Australia*

BUXTON, R., *CSIRO Division of Wildlife and Ecology, PO Box 2111, Alice Springs, NT 0871, Australia*

CANE, M.A., *Lamont-Doherty Earth Observatory of Columbia University, Route 9W, Palisades, New York 10964, USA*

CARBERRY, P., *Agricultural Production Systems Research Unit (APSRU), CSIRO Tropical Agriculture and Queensland Departments Primary Industries and Natural Resources, PO Box 102, Toowoomba, Qld 4350, Australia*

CARTER, J.O., *Queensland Department of Natural Resources, PO Box 631, Indooroopilly, Qld 4068, Australia*

CHAPMAN, S.C., *CSIRO Tropical Agriculture, 306 Carmody Rd, St. Lucia, Qld 4067, Australia*

CHARLES, S.P., *CSIRO Land and Water, Private Bag PO, Wembley, WA 6014, Australia*

CHIEW, F., *Department of Civil and Environmental Engineering, University of Melbourne, Parkville, Vic 3052, Australia*

CLEWETT, J.F., *Queensland Centre for Climate Applications, Queensland Departments Primary Industries and Natural Resources, PO Box 102, Toowoomba, Qld 4350, Australia*

CLIFFE, N.O., *Queensland Centre for Climate Applications, Queensland Departments Primary Industries and Natural Resources, PO Box 102, Toowoomba, Qld 4350, Australia*

COBON, D.H., *Queensland Department of Primary Industries, PO Box 102, Toowoomba, Qld 4350, Australia*

DANIELS, P., *CSIRO Australian Animal Health Laboratory, Private Bag 24, Geelong, Vic 3220, Australia*

DAY, K.A., *Climate Impacts and Natural Resource Systems, Queensland Department of Natural Resources, 80 Meiers Rd, Indooroopilly, Qld 4068, Australia*

DE HOEDT, G., *National Climate Centre, Bureau of Meteorology, PO Box 1289K, Melbourne, Vic 3000, Australia*

DROSDOWSKY, W., *Bureau of Meteorology Research Centre, PO Box 1289K, Melbourne, Vic 3001, Australia*

FREDERIKSEN, C., *Bureau of Meteorology Research Centre, PO Box 1289K, Melbourne, Vic 3000, Australia*

GEORGE, D., *Queensland Centre for Climate Applications, Queensland Departments Primary Industries and Natural Resources, PO Box 102, Toowoomba, Qld 4350, Australia*

HALL, W.B., *Climate Impacts and Natural Resources, Queensland Department of Natural Resources, 80 Meiers Rd, Indooroopilly, Qld 4068, Australia*

HAMMER, G., *Agricultural Production Systems Research Unit (APSRU), Queensland Departments Primary Industries and Natural Resources and CSIRO Tropical Agriculture, PO Box 102, Toowoomba, Qld 4350, Australia*

HIRST, A.C., *CSIRO Division of Atmospheric Research, PMB1 Aspendale, Victoria 3195, Australia*

HOCHMAN, Z., *Agricultural Production Systems Research Unit (APSRU), Queensland Departments Primary Industries and Natural Resources and CSIRO Tropical Agriculture, PO Box 102, Toowoomba, Qld 4350, Australia*

HUGHES, J.P., *Department of Biostatistics, University of Washington, Seattle, WA 98195, USA*

HUNT, B.G., *CSIRO Division of Atmospheric Research, PMB1 Aspendale, Victoria 3195, Australia*

IMRAY, R.J., *FarMarCo Enterprises (Aust.) Pty Ltd, Herries St, Toowoomba, Qld 4350, Australia*

JOHNSTON, P.W., *Sheep and Wool Institute, Queensland Department of Primary Industries, PO Box 3129, Brisbane, Qld 4001, Australia*

KLEEMAN, R., *Bureau of Meteorology Research Centre, PO Box 1289K, Melbourne, Vic 3000, Australia*

LIMPUS, C.J., *Queensland Parks and Wildlife Service, PO Box 155, Brisbane Albert St, Qld 4002, Australia*

LINDSAY, M., *Department of Microbiology, University of Western Australia, Nedlands, WA 6907, Australia*

MACKENZIE, J., *Department of Microbiology and Parasitology, University of Queensland, St. Lucia, Qld 4072, Australia*

MCINTOSH, P., *CSIRO Division of Oceanography, GPO Box 1538, Hobart, Tasmania 7000, Australia*

MCKEON, G.M., *Climate Impacts and Natural Resource Systems, Queensland Department of Natural Resources, 80 Meiers Rd, Indooroopilly, Qld 4068, Australia*

MCMAHON, T., *Department of Civil and Environmental Engineering, University of Melbourne, Parkville, Vic 3052, Australia*

MEINKE, H., *Agricultural Production Systems Research Unit (APSRU), Queensland Departments Primary Industries and Natural Resources and CSIRO Tropical Agriculture, PO Box 102, Toowoomba, Qld 4350, Australia*

NICHOLLS, N., *Bureau of Meteorology Research Centre, PO Box 1289K, Melbourne, Vic 3001, Australia*

O'REAGAIN, P.J., *Queensland Department of Primary Industries, PO Box 976, Charters Towers, Qld 4820, Australia*

O'SULLIVAN, D.B., *Queensland Centre for Climate Applications, Queensland Departments Primary Industries and Natural Resources, PO Box 102, Toowoomba, Qld 4350, Australia*

PARTRIDGE, I.J., *Queensland Centre for Climate Applications, Queensland Departments Primary Industries and Natural Resources, PO Box 102, Toowoomba, Qld 4350, Australia*

PAULL, C.J., *Queensland Centre for Climate Applications, Queensland Departments Primary Industries and Natural Resources, PO Box 102, Toowoomba, Qld 4350, Australia*

PIECHOTA, T., *Civil and Environmental Engineering Department, University of California, Los Angeles, California, USA*

PLANT, S., *Samarai Farming & Grazing Company, MS 918, Toowoomba Qld 4352, Australia*

SAAL, R.L., *Queensland Centre for Climate Applications, Queensland Departments Primary Industries and Natural Resources, PO Box 102, Toowoomba, Qld 4350, Australia*

SCANLAN, J.C., *Robert Wicks Research Centre, Queensland Department of Natural Resources, PO Box 318, Toowoomba, Qld 4350, Australia*

SMITH, I., *CSIRO Division of Atmospheric Research, PMB1 Aspendale, Victoria 3195, Australia*

STAFFORD SMITH, M., *CSIRO National Rangelands Program, PO Box 2111, Alice Springs, NT 0871, Australia*

STEPHENS, D.J., *Spatial Resource Information Group, Agriculture Western Australia, South Perth, WA 6151, Australia*

STONE R.C., *Queensland Centre for Climate Applications, Queensland Department Primary Industries, PO Box 102, Toowoomba, Qld 4350, Australia*

SUTHERST, R.W., *CSIRO Entomology, Long Pocket Laboratories, PMB No 3, Indooroopilly, Qld 4068, Australia*

WHITE, B., *Land and Water Resources R&D Corporation, PO Box 916, Indooroopilly, Qld 4068, Australia*

ZHOU, S., *Department of Civil and Environmental Engineering, University of Melbourne Parkville, Vic 3052, Australia*

THE IMPORTANCE OF CLIMATE VARIABILITY AND SEASONAL FORECASTING TO THE AUSTRALIAN ECONOMY

BARRY WHITE

Co-ordinator, Climate Variability in Agriculture R&D Program
Land and Water Resources R&D Corporation
PO Box 916, Indooroopilly, Qld 4068, Australia

Abstract

The underlying source of climate variability affecting the Australian economy is the fluctuations in agricultural production related to rainfall. Drought in the grain and extensive grazing industries is the prime contributor. As an introductory chapter in a volume on the Australian experience in applying seasonal climate forecasts, this chapter's role is firstly to provide context by describing the major impacts of climate variability on the agricultural sector. The importance of the impacts from the perspective of the Australian economy is then reviewed. Production variability in agricultural industries and regions is described together with the flow-on impacts to the economy generally. The 1994-95 drought, which had large impacts on farm output and exports, is used as an example. Notwithstanding the relatively small size of the farm sector, these impacts can have significant but infrequent flow-on effects to the rest of the economy, and on major macro-economic aggregates such as economic growth and the volume of exports. The strong linkages between the farm sector and the rest of the economy, and no significant dampening of agricultural instability by the non-farm sector, both contribute to the impacts of climate variability on the economy.

Seasonal climate forecasts are being increasingly used to benefit decision-making in the more climate-sensitive sectors of the economy. The second role of the chapter is to provide a broad research context for applications of seasonal forecasting to manage risk arising from climate variability. Farmers are the major group of potential users, and they have identified more confident use of forecasts as a priority for research. Extensive but neglected research on how risky decisions are made is reviewed to look for opportunities for alternative ways of presenting probability information. The relatively recent impact of probabilistic thinking on human affairs suggests that the concepts are not intuitive. Alternatives could take account of research on biases in intuitive approaches, and thus contribute to greater confidence in the use of seasonal forecasts. The paper concludes that the communication aspects of seasonal climate forecasts warrant greater priority if the potential importance of the forecasts in improved risk management is to be realised.

1

G.L. Hammer et al. (eds.), The Australian Experience, 1–22.

1. Introduction

The underlying source of climate variability affecting the Australian economy is the occasionally substantial fluctuations in agricultural production due to rainfall variability, notably major droughts. The importance of climate variability and seasonal forecasting to the Australian economy can therefore be restricted to two distinct perspectives. The first is the flow-on effects of the fluctuations in agricultural production to the rest of the economy. If these are significant, seasonal climate forecasts may have some value in macro-economic management, and benefit the community generally. The second perspective is the importance within agriculture, and this is essentially the benefits to individuals from applying seasonal forecasts for the improved management of the risks arising from climate variability. (Note that the perspectives are not seen as linked - even widespread adoption of seasonal forecasts by diverse decision-makers is unlikely to have a consistent or significant impact on either aggregated production or the flow-on effects.)

As an introductory chapter in a volume on the Australian experience in developing and applying seasonal climate forecasts, the chapter first provides a context by describing the more prominent impacts of climate variability on the agriculture sector and their importance from the perspective of the Australian economy. The regional and industry impacts of climate variability in the farm sector of the economy are summarised in the first part of the paper. Drought is the dominant and recurring feature in its impact on production and on the viability of farm businesses. Variability in prices of major commodities also needs to be considered to place climate variability in a broader risk management perspective. Flow-on effects to the Australian economy from variability in the farm sector have been the subject of a number of studies. The importance of the flow-on effects depends on the relative contribution of the farm sector to the national economy, the strength of linkages, and also on the relative variability of fluctuations in the farm sector compared with the rest of the economy.

The second and major focus of the paper then shifts to a broad and inter-disciplinary context to help shape and interpret research on applications of seasonal forecasting. The emphasis is on understanding and managing risk arising from climate variability. Seasonal forecasts are increasingly being used to benefit decision-making in the more climate-sensitive sectors of the economy, especially in agriculture. Farmers have identified research to enable more confident use of forecasts as a priority. Whilst improved forecasting skill will clearly contribute to more confident use, there is also ample scope to improve communication of existing forecasts. The scope includes research on valuing forecasts to demonstrate and promote potential applications in decision making, and opportunities to learn from the extensive research on biases in the use of probabilities. The research on biases might have predicted some of the confusion that accompanied the unfolding of the El Niño event of 1997. The final section of the paper presents some alternative approaches to presenting forecasts to better communicate probability information.

2. Volatility in Value of Farm Production

The shaping of the fundamentals of industry location and production volatility by climate variability has been the subject of many studies. Widespread droughts have been the catalyst for some of the more sporadic research. Campbell (1958) described the limited progress in coming to terms with climate variability - ' the most insidious and pervasive characteristic of our climate'. The comprehensive review of climate variability by Anderson (1979) was structured around a framework from farm to national and included responses to climate variability by farmers and by governments. The review highlighted the decline in the economic impacts of climate variability as the level of aggregation increased from farm to industry to national level. Drought experience and production risk have shaped many of the government policies relating to rural adjustment, and to economic efficiency and welfare in the farm sector (DPIE, 1997).

The Australian farm sector is extensive, and has a degree of diversity by regions and by industries, which might be expected to dampen much of the broader impact on the economy of specific and scattered shocks such as droughts. But some climatic influences such as ENSO (El Niño-Southern Oscillation) are near global-scale. As a result, droughts and good seasons tend to be widespread, and more continental than local phenomena. Gibbs and Maher (1967) used calendar year rainfall in the first decile as the criterion to define and map the spatial patterns of drought in Australia. The analysis was based on point rainfall from 1885-1965. The criterion gave a good correspondence with major widespread droughts as determined, for example, by media reports. On the basis of the spatial extent of the first decile range, more than a quarter of the continent was in drought in the 10% of years with the greatest extent of drought. For a geographically small state such as Victoria, the corresponding area was more than 60%. (The general limits are 10% of an area in the first decile range if spatial correlation was zero, and 100% with perfect correlation.) Thus, droughts have been widespread, spatial phenomena.

The spatial extent of droughts and favourable seasons is sufficient to have an occasional and major impact on farm production at the national level. Nicholls (1985) showed the scope to predict the gross value of production of Australian crops using an ENSO indicator. Over the period 1964 to 1983 over half the variance in gross value of crop production was attributable to sea surface temperature variations to the north of Australia.

In terms of diversity by major industry, crop and livestock sectors each typically account for about one half of the gross value of production. Crop and livestock production also respond on different time scales to drought and to favourable seasons. Crop yields are impacted more immediately than wool or meat production. Beef cattle slaughtering can increase in drought, resulting in lagged effects on future production. For the major industries, most variability in the volume of production in the short term is associated with rainfall variability. Whilst irrigation is an important contributor to stabilising production in some regions and industries, the contribution of irrigated agriculture to total farm value of production is only of the order of 20%.

Three major commodities, wheat, wool and beef, typically account for about half of the value of farm production. For the farm sector to have a significant impact on the rest of the economy, there needs to be a substantial fluctuation in either production or prices for at least one of the three commodities. Table 1 analyses the major sources of variation by presenting the two years with the largest increases, and the two with the largest decreases, in the gross value of farm production over the last two decades. Three of the four years of greatest year-on-year change have an exceptional wheat yield as a major contributing factor. A decline in wool price was the major factor in 1990-91. The upswing in beef price combined with increased wheat yields in 1978-79 resulted in an exceptional 43% increase in the gross value of farm production. Changes in Table 1 are in gross value and are smaller relatively than the impacts on the net value of farm production, a better indicator of the financial and social impact of boom and bust within the farm sector. The impact of the decreases is also greater when compared with the average annual growth of gross value over the period of about 6%.

TABLE 1. Years of greatest year on year change in gross value of Australian farm production and the main contributing factors - 1977-78 to 1996-97 (adapted from ABARE (1996)).

Change on previous year		Main contributing factors
Major annual increases		
1978-79	+ 43%	Value of livestock slaughtering up 58%
		Wheat yield up 88% from low base
1983-84	+ 32%	Drought recovery
		Weat yield up 120% from low base
Major annual decreases		
1982-83	- 8%	Widespread drought
		Wheat yield down 44%
1990-91	- 10%	Wool prices down over 25%

An indication of the relative contributions of price and production volatility to overall volatility in the gross value of agricultural production can be determined using indices of the volume of production and of prices received by Australian farmers. As shown in Table 2, after allowance is made for trends in the series, the volume of production appears slightly more volatile than prices. For the wheat industry, it has been shown that average temporal variability of national yields and of prices were similar at about 35% (Scoccimarro *et al.*, 1995). This is consistent with a broader and earlier analysis (Anderson 1979), which showed that about 40% of the temporal variation in the net value of agricultural production was attributable to climate risk, mainly rainfall-related.

Compensatory aspects, for example higher prices resulting from lower production in drought, are a further factor to consider for some commodities. There is considerable scope for research on ENSO-related impacts on prices, for example building on the exploratory analysis of impacts on world grain yields by Garnett *et al.* (1992).

TABLE 2. Sources of volatility in gross value of farm production (1975 - 1995) (adapted from ABARE (1996)).

Item	Annual average change (mean absolute) %	Annual trend %
Gross value of production	10	6
Index of volume of production	7	2
Index of prices received	7	5

TABLE 3. Australian farm income variability for major broadacre industries and regions (adapted from Scoccimarro *et al.* (1995)).

Broadacre industry	Australian region		
	South-west	South-east	Northern
	(Coefficient of variation class, %)		
Beef	25-30	25-30	>30
Sheep	20-25	20-30	25-30
Crops-livestock	<20	20-25	25-30

The influence of ENSO on the degree and regional pattern of climate variability has been demonstrated by Nicholls *et al.* (1997) by comparing Australian variability with the global average (used as a standard) for the same average rainfall. Variability is generally much higher than the standard value, particularly in the northeast and in the centre of Australia. Lower values occur in the southwest and southeast. Broad adaptations to climate variability, for example in terms of industry location in relation to production stability, irrigation development and as a driver of agricultural research, have long been established. The national pattern of farm income variability is, not surprisingly, also influenced by the pattern of climate variability. An ABARE study (Scoccimarro *et al.*, 1995) analysed the relative variability of the Receipts/Costs ratio (a desirably non-negative income measure) for farms classified by major regions and industries. A summary of the study is presented in Table 3. The industry pattern within regions is for income variability to be least for farms classified as crops-livestock. Many of these farms are well-diversified enterprises based on wheat and sheep. Farms classified as sheep have an intermediate level of variability between crops-livestock and beef, the most variable. The industry pattern reflects in part the location of industries within regions. Cropping is generally located in relatively higher rainfall areas that enjoy lower relative rainfall variability than the more extensive livestock farms. Across regions, there is also a pattern of increase from lowest in the southwest to increasing in the southeast, and with the highest value in the northern region of Australia. This regional pattern is consistent with the rainfall variability pattern, and thus reflects the impact of ENSO on income variability. A similar regional pattern was apparent in variability of wheat yields and wheat prices.

The broad national patterns of Table 3 also reflect many other factors, such as the capacity of soils to store moisture, rainfall seasonality and the cropping systems that have evolved. Low variability of wheat yields in the southwest of Australia is associated with low rainfall variability for crops grown on soils of low moisture storage in a winter rainfall environment. By contrast in the northeast where summer rain is dominant, the system is a more opportunist and complex winter wheat regime. Wheat yields are more variable. Yields there are more dependent on a fallow period to store highly variable summer rainfall, although yield variability is buffered to some extent by soils of high moisture storage potential.

3. Economy-Wide Impacts

Impacts on the Australian economy of fluctuations in the farm sector's value of production have been the subject of occasional studies extending back to droughts in the 1960s. The impact has been important to national policy makers because of potential impacts on policy settings for employment and economic growth objectives. Anderson's (1979) review concluded that the impact of a major drought was a macro-shock likely to be a 1-2% decline in Gross National Product. The widespread drought in 1982-83 compounded a severe recession with economic growth of only 0.3%. Without the drought, non-farm output would have grown by 1.3% (O'Mara, 1987). These economy-wide impacts, although infrequent, are considerable given that the farm sector is diverse, and now accounts for only a small share of the economy. They arise from both lower non-farm volatility and the substantial flow-on effects from the farm sector. The farm sector directly contributes about 20% of total exports, and the manufacturing sector includes major agricultural processing industries, which also contribute to exports.

In recognition of the farm sector's potential to be a source of short-term volatility in the Australian economy, ABARE analysed the impact of the 1994-95 drought on the Australian economy using the ORANI-E model (Hogan et al., 1995). Effects of drought were calculated from the production and prices estimated to have applied in the absence of the drought, compared with the forecasts for 1994-95 that included estimated drought impacts. The impact of the drought was estimated to reduce gross value of farm production by $2,400 million or 9.6%, a similar percentage decline to 1982-83. Net value of farm production and farm exports had declines of the order of $2,000 million.

The ABARE study used the forecast drought impacts in 1994-95, including for example a 47% decline in grain production, a 5% decline in wool production, and a 2% increase in beef production, to simulate the economy-wide impacts in the same financial year. Some of the lagged effects of drought in the rural sector were captured by continuing the analysis for 1995-96. This assumed a further 3% decline in wool production and a 10% decline in beef production. Table 4 summarises the economy-wide impacts. Consistent with previous major droughts, the impact on the national economy was about 1%. Impacts were substantial in industries involved in processing of agricultural products or in supply of inputs. The small impact of the drought on the mining industry was related in the simulation study to the appreciation of the Australian dollar of 2%. However, the ABARE study noted that the medium-term impact of the drought was likely to be a depreciation of the Australian dollar related particularly to increased debt repayments due to the drought. Other minor impacts of the drought included a greater reduction in employment for unskilled labour and increased inflation.

TABLE 4. Economy-wide impacts of the 1994-95 drought (Hogan et al.. (1995)).

Item	1994-95	1995-96
Total output ($ million)	-4,800	-1,800
Gross domestic product (%)	-1.1	-0.4
Employment (%)	-0.6	-0.3
Volume of exports (%)	-6.3	-2.6

McTaggart and Hall (1993) provide a further example of the significance of variations in farm output on the national economy. This study evaluated foreign (USA growth in GDP) and domestic disturbances (monetary policy and export supply) to determine their relative importance in shaping Australian economic growth and by implication, macro-economic policies. As a proxy for an aggregate export supply variable, the study used the SOI (Southern Oscillation Index) lagged by one quarter. The analysis concluded that 'the simple intuition of Australia as a small open economy in large world markets vulnerable to the vagaries of the world trade and the weather, proved correct'. Dominant influences on Australian economic activity were concluded to be economic growth in the USA and the SOI.

The studies on the economy-wide impacts of the major and infrequent droughts (such as 1982-83 and 1994-95) demonstrate the potential importance of seasonal climate forecasts in economic policy development. For example, with a forecast of major drought, economic growth would be likely to be about one percent less than expected. As a component of broader macro-economic policy development, this could imply a relaxing of monetary policy to maintain growth targets.

4. Increasing the Benefits from Seasonal Forecasting

The first part of the paper has demonstrated the historical impacts of climate variability on production instability in the agricultural sector and the occasional significant flow-on effects in the national economy. The following sections have the more difficult tasks of looking forward to how to manage the risks, and the challenges in communicating the potential benefits from the application of seasonal climate forecasts.

Although there is currently a high level of awareness of seasonal climate forecasts, or at least of El Niño, through for example regular media exposure, there is little consolidated information on how forecasts are being applied in decision making. Seasonal climate forecasts are beginning to be used more widely to benefit decision-making, particularly in the more climate-sensitive sectors of the economy. Farmers are potentially the major users. They have identified research to enable more confident use of forecasts as a priority (LWRRDC, 1998). The review of the Rural Adjustment Scheme (DPIE, 1997) stated that seasonal forecasts have not yet had a great impact on the risk management practices of Australian farmers. Poor links between the information available and on-farm decision making was seen as the principal reason.

The application of seasonal climate forecasts is a recent and rapidly evolving field of research. Many of the applications have been in northeastern Australia, a region of extreme ENSO impact, and a region with an exceptional number of droughts during the 1990s. Developments in forecasts based on sea surface temperature (Drosdowsky and Chambers, 1998) will add a new dimension with their potential for improved skill across southern Australia.

The skill in seasonal climate forecasting developed in the last decade is giving farmers new tools to manage some of the range of risks facing their businesses. Anderson's (1979) review recognised the ease with which the historical impacts of climate variability can be studied compared with research on how decision-makers might respond to manage the risk. The review also predicted that 'prospects for long-term forecasting of rainfall are so bleak that Australian farming will continue to be unstable and risky', which at a minimum, confirms the recency of research in this area.

As part of National Drought Policy aimed at improved self-reliance of farmers in managing climate-related risks, the Commonwealth Government has since 1992 been funding a research program currently named the Climate Variability in Agriculture R&D Program managed by the Land and Water Resources R&D Corporation. Recently the first full review of the program was undertaken (Hassall and Associates, 1997) to provide feedback to lay the foundations for further phases of the program. CVAP has been involved in much of the national effort on applications of seasonal climate forecasting in particular rural industries and in more generic research on improved forecasting skill. Many of the rural R&D Corporations have been involved as partners in the program. For projects that are industry-specific, there is a degree of concentration in grain and extensive grazing industries. In the previous sections, climate variability was shown to have a significant impact on production and income, particularly in the broadacre grain and grazing industries. There is also one major project directly relevant to the use of seasonal forecasts in water resources management. In geographic terms, the program has had activities relevant to most Australian agricultural regions and this has been achieved by a mix of projects with either regional or national objectives.

The CVAP objectives recognise that progress in applying forecasts depends on a comprehensive and integrated approach, and close interaction between researchers and potential users. The program review (Hassall and Associates, 1997) recognised that due to the short time frame since the program began, much of the applied research undertaken had yet to have a major impact on management decisions. More rapid adoption by farmers, the major user group, was seen to depend on their greater involvement in research. Further, research needed to be more focused on increased confidence in the use of forecasts, and demonstrations of the value of forecasts relevant to the decisions of individual farmers. Two neglected areas are taking advantage of forecasts of favourable seasons and improved management of natural resources. For example, Nicholls (1991) has interpreted some ENSO characteristics relevant to sustainable management of Australian soil and vegetation resources. Accordingly, the final four sections of this chapter will explore opportunities generally for research to achieve more effective adoption of current and emerging skill in seasonal climate forecasting.

5. Case Study on the Impacts of the 1997 El Niño on the Dairy Industry

As a result of growing concern on potential impacts on the Victorian dairy industry of the then developing El Niño event, a study was done over the period August to October 1997 to determine possible strategies (VCG Australia, 1997). The study reported on farmers' evolving responses to the El Niño event, some of the stress and confusion resulting in part from 'the almost universal media anticipation of catastrophic drought', and reactions of the sceptical following above average 1997 spring rainfall in some regions. Analogue El Niño years (13 since 1940) were used to determine potential impacts, particularly on pasture production, using a pasture simulation model. Increased feed requirements to maintain production were then determined and farm level economic analysis undertaken for a variety of situations. In comparison to the grain and extensive grazing industries, there was little impact expected on production because of the profitability of purchasing feed to maintain milk production. The 'Expected El Niño' scenario, based on the average of the 13 El Niño analogue seasons, would require additional feed purchases of $193 million or an average of $14,200 per farm.

The study, recognising that dairy farms were making little use of formal approaches to risk management, therefore used simple and introductory approaches to compare possible strategies in response to potential drought. Table 5 shows the simplest and standard approach for conveying the essentials of the decision-making process. The value of an alternative (Take Action) strategy to purchase or grow the additional feed required to maintain production is determined for two possible season types. The No-

Change strategy has a wider spread of possible cash flow change compared with the alternative of taking action. In the nature of insurance, the alternative action has a cost if the year turns out normal, but avoids some of the major loss if the outcome is an El Niño season. Based on the probabilities considered to operate in spring 1997, the expected value of the No Change strategy is $13,200 [(0.4 X $0) + 0.6 X $22,000)] less than under normal long-term weather circumstances. This compares with a reduction of only $9,200 [(0.4 X $5,000) + (0.6 X $12,000)] if the Take Action strategy is implemented. Thus, under the assumptions (and at least for a risk-indifferent decision maker), the Take Action strategy would be favoured in terms of the expected value of the pay-off for that season.

TABLE 5. Cash flow change outcomes for alternative strategies and seasons for a Victorian dryland dairy farm (adapted from VCG Australia, 1997).

Strategy	Cash flow change outcome ($)	
	Normal season (40% probability)	El Niño season (60% probability)
No change	0	-22,000
Take action (additional feed)	-5,000	-12,000

The presentation of possible outcomes in the dairy industry study recognised that risk and risk management are complex, difficult to communicate, and often poorly understood concepts in a formal sense. Acknowledged too, was that by and large farmers obviously do engage in a level of risk management, largely informal, but clearly sufficient for the majority of farm businesses to survive and grow.

6. Valuing Seasonal Climate Forecasts

The prospect of increased profit was seen by Ridge and Wiley (1996) as the practical key for unlocking an improved decision-making process based on seasonal forecasts. The potential for change was seen to depend more on the transparency of the argument than the sophistication of the analysis. The challenge is to cut through the complexity to develop robust approaches that can be communicated to decision-makers. The two ingredients for using a seasonal climate forecast in decision-making are (a) exploring outcomes for alternative decisions based on the forecast of the likelihood of possible outcomes, and (b) a comparative evaluation of the alternatives in terms of the decision maker's goals. The two under-pinning tools of choice from economic theory of information value are Bayesian revision of the probabilities and the expected utility framework. Johnson and Holt (1997) and Wilks (1997) have reviewed applications of the theory in valuing weather information. The principles also apply to seasonal climate forecasts. Many of the case studies reviewed have been in agriculture, covering decisions that are relevant to a large number of farmers. For example, Mjelde *et al.* (1988) valued seasonal forecasts for decisions on fertiliser amounts and timing for sowing corn. For simplicity, many studies have assumed decision-makers to be indifferent to risk. Maximising expected profit without concern for its variability is then the goal.

Marshall *et al.* (1996) in one of the few comprehensive Australian studies on value, used recursive stochastic programming to evaluate SOI-based forecasts in decision-making. The evaluation was based on the crop simulation study by Hammer *et al.* (1996) of the value of SOI phases in wheat crop management in southern Queensland. In decisions relating to nitrogen applications, cultivar choice and whether to harvest grains or graze, seasonal forecasts had mean values from $3.52 to $3.83 per hectare per year. These values were shown to be a little less than the benefits from a new wheat variety of $4.73. Stafford Smith *et al.* (1997) have evaluated SOI forecasts in grazing systems in northern Australia. The simulation studies coupled a pasture growth model, a model of a beef cattle herd, and a financial model to derive optimal stocking strategies. The studies suggested limited value of the forecasts for the simulations undertaken, but improved forecast skill had considerable potential to improve economic returns and to help protect the resource base.

Such studies are essentially optimising and prescriptive, and even further constrained by how well the model describes the system. They show potential value of forecasts for idealised users who respond according to Bayesian decision theory for example. Alternatively, descriptive studies can potentially show how users actually make decisions. Stewart (1997) argues for complementary and even convergent use of both approaches, whilst recognising that few descriptive approaches have gone beyond surveys and case studies. Further, no studies were located which had explored insights into the use and value of forecasts by comparing prescriptive and descriptive approaches. There is also unrealised scope for descriptive approaches to model and better understand the decision process using established tools from research on judgement and decision analysis.

Case studies on the value of forecasts have a primary role in demonstrations of value to potential users. The value of forecasts needs to embrace a range of possible benefits, including increased profitability and reduced risk to the enterprise and its resource base. Value studies also have an important role in guiding allocation of resources throughout the system from acquisition of climate data to extension of forecasts to the user. Although there is an extensive agricultural research literature on decision making under risk, there are few studies on research evaluation that also include the benefits from reduced income variability (Harrison *et al.*, 1991). The shaping of research priorities, the developing of a more general understanding of systems where forecasts are likely to have value, and feedback from users are related aspects that can benefit from a stronger emphasis on research evaluation. The wheat example (Hammer *et al.*, 1996) assumed monitoring of soil moisture and nitrogen at planting. Thus, the value of forecasts is likely to be dependent on monitoring systems being in place. In a water resource system, a low or no risk water release policy will constrain the value of a forecast whilst in extensive grazing operations, the frequency of mustering is likely to be a constraint. As case studies accumulate, it should be possible to use the experience to develop greater confidence in more intuitive assessments of where forecasts add value in changing farm management decisions.

One constraint on the scope for case studies is the development costs of the simulation models required for a rigorous comparison of outcomes from alternative decisions. The major research effort in developing models of cropping and grazing systems in northern Australia has been influenced by greater climate variability, and also the greater contribution from soil moisture in soils of high storage capacity. As a result, there is some understanding of interactions between research treatments and rainfall and there is potential to vary inputs to take advantage of seasonal forecasts. In contrast, agronomic research in lower variability areas of southern Australia has been able to rely on traditional approaches with replication by sites and seasons to simply uncover the mean response to inputs. The southern Australian approach may have also discriminated against risky inputs if there is a high variability in response. Opportunities to vary inputs to take advantage of seasonal forecasts may then be less clear.

The inherent predictability, a few months ahead, of many agricultural and hydrological systems compared to climate systems, also needs to be accounted for in studies of forecast value. Soil moisture storage for example can buffer the variability in agricultural production. A simulation study for a soil of high moisture capacity at Dalby in southern Queensland showed that 56% of the variation in wheat yields was accounted for by soil moisture at the time of planting (Berndt and White, 1976). This degree of predictability, which arises from persistence via the storage characteristics of the agricultural system, could be the base rate for comparison in the assessment of the value of a seasonal climate forecast for a relevant decision. For example, the study of Hammer et al. (1996) showed additional value of the forecast given knowledge of soil moisture at the time of planting.

The review by Wilks (1997) lists 25 studies, mostly in the United States, on weather and climate forecast value. The review demonstrated that the required decision-analytic tools can be applied to a wide range of real-world decisions. More emphasis has been put on research valuing forecasts to determine public investment in forecast services than on research to develop user confidence and increase adoption of the use of forecasts. Such an emphasis could be expected in a developing field of research. Many of the studies are complex involving several disciplines. Future studies in this vein are likely to be constrained to more general decisions relevant to a large number of users. Anderson (1991) considered cost to be a decreasing constraint as computing costs decrease, although more complex analyses remain as an alternative response to cheaper computing. Stewart (1997) recognised the opportunity for a stronger user perspective building on research on biases in judgement and decision-making, and on the role of intuition.

Various forms of cost/loss models, similar to the dairy industry example outlined earlier, were used by Katz and Murphy (1997) to explore theoretical relationships between quality and value of forecasts. Cost/loss models have been widely used to evaluate decisions of the 'umbrella problem' type. The decision is whether to incur a protection cost or alternatively withstand a loss if the weather outcome is adverse. The cost is incurred whatever the weather. One general result from both the simpler and more complex prototype studies was the existence of a threshold for quality below which the forecast value was zero. In the simplest case, the threshold depends on the cost/loss ratio and the adverse weather probability as determined by climatology. Katz and Murphy suggest the threshold may explain why long-range forecasts of low quality are apparently ignored by many decision-makers.

The cost/loss approach can also be simplified, and expressed as rules of thumb of some generality and intuitive appeal. For many decision-makers, the two seasonal outcomes of normal or drought are appropriate and a sufficient disaggregation of season types. The decision dilemma is often between continuing with a normal strategy or choosing an alternative. When drought is a possible outcome, one alternative could be reducing an input that is unlikely to be profitable in a drought. The cost savings from reduced inputs if a drought eventuates can be compared with the loss in profit if the season is normal. The cost/loss ratio determines the break-even odds. This approach, while traditional and straightforward, may be difficult to communicate because of the probability concepts. However, some conditional forecasts (as in the dairy El Niño example) will on occasions show a significant probability of drought in some regions. If the probability is not too far from an even chance then a simple, even obvious, rule of thumb can be used as follows. If there is a good chance of a drought, it would be worth further considering reducing inputs that show little response in a drought year, provided the cost saving is similar to the increased profit from the input in a normal year. Similar rules can be developed for evaluating other decisions, for example extra or reduced inputs for a favourable season or for insurance type strategies.

7. Decision Research

In this section, farmer decision-making with seasonal climate forecasts will be first considered in the broader contexts of decision research generally, and in relation to probability concepts. The review of research suggested imbalance between prescriptive approaches to the value of forecasts, compared with the more neglected and equally challenging area of descriptive studies on how users actually make decisions. The previous dairy industry example is likely to be representative in showing reliance by farmers and undoubtedly the community generally on less formal and therefore intuitive approaches to decision-making.

Extensive research in judgement and decision making generally (Dawes, 1988) has shown the superiority of even simple formal models over expert intuition (obviously only in cases where the outcomes can be compared). This finding is as consistent as it has been ignored. More positively, the research has shown that experts do excel in selecting the best criteria but not in their integration of the criteria. On that basis, good starting points are to at least support intuitive decision-making with a comprehensive decision framework, and to recognise that breadth is more important than depth.

The preceding is well aligned with Malcolm's (1990) defence of the continuing relevance and validity of simple, but comprehensive budgets, in farm management. The approach was seen to have stood tests of time and markets, despite the then largely unfulfilled promises of narrowly focused and more complex methodologies, for example incorporating systems simulation and expected utility analysis.

Another major body of relevant research on decision-making, largely within psychology, is on heuristics and biases as developed since the 1970s by Kahneman and Tversky (1996). The research has demonstrated that heuristics or 'rules of thumb' that violate the rules of probability theory are in widespread use. One consequence is decisions that are irrational in a fundamental sense - they contradict rules the decision-maker accepts. Expected utility is thus challenged in that it does not consistently describe how decisions are made or might best be made. The major sources of heuristics and biases influencing decisions are grouped under illusions relating to representativeness (what is similar), availability (what is easily recalled), and anchoring (what comes first). Dawes (1988) lists many related examples of various heuristics including the law of small numbers, base rate neglect and pattern illusions. Seeing patterns where none exist and missing patterns when they are present appear to be part of the human condition. Biases with compound events have been widely studied. The 'Linda' problem is a classic in psychology. An agricultural analogue would be asking participants the following question. Given that it is the beginning of summer and the Southern Oscillation Index is fluctuating, which one of the two following scenarios is more likely at the end of summer?
(a) there is a major El Niño event, or
(b) there is a major El Niño event associated with widespread drought.
Based on the Linda equivalent, a majority is likely to select the compound event (El Niño and drought). But the laws of probability decree a compound event to be less likely than one of its component events.

A further example involving Bayesian revision has been developed to show one frequent bias. The example draws on extensive experience in clinical diagnosis, broadly equivalent to forecasting. For major droughts with an assumed average frequency of occurrence of 10% of years and assuming (for illustration purposes) 70% accuracy in using an El Niño diagnostic test for the presence or absence of subsequent drought, what is the probability of drought given an El Niño? If experience in the failure of intuitive judgements in clinical diagnosis is a good analogy, then a majority of respondents to the above question are likely to produce estimates over 50% (Eddy 1982).

The answer (and using only Bayes Theorem) is 21% which compares to the base rate of 10% (a 50% probability of drought would require 90% accuracy). Dawes (1988) attributes this bias to confusion of the inverse, that is confusion with the probability of an El Niño given there is a drought (70%), which is effectively answering the wrong question correctly. The example exploited two frequent and related sources of confusion in communicating forecasts and their accuracy. Confusion of the inverse depends on the frequency of forecasts differing from the frequency of the event being forecast. (By most definitions, El Niño events are more frequent than major droughts.) The varying interpretations and alternative definitions of accuracy are also a common source of confusion and ambiguity. The El Niño example was artificially constructed to show how conditional probabilities can be over-estimated. A contrast is conservatism and inefficiency in revising probabilities given new information (Edwards 1968).

The above examples add to a pattern, which, at least for some, suggests a rather dismal view on the extent to which the mind functions as a competent intuitive statistician. Bernstein (1996) in documenting the remarkable history of how risk concepts evolved, has shown how probability theory is a relatively recent discovery. Many fields such as mathematics, insurance and gambling evolved well enough without an awareness of the concepts. Bernstein, in attributing the development of the principles of gambling and probability to Cardano in the sixteenth century, quotes Cardano's assessment - 'these facts contribute a great deal to understanding but hardly anything to practical play.'

The practicality and the generality of the research on biases in applying probability concepts have been challenged. For example Cohen (1981) argued that some of the experiments demonstrating bias simply test subjects on unusual (even conjured) tasks. Some tests may only be of intelligence and education, or even memory of Bayes Theorem. Biases are thus not necessary to explain difficulties in communicating some probability concepts. Economists have also and predictably been sceptical of the utility of laboratory research on biases where stakes and accountability are as low as they are unrealistic. Stewart (1997) in considering the application of decision research in weather forecasting, also cautions on the generality of the research, but with a hedge of not wishing to discount the possibility of flawed cognitive processes for coping with uncertainty. On the other hand, Gilovich (1991) is replete with real world confirmation of the significance of biases.

8. Domesticating Probability

The section title has its origins in the review by Daston (1987) on statistics and the domestication of risk in the insurance industry. During the last two centuries, the insurance industry was able to, in a sense, mature and reposition at a more comfortable distance from gambling with which it had been closely identified. Instead of a gamble on death insurance, the gamble became not investing in life insurance. Mathematically dignified probabilities replaced gambler's odds, which are calculated simply and intuitively from the relative frequencies of not winning and winning. (Ironically, gambling and lotteries are widely used instruments in research on biases in probability judgements.)

In clarifying the probabilistic revolution of the last century, Hacking (1987) provides a pertinent analogy, referring first to early textbooks on meteorology. Forecasters were then cautioned against forecasting an event as probable, on the basis that all forecasts are only probable anyway. Hacking then states: 'Now all is changed. When was the last time you heard a forecast not tricked out in probabilities?'. Nevertheless, for much of Australia, weather forecasts remain qualitatively hedged rather than probabilistic. The heuristics and biases of the previous section could be seen, in a sense, as some support for such a stance, although absence of probabilistic weather forecasts may well be hindering communication and understanding of probabilistic seasonal climate forecasts. For weather forecasts, Murphy (1991) has shown in the United States that users prefer, by a ratio of two or three to one, probabilistic forecasts over the more deterministic or qualitatively hedged forecasts they replaced. One argument that could be invoked in favour of probabilistic forecasts is the latitude in perception, and even the ambiguity, in terms such as 'probable'. In one survey (Lichtenstein and Newman 1967), where respondents were asked to specify an equivalent numerical probability, the terms 'probable' and 'possible' had mean associated probabilities of 0.71 and 0.37. The range of responses for each term was from 0.01 to 0.99, the maximum made available to the respondents.

Responses to the extensive body of research on heuristics and biases need not be dismal. Some of the challenges to the generality of the results actually help define approaches to minimise the biases. Research on debiasing has generated alternative approaches to presentation of information, which aid in understanding and creativity. Gigerenzer and Hoffrage (1995) have demonstrated that simply replacing probabilities with frequencies can substantially reduce bias. Feynman (1967) emphasised the stimulus to creativity from varied formulations of the same physical law and this reformulation underpins the frequency alternative. Plous (1993) in similar vein, reviews studies suggesting that simply, but explicitly considering alternative perspectives can be effective in debiasing. Miller (1985) presented a summary of biases and examples of their use and abuse in environmental judgements. He concluded that more attention should be given to their recognition and to minimising their detrimental effects. The exemplary record of weather forecasters in risk assessment and communication (National Research Council, 1989) formed the basis for a review by Monahan and Steadman (1996) on how analogies with meteorology might inform risk communication in mental health assessments. They concluded that there was a major imbalance in the effort expended on improved risk assessments compared with the equally challenging and creative research required to underpin effective risk communication.

As shown in a previous section on the dairy industry study, a common communication problem arises when a major El Niño event is occurring - 'the almost universal media anticipation of catastrophic drought'. This raises numerous other issues including media accountability and the extent to which the media leads or follows community views. A fundamental issue is the frequent dropping of the defining probabilistic terms by the media. Maybe this is (unintended) sympathy with the tradition of presenting weather forecasts in non-probabilistic terms. Whilst decision makers may, from long experience, be able to accommodate weather forecasts routinely delivered without probabilities, major El Niño forecasts are recent and rare.

The challenge for seasonal climate forecasting is to develop a richer variety of approaches that communicate information useful in decision making. As a starting point, an example is presented in Table 6 to take account of some of the biases detailed in previous sections. The presentation is in frequency format. For the simplest situation of a test to predict two outcomes, the format is the minimum able to convey the four possible outcomes. A frequency format is rarely used to show outcomes from seasonal forecasts. The example is for a forecast of summer rain at Emerald in Central Queensland using the SOI for the previous June and July. Emerald is in a region of strong ENSO influence (Partridge, 1994). The data are expressed in frequencies on a 100-year basis to reduce biases inherent in the use of probabilities (Gigerenzer and Hoffrage 1995). Following Simon (1969), the two digit integers may have the advantage over other representations of being retainable in immediate memory. The forecast could be applicable to management decisions on a summer crop. Based on farmer experience, a poor crop or drought might be more likely if summer rainfall is less than 350 mm. The base rate, or climatology, is 33 years in 100 (about 3 in 10) with summer rainfall less than 350 mm. Using an SOI < -5 as a predictor, the chances of a low rainfall year are 15 out of 25 (or 6 in 10), considerably more likely than the base rate.

TABLE 6. Summer (October to March) rainfall at Emerald (Qld) expressed as frequencies on a 100-year basis, in relation to the average SOI in the preceding winter (June-July). (adapted from Australian RAINMAN (Clewett et al., 1994))

Summer rainfall	SOI in preceding winter		
	SOI < - 5	SOI > - 5	Total years
> 350 mm	10	57	67
< 350 mm	15	18	33
Total years	25	75	100

A similar example for Emerald is presented in *Will It Rain* (Partridge, 1994) but using the percentage of summers with less than various amounts of rain, rather than frequencies, for various SOI classes. This form of presentation is widely used in tables and graphs to show how SOI classes or phases can have forecast skill. Shifts in the probability distributions have been emphasised, for example, Stone *et al.* (1996). In a similar manner, Table 6 could be re-presented in probabilities showing a 60% chance of a low rainfall year if the SOI is low, compared to only a 24% chance if the SOI is average to high; but this does not assist a decision-maker interested in outcomes and strategies for all seasons. A more complete and mature presentation is required when the purpose evolves and shifts from demonstrating forecast skill to communicating better understanding for decision-making. The decision-maker needs to know the number of misses (false negatives), which are the 18 low rainfall years when SOI was average to high. Also required are the 10 false alarms (false positives). The decision-maker's perspective is in accuracy in predicting outcomes, not the inverse.

A further bias that might be exploited relates to over reliance on recent experience- 'the mind is a prisoner of its recent past'. (Incidentally this bias could have unintended

18

advantages where trends exist.) The recent perspective, together with the value in using alternative presentations to remove bias, is the basis for the graphical presentation in Figure 1. The SOI relationship with summer rainfall is shown for only the last decade of the same data as summarised in Table 6. The decision maker can then more easily recall and visualise how the SOI forecast might have been applied each year recognising for example the point 93 (October 1993 to March 1994) as a false alarm with high rainfall associated with a low SOI. (Hindsight does incidentally reveal a rainfall of over 300 mm in March 1994.) With few exceptions, such as Australian Rainman (Clewett *et al.*, 1994), much of the information provided to decision-makers has been of current forecasts in a spatial format. For a decision-maker at a specific location, the recent performance of the forecast is likely to be a more relevant context. The technology is becoming available to make that feasible.

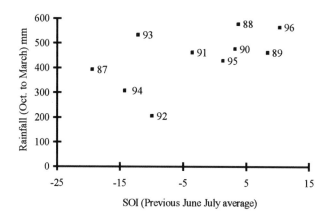

Figure 1. Relationship between the winter SOI (June-July average) and rainfall the following summer (October to March) for Emerald. The data labels are for winter. For example, the label 92 denotes the June-July SOI for 1992 and rainfall for the following summer, October 1992 to March 1993. (adapted from Australian RAINMAN (Clewett *et al.*, 1994)).

The graphical presentation in Figure 1 also suggests opportunities to vary the SOI criterion, for example to SOI < -10, even though there are fewer seasons meeting the stricter criterion. Would this increase the value of the forecast? There are likely to be fewer seasons correctly forecast with rain < 350 mm (even though the conditional probability of rain < 350 mm is likely to be higher than for SOI < -5). However, the reduced number of false alarms will compensate to some extent. Accuracy measures such as the proportions of true and false positives thus vary with the operating criteria. Given the benefits and costs of correct and incorrect decisions, the optimal criteria can be readily determined using the relative operating characteristic (ROC) approach. Despite its intuitive appeal in decision making generally, and the ease with which the optimal criteria can be communicated in a simple graph as shown by Swets (1992), the concept has rarely been applied. The concept does bridge the decision context from the safety of probability forecasts, which technically cannot be wrong, to risky decisions that can be. (Wrong decision here is in the common usage of a 'good' decision, which had a 'bad' outcome.) The elegance and power of the ROC are in the simple portrayal of the compensatory relationship between outcomes to define accuracy, and the demonstration that the optimal criterion is determined by value. The ROC approach appears to have considerable potential in understanding the links between accuracy and value, and in responsible disclosure to users of seasonal climate forecasts.

Conclusion

Widespread droughts with an average frequency of about one per decade have been the major impact of climate variability on Australian agriculture. Despite the diversity of Australian agriculture and adaptations that have evolved to buffer variability, the typical impact of a major drought has been of the order of a ten percent reduction in gross value of agricultural production. Although the farm sector is only a few percent of the national economy, lower non-farm volatility and the substantial flow-on effects from the farm sector can contribute to a significant impact on economic growth. Major droughts have typically reduced Gross National Product by about one percent and this can be significant given typical growth rates for the economy of up to a few percent. The potential to forecast major droughts offers opportunities to adjust macro-economic policies to reduce impacts on national economic objectives.

Although improved skill in seasonal forecasting is likely to be of benefit in a wide range of contexts, agricultural production and natural resource management will continue as the areas providing the most opportunities. Improved skill in southern Australia, using sea surface temperature forecasts, will open up a wide range of potential applications to complement SOI-based approaches being increasingly used to aid decisions in northeastern Australia. Communicating probability-based information in a form to encourage confident responses by decision-makers, is a continuing research challenge, and an integral component of a balanced research program. Research and extension to show the value of forecasts in specific applications is likely to be constrained by resources, the complexity of the analyses and the need for simulation models that can capture the biological and economic impacts of alternative decisions. Broader less intensive communication activities will also be required. These will need to draw on extensive research on decision-making in a range of disciplines. For example, biases in intuitive decision-making using probabilities may be avoided by using frequencies and alternative representations.

One broad but simple perception of forecasting skill is based on El Niño events being about twice as frequent as major Australian droughts. This alone suggests scope to place more emphasis on communicating forecasts in a broader decision making context to recognise and manage false alarms. The early emphasis has been to build on the exciting developments in ENSO understanding, and to convey that significant skill in forecasting is possible in some seasons and for some regions. Once decision-makers accept that there is a degree of skill, a more mature approach can be introduced to make false alarms and misses more explicit. Seasonal climate forecasting, and indeed this paper, can then avoid the Cardano conclusion: 'these facts contribute a great deal to understanding but hardly anything to practical play'.

Acknowledgments

Evolving experience in coordinating a national research program on managing climate variability in agriculture has provided a rich basis for this contribution. Interesting times with farmers pioneering applications of seasonal climate forecasts, and with a wide range of researchers spanning diverse disciplines, have provided the stimulus to further explore some of the issues that have emerged. The contributions of an anonymous reviewer, including other perspectives on biases in probability judgements, are gratefully acknowledged. The synthesis of all contributions and any residual biases are the author's responsibility.

References

ABARE (1996) Australian Commodity Statistics. Australian Bureau of Agricultural and Recourse Economics, Canberra.

Anderson, J.R. (1979) Impacts of climate variability in Australian agriculture: A review. *Review of Marketing and Agricultural Economics* **49**, 147-177.

Anderson, J.R. (1991) A framework for examining the impacts of climate variability, in R.C.Muchow and J.A. Bellamy (eds.), *Climatic risk in crop production: Models and management for the semi-arid tropics and sub-tropics*, CAB International, Wallingford, England. pp. 3-17.

Berndt, R.D. and White, B.J. (1976) A simulation-based evaluation of three cropping systems on cracking-clay soils in a summer-rainfall environment. *Agricultural Meteorology* **16**, 211-29.

Bernstein, P.L. (1996) Against the Gods: The remarkable story of risk. Wiley, New York.

Campbell, K.O. (1958) The challenge of production instability in Australian agriculture. *Australian Journal of Agricultural Economics* **2**, 3-23.

Clewett, J.F., Clarkson, N.M., Owens, D.T., and Abrecht, D.G. (1994) *Australian Rainman: Rainfall information for better management*. Department of Primary Industries, Brisbane.

Cohen, L.J. (1981) Can human irrationality be experimentally demonstrated? *The Behavioral and Brain Sciences* **4**, 317-70.

Daston, L.J. (1987) The domestication of risk: Mathematical probability and insurance, in L. Kruger, L.J. Daston, and M. Heidelberger (eds.), *The Probabilistic Revolution. Volume 1: Ideas in History*, MIT Press, Cambridge, Massachusetts. pp. 1650-1830.

Dawes, R.M. (1988) *Rational choice in an uncertain world*. Harcourt, Brace, Javonovich. New York.

DPIE (1997) Rural Adjustment-Managing Change. Department of Primary Industries and Energy, Canberra.

Drosdowsky, W. and Chambers, L. (1998) Near global sea surface temperature anomalies as predictors of Australian seasonal rainfall. BMRC Research Report 65, Bureau of Meteorology, Melbourne.

Eddy, D. (1982) Probabilistic reasoning in clinical medicine: Problems and opportunities, in D. Kahneman, P. Slovic, A Tversky, (eds.), *Judgement under uncertainty: Heuristics and biases*, Cambridge University Press, Cambridge, England. pp. 249-67.

Edwards, W. (1968) Conservatism in human information processing, in B. Kleinmuntz (ed.), *Formal Representation of Human Judgement*, Wiley, New York.

Feynman, R.A. (1967) *The character of physical law*. MIT Press, Cambridge, Massachusetts

Garnett, E.R. and Khondeker, M.L. (1992) The impact of large scale atmospheric circulations and anomalies on Indian monsoon droughts and floods and on world grain yields - a statistical analysis. *Agricultural and Forest Meteorology* **61**, 113-28.

Gibbs, W.J. and Maher, J.V. (1967) Rainfall deciles as drought indicators. Bulletin No 48, Bureau of Meteorology, Melbourne.

Gigerenzer, G. and Hoffrage, U. (1995) How to improve Bayesian reasoning without instruction: Frequency formats. *Psychological Review* **102** (4), 684-704.

Gilovich, T. (1991) *How do we know what isn't so: The fallibility of human reasoning in everyday life.* Free Press, New York.

Hacking, I. (1987) Was there a probabilistic revolution 1800-1930? in L. Kruger, L.J. Daston, and M. Heidelberger (eds.), *The Probabilistic Revolution. Volume 1: Ideas in History,* MIT Press, Cambridge, Massachusetts.

Hammer, G.L., Holzworth, D.P., and Stone, R. (1996) The value of skill in seasonal climate forecasting to wheat crop management in a region with high climatic variability. *Australian Journal of Agricultural Research* **47**, 717-37.

Harrison, S.R., Dent, J.B., and Thornton, P.K. (1991) Assessment of benefits and costs of risk-related research in crop production, in R.C. Muchow and J.A. Bellamy (eds.), *Climatic risk in crop production: Models and management for the semi-arid tropics and sub-tropics,* CAB International, Wallingford, England. pp 491-510.

Hassall and Associates (1997) Review of the National Climate Variability R&D Program. Report CV02/97. Land and Water Resources Research and Development Corporation, Canberra.

Hogan, L., Woffendon, K., Honslow, K., and Zheng, S. (1995) The impact of the 1994-95 drought on the Australian economy, in *Coping with Drought,* Occasional Paper No. 87, Australian Institute of Agricultural Science, Melbourne. pp. 21-33.

Johnson, S.R. and Holt, M.T. (1997) The value of weather information, in R. Katz and A. Murphy (eds.), *Economic Value of Weather and Climate Forecasts,* Cambridge University Press, Cambridge, England. pp 75-107.

Kahneman, D. and Tversky, A. (1996) On the reality of cognitive illusions. *Psychological Review* **103**, 582-91.

Katz, R. and Murphy, A. (1997) Forecast value: Prototype decision-making models, in R. Katz and A Murphy (eds.), *Economic Value of Weather and Climate Forecasts,* Cambridge University Press, Cambridge, England. pp 184-217.

Lichtenstein, S. and Newman, J.R. (1967) Empirical scaling of common verbal phrases associated with numerical probabilities. *Psychonomic Science* **9**, 563-4.

LWRRDC (1998) National Climate Variability Research and Development Program. Land and Water Resources R&D Corporation, Canberra.

Malcolm, L.R. (1990) Fifty years of farm management in Australia: Survey and review. *Review of Marketing and Agricultural Economics* **58**, 24-55.

Marshall, G.R., Parton, K.A., and Hammer, G.L. (1996) Risk attitude, planting conditions and the value of seasonal forecasts to a dryland wheat grower. *Australian Journal of Agricultural Economics* **40**, 211-33.

McTaggart, D. and Hall, T. (1993) Unemployment: Macroeconomic causes and solutions? Or, Are inflation and the current account constraints on growth? Conference on Unemployment, Causes, Costs and Solutions. Canberra, 16-17 February, 1993.

Miller, A. (1985) Psychological biases in environmental judgements. *Journal of Environmental Management* **20**, 231-43.

Mjelde, J.W., Sonka, S.T., Dixon, B.L., and Lamb, P.J. (1988) Valuing forecast characteristics of a dynamic agricultural production system. *American Journal of Agricultural Economics* **70**, 674-84.

Monahan, J. and Steadman, H.J. (1996) Violent storms and violent people - How meteorology can inform risk communication in Mental Health Law. *American Psychologist* **51**, 931-8.

Murphy, A. (1991) Probabilities, odds, and forecasts of rare events. *Weather and Forecasting* **6**, 302-7.

National Research Council (1989) Improving risk communication. National Academy Press, Washington, DC

Nicholls, N. (1985) Impact of the Southern Oscillation on Australian crops. *Journal of Climatology* **5**, 553-60.

Nicholls, N. (1991) The El Niño Southern Oscillation and Australian vegetation. *Vegetatio* **91**: 23-6.

Nicholls, N., Drosdowsky, W., and Lavery, B. (1997) Australian rainfall variability and change. *Weather* **52**, 67-72.

O'Mara, L.P. (1987) The contribution of the farm sector to annual variations in gross domestic product in Australia. *Economic Record* **63**, 255-69.

Partridge, I.J. (ed.) (1994) *Will it Rain? The effects of the Southern Oscillation and El Niño on Australia. 2nd ed.* Department of Primary Industries, Brisbane

Plous, S. (1993) *The psychology of judgement and decision making,* McGraw-Hill, New York.

Ridge, P. and Wylie, P. (1996) Farmers training needs and learning for improved management of climate risk. Proceedings of conference on managing with climate variability. Report CV03/96, Land and Water Resources Research and Development Corporation, Canberra. pp 126-30.

Scoccimarro, M., Mues, C., and Topp, V. (1995) Climate variability and farm risk. Report CV02/95, Land and Water Resources Research and Development Corporation, Canberra.

Simon, H.A. (1969) *The sciences of the artificial*, MIT Press, Cambridge, Massachusetts.

Stafford Smith, D.M., Clewett, J.F., Moore, A.D., McKeon, G.M., and Clark, R. (1997). DroughtPlan - Developing with graziers profitable and sustainable strategies to manage for rainfall variability. Report CV01/97, Land and Water Resources Research and Development Corporation, Canberra.

Stewart, T.R. (1997) Forecast value: descriptive decision studies, in R. Katz and A Murphy (eds.), *Economic Value of Weather and Climate Forecasts*, Cambridge University Press, Cambridge, England. pp 109-45.

Stone, R.C., Hammer, G.L., and Marcussen, T. (1996) Prediction of global rainfall probabilities using phases of the Southern Oscillation Index. *Nature* **384**, 252-55.

Swets, J.A. (1992) The science of choosing the right decision threshold in high-stakes diagnostics. *American Psychologist* **47**, 522-32.

Virtual Consulting Group, Australia. (1997) Dairy El Niño risk management strategy. Report to Dairy Research and Development Corporation, Melbourne.

Wilks, D.S. (1997) Forecast value: prescriptive decision studies, in R. Katz and A Murphy (eds.), *Economic Value of Weather and Climate Forecasts*, Cambridge University Press, Cambridge, England. pp 109-45.

THE RELEVANCE OF SEASONAL CLIMATE FORECASTING TO A RURAL PRODUCER

SID PLANT

Samarai Farming & Grazing Company
MS 918
Toowoomba, Qld 4352, Australia

Abstract

Weather and climate are the biggest single management considerations in primary production. Australian farms are still mostly family-operated, but demand a wide range of financial and other management skills. Under continued pressure in terms of trade, Australian farmers have become efficient, innovative and receptive to new technology.

In response to the extreme variability of the climate in Queensland, scientists have made good progress in seasonal forecasting and its practical application to decision-making. Current systems provide probabilistic forecasts. Uptake and use of modern seasonal forecasting technology to minimise risks and maximise opportunities has generally been greatest where there is a strong relationship between the El Niño–Southern Oscillation (ENSO) and local rainfall.

The recent long drought in Queensland has put climate forecasting on the political agenda - an opportunity to be maximised but needing reasoned scientific explanations. Climate and agricultural applications information is available from a range of sources including radio, TV, aviation meteorology and the internet. Statistical forecast systems such as that based on AUSTRALIAN RAINMAN are sufficiently accurate to be of use in rural Australia, and should be developed until General Circulation Models (GCMs) become useful at a district level. Much current attention to El Niño has been dramatised by media sources. Thus, some farmers who are starting to appreciate practical use of seasonal forecasting, are confused by different sources of information, and could become discouraged.

1. Major Features of a Farm Business

Australian farms often have an enterprise mix of cropping and livestock, making climate-related decisions complex. Farms are generally capital-intensive and likely to

G.L. Hammer et al. (eds.), The Australian Experience, 23–28.

be family-operated, although corporatisation is increasing. Financial and other management skills vary from basic to sophisticated.

Australian farmers have been under extreme cost/price pressure, largely because they cannot control, or even significantly influence, either costs or prices. As a result, farmers have become innovative and receptive to new technologies to increase efficiency and maintain economic competitiveness.

2. Impact of Climate Variability

Australia has a highly variable climate, and Queensland has the greatest variability in the continent. This has induced farmers to learn to manage climate variability and scientists to progress seasonal forecasting and modelling and analytical tools that facilitate its practical use.

Financial pressures over the years have pressured managers to push their resources to the limit. Over the last decade, this has been recognised and ecologically sustainable management has become a priority. There is a significant role for climate forecasting to help managers operate in an ecologically viable way and still survive financially.

2.1 BEEF INDUSTRY

Australia is a relatively small producer of beef, but is the largest beef exporter in the world. Fifty thousand producers sell to a small number of processors, and thus have little opportunity to command prices. The numbers of cattle suited to a particular market available at any particular time, is largely a function of climatic conditions over the preceding season. Thus, reliable seasonal forecasts provide a great opportunity to anticipate the likely market or the most likely place to find suitable cattle; this is often the difference between profit and loss.

Similarly, a reliable seasonal forecast can help in decisions of when or what fodder crops to plant to have cattle available for a foreseen opportunity in the market. Likewise, the feedlot industry would use seasonal forecasts when planning for cattle availability and likely market movements. They could also look further ahead when planning forward grain supplies. Good risk management requires knowledge of grain costs before the cattle are purchased to supply a particular market contracted for 300 days later.

I breed, buy and sell cattle, and grow crops. Some decisions affect product three to four years away, with climate acting on my entire operation. Thus better management of climatic risk can have a big impact on my production and on my costs, although my markets are also influenced by variations in exchange rates, interest rates, demand in international markets, production in competitor countries, food safety issues and climate variability in other countries.

I need good climate information to make good decisions and, as I am in competition with those other 49,999 producers, I need it first and I need to use it best. From an industry-wide point of view, intelligent use of climate information by all players would smooth out some of the wildest variations in availability of suitable cattle, and help us compete against other countries with less variable climates.

2.1.1 Livestock Transport

In the livestock transport industry, animals are regularly moved from breeding country to growing country to feedlots and finally to market. However, during and after droughts, there can be large migrations of cattle. Warning of these events would ease logistical problems for producers and transport operators.

2.1.2 Meat Processors

Meat processors have to tender for export markets for grass-fed cattle without knowing their future availability. Good seasonal forecasts can reduce this risk.

3. Cropping Industry

In the grain growing industry, the most important determinant of yield is moisture, both that stored in the soil during fallow and the rain that falls during crop growth. A full profile of stored moisture will reduce the risk of a low probability of in-crop rainfall.

For winter cereals, a frost during flowering can decimate yield. Thus a forecast of the likelihood of late frost is invaluable in decisions about when and which variety to plant. The chance of rainfall at harvest time can affect expensive decisions about early harvesting with artificial drying or hiring extra equipment to help get it off in time.

Good rainfall and moisture increase the potential yield of the crop. Additional expensive nitrogen fertiliser will be needed to achieve a premium protein level, but this decision has to be made before or at planting. Conversely, exceptionally wet conditions can reduce yield substantially through waterlogging or fungal disease.

3.1 COTTON INDUSTRY

The cotton industry is relatively new in this country, and has attracted young and progressive farmers with advanced management skills and a readiness to adopt available information. For a number of reasons, including the heavy use of expensive chemicals, cotton farmers stood a little apart from other broad-acre farmers. They have an excellent communication network and tend to consider the use of relevant technology to be paramount to their success. With their rapid adoption of fax and Internet, cotton farmers are a good example to other industries in their successful use of modern technology and climatic information sources.

Dryland cotton farmers have been quick to realise the implications of El Niño and seasonal forecasting to make maximum use of available moisture. The large financial implications of depending on in crop moisture has lead many cotton growers to establish irrigation systems using stored water. They then have to assess the significance of rainfall variability on their ability to harvest flood water for their storage dams

3.2 WEED AND PEST CONTROL

Weeds and pest problems are affected by climate variability. The populations of some insect pests explode under specific weather conditions, possibly resulting in too few aircraft available for the timely control of the outbreaks. Similarly, under wet conditions, weeds can flourish to compete with the crop if ground machinery can not get on the wet land. If the season can be predicted in advance, the number of spray applications can be planned.

4. Other Primary Industries – The Prawn Industry

A significant correlation has been demonstrated between El Niño and the prawn harvest in the Gulf of Carpentaria, (Love 1987). With poor stream flows into the Gulf, prawns breed less prolifically and mature later, resulting in reduced harvests. Seasonal forecasting can aid decisions related to length of the harvesting season, such as whether to invest in new equipment.

5. Related Industries

The effects of climate variability on farming reverberate on all rural suppliers. Examples of industry support include machinery manufacturers and dealers, stock and station agents, grain merchants, food processors and exporters. Rural services such as shops, banks and schools are also affected.

6. Potential Application of Seasonal Climate Forecasting

Attitudes to seasonal climate forecasting vary from folklore to a reasonable scientific understanding of the world's climate and weather systems. Climate is an important element in the life of rural Australia but there are many attitudes to forecasting. My experience is that some folk tales do seem to anticipate climatic events, but others reflect a frustration with 'scientific' forecasting in the past.

Probably half of the rural community could be called cynics or unconvinced, and they are the ones who could benefit most from good climate forecasting applications work. This involves pure and applied climate and agricultural science, information delivery, decision support development, facilitator training and workshop development, communication and other disciplines.

Many disbelievers have previously tried applying some scientific seasonal forecast information and failed for some reason. This could be either wrong information, incomplete information or information wrongly applied or misunderstood. For example, some farmers, having no knowledge of effects of the SOI at different times of the year or in different geographic locations, had been led to believe that an El Niño meant total drought, and so did not plant a crop, while a bank manager may have refused to extend finance.

Presenting forecasts as probabilities is the best method for the current level of forecasting skill. But explanations are required as some still interpret a "30 % probability of receiving median rain" as "you will probably get 30 % of the median rain". Most farmers have faith in the more obvious short-term weather forecasts from satellite cloud images or synoptic charts; these they have learned to interpret with a little skill.

The "converts" can see the value in seasonal climate forecasting and are doing their best to collect relevant information and apply it. Some simply interpret a negative SOI as predicting below average rainfall and manage accordingly. Others may have a detailed theoretical knowledge of the world's weather and climate; they may understand the sub-tropical ridge, Kelvin wave, 40-day wave, Walker cycle, SOI, El Niño Southern Oscillation (ENSO), and general circulation models, as well as a general understanding of meteorology. These people gather the climate and agricultural response information and then take responsibility for their own decisions. As long as they win more often than they lose, they receive a net gain from seasonal forecasting.

Most farmers have facsimile facilities and many make use of excellent weather service available on poll fax from the Bureau of Meteorology and the Queensland Department of Primary Industries. The farmer's nightly ritual includes the weather report on television, especially if they have some weather-related decision pending, such as cutting hay.

In my case, I studied meteorology when I trained for my pilot's licence in the 1960s and soon realised the potential to apply some climate science to farm decision-making. I have maintained my interest, and read widely on the subject. I now check the Internet daily for the satellite cloud chart, synoptic pressure analysis, 24-hour prognosis, table of current SOI data. I regularly check sea surface temperatures, GCMs, and information on "The Long Paddock" website (http://www.dnr.qld.gov.au/longpdk/) and Bureau of Meteorology website (http://www.bom.gov.au/). Other useful websites include SILO (http://www.bom.gov.au/silo/) for yesterdays rainfall, and that of the Agricultural Production Systems Research Unit (http://www.apsru.gov.au) for crop management implications.

I try to keep an up-to-date picture of the likely weather and climatic outlook for my farm for the coming period, and I use this as one source of relevant information on which I base my management decisions.

Statistical probability forecasts such as that based on AUSTRALIAN RAINMAN (Clewett *et al.*, 1994; Stone *et al.*, 1996) have sufficient accuracy to be of practical use in rural Australia. I would like to see them maintained and developed until General Circulation Models (GCMs) can give superior forecasts at a local level.

The level of adoption by farmers of the information available has been variable, with the best uptake where the SOI is a strong indicator of future rainfall. Here, there is greater opportunity to minimise risks and maximise profit opportunities with less confusion.

7. Political Implications and Media Relations

The 1997 El Niño attracted world-wide attention from the media, and introduced the phenomenon to many people for the first time. This has been a mixed blessing with many natural events being blamed on El Niño. This El Niño has caused major impact in other parts of the Pacific region, but not the one in a hundred and fifty year event that the media loudly forecast for Queensland. The dramatised signals received from the popular press have conflicted with the reasoned advice from services such as the DPI, and first-time users of climate information have been confused.

The extent of the drought in recent years has put climate forecasting on the political agenda to an extent never seen before and governments are investing significant funds in further research and development, such as with the establishment of the Queensland Centre for Climate Applications.

While some consider any publicity to be good publicity in keeping the attention of politicians and the public, 'crying wolf' may have a negative impact. In agriculture, the dramatic message was at best a nuisance. It may have contributed to a couple of the drought-related suicides in areas badly affected by the decade's long-term drought. Attempts must be made to better handle the media.

While rural Australia and agriculture are dependant on the scientific community for the means to manage climate variability more effectively in the future, the converse is also true. Research and development funding is dependant on community perception and support, and this will continue only as long as it is seen to be worthwhile or commercially relevant. There are some exciting times ahead for the research, development and extension of climatology and the business of farming in the future.

References

Clewett, J.F., Clarkson, N.M., Owens, D.T., and Abrecht, D.G. (1994) AUSTRALIAN RAINMAN: Rainfall Information for Better Management, Department of Primary Industries, Brisbane.
Love, G. (1987) Banana Prawns and the Southern Oscillation Index. *Aust. Met. Mag.* **35**, 47-49.
Stone, R.C., Hammer G.L., and Marcussen, T. (1996) Prediction of global rainfall probabilities using phases of the Southern Oscillation Index. *Nature* **384**, 252-55.

UNDERSTANDING AND PREDICTING THE WORLD'S CLIMATE SYSTEM

MARK CANE

Lamont-Doherty Earth Observatory of Columbia University
Route 9W
Palisades, New York 10964, USA

Abstract

This paper reviews the advances in understanding ENSO (El Niño and the Southern Oscillation) and the development of coupled ocean-atmosphere prediction systems. Prospects for improving forecasts of ENSO and its global consequences are considered in view of the factors limiting the skill of current forecasts: inherent limits to predictability, flaws in the models and data assimilation procedures, and gaps in the observing system. The possibilities for utilising other modes of climate variability, such as the North Atlantic Oscillation, are discussed. Experience with the 1997 warm event is highlighted.

1. Introduction

The weather this season is not the same as it was a year ago, and common experience leads us to expect that it will be different still a year hence. None of us, including the experts in long range forecasting, have a reliable idea of how it will differ.

Some of the year-to-year variations in climate are the result of random sequences of events, just as a series of coin flips will occasionally produce a long run of heads. A region may experience a dry spell because no storms happen to pass that way for a time. Prediction of such stochastic events is not possible. Many climatic variations, however, are part of patterns that are coherent on a large scale. Skilful prediction may then be possible, particularly if the patterns are forced by observable changes in surface conditions such as sea surface temperature (SST).

The most dramatic, most energetic, and best-defined pattern of interannual variability is the global set of climatic anomalies referred to as ENSO (El Niño and the Southern Oscillation). As recently as late 1982, little more than a decade ago, even many knowledgeable observers were unaware that the largest El Niño in at least a century was already well underway. In 1985 the international TOGA (Tropical Ocean-Global Atmosphere) program was launched with the goal of predicting ENSO, if possible. To

29

G.L. Hammer et al. (eds.), The Australian Experience, 29–50.

an unexpected degree that goal has been met: ENSO events are now predicted well in advance with useful (albeit less than perfect) skill.

The progress in ENSO prediction is built on advances in understanding, advances in coupled ocean-atmosphere modelling, and the development of an ocean observing system. These advances will be reviewed below, and prospects for improving forecasts of ENSO and its global consequences will be considered in view of the factors limiting the skill of current forecasts: inherent limits to predictability, model flaws, gaps in the observing system, and shortcomings in the ways the observations are used in a forecast system.

Our understanding of ENSO identifies the tropical Pacific as the genesis region, and ocean atmosphere interactions in that region as the generating mechanism. Other global aspects of the cycle are to be viewed as consequences of the changes in the tropical Pacific source region. Thus the ENSO prediction problem divides into two parts: the prediction of ENSO *per se* and the prediction of its global influence. We will also touch briefly on significant modes of variability in the climate system other than ENSO, but there is less to say because much less is understood. Beyond the changes in the physical climate system, one would like to predict impacts on agriculture, health, economics and so on, and then develop response strategies based on this knowledge. The mission of the newly formed IRI (International Research Institute for Climate Prediction; see the URLs iri.ldeo.columbia.edu and iri.ucsd.edu) spans these issues from end-to-end: from the prediction of variations in the climate system (SSTs in the tropical Pacific prominent among them), to the identification of social actions to mitigate the impacts of climate variations.

The next section discusses our understanding of ENSO, and the following one reviews the experience in predicting ENSO, lingering over the 1997 event. In Section 4 we move on to the global influence of ENSO, including a brief consideration of some other modes of climate variability with important impacts on local conditions. The final section summarises the present state of our forecasting skill and knowledge, and offers some speculations on future prospects.

2. The Nature of ENSO

On the average there is an ENSO warm event (an "El Niño") about every four years, but the cycle is highly irregular. Sometimes there are only two years between events, sometimes almost a decade. There are great variations in amplitude. Though each episode has its own peculiarities, all follow a general pattern. At an early stage, anomalously warm surface waters are found in the western equatorial Pacific. Associated with the warmer surface temperatures is an increase in convective activity, and at a certain stage, a persistent slackening of the normally westward flowing trade winds. Following this is a dramatic and expansive warming of the tropical Pacific Ocean from the dateline to the South American coast, and a further disruption of the trade winds. Very heavy rains fall in normally arid regions of Peru and Ecuador, while droughts are experienced in Australia and southern Africa, and anomalous tropical cyclones occur in regions such as French Polynesia and Hawaii. Farther away, there are often disruptions of the Indian monsoon, the seasonal rains of northeast Brazil, and regional climates over much of East Asia, North America, and Africa are usually affected.

The 1982/83 ENSO event was the most extreme in at least a century. Equatorial waters from the South American coast to the dateline warmed by an average of 2°C, with the warming along the coast exceeding 6°C. The trade winds actually reversed. The consequences of this event were often devastating. In Australia, the worst drought ever recorded spawned firestorms that incinerated whole towns; normally arid regions of Peru and Ecuador were inundated by as much as 3m of rain; the beaches of California were rearranged by the unusual winter storms; drastic changes in the tropical Pacific Ocean resulted in mass mortality of fish and bird life. All in all, it has been estimated that the 1982/83 ENSO event was responsible for $8 billion in damages and the loss of two thousand lives.

Historically, El Niño referred to a massive warming of the coastal waters off Ecuador and Peru. This warming leads to widespread mortality of fish and guano birds, crippling the local economy. The heavy rainfall associated with the event results in catastrophic flooding in coastal land areas. El Niño has been documented as far back as 1726 and it appears that El Niño rainfall more than a century earlier made it possible for the conquistadors to cross an otherwise impenetrable desert.

The atmospheric component of ENSO, the Southern Oscillation, is a more recent discovery. The seminal figure in delineating it was Sir Gilbert Walker, the Director-General of Observatories in India. Walker assumed his post in 1904, shortly after the famine resulting from the monsoon failure in 1899 (an El Niño year). He set out to predict the monsoon fluctuations, an activity begun by his predecessors after the disastrous monsoon of 1877 (also an El Niño year). Walker was aware of work indicating swings of sea level pressure from South America to the Indian-Australian region and back over a period of several years. In the next 30 years he added correlates from all over the globe to this primary manifestation of the Southern Oscillation. For example, he found that periods of low Southern Oscillation Index (Figure 1) are characterised by heavy rainfall in the central equatorial Pacific, drought in India, warm winters in southwestern Canada and cold winters in the southeastern US. No conceptual framework supported the patterns he found. Walker's methods were strictly empirical. Probably the very thoroughness of his search, together with the short duration of the records then available, made it easier for others to dismiss his findings as mere artefact.

Recently, Walker's global correlations have been re-examined with decades of new, independent data and found to hold. Walker did not consider El Niño, and although both El Niño and the Southern Oscillation had been known at the turn of the century, it was only in the 1960's that the close connection between the two (see Figure 1) was finally appreciated, principally through the work of Jacob Bjerknes.

Bjerknes (1969, 1972) did more than point out the empirical relation between the two; he also proposed an explanation that depends on a two-way coupling between the atmosphere and ocean. His ideas were prompted by observations of large-scale anomalies in the atmosphere and the tropical Pacific Ocean during 1957-58, the International Geophysical Year. A major El Niño occurred in those years, bringing with it all the atmospheric changes connected to a low Southern Oscillation Index. It is implausible that a warming confined to coastal waters off South America could cause global changes in the atmosphere, but the 1957 data showed that the rise in sea surface temperature (SST) was not confined to the coast. Bjerknes suggested that this feature was common to all El Niño events. He was correct, and the term "El Niño" is now most often used to denote basin-scale oceanic changes. In his account of the connection between the ocean and atmosphere, the coastal events constituting El Niño are incidental to the important oceanic change, the warming of the tropical Pacific over a quarter of the circumference of the Earth.

Bjerknes suggested a tropical coupling between El Niño and the Southern Oscillation. He also proposed that the changes in atmospheric heating associated with tropical Pacific SST anomalies cause changes in mid-latitude circulation patterns. This teleconnection idea is consistent with the global nature of Walker's Southern Oscillation. In his theory, however, the causes of ENSO are rooted solely in the coupling of the atmosphere and ocean in the tropical Pacific. They are entirely internal to the climate system, not responses to volcanic eruptions, solar variations, biological activity, etc.

Work in the past two decades, especially that under the auspices of the international TOGA Programme, has provided theoretical and observational support for Bjerknes' concept. An essential addition, equatorial ocean dynamics, was introduced by Klaus Wyrtki (1975, 1979) in the 1970s on the basis of data from a network of Pacific island tide gauges. The first model to successfully simulate ENSO, that of Zebiak and Cane (1987), was based explicitly on the Bjerknes-Wyrtki hypothesis.

This numerical ENSO model depicts in a simplified manner the evolution of the tropical Pacific Ocean and overlying atmosphere. It is a dynamical model, rather than a statistical one; that is, it relies on the governing physical equation rather than simply a sequence of observations. Such dynamical models also provide a means for physical interpretation and understanding of whatever they simulate. One of the most significant results of the model simulations was the recurrence of ENSO at irregular intervals as a result of strictly internal processes; that is, without any imposed perturbations. Analysis of the model helped in developing a now widely accepted theory that treats ENSO as an internal mode of oscillation of the coupled atmosphere-ocean system, perpetuated by a continuous imbalance between the tightly coupled surface winds and temperatures on the one hand, and the more sluggish subsurface heat reservoir on the other.

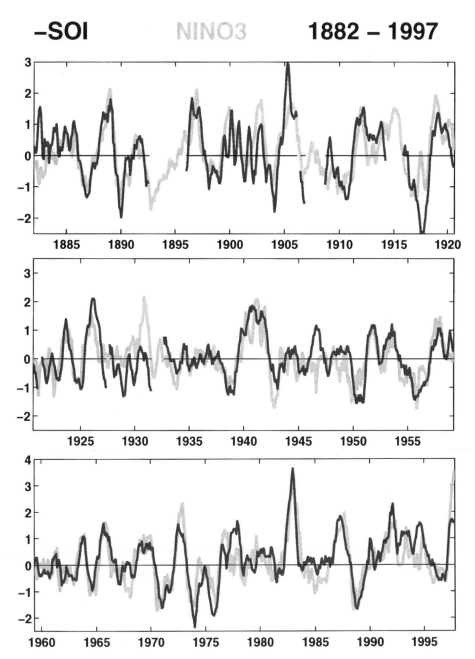

Figure 1. Historical trends in the sea surface temperature (SST) anomaly in the NINO3 region of the eastern equatorial Pacific (90°W-150°W, 5°S-5°N), which is a commonly used index of El Niño (grey line), and the negative of the Southern Oscillation Index (SOI), which is the normalised sea level pressure difference between Tahiti and Darwin, Australia (black line). Note the close relationship between the two indices.

3. Prediction of ENSO

This theory has a number of implications for the prediction of El Niño events. First, since the essential interactions take place in the tropical Pacific, data from that region alone may be sufficient for forecasting. Second, the memory of the coupled system resides in the ocean. Anomalies in the atmosphere are dissipated far too quickly to persist from one El Niño event to the next. The surface layers of the ocean are also too transitory. Hence the memory must be in the subsurface ocean thermal structure. The crucial set of information for El Niño forecasts is the spatial variation of the depth of the thermocline in the tropical Pacific Ocean. (The thermocline is the thin region of rapid temperature change separating the warm waters of the upper ocean from the cold waters of the abyssal ocean.)

Starting with experimental studies in 1985, our group at Lamont has used the Zebiak-Cane model not only to simulate, but also to predict El Niño (Cane *et al.*, 1986). As noted above, theory argued for a deterministic origin of ENSO, that is, a systematic evolution throughout the cycle rather than a sequence of random events. However, even deterministic systems that are chaotic have limited predictability, and in this case the situation was made worse by a very poor observational base over vast regions of the tropical Pacific. It was with some scepticism that we first undertook retrospective forecasts.

Forecasts were made by utilising observations of surface winds over the ocean, beginning in 1964. On the basis of these wind data we ran the ocean component of the model to generate currents, thermocline depths, and temperatures that served as initial conditions for forecasts - a necessary step because of the lack of direct observations of oceanic variables. Each forecast then consisted of choosing the conditions corresponding to a particular time, and running the coupled model ahead to predict the evolution of the combined ocean-atmosphere system. By making predictions based on past periods we could compare forecasts directly with reality. This we did, starting with 1970. The results clearly demonstrated predictive skill at lead times longer than one year. This set the stage for the first predictions of the future, made in early 1986, which called unambiguously for an El Niño occurrence later that year.

That the forecast was made public did not receive universal approval at the time, as even the notion of climate predictability on the time scale of a year was not generally accepted. After what seemed like imminent failure during mid-year, conditions evolved rapidly into El Niño after that. Figure 2 shows the sea surface temperature anomalies for January 1987 as observed, and as predicted, in early 1986. The appearance of a moderate El Niño in both allow us to count the forecast as a success, although differences in timing and other details show that the prediction scheme was far from perfect.

Prior to the 1997 event, the model was generally successful in forecasting the major events (1972, 76, 82, 86 and 91) a year or more ahead but demonstrated little skill in predicting smaller fluctuations (some of which may influence climate elsewhere on the globe). Figure 3 illustrates the overall performance of the Lamont forecasts in terms of a widely used index of ENSO events, the sea surface temperature anomaly in the NINO3 region of the eastern equatorial Pacific (90°W-150°W, 5°S-5°N). The curve labelled "standard" represents the scheme that made the forecasts shown above. The one labelled "new" uses the same model and the same data but it makes better use of the same wind data by combining it with a model estimate of the wind (Chen *et al.*, 1995). It is clearly an improvement, but primarily for the 80s not the 70s. Apparently, some periods are harder to predict than others.

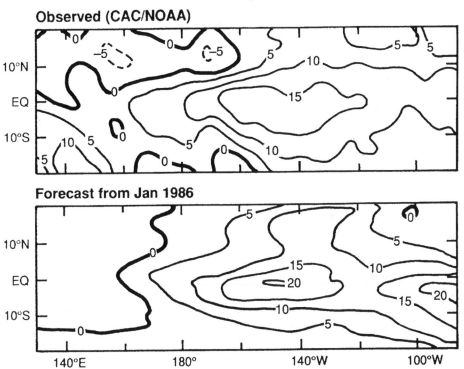

Sea Surface Temperature Anomalies (0.1°C)
January 1987

Figure 2. Prediction of El Niño one year in advance: Comparison of the observed sea surface temperature anomaly field (0.1°C) for January 1987 and that predicted by the Lamont atmosphere-ocean model from one year ahead. This forecast, which was published in *Nature* in June of 1986, is the first true forecast of El Niño with a dynamical model.

In the 1980s the Lamont model was the only physically based forecasting system with this level of skill, though there were (and are) comparably skilful statistical schemes. More recently a number of other models have been developed for ENSO forecasting, many of them based on the comprehensive physical models of the atmosphere and ocean known as general circulation models (GCMs). The GCMs are generally agreed to offer the greatest promise for accurate prediction, because they simulate the climate system with the greatest verisimilitude (see Hunt and Hirst, 2000). In addition, unlike models of the tropical Pacific alone, they make it possible to predict the global impacts of ENSO (Section 4). Their complexity, however, makes them less forgiving of errors and imperfections than the more simplified models (see Nicholls *et al.*, 2000). It was thus a considerable achievement to bring them to the same general skill level as the Lamont model and the statistically based schemes.

Though the reaction to the first ENSO forecasts may have been surprise that it could be done at all, at this point the natural question to ask is why it isn't being done better (viz. Figure 3). The factors limiting the current skill of forecasts are: inherent limits to predictability; flaws in the models; gaps in the observing system; flaws in the data assimilation systems used to introduce the data into the models. We do not know enough to make a precise attribution of loss of skill factor by factor, and in any case the effects are not simply additive: the different factors can combine in unpleasant ways. However, a rough assessment is possible.

There is no question that the predictability of ENSO is limited, although there is disagreement as to whether this is due to chaos (eg. Tziperman *et al.*, 1994; Jin *et al.*, 1994) or the effect on the ENSO system of "noise" such as intraseasonal oscillations or weather disturbances intruding from mid-latitudes (eg. Penland and Sardeshmukh, 1995). Either way, the ENSO system exhibits "sensitivity to initial conditions" so that small inaccuracies in the estimate of the initial state of the climate system can grow into substantial errors. Our best guess, based on numerical experimentation with models such as that of Zebiak and Cane (1987), is that the inherent limit to predictability is several years. That is, on average it will take several years for a small initial error to grow to a size comparable to a typical ENSO anomaly (cf. Zebiak, 1989; Goswami and Shukla 1991; Xue *et al.*, 1997a,b). Equivalently, only in a small percentage of cases will the forecasts at a lead-time less than a year be substantially impacted. At present the primary problems lie elsewhere.

Experience has shown that imperfections that seem acceptable when atmosphere and ocean models are run independently become unacceptable when the two models are coupled together. Small flaws apparently reinforce each other. Even the most complete models necessarily include only approximate treatments of important physical processes. The most notable examples of processes known to be inadequately represented are clouds and atmospheric convection, which are intimately tied to the heating of the atmosphere. As discussed above, variations in atmospheric heating are crucial both in the working of the ENSO cycle and in generating its global effects. Model shortcomings are surely a major limitation on forecast skill.

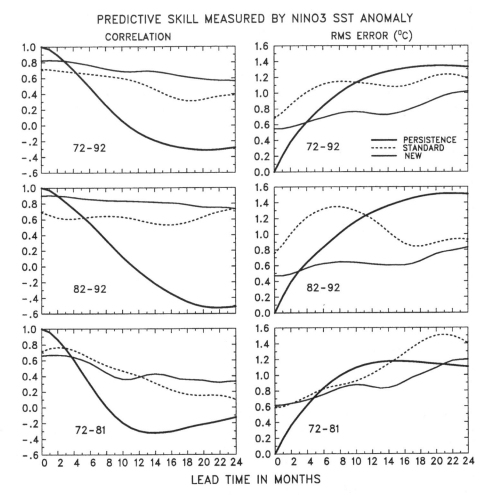

Figure 3. Correlations and root mean square (RMS) errors between predicted and observed NINO3 SST anomalies for four different time periods. In each panel results from the standard, new, and persistence forecasts are shown for comparison. Persistence forecasts are obtained by assuming initial SST anomalies remain constant. (from Chen *et al.*, 1995.)

Though the observing system was certainly inadequate in the past, part of the legacy of the TOGA program is an array of highly instrumented buoys in the tropical Pacific. The experience of others (Leetmaa, private communication) confirms our own estimates that the present observing system provides adequate initial conditions. Two caveats apply. First, the 1997 experience may show us that this optimistic assessment is wrong (see below). Second, maintenance of the observing system at an adequate level is not guaranteed, so interested parties must exercise due vigilance.

There are, however, important shortcomings in the way the observational data is used. In the decades of experience with numerical weather prediction, data assimilation has been developed into a sophisticated and effective technology. There is only a decade of experience with coupled model ENSO forecasting and our understanding of data assimilation for this purpose is in its infancy. Figure 3 illustrates that even a fairly simple improvement in data assimilation methodology (Chen *et al.*, 1995) may improve forecasts noticeably. Work with coupled GCMs at NCEP (National Center for Environmental Prediction in Washington DC) also shows a substantial increase in skill from fairly small changes in the data assimilation system (Ji *et al.*, 1995).

There is certainly room for improvement, and this analysis of the limits to ENSO forecast skill is hopeful. The inherent limits to predictability and the gaps in the observing systems are, respectively, the most intractable and most expensive of the limiting factors. The best estimate is that these are not now the major limiting factors, though they will have to be faced if the other factors are reduced. The major problems at the moment are shortcomings in the simulation models and in the systems for introducing data into these models. Both should diminish in the face of concerted efforts by forecasting centres such as IRI and NCEP. The difficulty of the work is increased by the data scarcity for past events, which restricts the time period suitable for evaluating putative improvements in the model and data assimilation system. Rapid gains in data assimilation should result by borrowing techniques developed for numerical weather prediction. Fixing the model flaws appears to be more difficult, though progress should be expected.

3.1 THE 1997 EXPERIENCE.

The highly unusual warm event of 1997 holds useful lessons for ENSO forecasting. Figure 4 shows the strongest events of the period of instrumental data, approximately 140 years. Not only is the 1997 El Niño at least comparable to the strongest of these, but the rate at which the ocean warmed is unprecedented.

Both versions of the Lamont forecasting system utterly failed to forecast this event (viz. Figure 5a,c). Neither shows any tendency to warm until well into 1997, when the warm event was well underway. If anything, the performance of LDEO2, the improved version (Chen *et al.*, 1995), is even worse than the standard version, LDEO1. However, Figure 5b shows that the forecast is much improved (cf. Figure 5a) by initialising the forecasts using wind fields derived from the space borne NSCAT scatterometer rather than the FSU (Florida State University) wind fields used in the standard procedure (Chen *et al.*, 1998). The ADEOS satellite containing the NSCAT instrument was launched in August 1996, and the winds are available from October. It is notable that this data had the impact it did given how short a time it was used. (Unfortunately, the NSCAT data ceases in June 1997 when the satellite's solar panel failed.)

Figure 5d shows the impact of another data source. In this case, we assimilated a field of thermocline depths derived from an analysis of tide gauge data (Cane *et al.*, 1996).

The forecasts of Figure 5d were made with the improved wind analysis procedure of Chen *et al.* (1995). There is a marked improvement compared to the forecasts of Figure 5c, which were made with the same wind analysis but without thermocline depth information. Even with the older wind analysis procedure, the forecasts using the thermocline depths derived from observations are much like those shown in Figure 5d.

NINO3 SSTA of Major El Nino Events

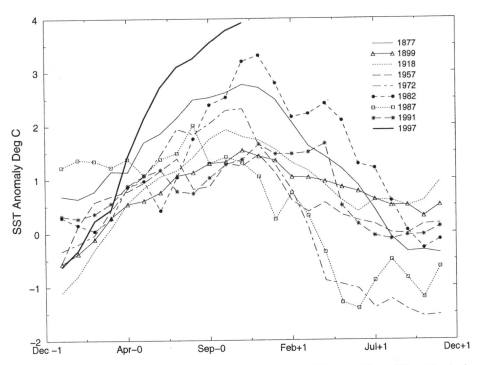

Figure 4. NINO3 SST anomalies for the strongest events in the past 130 years. The rapidity of the rise in 1997 is unprecedented in the period for which we have instrumental data. Figure courtesy of the IRI.

It is unclear why after so many years of success using the FSU winds and nothing else, this year demands more. Chen *et al.* (1998) show that the principal difference from the NSCAT winds is in the southeastern Pacific, a region not considered as highly important in ENSO evolution. Most centres forecasting ENSO with physical models have long found it necessary to use ocean thermal data in their initialisation procedure, but previous experience did not show it to be essential for the Lamont model. The suggestion is that some differences in the evolution of the 1997 event impacted performance, highlighting the need for (at least) better data assimilation procedures. In this regard, note the unprecedented rapid rise of SST in early 1997.

It should be added that many, perhaps most, of the models, did a credible job of predicting the 1997 event. Examples of some of the better forecasts, from the NCEP model, are shown in Figure 6. Forecasts from both November 1996 and February 1997 both indicate a warm event in mid-year 1997. Still, both underestimate the amplitude of the warming, especially the forecast from late 1996. In fact, few, if any, of the forecasts made before the event became evident in early 1997 predicted the outsized nature of the event to come (viz. NOAA's *Experimental Long Lead Forecasting Bulletin* for late 1996 and early 1997). Future analysis of this event should identify the reasons for this, leading to improvements in the forecasting procedures.

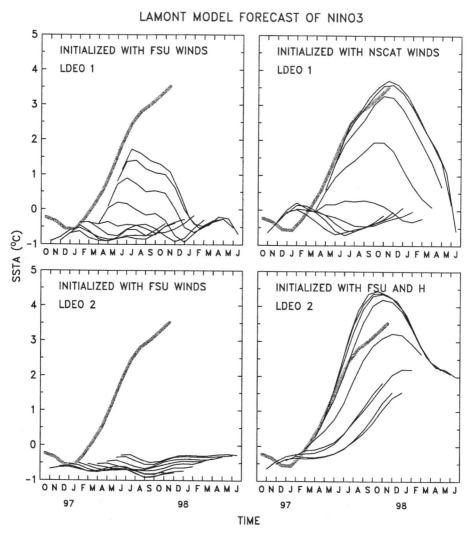

Figure 5. LDEO forecasts of SST anomalies (SSTA) for 1997 (dark line traces) compared with measured values (grey line) for: (a) LDEO1, the standard procedure described in Cane *et al.* (1986) using the FSU wind product; (b) LDEO1 but with the NSCAT wind product; (c) LDEO2, the new data assimilation procedure of Chen *et al.* (1995), with FSU winds; (d) LDEO2 with assimilation of sea level (thermocline depths).

SST ANOMALIES

6 and 9 MONTH LEAD FORECAST FOR JUN–AUG97

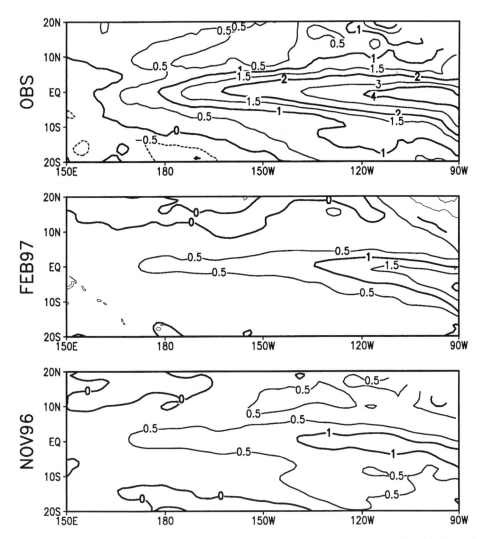

Figure 6. Comparison of NCEP forecasts of SST anomalies for June, July and August 1997 with observed values. The top panel shows the observed anomalies; the middle panel shows the 6-month lead forecast from NCEP using data through February 1997; and the bottom panel shows the 9-month lead forecast from NCEP using data through November 1996.

4. Global Impacts

Figure 7 is a version of the well-known diagram of the global influence of an ENSO warm event (after Ropelewski and Halpert, 1987; all of the relationships discussed below may be found in that paper or Ropelewski and Halpert, 1996). As a first approximation, one may say that ENSO cold events (ie. "La Niña" events, where the eastern tropical Pacific is colder than normal) have the opposite effects, but there are significant exceptions. As a general rule, the effects of an ENSO event are strongest and most reliable in the tropical Pacific genesis region and contiguous continents. When there is a warm event one can be fairly certain of heavy rains in Peru, drought in Indonesia (Figure 8a). Typical consequences are somewhat less reliable in the global tropics, but still highly likely. Thus, there is frequent concurrence of ENSO warm events and a poor monsoon in India, poor maize yield in Zimbabwe (Cane et al., 1994), drought in the Nordeste of Brazil, and reduced numbers of hurricanes in the Atlantic.

There is no question that ENSO has an influence in extra tropical latitudes, but the response is less certain than in the tropics. Other factors may intervene, and, characteristically, the extra tropical climate system is more chaotic and thus less predictable. Thus in these latitudes an ENSO event should be thought of as putting a bias in the system rather than as a certain cause. With warm (El Niño) events, heavy rains in the Great Basin region of the USA are more likely, and with cold (La Niña) events, midwestern drought (1988, for example) and lower corn yields are more likely. Certain patterns are more likely to persist, altering the paths of hurricanes, typhoons, and winter storms.

Predictions of the global impacts of ENSO are currently made with physical models, statistical procedures, and other empirical methods. If the physical models are global coupled GCMs then they predict global impacts along with the prediction of core changes in the tropical Pacific. Alternately, one may take a "two-tiered" approach in which a simpler model predicts tropical Pacific SSTs and then a global atmospheric GCM uses the predicted SSTs as boundary conditions to calculate global climate variations (Barnett et al., 1994; Hunt et al., 1994). Empirical approaches may also be two-tiered, deriving forecasts by combining a predicted ENSO index such as SOI or NINO3 with the historical relationship of a local climate variable (eg. rainfall in Goondiwindi). They may also do the entire prediction at once, using observed values of an ENSO index to predict a future local condition. For example, the prediction of global rainfall of Stone et al. (1996), based on the phase system uses values of the SOI at 2 different times to predict rainfall a season or more ahead.

Physical models are not now demonstrably better for this purpose than statistical ones, but as with ENSO itself, there are sound reasons to believe the physical models will ultimately prove superior and thus that the greater effort needed to develop them is justified. To begin with, empirical methods require a historical data set to train on and this is often too short or missing altogether. Since ENSO events occur about every 4 years, even a 50 year long record will contain only a dozen or so examples of warm and cold extremes. Since the response to ENSO events is at least somewhat nonlinear, the most straightforward statistical methods may be inadequate. More sophisticated methods typically demand even more data for training.

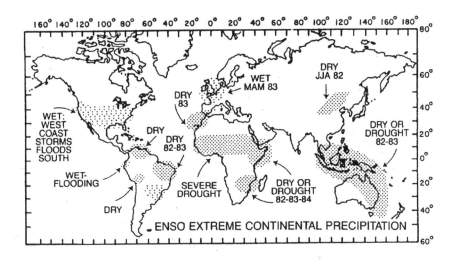

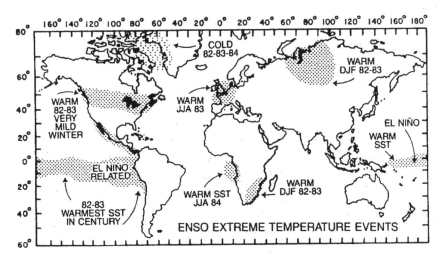

Figure 7. Global influence on precipitation and temperature of an ENSO warm event (after Ropelewski and Halpert, 1987)

It was already noted that not all ENSO connections are equally strong and reliable. A more general statement is that the global impacts of each ENSO event are different. Not every El Niño is accompanied by the same global variations, nor is the magnitude of what variations there are simply related to the strength of the El Niño event. Understanding of these differences is limited: they have hardly been classified satisfactorily, let alone explained in physical terms. A corollary is that the differences between events are not well predicted.

There are a number of reasons why this might be so. Perhaps the prediction schemes fail to respond to the idiosyncrasies of each event such as the subtle (and not so subtle) differences in the pattern of its SST anomalies. It is not clear what features we should pay attention to. It is known that the global response is sensitive to the location and strength of the atmospheric heating in the tropics (eg. Hoerling *et al.*, 1997).

That the atmosphere is a chaotic dynamical system means that small changes in boundary and initial conditions can be amplified to give very large differences in the future state of the atmosphere. One result of this chaos is "weather"; that is, variability on a timescale much shorter than the timescale of what we call "climate variations". We may think of weather (and other short time variability) as "noise" in the climate system, random noise that makes the system unpredictable. Whether one's conceptual view favours "chaos" or "noise" the system is unpredictable: even with SST and other boundary conditions fixed, the atmosphere may evolve into very different states if started with only slight differences in its initial state. Suppose, for example, one used an atmospheric model with correctly specified SST boundary conditions to make 10 simulations initialising the model with our best estimate of the state of the atmosphere on October 1, 1982, October 2,1982,..., October 10, 1982 . The outcomes for, say, March 1983 over North America would be very different (Barnett, 1995). With a good model one or more might resemble the actual outcome, but Nature only runs this experiment once and it is not possible to say *a priori* which it will be. In order to be accurate, the prediction has to be stated as a probability distribution of possible states. This is a consequence of the nature of the climate system and would be true if the model were perfect: forecast skill is known to be limited by this intrinsic loss of predictability in midlatitudes and elsewhere.

ENSO is not the only mode of climate variability with large-scale impacts. There is enormous current interest in the North Atlantic Oscillation (NAO), which is usually defined by an oscillation in sea level pressure between stations in Iceland and the Azores because of well identified connections to climate anomalies in Europe, North Africa, the Middle East, and eastern North America (Hurrell, 1995). SST variations in the tropical Atlantic have been related to droughts in the Sahel region (Folland *et al.*, 1986) and the Nordeste region of Brazil (Nobre and Shukla, 1996). Climate anomalies in Australia have relationships to Indian Ocean SST anomalies that are independent of ENSO (Nicholls, 1989). Perhaps these interact with ENSO, and our current inability to take account of these interactions limits forecast skill. Moreover, we have yet to establish that any of these modes can be predicted several seasons ahead -- or even that such a prediction is theoretically possible. So even if the interaction with ENSO were understood thoroughly, this knowledge might not be much help in improving actual forecasts of global impacts.

A brief global tour through the historical record and the 1997-98 event will illustrate the range of possibilities in impact forecasting. In Indonesia and New Guinea it is a virtual certainty that El Niño years are drought years and La Niña years bring excess rain (Figure 8a). The 1997 forest fires in Indonesia and famine inducing drought in Papua New Guinea fit the pattern. In Australia the expected rainfall anomalies are in the same sense, but are not as reliable (Figure 8b). Drought in Australia in 1997-1998 was not nearly as severe as the size of the event would have suggested. In Zimbabwe there is a very strong connection in the same sense between ENSO and rainfall and an even stronger connection to the maize crop, which integrates rainfall and temperature effects (Cane et al., 1994). However, the relationship is not entirely reliable or straightforward: 1992 was the most severe drought in at least the last 150 years in Southern Africa, but only a moderate El Niño.

While the Nordeste region of Brazil usually experiences drought in El Niño years, the relationship to tropical Atlantic SSTs is even stronger (Figure 9). Numerous atmospheric GCM experiments have shown that if the SSTs in the tropical oceans are known, then Nordeste rainfall is predictable. That is, our ability to forecast rainfall there is not limited by the intrinsic limits to predictability in the atmosphere. Unfortunately, we cannot yet predict the necessary tropical Atlantic SSTs, and do not even know in principle how predictable they are.

The Sahel is another region whose rainfall has links to tropical SSTs in both the Atlantic and Pacific with the Atlantic outweighing the ENSO impact. Moreover, Sahel rainfall is related to the North Atlantic Oscillation (NAO). Presumably, predictions of Sahel rainfall would be improved if one could know the future state of the NAO. Since much of the NAO -- Sahel connection is at interdecadal timescales, there is something to be gained by taking note of the present state of the NAO. Beyond that, the causes of the NAO are poorly understood, and its potential predictability is not known.

Figure 10 shows the relationship between ENSO and a measure of the quality of the Indian monsoon, the All India Rainfall index. It is obvious that poor monsoons are generally associated with El Niño events and excess rain with La Niña events, but the connection is far from perfect. Sometimes El Niño year rainfall is average, and sometimes there is a poor monsoon without an El Niño event. Based on this history, if one had been asked early in 1997 what sort of monsoon to expect, the forecast would have to have been that a poor monsoon was likely. Indeed, two of the best atmospheric GCMs used for global prediction (NCEP and ECHAM; see the IRI website) predicted significantly below average June to September rainfall for India. In the event, the rainfall turned out to be indistinguishable from the climatological normal. It should be mentioned that the forecast of the Indian Meteorological Service was somewhat better, predicting only a marginally below normal year. This forecast is finally issued after a subjective evaluation, but begins with a statistical prediction scheme which includes El Niño information and factors other than El Niño -- factors seemingly included, inter alia, in the GCM's initial and boundary conditions.

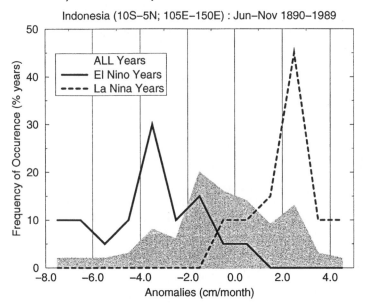

a) ENSO Impacts : Rainfall Anomalies

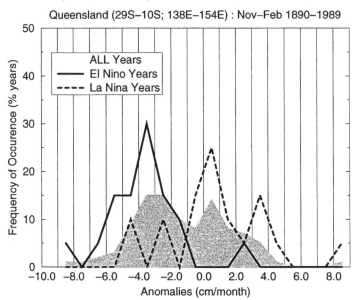

b) ENSO Impacts : Rainfall Anomalies

Figure 8. Impacts of ENSO on frequency of occurrence of rainfall anomalies (departures from climatological normal) in (a) Indonesia (10S-5N; 105 E-150E) and (b) Queensland, Australia (29S-10S; 138E-154E). Frequencies are calculated from rainfall records for the period 1890-1989. The frequency distribution of rainfall anomalies based on all years in the period is shaded, and the distribution during El Niño (La Niña) years only is the solid (dashed) line. Courtesy of the IRI.

NEB RAINFALL vs. TROPICAL SST

SEASONAL (February-May) CORRELATIONS

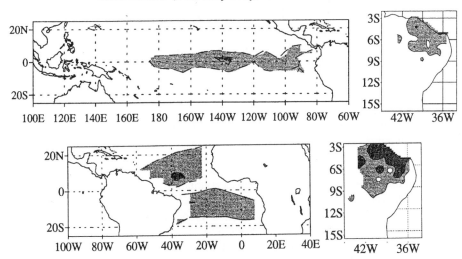

Figure 9. Association of rainfall in the Nordeste region of Brazil with tropical SST patterns in the Pacific and Atlantic oceans. Shown are the singular value decomposition patterns of SSTs and Northeast Brazil rainfall for the rainy season months of February through May. The top panels show the relation with tropical Pacific SST, the bottom panels with tropical Atlantic SST. Shaded correlations are above 0.4; dark shading is above 0.6. Courtesy of Y. Kushnir

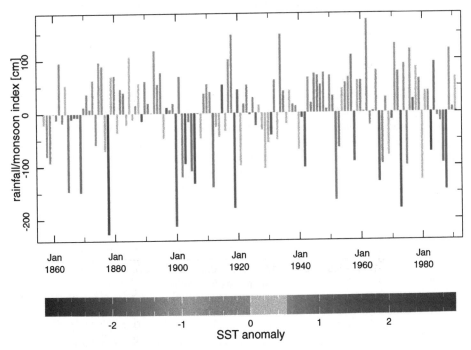

Figure 10 (A). All India rainfall index for 1856-1988. The bars are colour coded to indicate the strength of the SST anomaly in the NINO3 area. (This figure is reproduced in the colour section at the end of the book as plate A)

5. Discussion: Future Prospects

The towering landmark in the understanding of ENSO is the work of Jacob Bjerknes (1969, 1972). Among the consequences of the Bjerknes hypothesis is the idea that ENSO events can be predicted. Indeed, ENSO events now have been predicted, using models built to include the ocean-atmosphere physics Bjerknes singled out as essential. Moreover, the essence of ENSO is now understood: ENSO is a coupled instability of the atmosphere-ocean system in the tropical Pacific. Our theory for the ENSO mechanism is a combination of the Bjerknes hypothesis and linear shallow water ocean dynamics. In this understanding, the ENSO variations beyond the tropical Pacific are not integral to the generation of the ENSO cycle. Rather, they are consequences of the alterations in atmospheric heating due to the changes in tropical Pacific SST patterns.

As a general rule, the effects of an ENSO event are strongest and most reliable in the tropical Pacific genesis region and contiguous continents. When there is a warm event one can be fairly certain of heavy rains in Peru and drought in Indonesia. Typical consequences are somewhat less reliable in the global tropics, but still highly likely. The inactive 1997 Atlantic hurricane season is a striking confirmation of the expected pattern, while the normal 1997 Indian monsoon is a striking exception.

ENSO influence in extra tropical latitudes is less certain than in the tropics. Being more chaotic, the extra tropical climate system is intrinsically less predictable. Thus in these latitudes an ENSO event should be thought of as putting a bias in the system rather than as a certain cause. We are only beginning to learn the limits to predictive skill in such cases. Finally, there are patterns other than ENSO -- most prominently the North Atlantic Oscillation -- which influence climate, but the potential for predicting these patterns is not known.

Over the past years, truly remarkable strides have been made in mobilising climate prediction. There are presently many research groups doing routine ENSO prediction using a variety of methods. Regular observational updates for the tropical Pacific and summaries of forecast results are published monthly in the Climate Diagnostics Bulletin (available from the Climate Analysis Center of NCEP) and are available at a number of websites. This information has been used by groups in Peru, Northeast Brazil, India, China, Ethiopia, the United States, Australia and elsewhere to suggest actions to mitigate the effects of the local climatic variations associated with ENSO. The 1997 event has brought a high level of awareness to the public around the world, creating a new context for the task of designing and implementing response strategies for ENSO impacts.

More sophisticated and effective prediction procedures are emerging rapidly. The coupled general circulation model being run at NCEP is an example. The atmospheric component of this model is taken from the NCEP weather prediction system. A state-of-the-art procedure for creating fields of oceanic initial conditions takes advantage of the vastly expanded tropical Pacific Ocean data sets available from the observational network brought into being by the TOGA program (whose primary objective was the prediction of climate on time scales of months to years). Although it under predicted the amplitude somewhat, overall this prediction scheme can be credited with a successful forecast of the 1997 event.

It is unquestionably a great advance to have demonstrated the possibility of long range climate prediction. The forecasts are certainly far from perfect, especially so for the connections to local conditions with the greatest human consequences. We know that the nature of the climate system does not allow unlimited predictability, so that even a perfect forecasting system would not be able to deliver precise forecasts. Thus to be correct a forecast must be phrased as a probabilistic statement. Some aspects of the climate are more predictable than others. For example, given an ENSO event for some climate features the probability distribution will differ greatly from the climatological expectation, while for others it will not. For the former knowing that there is an ENSO event adds information. The forecast should be useful, all the more if the probability conditioned on ENSO is sharply peaked so that a rather specific outcome may be expected. Our current knowledge of the predictability of different features of the global climate is quite incomplete, but research over the next few years will add to it rapidly.

There is much room for improvement in prediction models, and work is underway to adapt existing technologies from weather prediction and elsewhere. As a legacy of the TOGA program an adequate observing system has been built, leaving us the simpler but still essential task of maintaining and improving it. Finally, the establishment of the International Research Center (IRI) for Seasonal to Interannual Climate Prediction provides an international focal point for efforts to capitalise on the progress in climate prediction and make it responsive to the needs of all impacted nations.

Acknowledgments

My thanks to Virginia DiBlasi-Morris, Larry Rosen, Dake Chen, Yochanan Kushnir, Steve Zebiak, Krishna Kumar, Ming Ji and Lisa Goddard for help with the manuscript preparation and useful advice. This work was supported by NOAA grant number MA56GPO221 and NASA- JPL grant JPL 957647.

References

Barnett, T.P. (1995) Monte Carlo climate forecasting. *J. Climate* **8**, 1005-1022.
Barnett, T.P., Bengtsson, L., Arpe, K., Flugel, M., Graham, N., Latif, M., Ritchie, J., Roeckner, E., Schlese, U., Schulzweida, U., and Tyree, M. (1994) Forecasting global ENSO-related climate anomalies. *Tellus* **46A**, 381-397.
Bjerknes, J. (1969) Atmospheric teleconnections from the equatorial Pacific. *Mon. Wea. Rev.* **97**, 163-172.
Bjerknes, J. (1972) Large-scale atmospheric response to the 1964-65 Pacific equatorial warming. *J. Phys. Oceanogr.* **15**, 1255-1273.
Cane, M.A., Eshel, G., and Buckland, R.W. (1994) Forecasting maize yield in Zimbabwe with eastern equatorial Pacific sea surface temperature. *Nature* **370**, 204-205.
Cane, M.A., Zebiak, S.E., and Dolan, S.C. (1986) Experimental forecasts of El Niño. *Nature* **321**, 827-832.

Cane, M.A., Kaplan, A., Miller, R.N., Tang, B., Hackert, E.C., and Busalacchi, A.J. (1996) Mapping tropical Pacific sea level: data assimilation via a reduced state space Kalman filter. *J. Geophys. Res.* **101**, 22599-22617.

Chen, D., Cane, M.A., and Zebiak, S.E. (1998) The impact of NSCAT winds on predicting the 1997/98 El Niño: a case study with the Lamont model. *J. Geophys. Res.* submitted.

Chen, D., Zebiak, S.E., Busalacchi, A.J., and Cane, M.A. (1995) An improved procedure for El Niño forecasting. *Science* **269**, 1699-1702.

Folland, C.K., Palmer, T.N., and Parker, D.E. (1986) Sahel rainfall and worldwide sea temperatures: 1901-85. *Nature* **320**, 602-606.

Goswami, B.N. and Shukla, J. (1991) Predictability of a coupled ocean-atmosphere model. *J. Climate* **4**, 3-22.

Hoerling, M.P., Kumar, A., and Zhong, M. (1997) El Niño, La Niña, and the nonlinearity of their teleconnections. *J. Climate* **10**, 1769-1786.

Hunt, B.G. and Hirst, A.C. (2000) Global climate models and their potential for seasonal climate forecasting, in G.L. Hammer, N. Nicholls, and C. Mitchell (eds.), *Applications of Seasonal Climate Forecasting in Agricultural and Natural Ecosystems – The Australian Experience.* Kluwer Academic, The Netherlands. (this volume)

Hunt, B.G., Zebiak, S.E., and Cane, M.A. (1994) Experimental predictions of climatic variability for lead time of twelve months. *Int. J. Climate* **14**, 507-526.

Hurrell, J. (1995) Decadal trends in the North-Atlantic oscillation – Regional temperatures and precipitation. *Science* **269**, 676-679.

Ji, M., Leetmaa, A., and Derber, J. (1995) An ocean analysis system for seasonal-to-interannual climate studies. *Mon. Wea. Rev.* **123**, 460-481.

Jin, F.F., Neelin, J.D., and Ghil, M. (1994) El Niño on the devils staircase – annual subharmonic steps to chaos. *Science* **264**, 70-72.

Nicholls, N. (1989) Sea surface temperature and Australian winter rainfall. *J. Climate* **2**, 965-973.

Nicholls, N., Fredericksen, C., and Kleeman, R. (2000) Operational experience with climate model predictions, in G.L. Hammer, N. Nicholls, and C. Mitchell (eds.), *Applications of Seasonal Climate Forecasting in Agricultural and Natural Ecosystems – The Australian Experience.* Kluwer Academic, (The Netherlands. (this volume)

Nobre, P. and Shukla, J. (1995) Variations of sea surface temperature, wind stress, and rainfall over the tropical Atlantic and South America. *J. Climate* **9**, 2464-2479.

Penland, C. and Sardeshmukh, P.D. (1995) The optimal-growth of tropical sea surface temperature anomalies. *J. Climate* **8**, 1999-2024.

Ropelewski, C.F. and Halpert, M.S. (1987) Global and regional scale precipitation patterns associated with the El Nino/Southern Oscillation. *Mon. Wea. Rev.* **115**, 1606-1626.

Ropelewski, C.F. and Halpert, M.S. (1996) Quantifying southern oscillation – precipitation relationships. *J. Climate* **9**, 1043-1059.

Stone, R.C., Hammer, G.L., and Marcussen, T. (1996) Prediction of global rainfall probabilities using phases of the southern oscillation index. *Nature* **384**, 252-255.

Tziperman, E., Stone, L., Cane, M.A., and Jarosh, H. (1994) El Niño chaos: Overlapping of resonances between the seasonal cycle and the Pacific Ocean-atmosphere oscillator. *Science* **264**, 72-74.

Wyrtki, K. (1975) El Niño - The dynamic response of the equatorial Pacific Ocean to atmospheric forcing. *J. Phys. Oceanogr.* **5**, 572-584.

Wyrtki, K. (1979) The response of sea surface topography to the 1976 El Niño. *J. Phys. Oceanogr.* **9**, 1223-1231.

Xue, Y., Cane, M.A., Zebiak, S.E. (1997a) Predictability of a coupled model of ENSO using singular vector analysis. Part I: Optimal growth in seasonal background and ENSO cycles. *Mon. Wea. Rev.* **125**, 2043-2056.

Xue, Y., Cane, M.A., Zebiak, S.E., and Palmer, T. (1997b) Predictability of a coupled model of ENSO using singular vector analysis. Part II: Optimal growth and ENSO forecast skill. *Mon. Wea. Rev.* **125**, 2057-2073.

Zebiak, S.E. (1989) On the 30-60 day oscillation and the prediction of El Niño. *J. Climate* **2**, 1381-1387.

Zebiak, S.E. and Cane, M.A. (1987) A model El Niño/Southern Oscillation. *Mon. Wea. Rev.* **115**, 2262-2278.

A GENERAL SYSTEMS APPROACH TO APPLYING SEASONAL CLIMATE FORECASTS

GRAEME HAMMER

Agricultural Production Systems Research Unit (APSRU)
Queensland Departments Primary Industries and Natural Resources and
CSIRO Tropical Agriculture
PO Box 102, Toowoomba, Qld 4350, Australia

Abstract

Agricultural and natural ecosystems and their associated business and government systems are diverse and varied, ranging from farming systems to water resource systems to species population systems to marketing and government policy systems, among others. These systems are dynamic and responsive to fluctuations in climate. Production, profit, conservation, and policy issues provide the major focus for intervention in these systems. Risk, or the chance of incurring a financial or environmental loss, is a key factor pervading decision-making. Skill in seasonal climate forecasting offers considerable opportunities to managers via its potential to realise system improvements (ie. increased profits and/or reduced risks). Realising these opportunities, however, is not straightforward as the forecasting skill is imperfect and approaches to applying the existing skill to management issues have not been developed and tested extensively.

The concepts associated with a systems approach to management are presented as a suitable means to apply seasonal climate forecasting within a decision-making context. An effective application of a seasonal climate forecast is defined as use of forecast information leading to a change in a decision that generates improved outcomes in the system of interest. A simple example of tactical management of row configuration in a cotton crop on the Darling Downs, Queensland, is presented to demonstrate an effective application. The profit outcomes with a fixed (all years the same) decision, or a tactical decision based on the SOI phase, or perfect knowledge of the season were compared using a crop simulation study. Over the complete historical climate record, the tactical approach gave an average profit increase of 11%. Adopting a tactical approach, however, did not give increased profit in every year. In 80% of years adopting a tactical approach was as good as or better than not adopting it, but in 20% of years the manager would have been worse off. It is suggested that effective implementation requires understanding of these risks and highlights the point that although tactical response to a forecast may pay off on average over a period of years, there can be no guarantees for the ensuing season.

G.L. Hammer et al. (eds.), The Australian Experience, 51–65.

The systems approach to applying climate forecasts in decision-making across the range of agricultural and natural ecosystems can be generalised to -

- understand the system and its management
- understand the impact of climate variability
- determine opportunities for tactical management in response to seasonal forecasts
- evaluate worth of tactical decision options
- participative implementation and evaluation
- feedback to climate forecasting

The nature of the interdisciplinary approach needed to pursue this systems approach to applying seasonal climate forecasts is discussed.

1. Introduction

Agricultural and natural ecosystems are diverse and varied, ranging from farming systems to water resource systems to species population systems, among others. These ecosystems are dynamic and responsive to fluctuations in climate. Production and conservation issues in these systems provide the major focus for management intervention. While management focus will largely depend on the specific objectives for intervening in any particular system, in general, a short term outlook concerned with production and profitability issues needs to be combined with a longer term outlook concerned with resource conservation and biodiversity issues. Climate variability generates risks for management decision-making on both short and long time horizons because outcomes of decisions cannot be predicted with any surety, be they decisions on crop management, stocking rate, water allocation, or fish or insect population management. Risk, or the chance of making a financial or environmental loss, is a key factor pervading decision-making in management of agricultural and natural ecosystems (Hardaker *et al.*, 1997).

Beyond the biophysical ecosystems, there is a range of other systems that effect, or are affected by, management of those ecosystems. These are the business and government systems that produce inputs required for management, market outputs, and determine policies influencing ecosystem management. These business and government systems, although operating at a different scale, are also dynamic and responsive to fluctuations in climate. Climate variability generates risks for decision-making in these systems in much the same way, because again, outcomes of decisions cannot be predicted with any surety, be they decisions on inventories, marketing strategies, or drought or taxation policy.

The introduction over the last decade of seasonal climate forecasts based on the El Niño - Southern Oscillation (ENSO) phenomenon and other research on global climate forcing factors (Cane, 2000), has provided the basis to consider taking advantage of climate variability, rather than passively accepting the risks it generates (Hammer and Nicholls, 1996). The possibility of adjusting management to what the next season is predicted to be offers considerable opportunities to managers of agricultural and natural ecosystems and the associated business and government systems. Realising these opportunities, however, is not straightforward as the forecasting skill is imperfect and approaches to applying the existing skill to management issues have not been developed extensively.

Hence, the aim of this paper is to
- present the concepts of a systems approach to management issues in agricultural and natural ecosystems, and their associated business and government systems,
- consider what is an effective application of seasonal climate forecasting in these systems, and
- via use of a simple example of an application using the systems framework, set out a general framework for approaching applications

2. What is a Systems Approach?

A system can be defined as a network of interacting elements receiving certain inputs and producing certain outputs. The dynamics of a system is implicit in this definition, which can be applied not only to the full range of agricultural and natural ecosystems and their associated business and government systems, but also to climate systems or economic systems. It is important to consider boundaries clearly when defining a system. The scale of the system and what is internal and external to the system are key issues to be clarified in the systems approach. These issues influence the perspective adopted when considering system function and management. A systems approach seeks and utilises understanding of system composition and dynamics derived from relevant research to enable prediction of the responses or behaviour of a system. In many cases, this capacity to predict utilises system models, which are a simplified representation of the system, often expressed mathematically. These concepts originated in the 1960's in industrial systems (eg. Forrester, 1961) and were adapted to agricultural and natural systems shortly after (eg. Duncan *et al.*, 1967; Patten, 1971)

The range of systems relevant to agricultural and natural ecosystems and their associated business and government systems includes
- a crop field and its management
- a farm and its management
- the national wheat crop and its marketing
- the national drought policy and its development
- a catchment and its management
- a species population and its management

For example, a crop field and its management consists of the crops, the soil and its physical and chemical attributes, and the manager undertaking a range of management practices, such as planting, cultivating, spraying and harvesting (Figure 1). Inputs to this system include daily weather, fuel, fertiliser, and pesticide. Outputs include harvested products, runoff, and soil loss. One could also look at balances for elemental C and N in compounds entering and leaving the system. The dynamics involve crop

54

growth and development processes and soil chemical and physical processes given the specific conditions of the system. The other systems listed could be similarly described in terms of their components, inputs, outputs and dynamics.

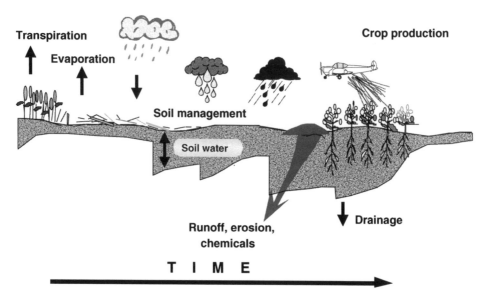

Figure 1. Schematic of a cropping system showing transitions of the system state through time in response to management activities. Some system inputs and outputs are indicated.

Management of a system is the manipulation of that system to achieve specific objectives. Hence, managers must be considered as an element of the system, or at least their actions considered as inputs to the system. The consequences of management manipulations may be predicted using system models, although the adequacy of such models for this task must be examined critically.

This systems approach can be applied to the full range of management and policy decisions associated with agricultural and natural ecosystems and their associated business and government systems. It is necessary, however, to place the systems approach within a problem solving context to realise its potential in application to decision-making. This concept is not new and stems from the rational approaches to problem solving derived from operations research methods developed in industry in the 1960's (Robertshaw *et al.*, 1978). It has formed the basis for the development of decision analysis procedures in agriculture (Dent and Anderson, 1971; Anderson and White, 1991).

A general approach to problem solving takes the following steps
• define the issue or problem in the system of interest
• determine objectives and criteria for measuring them
• gather information and knowledge about the system of interest

- identify and evaluate decision options
- implement preferred decision
- monitor the outcome and feedback new knowledge

Clearly defining the system of interest and the relevant issue and objective are critical components in problem solving. Credible system models facilitate problem solving in that they provide the means to evaluate alternatives via simulation analyses. Considerable scientific effort has been put to developing a range of system models. It has been argued that this effort has been most effective where connection between scientific rigour and predictive capability has been at the fore in driving model development (Hammer, 1998).

A systems approach in a problem-solving context requires on-going connections between decision-makers, advisors, modellers and researchers for effective outcomes. This integration of skills is required to achieve the balance needed between practicalities of system management, needs of decision-makers, and development and use of system simulation or expert knowledge to evaluate options. Checkland (1983) noted that the failure to be more aware of the human factor had resulted in a general failure of "hard" systems analysis approaches to influence what practitioners do. The failure of the "hard" methods to adequately address the human factor has lead to the emergence of different methodologies for dealing with the "soft" people-related issues, such as learning processes (eg. Bawden and Packham, 1991). McCown *et al* (1994) argued that the best prospects for developing better policy and management strategies lie with skilful use of "hard" system tools within a "soft" systems philosophical framework. An interdisciplinary and participative approach facilitates this combination and has been found effective in farming systems activities (McCown *et al.*, 1994).

3. What is an Application of a Seasonal Climate Forecast?

An effective application of a seasonal climate forecast is the use of forecast information that leads to the change in a decision that generates improved outcomes. A seasonal forecast has no value if it does not generate changed decisions. But to be effective the decision changes must produce positive changes in value by improving the relevant aspect of system performance targeted. In agricultural and natural ecosystems this most often relates to management decisions in those systems pertaining to their profitability, use, and conservation. In the associated business and government systems changed decisions may relate to profitability of the business and policy positions that are designed to influence ecosystem management indirectly.

An effective application differs considerably from the impact of a seasonal forecast. Impact measures the extent of effect, whereas application is the management response to change the anticipated impact. There have been many reports of significant impacts of predictions of certain season types (White, 2000; Nicholls, 1986; Rimmington and Nicholls, 1993; Meinke and Hammer, 1997) and while this is of interest in establishing the relevance of seasonal forecasting, it is of no value if nothing can be done to change from passive acceptance of impact to active response. Some studies have pursued effective applications by examining tactical changes in decisions associated with a seasonal forecast (Mjelde *et al.*, 1988; Hammer *et al.*, 1991; Keating *et al.*, 1993; Hammer *et al.*, 1996; Marshall *et al.*, 1996; Meinke and Stone, 1997).

3.1 EXAMPLE APPLICATION - COTTON CROP MANAGEMENT

A simple example of potential tactical management of row configuration in a cotton crop on the Darling Downs, Queensland, demonstrates an effective application of seasonal climate forecasting. Dryland cotton is planted as a row crop and can be configured with all rows planted (solid), every third row missed (single skip), or two rows planted and two rows missed (double skip). The rationale for manipulating row configuration is that although yield potential in good seasons is lost as rows are deleted, the slower rate of water use can give better yield in low rainfall seasons and, thus, reduce production variability risks associated with climatic variability. In addition, pesticide spray and some other management costs are reduced as rows are missed. With the solid row configuration, aerial spraying of insecticides is required, whereas more frequent use of less costly ground spraying technologies is feasible with the skip row configurations.

Is it possible to improve profitability by tactically manipulating row configuration in dryland cotton in response to a seasonal climate forecast? A simulation analysis using the cotton crop model, OZCOT (CSIRO Division of Plant Industry), in the agricultural production system simulator, APSIM (McCown *et al.*, 1996) was used to compare the management options over all years of the historical climate record (1887 - 1993) for Dalby, Queensland. Options were then compared for profitability over either all years or over sets of analogue years associated with a seasonal climate forecast. The historical analogue sets were defined using the SOI phase analysis system of Stone *et al.* (1996). By taking the SOI phase immediately preceded sowing (August-September), the climate of the subsequent growing season (October-April) could be allocated to one of five types for each year of the historical record. The best tactical management system was determined by selecting the row configuration giving maximum profit in each of the five season types associated with the SOI phases. This was compared with the row configuration giving maximum profit over all years to determine the degree of profit improvement associated with responding to the climate forecast. Both the fixed and tactical management approaches were compared to the "perfect knowledge" case, where the management system giving highest profit each year was implemented.

The scenario simulated was for a Siokra-type cotton planted with row configuration either solid, single skip, or double skip on 1 October on a Brigalow soil type with 260mm plant available water capacity to 1.8m depth and with 100mm of stored water in the soil profile at planting. These starting conditions were reset every year of the simulation so that the outcome reflected only variability associated with row configuration and seasonal climate conditions. Standard management practices for the region were assumed so that variable costs of production were $850, $790, and $750 per hectare for solid, single, and double skip, respectively (DPI Crop Management Notes).

Returns were assumed to be $400 per bale of cotton produced after allowing for returns from lint and seed and costs of ginning. Gross margin was calculated as the product of return per bale and bales produced per hectare less costs of production per hectare.

The simulated yield time series for the solid planting configuration (Figure 2) showed extreme variability in annual production, which emphasises the riskiness associated with growing dryland cotton. The variability associated with single or double skip row configurations was less but remained high (data not shown). Average simulated production over all years was 3.4, 3.2, and 2.8 bales per hectare for solid, single, and double skip configurations, respectively (Figure 3).

Average production varied among sets of years defined by SOI phases (Figure 3). When the SOI phase in August-September was consistently positive, average yield was greater. In contrast, if the SOI phase was rapidly falling, average yield was lower. In general, average yield was greatest with solid planting (Figure 3). It was only in years with consistently negative or rapidly falling SOI phase in August-September that average yield for single skip exceeded solid planting. However, the production averages do not reflect the economic outcome as costs differ among row configurations. They also contain no information about the risk associated with the high year-to-year variability (Figure 2).

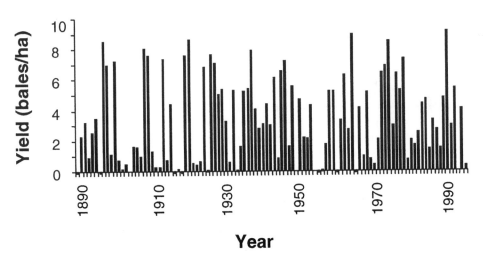

Figure 2. Simulated cotton yield versus year of planting from 1887 to 1992 for crops planted at Dalby, Queensland, on 1 October each year.

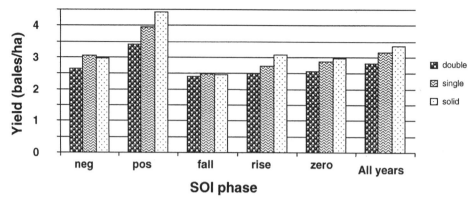

Figure 3. Average simulated cotton yield for three row configurations – solid, single skip row, and double skip row. Averages are given for the entire period 1887-1992 (all years) or for subsets of years within that period derived from the SOI phase preceding planting (August-September). The five phase types are consistently negative (neg), consistently positive (pos), rapidly falling (fall), rapidly rising (rise), and near zero (zero).

Average gross margin over all years was greatest with solid planting ($495/ha), but this was not consistent across groups of years associated with SOI phases (Figure 4). Assuming the simple criterion of maximising gross margin or profit, the best tactical decision making use of the forecast information would be to plant single skip, solid, double skip, solid, or single skip when the SOI phase preceding planting was consistently negative, consistently positive, rapidly falling, rapidly rising, or near zero. Ignoring the forecast information, the profit maximising strategy would be to plant solid in all years. In years when the August-September phase was consistently positive, average gross margin was much higher ($919/ha for solid planting). While this is a substantial *impact* on returns, associated with identifying high yield years, it had *no consequence on the decision* in those years, which remained as solid planting. This contrasted with the situation when the August-September SOI phase was either consistently negative or rapidly falling. In these season types, *the application of the forecast was effective as it indicated benefit from a change of decision*, as single and double skip options (average gross margins $427/ha and $206/ha) were more profitable than the solid planting configuration in those years ($340/ha and $129/ha, respectively). This was associated with the likely lower rainfall in those season types, which caused more favourable yield due to the slower water use of the single and double skip systems, combined with the cost savings of the single and double skip systems.

Comparing the tactical and fixed management approaches over the complete historical climate record, gave an average gross margin increase of $28/ha for the tactical approach (gross margins $523/ha and $495/ha, respectively). This is an increase of about 6% in gross margin or 11% in profit (calculated by deducting fixed costs). The "perfect knowledge" case gave an average gross margin of $819/ha, which is an increase of 65% in gross margin or 130% in profit over the fixed management approach. Hence, the seasonal climate forecasting system used yielded about 10% of the value contained in a perfect forecast. These percentage increases for value of the forecast information and its relationship to perfect knowledge are similar to that reported for wheat in NE Australia (Hammer *et al.*, 1996; Marshall *et al.*, 1996). Although the "perfect knowledge" situation is an unrealistic target, this indicates that considerable scope remains for improving forecasts. This methodology provides a general means to evaluate forecasts using the extent of the shift in value from no use of forecasts to the perfect knowledge case.

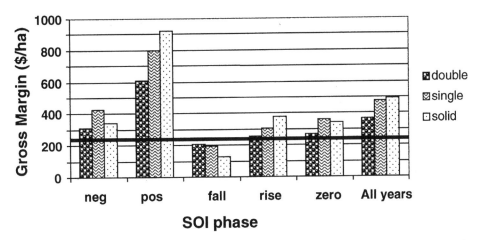

Figure 4. Average simulated gross margin for cotton grown using three row configurations – solid, single skip row, and double skip row. Averages are given for the entire period 1887-1992 (all years) or for subsets of years within that period derived from the SOI phase preceding planting (August-September) as defined in Figure 3. The solid line represents the level of fixed costs ($250/ha).

In this example application, a number of simplifications were maintained for ease of communication. For example, the risk, or chance of making a loss, associated with each management option was not considered. The simple criterion of maximising gross margin or profit, independent of risk, was used. The trade-off between profit and risk, however, is a vital component in decision analysis in this type of situation (Hardaker *et al.*, 1997) and has been an important aspect of previous studies (Hammer *et al.*, 1996). In this example, if risk is defined as the chance of making a loss (ie. not recovering fixed costs) in any year, then there was a shift in risk with season type (Figure 4). There was a considerably increased chance of making a loss in years with rapidly falling SOI (as the average gross margin was less than the fixed costs) and there was a considerably decreased chance of making a loss in years with SOI consistently positive. These shifts in risk can have implications on tactical management responses if the risk differs among management options within season types. This will not be examined in this case study as it is dealt with in greater detail in subsequent chapters (eg Carberry *et al.*, 2000), but, in general, it should form part of the analysis procedure.

60

A further simplification is that this example related only to a specific situation – a specific combination of soil condition, location and planting date. The outcome will vary with changes in these conditions and cannot be generalised without further analysis. To implement an application of this nature would require such further analysis and direct interaction with decision-makers. The analysis presented provides only a basis for discussion as a starting point in this process.

One of the key issues in discussing such analyses with decision-makers concerns the chance of coming out ahead in any one year if the forecast information was used to adjust decisions as suggested. This *average* increase found for the tactical approach is derived from a distribution of differences between the two approaches (ie. using or not using the forecast information) on a year-to-year basis. Adopting a tactical approach did not give increased profit in every year. In some years it was substantially better, whereas in others it was substantially worse (Figure 5). Overall, there were about 40% of years in which the tactical approach gave higher profit, 40% of years with no difference, and 20% of years with lower profit. This distribution is what the manager faces each year and the risks must be understood for effective implementation. Statistical testing of the time series of the benefit of the tactical over the fixed decision strategy showed no significant patterns or trends in these data over the last century (data not shown), indicating that outcomes from this distribution are equally likely in any year. While it is clear that, on average, outcomes can be improved in the longer term, there can be no guarantees for the ensuing season. The question a manager faces in deciding whether or not to apply a seasonal climate forecast is – am I content with an 80% chance of doing as well as or better than I would do without applying the forecast? This percentage is the targeted piece of information needed and cannot be derived by simple examination of the forecast. Dealing with probabilistic information remains an important consideration in implementing such applications. This issue is addressed in subsequent chapters (Meinke and Hochman, 2000; Nicholls, 2000).

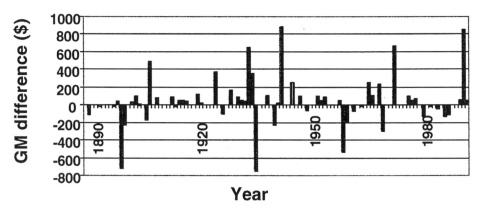

Figure 5. Difference in gross margin between tactical (responsive to seasonal climate forecast) and fixed (non-responsive) row configuration management strategies for each year of the cotton simulation study.

4. Using the Systems Approach in Applying Seasonal Climate Forecasts

To generalise from the cotton management example, the emphasis in a systems approach is to develop targeted information for influencing the most relevant decisions in the system of interest. This concept is relevant across the range of scale and issues associated with agricultural and natural ecosystems and their associated business and government systems (Figure 6). Whether the key issue is short-term profitability or long-term resource conservation, the emphasis needs to be on the analysis required to target the forecast information to the issue and the decision-maker. Generalised seasonal forecasts, which have information relevant across all systems, are likely to have little value if their targeting is not considered. The relevant decision-maker at each scale must be included as part of the systems approach to ensure clear problem definition and understanding of relevant decisions and information needs. The systems approach will often involve systems modelling as a means to move from general to targeted information. A number of subsequent chapters will pursue this systems approach for this diverse range of systems.

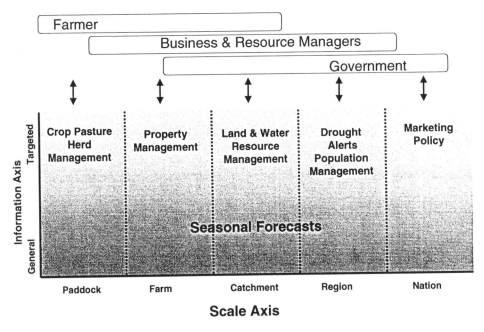

Figure 6 (B). The relationships between scale, information content, and decision-makers in defining relevant systems and the systems approach to applying seasonal climate forecasts in agricultural and natural ecosystems. (This figure is reproduced in the colour section at the end of the book as plate B)

The systems approach to applying climate forecasts to decision-making in agricultural and natural ecosystems can be generalised to a sequence of steps that are relevant to the wide range of types of decision in these systems. The steps can be summarised as-

1. Understand the system and its management - Whether the system is a paddock, a farm, or the national wheat crop it is essential to understand its dynamics and management and the points at which significant decisions influence desired system performance. Such understanding can only be gained by close interaction with system managers or their advisers. Developing predictive system models is an excellent means to learn about the system as well as generate a suitable tool for subsequent use.

2. Understand the impact of climate variability - It is essential to know just where in the system climate variability is an issue. Only at those points will a seasonal forecast be worth pursuing.

3. Determine opportunities for tactical management in response to seasonal forecasts - If a seasonal forecast was available, what possible options are there at relevant decision points? That is, how might decisions be changed in response to a seasonal forecast? What nature of forecast would be most useful? What lead-time is required for management response?

4. Evaluate worth of tactical decision options - Given a range of potential tactical management responses to a seasonal forecast, an evaluation of their expected outcomes and risks is required as a basis for discussion with decision-makers. Simulation analyses using system models provide a means to evaluate tactical options.

5. Participative implementation and evaluation - Working with managers/decision-makers generates valuable insights and learning throughout the entire process. An evaluation of outcomes with respect to changed decisions or processes highlights issues for subsequent focus.

6. Feedback to climate forecasting - Rather than just accept a given climate forecast, consider what specific improvements would be of greatest value in the system. This can provide some direction for the style of delivery of forecasts and for climate research.

5. Discussion

Connecting forecasts into applications remains an area where we have a lot to learn. This paper argues that the systems and operational research approach to applying seasonal climate forecasts provides an appropriate and general methodology for integrating forecasts into decision-making. Implementing this approach, however, imposes requirements on design of forecasts. Using agricultural and natural system models to analyse and evaluate decision options, usually requires input streams of daily climate data that sample the seasonal variability likely to be experienced. Many seasonal climate forecasts provide only general predictions and, hence, are difficult or impossible to use in the systems approach to applications. A forecasting scheme that identifies historical analogue seasons or years, as in the cotton example given above, is perhaps the most useful concept to date for facilitating applications. The use of this style of forecasting scheme in many of the case studies to follow in this book is testament to this. If seasonal forecasts are intended for targeted use as distinct from general information (Figure 6) then they must be designed to accommodate this requirement.

The analogue seasons or years concept does not, however, need to be confined to partitioning the historical record as in the cotton case study. It may be possible to derive synthetic analogues in some way via stochastic weather generators, for example. This would facilitate connection of forecasts to applications for forecasting systems that do not generate historical analogues, such as forecasts derived from dynamic ocean and/or atmosphere models. Developments in weather generators would also facilitate inclusion of effects of potential climate change, which may underlie trends in historical data that need to be considered when using historical analogues. The point remains, however, that development of appropriate methods to derive analogue seasons or years in order to connect forecasts and applications needs to be viewed as an essential component of forecasting research and development. It is an area where increased effort is needed.

Quantifying the unexplained variability in a seasonal climate forecast is as important in applications as the predicted shift in mean climate. This unexplained variability is what drives decision-making risks and opportunities. It needs to be communicated in a useful way, not hidden in an average, median, or simple probability statement. Generating an average forecast from some ensemble may simplify forecast delivery, but it removes the unexplained variability from the forecast. The issuing of forecasts as simple probability statements is better, but provides only general information. Defining analogue seasons or years captures this variability in a way that enables riskiness of alternative decisions to be evaluated by examining each year in the analogue set separately (eg. Figure 3).

Separate analysis of each analogue year is essential, as is careful consideration of how to condense the resulting output to the most useful form. Averages are often a far less meaningful statistic of the probability distribution of outcomes than some consideration of the likelihood of exceeding (or not exceeding) some critical system state, be it profit or land condition. Given this approach, outcomes of decision options can then be assessed in terms of both their expected outcome and the associated risk. Both aspects are important to the decision-maker when having to decide among competing options. One of the most important features of a useful forecast may be the extent to which it reduces risk of making a loss (Hammer *et al*, 1996).

Understanding targeted information about effects of seasonal climate forecasts on likely outcomes of decision-making options is the key to successful application of forecasts. The information is probabilistic in nature and is relevant to managing risks over a number of years. In any one year, responding to a forecast may or may not pay-off. Over a number of years, however, responding to a forecast results in significant advantages (eg. Figure 5). The notions of a "win" or a "loss" or "getting it right" or "getting it wrong" on a single occasion are not really appropriate. The shift in outcome probabilities associated with application of a forecast is valuable in risk management if used with such understanding. This requires close interaction with decision-makers as part of the development process. The participative approach facilitates learning by the decision-maker in relation to the nature of the information and how to use it effectively, while facilitating learning by the researcher in relevant decisions to target and in communication process. This approach provides case studies that are useful for broader educative activities with decision-makers and researchers. The difficulties associated with using this type of information are explored in some depth in subsequent chapters (Meinke and Hochman, 2000; Nicholls, 2000).

Converting general information on seasonal likelihood to targeted information on likely outcomes of decision options requires robust and reliable models of the systems in question. This is not a trivial task and should not be taken for granted. The models used in the cotton example have taken many years of intensive research, development, and testing to get to their current state. There are many possibilities for improvement in models (eg. Hammer, 1998) making it important to maintain on-going effort in this area.

The systems approach to applying seasonal climate forecasts offers considerable opportunities to improve decision-making in our agricultural and natural ecosystems. It does, however, require interdisciplinary action. Continual interaction of forecast application R&D, forecast development R&D, and decision makers managing the target systems is essential to ensure that forecasts are produced in a way that maximises their utility and relevance. Such interaction also provides opportunities to generate targets for improvement of forecasts as well as a means to capture quickly any innovations in forecasting.

Acknowledgments

The use of the OZCOT cotton model developed by CSIRO Division of Plant Industry, Cotton Research Unit, Narrabri and assistance from Dr P Carberry and Mr D Holzworth (APSRU) in conducting the simulation analysis are gratefully acknowledged.

References

Anderson, J.R. and White, D.H. (1991) Systems thinking as a perspective for the management of dryland farming, in V. Squires and P.G. Tow (eds.), *Dryland Farming - A Systems Approach.* Sydney University Press, Sydney. pp.16-23.

Bawden, R.J. and Packham, R.G. (1991) Improving agriculture through systemic action research, in V. Squires and P.G. Tow (eds.), *Dryland Farming - A Systems Approach.* Sydney University Press, Sydney. pp.261-270.

Cane, M.A. (2000) Understanding and predicting the world's climate system, in G.L. Hammer, N. Nicholls, and C. Mitchell (eds.), *Applications of Seasonal Climate Forecasting in Agricultural and Natural Ecosystems – The Australian Experience.* Kluwer Academic, The Netherlands. (this volume)

Carberry, P.S., Hammer, G.L., Meinke, H. and Bange, M. (2000) The potential value of seasonal climate forecasting in managing cropping systems, in G.L. Hammer, N. Nicholls, and C. Mitchell (eds.), Applications *of Seasonal Climate Forecasting in Agricultural and Natural Ecosystems – The Australian Experience.* Kluwer Academic, The Netherlands. (this volume)

Checkland, P.B. (1983) O.R. and the systems movement: Mappings and conflicts. *J. Operational Research Society* **34**, 661-675.

Dent, J.B. and Anderson, J.R. (1971) *Systems analysis in agricultural management*. John Wiley, Sydney. pp.394.

Duncan, W.G., Loomis, R.S., Williams, W.A. and Hanau, R. (1967) A model for simulating photosynthesis in plant communities. *Hilgardia* **38**, 181-205.

Forrester, J.W. (1961) *Industrial dynamics*. Massachusetts Institute of Technology Press, Cambridge, Massachusetts, pp.464.

Hammer, G.L. and Nicholls, N. (1996) Managing for climate variability - The role of seasonal climate forecasting in improving agricultural systems. Proc. Second Australian Conference on Agricultural Meteorology. Bureau of Meteorology, Commonwealth of Australia, Melbourne. pp.19-27.

Hammer, G.L. (1998) Crop modelling: Current status and opportunities to advance. *Acta Horticulturae* **456**, 27-36.

Hammer, G.L., McKeon, G.M., Clewett, J.F. and Woodruff, D.R. (1991) Usefulness of seasonal climate forecasts in crop and pasture management. Proc. First Australian Conference on Agricultural Meteorology. Bureau of Meteorology, Commonwealth of Australia, Melbourne. pp.15-23.

Hammer, G.L., Holzworth, D.P. and Stone, R. (1996) The value of skill in seasonal climate forecasting to wheat crop management in a region with high climatic variability. *Aust. J. Agric. Res.* **47**, 717-737.

Hardaker, J.B., Huirne, R.B.M. and Anderson, J.R. (1997) *Coping with risk in agriculture*. CAB International. pp.274.

Keating, B.A., McCown, R.L. and Wafula, B.M. (1993) Adjustment of nitrogen inputs in response to a seasonal forecast in a region of high climatic risk, in F.W.T. Penning de Vries *et al.* (eds.), *Systems Approaches for Agricultural Development*. Kluwer, The Netherlands. pp.233-252.

Marshall, G.R., Parton, K.A. and Hammer, G.L. (1996) Risk attitude, planting conditions and the value of seasonal forecasts to a dryland wheat grower. Aust. *J. Agric. Econ.* **40**,211-233.

McCown, R.L., Cox, P.G., Keating, B.A., Hammer, G.L., Carberry, P.S., Probert, M.E. and Freebairn, D.M. (1994) The development of strategies for improved agricultural systems and land-use management, in P. Goldsworthy and F.W.T. Penning de Vries (eds.), *Opportunities, Use, and Transfer of Systems Research Methods in Agriculture to Developing Countries*. Kluwer, The Netherlands. pp.81-96.

McCown, R.L., Hammer, G.L., Hargreaves, J.N.G., Holzworth, D.P. and Freebairn, D.M. (1996) APSIM: A novel software system for model development, model testing, and simulation in agricultural research. *Agric. Sys.* **50**, 255-271.

Meinke, H. and Hammer, G.L. (1997) Forecasting regional crop production using SOI phases: a case study for the Australian peanut industry. *Aust. J. Agric. Res.* **48**, 231-240.

Meinke, H. and Stone, R.C. (1997) On tactical crop management using seasonal climate forecasts and simulation modelling - a case study for wheat. *Sci. Agric., Piracicaba* **54**, 121-129.

Meinke, H. and Hochman, Z. (2000) Using seasonal climate forecasts to manage dryland crops in northern Australia, in G.L. Hammer, N. Nicholls, and C. Mitchell (eds.), *Applications of Seasonal Climate Forecasting in Agricultural and Natural Ecosystems – The Australian Experience*. Kluwer Academic, The Netherlands. (this volume)

Mjelde, J.W., Sonka, S.T., Dixon, B.L., and Lamb, P.J. (1988) Valuing forecast characteristics in a dynamic agricultural production system. *American Journal of Agricultural Economics* **70**, 674-684.

Nicholls, N. (1986). Use of the southern oscillation to predict Australian sorghum yield. *Agric. For. Meteorol.* **38**, 9-15.

Nicholls, N. (2000) Impediments to the use of climate predictions, in G.L. Hammer, N. Nicholls, and C. Mitchell (eds.), *Applications of Seasonal Climate Forecasting in Agricultural and Natural Ecosystems – The Australian Experience*. Kluwer Academic, The Netherlands. (this volume)

Patten, B.C. (1971) *Systems Analysis and Simulation in Ecology*. Academic Press, New York.

Rimmington, G.M. and Nicholls, N. (1993) Forecasting wheat yields in Australia with the southern oscillation index. *Aust. J. Agric. Res.* **44**, 625-632.

Robertshaw, J.E., Mecca, S.J. and Rerick, M.N. (1978) *Problem Solving: A Systems Approach*. Petrocelli Books Inc., New York. pp.272.

Stone, R.C., Hammer, G.L. and Marcussen, T. (1996) Prediction of global rainfall probabilities using phases of the Southern Oscillation Index. *Nature* **384**, 252-256.

White, B. (2000) The importance of climate variability and seasonal forecasting to the Australian economy, in G.L. Hammer, N. Nicholls, and C. Mitchell (eds.), *Applications of Seasonal Climate Forecasting in Agricultural and Natural Ecosystems – The Australian Experience*. Kluwer Academic, The Netherlands. (this volume)

THE DEVELOPMENT AND DELIVERY OF CURRENT SEASONAL CLIMATE FORECASTING CAPABILITIES IN AUSTRALIA.

ROGER C. STONE[1] AND GRAHAM C. DE HOEDT[2]

[1] Queensland Centre for Climate Applications
Queensland Dept. Primary Industries, PO Box 102
Toowoomba, Qld 4350, Australia

[2] National Climate Centre
Bureau of Meteorology, PO Box 1289K
Melbourne, Vic 3000, Australia

Abstract

Current seasonal climate forecasting capabilities in Australia are the result of many decades of research and development, much of which has been carried out in Australia. Current forecasting activity is focused on statistical forecasting techniques that use the Southern Oscillation Index or certain sea-surface temperature anomaly patterns. More recent research in seasonal climate forecasting is being directed towards the use of atmospheric general circulation models or more complex fully coupled ocean-atmosphere models. Climate forecast systems in Australia have now been integrated into crop and pasture simulation models to aid the development of farming decision-support systems.

1. Development of Seasonal Climate Forecasting in Australia

As early as 1893, the relevance and value of probabilistic climate forecast information for Australia's rural industries had been recognised (Todd, 1893). The Australian Bureau of Meteorology has been generating seasonal climate forecasts since 1929. This work followed analyses by meteorologist H.A. Hunt (Hunt, 1913; 1914; 1916; 1918) who provided much initial understanding of rainfall variability in Australia. E.T Quayle, a Bureau of Meteorology scientist, published the first scientific method for Australian climate prediction in 1929 (Quayle, 1929).

'Good rains' during the 20 years following the Second World War may have resulted in less emphasis and importance being placed on climatic analyses and research in Australia (Nicholls, 1997). Re-emergence of drought conditions in many areas of Australia in the 1960s along with a renewed awareness of the extremely high variability

G.L. Hammer et al. (eds.), The Australian Experience, 67–75.

of Australian rainfall (Gibbs, 1963) coincided with an upsurge in interest in mechanisms responsible for equatorial Pacific Ocean atmospheric variability, especially in the Southern Oscillation (eg. Troup, 1965). Importantly, Berlage (1961, 1966) developed a simple two-station monthly pressure difference to represent the two 'centres of action' of the Southern Oscillation. Troup's analysis re-confirmed the seasonal structure of global sea-level pressure correlations and the nature of persistence in lag correlation patterns in the Southern Oscillation (Allan *et al.*, 1996).

Not surprisingly, many of the recently employed seasonal climate forecasting systems in Australia have drawn upon the findings of the earlier correlation and regression analyses that used an index of the Southern Oscillation in their compilation (eg. Troup 1965; Berlage, 1966). Efforts by Nicholls (1978, 1983, 1984a,b,c, 1985a,b, 1986a,b, 1988a,b, 1989, 1991a,b, 1992), Nicholls *et al.* (1982), and McBride and Nicholls (1983) provided the lead in Australia for much of the current concerted activity in seasonal climate forecasting. Nicholls' work described the links between the El Niño/Southern Oscillation (ENSO) and phenomena such as tropical cyclone activity, onset of the northern wet season, Australian droughts, crop yields, disease outbreaks, patterns of Australian vegetation, and rainfall patterns. Nicholls' work also evaluated the stability of Southern Oscillation/Australian rainfall relationships over time (Nicholls and Woodcock, 1981; Nicholls, 1984c), which was important as evidence for assessing the long-term feasibility of using the Southern Oscillation Index (SOI) in operational climate forecasting systems. Important work by Williams (1986) and Drosdowsky (1988) greatly facilitated the understanding of more complex relationships between the Southern Oscillation and the troposphere over Australia.

Pittock (1975) showed that the dominant principal component of rainfall variability in Australia was related to the Southern Oscillation. However, Pittock also demonstrated that the latitude of the sub-tropical ridge was an important additional contributor to variability in Australian rainfall. Work by Priestley (1962, 1963) had earlier established relationships between Darwin pressure and Australian rainfall and the marked persistence in eastern Australian spring and summer rainfall. Coughlan (1979) identified patterns and trends in temperatures over Australia that could be related to trends in circulation features attached to either the mean latitude of the sub-tropical ridge or to the Southern Oscillation. However, it was Streten (1975, 1981, 1983, 1987) and Lough (1991, 1992, 1993, 1994) who re-emphasised the need to identify relationships with Pacific Ocean sea-surface temperature (SST) anomalies (including Coral Sea SST anomalies). Particularly useful overviews of ENSO and climatic fluctuations in Australia are found in Allan (1985, 1988, 1989, 1993) and especially Allan *et al.* (1996).

More recent attention in statistical climate forecasting research in Australia is being directed towards investigation of propagation characteristics of Indian Ocean and Pacific Ocean pressure fields (which may evolve in response to an interaction between the annual cycle and quasi-biennial and lower frequency components). In particular, core regions in the Indo-Pacific appear to be influenced by decadal and interdecadal pressure signals which may be significant in the further development of seasonal climate forecasting for Australia (Allan *et al.*, 1996; Power *et. al.*, 1999; White, 1999; Drosdowsky and Allan, 2000).

Linear, lagged regression techniques have featured as the basis for the *Seasonal Climate Outlooks* provided by the National Climate Centre of the Bureau of Meteorology in Australia. Multiple linear regression techniques and linear discriminant analyses have been routinely applied to provide seasonal forecasts of rainfall based on the SOI or, more recently, Pacific and Indian Ocean SST. Since 1995, trends and 'phases' of the SOI have also been employed to provide seasonal climate outlook maps and conditional rainfall probability distributions. Output from these methods include forecast maps that describe the chances of rainfall in the 'above, average, or below average' range or the probability of exceeding the climatological median (Figure 1). Climate forecast output is also available as cumulative probability distributions for individual locations, based on lagged relationships between SOI patterns or phases and rainfall (Zhang and Casey; 1992; Stone and Auliciems, 1992; Stone and Hammer, 1992; Stone *et al.*, 1996a,b; Nicholls, 1997) (Figure 2). This type of climate forecast output is available through computer programs, such as *Australian Rainman* (Clewett *et al.*, 1994).

Seasonal climate forecasts of temperature and frost have not yet been developed to a level where routine climate outlooks are disseminated. However, early work by Coughlan (1979) and more recently Jones (1998) suggests seasonal climate forecasts of temperature that use the SOI or certain sea-surface temperature components as predictors should be relatively straightforward to produce. Stone *et al.* (1996b) suggest forecasts of frost season length and frost frequency for many regions of Australia could also be feasible.

A more recent development in seasonal climate forecasting research and development is the use of empirical orthogonal functions (EOFs) to isolate significant regions of global sea-surface temperature that affect rainfall patterns in Australia. This statistical forecasting system has been developed to replace climate forecasting systems that use the SOI. This method also captures effects that are considered to be separate from ENSO, such as Indian Ocean sea-surface temperatures, which can provide forecasting skill in those parts of the country where ENSO impacts are considered minimal (Nicholls, 1989; Drosdowsky, 1993a,b,c; Drosdowsky and Chambers, 1998; Drosdowsky and Allan, 2000).

A current major focus of research and development in seasonal climate forecasting activity in Australia is towards the use of both 'simple' atmospheric general circulation models (GCMs) and more complex coupled ocean-atmosphere models (Hunt and Hirst, 2000). The development of so-called limited area models has been aimed at capturing the essence of ENSO and therefore are aimed at producing accurate forecasts of El Niño and La Niña (Latif *et al.*, 1994; Allan *et al.*, 1996). Some 'less complex' models developed at the Bureau of Meteorology Research Centre (BMRC) possess minimal physical representation of the atmosphere by usually representing just two layers and are developed to resolve the basic dynamics of ENSO (see Kleeman, 1991, 1993). More recent developments have seen an improvement in the ability to construct more

70

complex ocean-atmosphere models and in the potential to predict both ENSO and Australian rainfall. Both BMRC and the CSIRO are continuing to improve the representation of the coupling between ocean and atmosphere: the Australian Community Ocean Model (Frederiksen *et al.*, 1995; Power *et al.*, 1995; Allan *et al.*, 1996; Nicholls, 1997). While useful output from coupled ocean-atmosphere models that is suitable for operational seasonal climate forecasts may still be some years from being produced, some results from this research could be applied to practical forecasting systems in the near future. For example, the SOI can be predicted quite readily from these atmospheric GCMs (I. Smith, pers. comm.). This forecast-SOI may then be used to provide probability distributions of rainfall for locations and districts throughout Australia (Stone *et al.*, 2000).

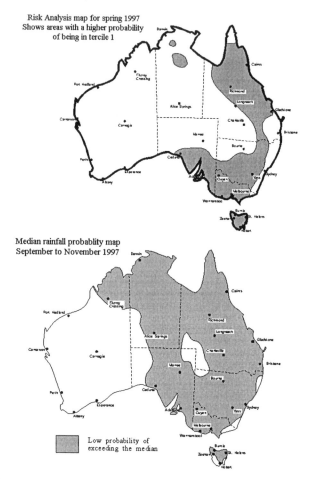

Figure 1. Example of seasonal climate forecast output recently generated by The Bureau of Meteorology. Maps indicate the probability of rainfall being within the lowest tercile and the probability of exceeding the climatological median, respectively, for the September to November period, 1997.

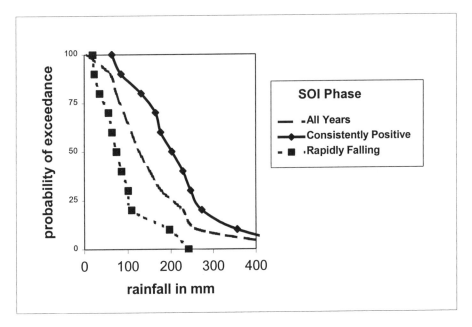

Figure 2. Example of cumulative probability distributions of forecast rainfall for a selected individual location in Australia. Shown are the rainfall probability distributions for the October to December period for Charters Towers, Queensland associated with two of the phases of the SOI through the preceding August/September. Non-parametric tests have been applied to test the significance of the shift in the distributions. K-W test p=.05; K-S test 'SOI-phases'(2,3) p=.01

2. Delivery of Current Seasonal Climate Forecasting in Australia

Early experimental seasonal forecasts produced internally by the Bureau of Meteorology were based on certain aspects of persistence of anticyclonicity, blocking patterns, and the movement of large-scale anomalies (Nicholls, 1997). However, since 1989, following research developments based on the SOI described above, seasonal climate forecasts have been issued publicly by the Bureau of Meteorology. Another key aspect of the compilation of these national seasonal climate forecasts has been a monthly climate forecast policy meeting hosted by the Bureau of Meteorology National Climate Centre (NCC). This meeting forms a key part of the process of routine seasonal climate forecast generation. It involves scientists from organisations such as the Bureau of Meteorology National Climate Centre, the Bureau of Meteorology Research Centre (BMRC), CSIRO Division of Atmospheric Research, the Queensland Department of Primary Industries and other State and Federal organisations. This meeting is similar to that conducted by the United States Climate Prediction Centre (CPC) in that it provides a useful 'sounding-board' in which many aspects of seasonal climate forecasting research, development, and delivery can be discussed. The relevant current 'forecast policy' is developed at this meeting.

The National Climate Centre currently issues the *Seasonal Climate Outlook* publication that provides the national seasonal climate forecast plus relevant background information. The NCC also provides extensive output relating to global atmospheric and oceanographic states, histories of the SOI, rainfall figures and maps, and statistical summaries of rainfall, temperature, and drought. The Queensland Department of Primary Industries, in conjunction with the Bureau of Meteorology, provides extensive climate forecast information targeted towards rural clients. This activity includes 'Managing for Climate Workshops' (workshops that facilitate linkages between climate forecasting and farm management) and production of computer programs such as Australian RAINMAN (Clewett *et al.*, 1994). Australian RAINMAN provides rainfall probability distributions associated with certain SOI classes or SOI phases which allows users to develop their own 'tailor-made' climate forecasts for their location. Internet services provided by the Bureau of Meteorology are particularly extensive and these include climate information on the 'Met Net' site. The Queensland Departments of Primary Industries and Natural Resources also provide climate forecasting and climate information services through their 'The Long Paddock' internet site. The New South Wales Agriculture Department has developed a web site that provides access to climate forecast models and output and the latest drought summaries. The Bureau of Meteorology and the Queensland Departments of Primary Industries and Natural Resources have also developed the 'SILO' internet site that provides extensive climate data and information for users.

Users' needs in seasonal climate forecasting have been continually assessed since seasonal climate forecasting systems and products were introduced in the late 1980s. Results obtained from workshops conducted in Queensland and New South Wales suggest many users of seasonal climate forecasts may have had difficulty in understanding the value of seasonal climate forecasting unless they were able to identify direct application of seasonal climate forecasting systems to particular management needs. Numerous workshops conducted over the past 6 years in rural areas of Australia have allowed participants to gain ownership of seasonal climate forecasting techniques and become aware of key linkages between climate variability, seasonal climate forecasting, and cropping and pasture management decisions (eg. Hammer *et al.*, 1996; Meinke and Hochman, 2000).

Summary

Seasonal climate forecasting in Australia is based on many years research and development conducted both in Australia and internationally. Research work conducted since the 1970s primarily focused on identifying empirical relationships between the Southern Oscillation, ENSO, sea-surface temperature patterns, and rainfall. A remarkable amount of effort by scientists at the Bureau of Meteorology and the CSIRO has led the way for much of the more recent and innovative research work in Australia into ENSO rainfall relationships. This research effort has also formed the basis for much of the seasonal climate forecasting output in this country produced by the National Climate Centre and such organisations as the Queensland Department of

Primary Industries. However, more recent research and development work in Australia has focused on development of complex coupled ocean-atmosphere models (that may predict future regional rainfall patterns as well as sea-surface temperature values) and on integrated climate forecasting/farming systems models that greatly facilitate the linkages between climate forecasting and on-farm decisions.

References

Allan, R.J. (1985) The Australian summer monsoon, teleconnections, and flooding in the Lake Eyre basin. South Australian Geographical Papers No 2, Roy. Geogr. Soc. Australasia (South Australian Branch). pp.47.

Allan, R.J. (1988) El Nino Southern Oscillation influences in the Australasian region. *Prog. Phys. Geogr.* **12**, 4-40.

Allan, R.J. (1989) ENSO and climatic fluctuations in Australasia, in Donnelly, T.H. and Wasson, R.J. (eds.), Proceedings CLIMANZ III Symposium, University of Melbourne. pp.49-61.

Allan, R.J. (1993) Historical fluctuations in ENSO and teleconnection structure since 1879: Near-global patterns. *Quat . Aust.* **11**, 17-27.

Allan, R.J., Lindesay, J. and Parker, D. (1996) *El Niño Southern Oscillation and Climatic Variability.* CSIRO Publishing, Collingwood, Victoria, Australia. pp.405.

Berlage, Jr., H.P. (1961) Variations in the general atmospheric and hydrospheric circulation of periods of a few years duration affected by variations in solar activity. *Annals of the New York Academy of Sciences* **95**, 354-367.

Berlage, Jr., H.P. (1966) The Southern Oscillation and world weather. Mededlingen en Verhandelingen No. 88, Koninklijk Meteorologische Institut, Staatsdrukkerijs-Gravenhage, Netherlands. pp.152.

Clewett, J.F., Clarkson, N.M., Owens, D.T. and Arbrecht, D.G. (1994) Australian Rainman: Rainfall Information for Better Management. Department of Primary Industries, Brisbane, Queensland, Australia.

Coughlan, M.J. (1979) Recent variations in annual-mean maximum temperatures over Australia *Q. J. R. Meteor. Soc.* **105**, 707-719.

Drosdowsky, W. (1988) Lag relations between the Southern Oscillation and the troposphere over Australia. BMRC Research Report No. 13, Bureau of Meteorology, Melbourne. pp.201.

Drosdowsky, W. (1993a) An analysis of Australian seasonal rainfall anomalies: 1950-1987. I: Spatial patterns. *Int. J. Climatol.* **13**, 1-30.

Drosdowsky, W. (1993b) An analysis of Australian seasonal rainfall anomalies: 1950-1987. II: Temporal variability and teleconnection patterns. *Int J. Climatol.* **13**, 111-149.

Drosdowsky, W. (1993c) Potential predictability of winter rainfall over southern and eastern Australia using Indian Ocean sea-surface temperature anomalies. *Aust. Meteor. Mag.* **42**, 106.

Drosdowsky, W. and Allan, R. (2000) The potential for improved statistical seasonal climate forecasts, in G.L. Hammer, N. Nicholls, and C. Mitchell (eds.), *Applications of Seasonal Climate Forecasting in Agricultural and Natural Ecosystems – The Australian Experience.* Kluwer Academic, The Netherlands. (this volume).

Drosdowsky, W. and Chambers, L. (1998) Near global sea surface temperature anomalies as predictors of Australian seasonal rainfall. BMRC Research Report No. 65, Australian Bureau of Meteorology, Melbourne, Victoria, Australia.

Frederiksen, C.S., Indusekharan, P., Balgovind, R.C. and Nicholls, N. (1995) Multidecadal simulations of global climate trends and variability. Proceedings TOGA95 International Scientific Conference, Melbourne, Australia, 2-7 April, 1995.

Gibbs, W.J. (1963) Some notes on rainfall in Australia. Working paper No. 63, Bureau of Meteorology, Melbourne, Australia. pp.8.

Hammer, G.L., Holzworth, D.P. and Stone, R.C. (1996) The value of skill in seasonal climate forecasting to wheat crop management in north-east Australia. *Aust. J. Agric. Res.* **47**, 717-737.

Hunt, H.A. (1913) Australian Monthly Weather Report and Meteorological Abstract, Vol. 4. Government Printer, Melbourne.

Hunt, H.A (1914) Results of rainfall observations made in Queensland. Commonwealth Government Printer, Melbourne. pp.285.

Hunt, H.A. (1916) Results of rainfall observations made in New South Wales during 1909-1914. Commonwealth Government Printer, Melbourne. pp.224.

Hunt, H.A. (1918) Results of rainfall observations made in South Australia and the Northern Territory. Commonwealth Government Printer, Melbourne. pp.421.

Hunt, B.G. and Hirst, A.C. (2000) Global climatic models and their potential for seasonal climate forecasting, in G.L. Hammer, N. Nicholls, and C. Mitchell (eds.), *Applications of Seasonal Climate Forecasting in Agricultural and Natural Ecosystems – The Australian Experience.* Kluwer Academic, The Netherlands. (this volume).

Jones, D.A. (1998) The prediction of Australian land-surface temperatures using near-global sea-surface temperatures. Proceedings 23rd Annual Climate Diagnostics and Prediction Workshop, Rosenstiel School of Marine and Atmospheric Science, October 26-30, 1998, Climate Prediction Center, NCEP/NWS/NOAA, Maryland, Virginia.

Kleeman, R (1991) A simple model of the atmospheric response to ENSO sea surface temperature anomalies. *J. Atmos. Sci.* **48**, 3-18.

Kleeman, R. (1993) On the dependence of hindcast skill on ocean thermodynamics in a coupled ocean-atmosphere model. *J. Climate* **6**, 2012-2033.

Latif, M., Barnett, T.P., Cane, M.A., Flugel, M., Graham, N.E., von Storch, H., Xu, J-S. and Zebiak, S.E. (1994) A review of ENSO prediction studies. *Clim. Dyn.* **9**, 167-179.

Lough, J.M (1991) Rainfall variations in Queensland, Australia: 1891-1986. *Int. J. Climatol* **12**, 745-764.

Lough, J.M. (1992) Variations of sea-surface temperatures off north-eastern Australia and associations with rainfall in Queensland: 1956-1987. *Int. J. Climatol* **12**, 765-782.

Lough, J.M. (1993) Variation of some seasonal rainfall characteristics in Queensland, Australia: 1921-1987. *Int. J. Climatol* **13**, 391-409.

Lough, J.M. (1994) Climate variation and El Nino-Southern Oscillation events on the Great Barrier Reef: 1958 to 1987. *Coral Reefs* **13**, 181-195.

McBride, J.L. and Nicholls, N. (1983) Seasonal relationships between Australian rainfall and the Southern Oscillation. *Mon. Wea. Rev.* **111**, 1998-2004.

Meinke, H. and Hochman, Z. (2000) Using seasonal climate forecasts to manage dryland crops in northern Australia – Experiences form the 1997/98 seasons, in G.L. Hammer, N. Nicholls and C. Mitchell (eds.), *Applications of Seasonal Climate Forecasting in Agricultural and Natural Ecosystems – The Australian Experience.* Kluwer Academic, The Netherlands. (this volume).

Nicholls, N. (1978) A possible method for predicting seasonal tropical cyclone activity in the Australian region. *Mon. Wea Rev.* **107**, 1221-1224.

Nicholls, N. (1983) Predictability of the 1982 Australian drought. *Search* **14**, 154-155.

Nicholls, N. (1984a) The Southern Oscillation, sea surface temperature, and interannual fluctuations in Australian tropical cyclone activity. *J. Climatol* **4**, 661-670.

Nicholls, N. (1984b) El Niño Southern Oscillation and North Australian sea-surface temperature. *Trop. Ocean-Atmos. News* **24**, 11-13.

Nicholls, N. (1984c) The stability of empirical long-range forecast techniques: a case study. *J. Clim. Appl. Met.* **23**, 143-147.

Nicholls, N. (1985a) Impact of the Southern Oscillation on Australian crops. *J. Climatol.* **5**, 553-560.

Nicholls, N. (1985b) Towards the prediction of major Australian droughts. *Aust Meteor Mag.* **33**, 161-166.

Nicholls, N. (1986a) Use of the Southern Oscillation to predict Australian sorghum yield. *Agric. For. Met.* **38**, 9-15.

Nicholls, N. (1986b) A method for predicting Murray Valley Encephalitis in southeast Australia using the Southern Oscillation. *Aust. J. Exp. Bio. Med. Sci.* **64**, 587-594.

Nicholls, N. (1988a) More on early ENSOs: Evidence from Australian documentary sources. *Bull. Amer. Meteor. Soc.* **69**, 4-6.

Nicholls, N. (1988b) El Niño-Southern Oscillation impact prediction. *Bull. Amer. Meteor. Soc.* **69**, 173-176.

Nicholls, N. (1989) Sea-surface temperatures and Australian winter rainfall. *J. Climate* **2**, 965-973.

Nicholls, N. (1991a) The El Niño-Southern Oscillation and Australian vegetation. *Vegetatio* **91**, 23-36.

Nicholls, N. (1991b) Advances in long-term weather forecasting, in Muchow, R.C. and Bellamy, J.A. (eds.), *Climatic risk in crop production: Models and management for the semiarid tropics and subtropics,* CAB International. pp.427-444.

Nicholls, N. (1992) Recent performance of a method for forecasting Australian seasonal tropical cyclone activity. *Aust. Meteor. Mag.* **40**, 105-110.

Nicholls, N. (1997) Developments in climatology in Australia: 1946-1996. *Aust. Meteor. Mag.* **46**, 127-135.

Nicholls, N. and Woodcock, F. (1981) Verification of an empirical long-range weather forecasting technique. *Q. J. R. Meteor. Soc.* **107**, 973-976.

Nicholls, N., McBride, J.L. and Ormorod, R.J. (1982) On predicting the onset of the Australian wet season at Darwin. *Mon. Wea. Rev.* **110**, 14-17.

Pittock, A.B. (1975) Climatic change and the patterns of variation in Australian rainfall. *Search* **6**, 498-504.

Power, S., Casey, T., Folland, C., Colman, A. and Mehta, V. (1999) Decadal modulation of the impact of ENSO on Australia. *J. Climate* (submitted).

Power, S.B., Kleeman, R., Tsietkin, F. and Smith, N.R. (1995) A global version of the GFDL Modular Ocean Model for ENSO studies. BMRC Technical Report , Bureau of Meteorology, Australia. pp.18.

Priestley, C.H.B. (1962) Some lag associations in Darwin pressure and rainfall. *Aust. Meteor. Mag.* **38**, 32-41.

Priestley, C.H.B. (1963) Some associations in Australian monthly rainfalls. *Aust. Meteor. Mag.* **41**, 12-21.

Quayle, E.T. (1929) Long-range rainfall forecasting from tropical (Darwin) air pressures. *Proc. Roy. Soc., Victoria* **41**, 160-164.

Stone, R.C. and Auliciems, A. (1992) SOI phase relationships with rainfall in eastern Australia. *Int. J. Climatol.* **12**, 625-636.

Stone, R.C. and Hammer, G.L. (1992) Seasonal climate forecasting in crop management. Proceedings 6th Australian Agronomy Conference, Armidale, February, 1992. Australian Society of Agronomy, Parkville, Victoria. pp.218-221.

Stone, R.C., Hammer, G.L. and Marcussen, T. (1996a) Prediction of global rainfall probabilities using phases of the Southern Oscillation Index. *Nature* **384**, 252-255.

Stone, R.C., Nicholls, N. and Hammer, G.L. (1996b) Frost in northeast Australia: Trends and influences of phases of the Southern Oscillation. *J. Climate* **9**, 1896–1909.

Stone, R.C., Smith, I. and McIntosh, P. (2000) Statistical methods for deriving seasonal climate forecasts from GCMs, in G.L. Hammer, N. Nicholls and C. Mitchell (eds.), *Applications of Seasonal Climate Forecasting in Agricultural and Natural Ecosystems – The Australian Experience.* Kluwer Academic, The Netherlands. (this volume).

Streten, N.A. (1975) Satellite derived inferences to some characteristics of the South Pacific atmospheric circulation associated with the Niño event of 1972-73. *Mon. Wea. Rev.* **103**, 989-995.

Streten, N.A. (1981) Southern Hemisphere sea surface temperature variability and apparent associations with Australian rainfall. *J. Geophys. Res.* **86**, 485-497.

Streten, N.A. (1983) Extreme distributions of Australian annual rainfall in relation to sea-surface temperature. *J. Climatol.* **3**, 143-153.

Streten, N.A. (1987) Sea-surface temperature anomalies associated with the transition to a year of widespread low rainfall over southern Africa. *Aust. Meteor. Mag.* **35**, 91-93.

Todd, C. (1893) Meteorological work in Australia: A review. Report of the Fifth Meeting of the Australasian Association for the Advancement of Science, Adelaide. pp.246-270.

Troup, A.J. (1965) The Southern Oscillation. *Q.J. R. Meteor. Soc.* **91**, 490-506.

White, W.B. (1999) Influence of the Antarctic Circumpolar Wave on Australian Precipitation from 1958-1996. *J. Climate* (submitted).

Williams, M. (1986) Relations between the Southern Oscillation and the troposphere over Australia. BMRC Research Report No. 6, Bureau of Meteorology, Melbourne. pp.201.

Zhang, X.-G. and Casey, T.M. (1992) Long-term variations in the Southern Oscillation and relationships with Australian rainfall. *Aust. Meteor. Mag.* **40**, 211-225.

THE POTENTIAL FOR IMPROVED STATISTICAL SEASONAL CLIMATE FORECASTS

WASYL DROSDOWSKY[1] AND ROB ALLAN[2]

[1] *Bureau of Meteorology Research Centre*
PO Box 1289K
Melbourne, Vic 3000, Australia

[2] *CSIRO Division of Atmospheric Research*
PMB1 Aspendale
Vic 3195, Australia

Abstract

From its beginning, statistical climate prediction has been hampered by poor observational data coverage, both spatially and temporally; incomplete theoretical understanding of the climate system; availability of only basic statistical techniques and limited computational capabilities. Progress in recent years, and the potential for further improvement in the future is the result of improvements in all these areas. Obviously these topics are all connected, and improvement in one area leads to, or requires, improvement in the others. This paper examines these issues as a basis to discuss potential to improve statistical seasonal climate forecasts.

Until recently, most seasonal climate outlooks, such as that provided by the Bureau of Meteorology National Climate Centre (NCC) have used the Southern Oscillation Index (SOI) as the sole or major predictor, and simple forecast techniques such as linear regression. Now seasonal forecasts are being issued based on large scale sea surface temperature (SST) and global circulation anomalies. The availability of large multi-variable data sets has and will continue to increase the use of more advanced statistical techniques beyond the simple linear regression approach. The availability of many additional potential predictors, and different statistical forecasting systems greatly increases the possibility of artificial skill, which can be minimised by rigorous cross validation techniques.

Much of the potential predictability in Australian seasonal rainfall one season ahead, may be realised by a system utilising global SST patterns as predictors. However, climate variability encompasses other parameters, and occurs on a variety of time scales ranging from intra-seasonal through inter-annual through to decadal. This leaves considerable scope for improvement, or expansion of seasonal climate prediction of other parameters besides rainfall, and with increased lead times, and different target season length.

G.L. Hammer et al. (eds.), The Australian Experience, 77–87.

1. Introduction

An historical account of the current seasonal climate outlook service provided by the Bureau of Meteorology National Climate Centre (NCC) is given by Stone and de Hoedt (2000). This service currently uses the Southern Oscillation Index (SOI) as its sole predictor. Elsewhere, particularly at the National Center for Environmental Prediction (NCEP) Washington, and the Hadley Centre for Climate Prediction and Research of the United Kingdom Meteorological Office (UKMO) empirical seasonal forecasts are being issued based on large scale sea surface temperature (SST) anomalies, amongst other predictors. A similar scheme is under development, and has been run in quasi-operational mode for the past twelve months, in the Bureau of Meteorology Research Centre (BMRC) (Drosdowsky and Chambers, 1998).

In order to assess the potential for improved statistical seasonal climate forecasts for Australia beyond that provided by either the current system, or the BMRC SST based forecast system, we need to consider the basic requirements for such a forecast system. These include;
(a) long time series of reliable data, of both the predictand and any potential predictors,
(b) appropriate statistical methodologies to analyse and find relationships in these data,
(c) the computing power, including data storage capacity, and ultimately,
(d) a knowledge of the science or the theoretical background to verify the physical plausibility of the predictor - predictand relationships.
Until very recently, statistical climate prediction has been restricted in all these requirements.

2. Data

Most observational data networks have been designed for weather forecasting, and while these may have adequate spatial coverage at present, this has not always been the case. As a result very few stations have the long, continuous and homogeneous records necessary for seasonal prediction. In the past 10 to 15 years much effort has been directed towards the compilation of historical data, for example the global SST data in the Comprehensive Ocean Atmosphere Data Set (COADS) (Slutz et al., 1985). This compilation of ship data has been further refined to interpolate missing values, and include other sources of SST such as satellite radiances, resulting in the UKMO Global sea-Ice and Sea Surface Temperature (GISST) data set (Rayner et al., 1998), and the NCEP SST reanalysis (Reynolds and Smith, 1994). Other global surface and upper air data sets being developed include the Global Historical Climate Network (GHCN) data set (Vose et al., 1992), the Global Mean Sea Level Pressure (GMSLP) data set from the UKMO (Allan et al., 1996; Basnett and Parker 1997), and the Comprehensive Aerological Reference Data Set (CARDS) (Eskridge et al., 1995). Over the past few years this array of observed data has been subjected to comprehensive global reanalyses using current, state-of-the-art analysis systems at NCEP (Kalnay et al., 1996) and the European Centre for Medium Range Weather Forecasts (ECMWF) (Gibson et al., 1997).

3. Computing Power

The global reanalysis projects at NCEP and ECMWF are feasible due to the enormous computing and data storage facilities available at these centres. The full potential of these large data sets can now also be realised by national meteorological services. In Australia, access to historical data has been greatly enhanced with the introduction of the Australian Data Archive for Meteorology (ADAM) (Lee, 1994) system within the National Climate Centre. This now allows virtually the entire historical data base to be kept on disk and accessed easily from anywhere within the Bureau of Meteorology and by external users.

4. Statistical Methods

The availability of these large multi-variable data sets and increased computing capabilities has and will increase the use of more advanced statistical techniques, beyond the simple traditional methods such as linear correlation and regression. These statistical techniques are used for a number of purposes. Multivariate analysis techniques such as Principal Component Analysis (PCA) or Empirical Orthogonal Functions (EOF) analysis, Singular Value Decomposition (SVD), cluster analysis, and Canonical Correlation analysis (CCA), are now commonplace in climate research and seasonal climate forecasting, being used as data reduction techniques, and also to explore patterns or modes of variability in the data (see reviews of these techniques in Mann and Park, 1998). Potentially more powerful forecast techniques such as linear discriminant analysis (LDA) and non linear methods such as neural networks are being applied to seasonal forecasts.

An essential requirement of all forecast schemes, whether statistical or dynamical, is an estimate of the skill to be expected by the scheme. Here we need to distinguish between the in-sample skill estimated by using all the data (ie the model fit) and true out-of-sample skill obtained by testing the model on independent data. This latter skill is usually obtained through a cross-validation or "leave-one-out" procedure. The difference between these two estimates is sometimes referred to as the artificial skill.

A more important source of possible artificial skill in seasonal climate predictions is the availability of many additional potential predictors, advanced statistical techniques and forecast methodologies and the computing power to apply them. To obtain a true out-of-sample skill estimate, when selecting among different potential predictors or different forecast models a nested cross validation procedure is necessary. Alternatively, an improved theoretical understanding of the climate system, and experience with global climate models, may lead to the formulation of appropriate *a priori* hypotheses, which can be tested via the improved data base. This is preferred to engaging in haphazard "fishing expeditions", which will almost always find some statistical relationship between the predictand and a large number of potential predictors. Simply selecting the best set of predictors or best model will lead to increased artificial skill.

5. The BMRC SST-based Prediction System

To illustrate the interaction between the various factors described above we present a brief description of the BMRC SST-based statistical system. The predictand is the Australian seasonal rainfall anomaly, on a uniform one degree grid over the entire continent. This grided data set was developed by NCC using all the available monthly rainfall records in the ADAM data base (Jones and Weymouth, 1997). The predictand to be used is the global or near global SST anomaly field. Both these data sets consist of many hundreds of individual data gridpoints. To relate one field to another directly would mean examining many thousands of possible predictor - predictand combinations. Clearly these will not be all independent as there is appreciable spatial correlation in both fields. To reduce the number of predictors and predictands to manageable size, and at the same time produce largely uncorrelated predictors, the coherent variability of both fields is summarised by rotated principal component analysis. For the rainfall data, retaining the first nine components (Figure 1) accounts for about 60% of the variance, while the first twelve components (Figure 2) account for about 50% of the SST data variance. This procedure reduces the number of potential predictors to around twenty to thirty, if not all SST components are considered, but some lagged values (at sufficiently large lags to reduce the autocorrelations to insignificant levels) are also used. The number of actual forecasts required for each season is then reduced from over 600 to just the nine rainfall principal components. An additional benefit is that the resulting forecast, when interpolated back to the original gridpoints in some appropriate manner will be much smoother spatially than that produced by forecasting for each individual gridpoint.

Having reduced the predictors and predictands to a small number of uncorrelated principal components we need to decide on a prediction methodology; the method chosen for this forecast scheme is LDA, mainly due to its ability to produce probability forecasts. In LDA the rainfall for each season in the training set is categorised according to its rank in the cumulative frequency distribution, with the driest third of years assigned to tercile 1 (Below normal), the middle third to tercile 2 (Near normal), and the wettest third to tercile three (Above normal). LDA then uses Bayes Theorem to estimate the probability of a new observation of SSTs belonging to a particular category, based on the distribution of the observations in the training set. This approach is identical to that employed by Ward and Folland (1991) to predict rainfall in Northeast Brazil using eigenvectors of global SST.

The estimate of hindcast skill (ie based on the training set of data) can be obtained in a number of ways. The two most common methods are the "Holdout Method" in which the data is split into the development set to which the LDA model is fitted, and a test sample (usually 1/4 to 1/3 of the total) to which the model is applied and from which the hindcast skill is calculated. The second method, and that used here, is the "Leave-One-Out" or cross validation method. This is similar to the Holdout method except that the split is into N-1 cases in the development sample, and only one test case. However this procedure is then repeated over all N cases, leaving one out each time. Skill is then evaluated over all N hindcasts. In either method the development and test samples must be kept independent to prevent the introduction of artificial skill.

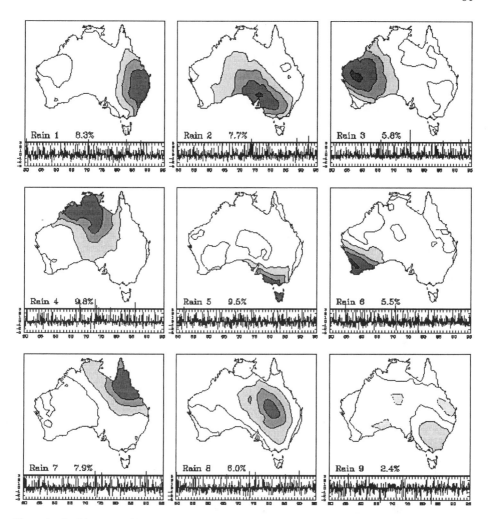

Figure 1. Spatial pattern of loadings and associated scores (time series) of the first nine VARIMAX rotated principal components of the standardised monthly anomalies of grided Australian rainfall. Contour interval is 0.2, with zero contour heavy, negative contours dashed and areas above +0.2 and below -0.2 shaded.

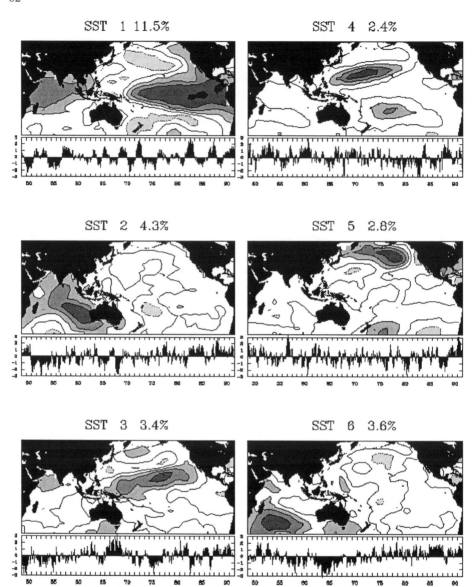

Figure 2. Spatial pattern of loadings and associated scores (time series) of the first twelve VARIMAX rotated principal components of the standardised monthly anomalies of the GISST data set. Contour interval is 0.2, with zero contour heavy, negative contours dashed and areas above +0.2 and below -0.2 shaded.

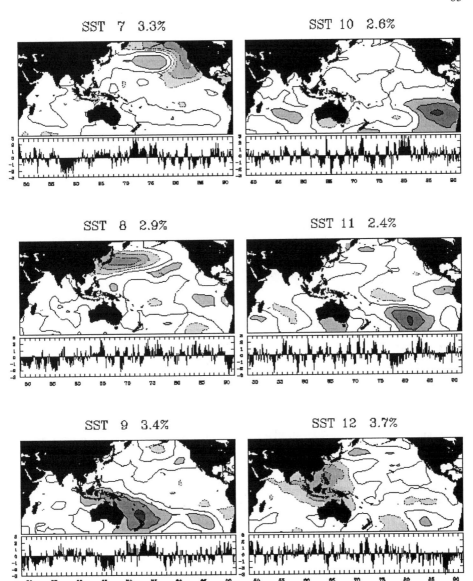

Figure 2 (cont'd). Spatial pattern of loadings and associated scores (time series) of the first twelve VARIMAX rotated principal components of the standardised monthly anomalies of the GISST data set. Contour interval is 0.2, with zero contour heavy, negative contours dashed and areas above +0.2 and below -0.2 shaded.

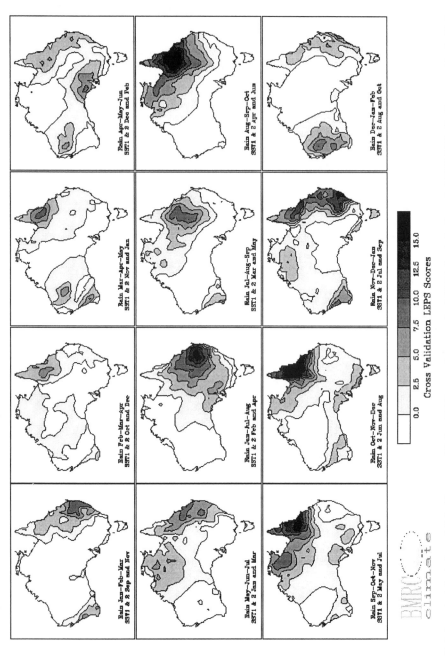

Figure 3. Independent, "out of sample" double cross-validation LEPS scores for seasonal rainfall hindcasts for the period 1950-1993. Predictors used in the hindcasts are the best combination of any size selected from a pool of four potential predictors, these being the first two SST principal components shown in Figure 2 lagged by one and three months.

This discussion applies to the simple case where the form of the LDA model (ie. the predictors to be used) are known or specified in advance. When the structure of the model is not known (ie. when we need to select the best subset of predictors from a large pool of potential predictors) then the selection procedure itself must also be cross-validated, as described in detail by Elsner and Schmertmann (1994). For the BMRC SST forecast system, we have examined the use of up to ten SST principal components, lagged by one and three months for a total of twenty potential predictors (Drosdowsky and Chambers, 1998). With such a large number of potential predictors it is not feasible to include every possible subset of size up to 20 predictors; in this case the search has been restricted to the best model involving a subset of at most two predictors. This still results in over 200 possible forecast models, and, as shown by Drosdowsky and Chambers (1998), an enormous potential for artificial skill as measured by the difference between the "in-sample" and the cross-validated "out of sample" skill. Nevertheless, the true cross-validated skill of this system exceeds that obtained from simple correlation with the SOI alone, particularly through the "autumn predictability barrier" when El Nino Southern Oscillation (ENSO) events typically change phase.

There are two major problems with the use of this scheme in operational seasonal prediction. Firstly, some of the predictor choices are difficult to justify on physical grounds, being simply the best in a statistical sense. Secondly, there is no requirement for temporal continuity of the selected predictors, which can result in major shifts in the forecasts from month to month. To overcome these problems the operational scheme has been restricted to the use of the first two SST components, which are known to be related to Australian rainfall variability. Retaining the one and three month lags results in four potential predictors, and it is feasible to examine the 15 possible subsets of any size. The resulting cross-validated skill is shown in Figure 3. Again this shows some improvement over the SOI alone, especially at the same lead time, with most of the increased skill due to the inclusion of the second, Indian Ocean, SST component.

6. Improved Future Forecast Systems

The potential for improvement of current seasonal climate prediction systems, and specifically the NCC system, depends on what is meant by "improved". In the narrow sense of more accurate or skilful forecasts in the current format, there may be only limited potential for improvement. Much of the potential predictability in Australian seasonal rainfall one season ahead, may be realised by the system utilising global SST patterns as predictors. From the vast array of new potential predictands, which will be available in the near future, only a small portion will display variability truly independent of the SST data.

Climate variability occurs on a variety of time scales ranging from intra-seasonal through inter-annual (ENSO) through to multi-decadal. More significantly the relationship between predictor and predictand can also vary on decadal to multi-decadal time scales, for example the changing relationships over time between the SOI and seasonal rain over most of Australia (Nicholls *et al.*, 1996, 1997). Wider physical evidence for these changes is now coming from concerted studies of the climate system. One line focusing on detailed global analyses of historical atmospheric pressure and SST compilations using techniques such as EOFs and SVD has isolated several modes of variability operating on decadal to secular time scales (Allan *et al.*, 1999; Mann and Park, 1998). This research indicates that not only do these climatic modes display ENSO-like structure and rainfall relationships at low frequencies (Allan, 1999), but that they interact with interannual ENSO signals to provide important modulations of the phenomenon. Consequently, protracted El Nino and La Nina event sequences, such as the 1990-1995 El Nino period, are manifest through the superposition of interannual ENSO and decadal ENSO-like modes in the climate system. In addition, long-term changes in ENSO, such as the climate shift in its characteristics since the 1970s, are seen to result from the operation of multidecadal ENSO-like fluctuations.

This low frequency variability can confound statistical forecasts for average conditions over a three-month season. If, however, we regard seasonal climate prediction in a wider sense including the intraseasonal and decadal variability, then there is considerable scope for improvement, or expansion of seasonal climate prediction.

Areas of possible expansion of seasonal climate prediction include;
(a) Introduction of different target season length. Many parameters, particularly in the tropics, display significant variability on the 30 to 90 day "intraseasonal oscillation" time scale. While this variability can be aliased onto the seasonal time scale and confound seasonal predictions, it may itself be predictable if the target "season" is much shorter than the traditional three months. Closely related to shorter target seasons is the prediction of significant events such as dates of last frost or seasonal changes such as monsoon onset in northern Australia or the "winter break" in southern Australia.

(b) Increased lead times for seasonal forecasts. The current system based on the SOI has essentially zero lead time, while the SST-based system has been designed with a one month lead. Forecasts issued by NCEP for the United States have lead times ranging from one to twelve months ahead. Appropriate realisable lead times for seasonal (and subseasonal) forecasts need to be established by consultation with the users of the forecasts.

(c) Prediction of other parameters besides rainfall. BMRC is currently developing a system for prediction of seasonal maximum and minimum temperatures, and has explored the feasibility of the prediction of seasonal extremes. Many other agriculturally useful potential predictands have also been identified.

Realisation of these extensions will require the availability of data sets of both the predictands and potential predictors on the appropriate time scales. For example, the introduction of sub-monthly target seasons may require daily or pentad (five day) resolution data. This in turn requires much greater computing and data storage capabilities than required for monthly or seasonal data. Some research effort has already been directed at these possible improvements by BMRC, especially the extension to other parameters such as temperature as mentioned earlier, and examination of intraseasonal variability of Australian rainfall. The SST-based season prediction scheme will also be tested with much longer lead times.

References

Allan, R.J. (1999) ENSO and climatic variability in the last 150 years, in Diaz, H.F. and V. Markgraf (eds.), *El Nino and the Southern Oscillation: Multiscale variability and its impacts on natural ecosystems and society.* Cambridge University Press, Cambridge, UK in press.

Allan, R.J., Lindesay, J.A. and Parker, D.E. (1996) *El Nino Southern Oscillation and Climatic Variability,* CSIRO Publishing, Melbourne. pp.405.

Allan, R.J., Folland, C.K., Mann, M.E., Parker, D.E., Smith, I.N., Basnett, T.A. and Rayner, N.A. (1999) ENSO and decadal-multidecadal modes of climatic variability in global instrumental data. *Climate Dynamics* (Submitted).

Basnett, T.A. and Parker, D.E. (1997) Development of the Global Mean Sea Level pressure data set GMSLP2. Climate Research Technical Note CRTN79, Hadley Centre, Meteorological Office, Bracknell, U.K. pp.16.

Drosdowsky, W. and Chambers, L.E. (1998) Near global sea surface temperature anomalies as predictors of Australian seasonal rainfall. BMRC Research Report No. 65, Bureau of Meteorology, Melbourne.

Elsner, J.B. and Schmertmann, C.P. (1994) Assessing forecast skill through cross validation. *Wea. and Forec.* **9**, 619-624.

Eskridge, R.E., Alduchov, O.A., Chernykh, I.V., Panmao, Z., Polansky, A.C. and Doty, S.R. (1995) A Comprehensive Aerological Reference Data Set (CARDS). Rough and systematic errors. Bull. *Amer. Meteor. Soc.* **76**, 1759-1775.

Gibson J.K., P. Kallberg, S. Uppala, A. Nomura, A. Hernandez and E. Serrano (1997) ERA Description. ECMWF Re-Analysis Project Report Series, 1. European Centre for Medium Range Weather Forecasts.

Jones, D.A., and Weymouth, G. (1997) An Australian monthly rainfall dataset. Technical Report No. 70, Bureau of Meteorology, Melbourne, Australia. 19pp.

Kalnay, E. and co-authors (1996) The NCEP/NCAR 40-year reanalysis project. *Bull. Amer. Meteor. Soc.* **77**, 437-471.

Lee, D.M. (1994) Australian Data Archive for Meteorology (ADAM) – Manual. Unpublished Report, National Climate Centre, Bureau of Meteorology, Melbourne. pp.64.

Mann, M.E. and Park, J. (1998) Oscillatory spatiotemporal signal detection in climate studies. *Adv. Geophys* (in press).

Nicholls, N., Drosdowsky, W. and Lavery, B. (1997) Australian rainfall variability and change. *Weather* **52**, 66-72.

Nicholls, N., Lavery, B., Frederiksen, C. and Drosdowsky, W. (1996) Recent apparent changes in relationships between the El Nino Southern Oscillation and Australian rainfall and temperature. *Geophys. Res. Lett.* **23**, 3357-3360.

Rayner, N.A, Horton, E.B., Parker, D.E. and Folland, C. K. (1998) The GISST2.3 and GISST3.0 data sets. Climate Research Technical Note CRTN??, Hadley Centre, Meteorological Office; Bracknell, U.K (in press).

Reynolds, R.W. and Smith, T.M. (1994) Improved global sea surface temperature analyses using optimal interpolation. *J. Climate* **7**, 929-948.

Slutz, R.J., Lubker, S.J., Hiscox, J.D., Woodruff, S.D., Jenne, R.L., Joseph, D.H., Steurer, P.M. and Elms, J.D. (1985) Comprehensive Ocean Atmosphere DataSet; Release 1. NOAA Environmental Research Laboratories, Climate Research Program, Boulder, CO. (NTIS PB86-105723). pp.268.

Stone, R.C. and de Hoedt, G. (2000) The development and delivery of current seasonal climate forecasting capabilities in Australia, in G.L. Hammer, N. Nicholls, and C. Mitchell (eds.), *Applications of Seasonal Climate Forecasting in Agricultural and Natural Ecosystems – The Australian Experience.* Kluwer Academic, The Netherlands. (this volume).

Vose, R.S., Schmoyer, R.L., Steurer, P.M., Heim, R., Karl, T.R. and Eischeid, J.K. (1992) The Global Historical Climatology Network: Long-term monthly temperature, precipitation, sea level pressure, and station pressure data. ORNL/CDIAC-53, NDP-041. CDIAC, Oak Ridge, Tennessee. pp.315.

Ward, N.M. and Folland, C.K. (1991) Prediction of seasonal rainfall in the north Nordeste of Brazil using eigenvectors of sea surface temperature. *Intl. J. Climatol.* **11**, 711-743.

GLOBAL CLIMATIC MODELS AND THEIR POTENTIAL FOR SEASONAL CLIMATIC FORECASTING

BARRIE HUNT AND A.C. HIRST

CSIRO Division of Atmospheric Research
PMB1 Aspendale, Vic 3195, Australia

Abstract

Currently there are major initiatives in a number of research institutions to develop operational long-lead climatic prediction schemes. These schemes are all based on the intrinsic predictability of El Niño/Southern Oscillation events occurring in the low latitude Pacific Ocean. The physical processes that provide this long-lead predictability are identified and briefly discussed here. These processes have now been incorporated into a range of climatic models, extending from highly parameterised two-level models to extremely complicated coupled atmospheric-oceanic global climatic models. The manner in which the various models are used for predictions is discussed, and some results from an extensive series of hindcasts are presented to illustrate present levels of predictability, particularly for rainfall. The paper concludes with a list of major problems that need to be resolved in order to enhance predictability, especially the critical requirement to extend oceanic feedbacks beyond the low latitude Pacific Ocean.

1. Introduction

In the last 5 years or so, rapid progress has been achieved in the area of long-lead time climatic predictions using numerical models. The current state of the art is demonstrated by the various predictions routinely presented in the NOAA "Experimental Long-Lead Forecast Bulletin". The range and dispersion of the results is illustrated by the fact that for the 1997 El Niño event, the Bulletin listed predictions issued at the beginning of 1997 that varied from El Niño to La Niña events!

Excluding statistical methods, there are three basic modelling approaches currently in vogue. For the first two approaches the only predicted output is sea surface temperature (SST) anomalies for the low latitude Pacific Ocean. The first, and the one with the longest history, is the intermediate coupled model attributable to Zebiak and Cane (1987), which uses highly simplified atmospheric and oceanic components. The second, or hybrid coupled model, approach is based on a more complete oceanic general circulation model with a statistical or simplified atmospheric component (Neelin, 1990).

G.L. Hammer et al. (eds.), The Australian Experience, 89–107.

The SST predictions from these two approaches are used to force an atmospheric global climatic model in what is called the two-tiered method, with SST outside of the Pacific domain being specified from climatology (Hunt *et al.*, 1994). This method is discussed in some detail below. The third method uses a coupled atmospheric-oceanic global climatic model and replicates all relevant climatic variables as well as SST. An example of this approach is the NCEP model of Ji *et al.* (1994). Until very recently this method was also restricted to predicting SST for the low latitude Pacific Ocean domain, although interactive global oceanic domains are now being evaluated (Schneider *et al.*, 1997; Stockdale *et al.*, 1998).

An additional method that can be used for short term predictions, 1-3 months in advance, is to superimpose current, observed SST anomalies onto a global climatological SST distribution and use this hybrid SST data set to force an atmospheric model. The SST anomalies are held fixed for the duration of the prediction. This method has been used with some success for predictions of Sahelian rainfall (Rowell *et al.*, 1992).

Because of chaotic influences in the climatic system multiple runs are usually made to generate an ensemble of results, regardless of which approach is used.

The emphasis of current prediction schemes on the low latitude Pacific Ocean is because El Niño/Southern Oscillation (ENSO) events are generated in this oceanic region. The long timescale of these events (~18 months) implies that there is a controlling mechanism associated with the complex air-sea interactions that occur. The aim of the models is to replicate this mechanism, and thereby to exploit the long-lead times of ENSO events to produce corresponding predictions.

The genesis of the model prediction schemes lies in earlier studies of ENSO mechanisms, which were based on conceptual or mechanistic models. This review therefore commences with a description of such models and then progresses to the coupled GCMs. The application of the two-tiered intermediate approach is described in some detail, as well as the need to develop the oceanic component of the model for predictions. The practical requirements for typical forecasts involving running an ensemble of predictions, the interpretation of such ensembles, the problems in specifying initial conditions, and the optimal method for presenting the predictions, are discussed.

Before reviewing progress it is worthwhile highlighting why it is possible to consider predictions up to a year in advance, when daily weather predictions struggle to achieve credibility for only a few days ahead. The difference derives from the critical importance of *boundary* conditions in long range predictions compared to *initial* conditions in daily forecasts. The latter require an accurate specification of critical atmospheric variables to be available at the initialisation of the forecasts. Even then, imperfections in such specifications compound owing to the intrinsic chaotic nature of the controlling mathematical equations used for the forecasts, and result in a breakdown in predictability after about 10 days. In contrast, in the long range predictions the boundary conditions are provided by the SST predictions derived from the oceanic component of the model. As has been demonstrated in many GCM experiments, specification of SST variability is sufficient to permit broad scale, interannual climatic variability to be simulated. Hence the prediction of SST provides a constantly evolving boundary condition which serves to keep the climatic prediction on-line, even though chaos is still an issue.

2. What Makes ENSO (Partly) Predictable?

Many statistical and physical models have been developed in the attempt to forecast ENSO. These models show that the state of ENSO is somewhat predictable up to several seasons in advance. There are two aspects of the tropical ocean and atmosphere that contribute to this predictability. The first aspect is that the SST in the eastern equatorial Pacific is largely controlled by patterns of wind over the remainder of the tropical Pacific during the *previous* months and years. The drag of the winds cause changes in the depth of the layer of warm surface water, which lies throughout the tropical Pacific atop a deep layer of much colder water. These changes in upper layer depth then move slowly away from their formation regions. The pattern of movement is westward in the open ocean away from the equator, towards the equator along the western ocean boundary, and then eastward along the equator, to eventually affect the upper layer depth in the eastern equatorial Pacific. In the eastern equatorial Pacific, there is strong upwelling of the cold substrate water into the overlying warm water layer, and the SST there is strongly dependent on the rate of this upwelling. When the upper layer is relatively thick, there is little upwelling of cold water and the surface water is warm (as in El Niño). Conversely, when the upper layer is relatively thin, there is much upwelling of the cold water and the surface water is cooler (as in La Niña). Therefore, by controlling the depth of the upper water layer in the eastern equatorial Pacific, the prior winds through the tropical Pacific control the surface temperature in the eastern equatorial Pacific, and hence the state of ENSO. Consequently, at any given time, signals in the upper layer depth are present in the tropical Pacific which will affect the SST in the eastern equatorial Pacific over the following months and years. Hence, a model of the Pacific ocean, when forced by winds observed during the several years up to the present, and then run on into the future using 'normal' climatological wind for forcing, will still yield information about the future SST in the eastern equatorial Pacific. This was the basis for one early ENSO prediction scheme, developed at Florida State University (Inoue and O'Brien, 1984).

However, ENSO is a phenomenon involving the *coupled* ocean and atmosphere, in that changes in the pattern of ocean surface temperatures affect the winds, which then drive further changes in the oceanic fields and surface temperature. The second important aspect adding to predictability of the state of ENSO is that the winds over the tropical Pacific respond to changes in surface temperature in the eastern equatorial Pacific in a fairly systematic manner. These are just the winds needed to generate the signals that will change the surface temperature in the eastern equatorial Pacific at a later date. Therefore, if an ocean model is forced with observed winds up to the present, and then coupled with an atmospheric model, the winds generated by the model in response to the surface temperatures in the eastern equatorial Pacific will cause changes in surface temperatures in that region at times still further into the future. Clearly, the ability of such a coupled model to predict ENSO is potentially much greater than that of an ocean model alone.

3. Low-order Coupled Models for ENSO Predictability

The essential factors in ENSO predictability, noted above, were realised by the mid 1980s, and investigators then began experimenting with models of differing complexity to develop practical forecasting schemes. It was clear from the start that the factors could be captured by use of very simple models. For example, the slow transmission of the ocean signals is simulated fairly accurately by a very simple ocean model consisting of just two layers of fluid (representing the upper and deeper layers of the ocean) and with most of the complicated (nonlinear) mathematical terms discarded (eg. McCreary, 1983). Further, the atmospheric response to eastern equatorial Pacific temperatures was shown to be fairly well simulated by a remarkably simple physical model (Zebiak, 1982; Neelin, 1989), which treats the atmospheric boundary layer as a single layer and considers only the effect of surface temperature and uplift associated with convection (ie. thunderstorms) on the surface wind field. These simplifications led rapidly to the development of 'low order' (ie. simple) coupled models for the purpose of studying the essentials of ENSO behaviour (eg. Anderson and McCreary, 1985; Cane and Zebiak, 1985; Hirst, 1986; Neelin, 1991).

A 'low order' coupled model consists of a set of several mathematical equations for currents, upper ocean layer depth, ocean surface temperature, surface air pressure and surface winds. The ocean surface temperature determines the pattern of surface air pressure, and hence the surface winds. The surface winds in turn exert a drag on the ocean, affecting the surface currents and upper ocean layer depth, and ultimately, the SST field as discussed above. The equations are used to advance the model variables forward in time by discrete time increments, called 'time steps'. Values are calculated at each time step for points on a horizontal grid, with grid points spaced at intervals of several tens of kilometres. The ocean model grid typically covers the tropical and subtropical Pacific Ocean. The atmospheric model grid typically extends further to cover the entire tropics and subtropics of the globe. Low order models stand in contrast to the much more complicated general circulation models (GCMs), which include, for example, temperatures and currents at many depths in the ocean, winds and pressure at many heights in the atmosphere, and detailed computation of atmospheric radiation transfer and atmosphere-ocean heat exchange. Low order models include only the processes believed to be essential for ENSO predictability, and even some of these are represented in idealised or semi-empirical form.

Particularly realistic ENSO-like behaviour occurred in the model of Cane and Zebiak (1985) which was comprised of the simple ocean and atmospheric models discussed above, and a complicated equation for determining ocean surface temperature based on the model ocean's upper layer thickness and currents. Analysis of several versions of this model led to many insights into the dynamics of ENSO (Zebiak and Cane, 1987; Battisti, 1988; Munnich *et al.*, 1991; Jin and Neelin, 1993), and was important in establishing the predominant theory of ENSO oscillation (Suarez and Schopf, 1988; Battisti and Hirst, 1989). Cane *et al.* (1986) showed that their low order model had significant predictive capability for ENSO. This came as a surprise, for although the Cane and Zebiak (1985) model included all the processes essential for ENSO predictability, all investigators had believed that their representation was too simple for the model to be used for practical forecasting. In fact, the forecasting system based on this model proved to be the most successful of any physically-based system, at least through to the early 1990s. This success underscores the simplicity of many of the processes underlying ENSO predictability. Further general reading on ENSO mechanisms may be found in Hirst (1989), Philander (1990), McCreary and Anderson (1991), and Cane (2000).

4. Two-tiered Approach

The two-tiered approach is probably the simplest approach to forecasting climatic impacts of future ENSO states, which still remains viable despite the development of fully coupled models.

In this approach (Hunt *et al.*, 1994; Barnett *et al.*, 1994) a SST prediction of the low latitude Pacific Ocean is used to force an atmospheric GCM. Outside of the domain of the predicted Pacific SSTs the required SSTs are specified from climatology as monthly varying values. The attraction of this approach is that anyone with an atmospheric GCM can make climatic predictions by requesting an appropriate SST prediction made by another group. For example, Hunt *et al.* (1994) used SST predictions derived from the Cane and Zebiak model (Zebiak and Cane, 1987) to make rainfall *predictions* for 1990 and 1991 in real time. Barnett *et al.* (1994) used a very similar approach based on SST predictions from a hybrid coupled model to force an atmospheric GCM. They then conducted retrospective predictions or *hindcasts* for a number of climatic variables for selected ENSO years. It should be noted that the Cane and Zebiak model performed very poorly in predicting SST anomalies for the 1997 El Niño event (see Cane, 1999), hence any choice of SST predictions needs to be assessed against the range of outcomes documented in the NOAA Bulletin.

A limitation of the two-tiered approach as used by Hunt *et al.* (1994) and Barnett *et al.* (1994) has been the lack of SST anomalies outside the low latitude Pacific Ocean for forcing the GCM in predictive mode. This limitation, of course, also applies to many coupled models discussed below. However, coupled models with global oceans are now being applied to seasonal predictions (Schneider *et al.*, 1997; Stockdale, *et al.*, 1998), hence, this restriction will eventually be removed.

Figure 1 provides an indication of the potential utility of this approach (from Barnett *et al.*, 1994). The figure shows the correlation between the model and observation for three climatic variables for a 6-month lead time averaged over the period December/February. The results are hindcasts (not predictions) for a composite of seven major El Niño/La Niña events based on an ensemble of three runs. Substantial correlations, up to 0.8, were achieved for all three variables for a number of regions in Figure 1, indicating encouraging skill. The most extended spatial significance was associated with the 500 mb geopotential height, which is a climatic variable having large scale characteristics. The rainfall hindcasts show maximum response over regions usually influenced by ENSO events. Since a composite of seven major events was used the results exaggerate the likely outcome compared to that expected for a single event – which is the practical situation required to be predicted. A more utilitarian test of this approach would have been to show time series of hindcasts and observations, as presented below. The skill shown in Figure 1 is related to both the hindcast of the SST anomalies used and to the adequacy of the GCM employed. As such, improved outcomes can be expected as techniques are developed further.

To illustrate more clearly the possibilities of the two-tiered method a lengthy hindcast for the period 1972-1992 was undertaken. These hindcasts include a range of *individual* El Niño/Southern Oscillation (ENSO) events and thus provide a comprehensive data set for assessment. Given the unique nature of the data set the results will be presented in some detail. The SST anomaly hindcasts used were for the Pacific region between 30°S and 30°N and were produced in the 'reanalysis' run of the Cane and Zebiak model reported by Chen *et al.* (1995). These SST data were used to force the T63 version of the Mark2 version of the CSIRO9 atmospheric GCM. Each hindcast for 1972-1992 was for a 12-month lead time starting on 1st January of the 21 individual years. An ensemble of 10 hindcasts was made for each year, to allow for chaotic effects in the climatic system. Starting conditions were taken from a lengthy control run of the atmospheric model, with each hindcast year using the same set of 10 initial conditions.

Coupled atmospheric-oceanic models (see below) use a similar ensemble approach. Normally, the coupled model generates a single SST prediction, which is subsequently used to force the atmospheric component of the coupled model in order to produce an ensemble of predictions. Hence, the basic difference between a two-tiered and a coupled model hindcast is the manner in which the SST results were generated.

The Southern Oscillation Index (SOI) is considered to be a broad indicator of likely climatic changes associated with ENSO events. In Figure 2 the hindcast SOI for 1972-1992 is compared with observations. The two datasets have been smoothed with a 10-point filter. The hindcast results, in this and subsequent time series, consisted of 21 years of monthly mean values that were combined into a single time series. The hindcasts replicate the observations in broad characteristics, especially, and importantly, as regards transitions between El Niño and La Niña events and vice versa. Such transitions are, of course, the most important feature that is required to be predicted in the SOI. The model does not reproduce many of the extreme values of the SOI, but this is presumably an indication of the limitations of current models, rather than an intrinsic problem. The correlation between the two time series in Figure 2 is 0.79.

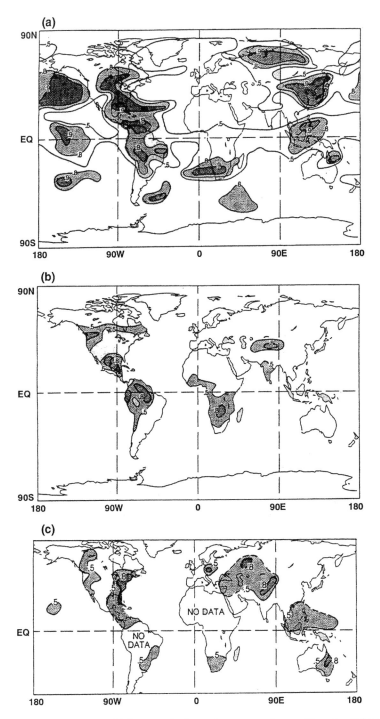

Figure 1. Correlations between hindcasts and observations averaged over December-February for a composite of seven large ENSO events for (a) 500 mb height anomalies, (b) near surface air temperature, and (c) precipitation. Light (heavy) stippling indicates 0.10 (0.05) confidence level. (From Barnett *et al.*, 1994).

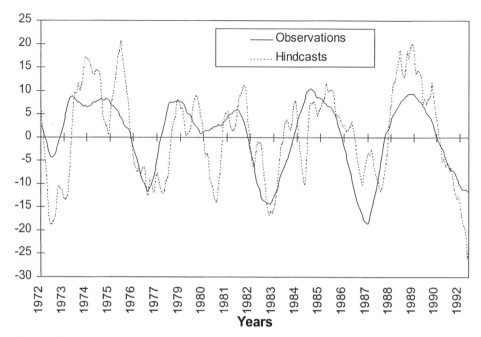

Figure 2. Comparison of model hindcasts and observations for the years 1971-1992. The hindcasts are for 12 months in advance starting on 1st January of each year. The mean of a 10-member hindcast ensemble is shown. Both time series in the figure have been smoothed with a 10-point filter.

Given that rainfall is probably the single most critical climatic variable for many situations, results are presented in Figure 3 for three distinct regions. It should be noted that rainfall is the most difficult climatic variable to replicate in a model, hence comparison of model and observed rainfall outcomes is a very stringent test of a model's capabilities. Hindcasts are compared with observations for the central Pacific, northeast Australia and southeast USA for the period 1979-1992. These regions were selected as examples where useful predictions appear to be possible. The NOAA global rainfall data set which was used (see Xie *et al.*, 1996 for some partial details of these data) only commenced in 1979, hence the whole hindcast period could not be compared with observations in Figure 3.

The hindcast results for the *individual* years were combined into a single 21 year time series of monthly accumulated values. A single gridbox (1.875° x 1.875°) was used in Figure 3(a) and area averages were used in Figures 3(b) and 3(c) to improve signal-to-noise ratios. No additional post-processing or removal of systematic model errors has been made. The full seasonal cycle, rather than just anomalies, is given for the rainfall in order to emphasise the close relationship between the rainfall amplitudes for the model and observation.

(a) Central Pacific 0°, 180°E

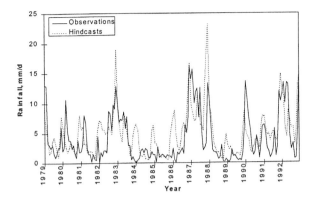

(b) USA/Gulf of Mexico (25°-35°N, 255°-290°E)

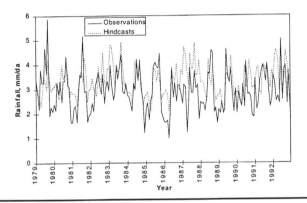

(c) Northeast Australia (20°-26°S, 144°-148°E)

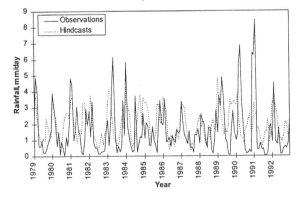

Figure 3. Hindcast and observed rainfall for three selected regions. Hindcast rainfall is the mean of the 10-member ensemble. A single model gridbox is used for the central Pacific (a). Averages over the identified areas are used for the other two locations ((b) and (c)).

For the Equatorial Pacific comparison (Figure 3(a)), the use of a single model gridbox constitutes an extreme test of the model's performance. Such a test could be considered, as the influence of hindcast SST anomalies was strongest in this location. The hindcast rainfall agreed with the observations in reasonable detail over the 13 years of the intercomparison; in particular, the interannual variability was well captured. The major deficiency is that the model over predicted rainfall in some years, particularly those with minimal rainfall. The correlation between the anomalies of the two time series in Figure 3(a) is 0.62, significant at the 99.9% level.

The time series in Figure 3(b) is spatially averaged over a region including the southeast states of the USA and the adjacent Gulf of Mexico. This is the same region used by Stern and Miyakoda (1995) in their simulation study with a global observed SST distribution. They found that reasonably high predictability existed for this region. The hindcast in Figure 3(b) reproduced many of the observed temporal rainfall variations: note only low latitude Pacific Ocean SST forcing was used in the hindcast compared with the global SST forcing of Stern and Miyakoda. The hindcasts for 1984 and 1985 were particularly good, in contrast to 1987 and 1988. Overall, these results suggest that this region has considerable potential predictability, which may be achievable with model improvements. The anomalies of the two time series in Figure 3(b) had a correlation coefficient of 0.22, significant at the 99.0% level.

The final panel (Figure 3(c)) is for northeast Australia, which is influenced strongly by ENSO events. The hindcast captured much of the observed *interannual* variability of the rainfall, especially the major 1982/83 drought situation. However, after 1989 the hindcast deteriorated, with a noticeable underestimation of the observed rainfall. The overall correlation of the anomalies of the two time series was only 0.07, which was not statistically significant. The causes of this rather poor outcome include inadequacies in the SST hindcasts used, deficiencies in the GCM and the restriction of the comparison to a relatively small region. Correlations tend to improve as the area over which results are averaged is expanded, but this then reduces the regional detail possible.

Comparisons of time series of hindcast and observed rainfall were made for a number of other regions. For central China hindcasts comparable to those shown in Figure 3 were obtained. As discussed in Hunt (1997) this region appears to have quite high predictability. In contrast, hindcasts for northeast Brazil, southern Africa and the Indian Ocean were very poor. These outcomes are attributable to the lack of SST forcing from the Indian and Atlantic Oceans, as simulations made with global SST datasets indicate that considerable predictability exists for these regions (see for example, Stern and Miyakoda (1995), Kirtman et al. (1997) and Rowell et al. (1995)).

5. Coupled Models for Prediction Purposes

Coupled atmospheric-oceanic global climatic models developed for long-lead predictions are in 'operational' use at a number of institutions (Ji et al., 1994; Kirtman et al., 1997; Ineson and Davey, 1997). Outputs from these models are regularly reported in the NOAA Experimental Long-Lead Forecast Bulletin. Other research groups are also developing such models. The oceanic region used in most of these predictions is restricted to the Pacific Ocean, but models with global oceans are now also being developed for routine use, see for example Stockdale *et al.* (1998). Details of many of these models are given in the Monthly Weather Review 1997, vol. 125, no.5. Hence, only very brief comments will be made here.

The capabilities of the models are demonstrated by predictions documented in the NOAA Bulletin in December 1996 that an El Niño event could be expected in 1997. While none of the models predicted the severity and rapid development of this El Niño event, they are still to be commended as regards the correctness of the basic prediction.

As would be expected, there is considerable diversity in the SST patterns and amplitudes predicted by the various models, and much development will be needed to improve the quality of the predictions and to demonstrate that the models can cope with decadal variability. An example of the current status of predictions is given in Figure 4 for 6 months in advance for two coupled models, from the NOAA Experimental Long-Lead Forecast Bulletin.

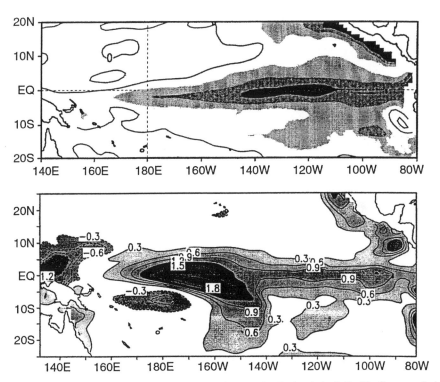

Figure 4. Comparison of SST predictions for March-May 1998 for the low latitude Pacific Ocean made in late 1997. The top panel shows predictions from the NCEP model and the bottom panel shows predictions from the COLA model. (Redrawn from the NOAA Experimental Long-lead Forecast Bulletin, 1997, Vol.6, no.3).

The SST predictions produced by the coupled models can be used 'off line' to force alternative atmospheric models, or the atmospheric component of the coupled model, to generate the ensemble of predictions needed. The output from the coupled models presented in the NOAA Bulletin is usually restricted to SST anomaly patterns for the Pacific Ocean. There appears to be rather little systematic documentation of critical outcomes, such as predictions of rainfall for seasons in advance, as presented by Hunt *et al.* (1994).

6. Future Issues in Long Lead Predictions

6.1 EMPIRICISM

A considerable number of both *a priori* and *a posteriori* corrections are currently applied to operational long-lead predictions. For example, Kirtman *et al.* (1997) extrapolate model winds from 850 mb to produce a zonal surface wind stress in their model. An iterative correction procedure between SST anomaly error and zonal wind stress is also applied in the oceanic initialisation procedure, while systematic errors are removed before processing prediction outcomes. Ji *et al.* (1994) apply a flux correction of the order of 10 W m^{-2} during their prediction experiments, and a model output statistics procedure is used to correct for wind stress anomalies. Both groups also use anomaly coupling between the atmospheric and oceanic models to overcome the problem of climatic drift.

While these procedures may be considered to be somewhat extreme, and to prejudice the model predictions, they are an inevitable consequence of the intrinsic problems associated with coupling two highly non-linear model systems. Both the atmospheric and oceanic models are remarkably sensitive to exchanges across their mutual interface and any error in either model rapidly propagates within the coupled system.

The need for this empiricism is an indication of the subtleties and sensitivities of the interactions. The model problems to be resolved are extremely difficult and rapid solutions are not likely to be achieved. In the meantime the pragmatism being used to achieve useful predictions at least provides valuable guidance as to future climatic variability.

6.2 MODEL INITIALISATION

The major problem lies in the initialisation of the oceanic model in view of the perceived climatic 'memory' that resides in the ocean. The problem is difficult because of the scarcity of three-dimensional oceanic data (primarily temperature) and the intermittent temporal nature of such data.

Oceanic data assimilation systems are being developed (Ji and Smith, 1995), while 'nudging' techniques are also used in which wind and/or SST are relaxed towards observations while the *coupled* model is spun up (Rosati *et al.*, 1997). An SST-wind nudging system has some appeal because of its basic simplicity compared with a three-dimensional oceanic data assimilation system. The experiments of Rosati *et al.* suggest that the former is competitive with the latter, although they used very strong relaxation coefficients in their nudging experiments.

In order to avoid 'shocks' when the oceanic and atmospheric models are used in coupled mode, it is undesirable that the models be initiated separately and then combined. In this regard the nudging technique within a coupled model is superior to the use of an oceanic data assimilation scheme. The latter is specific to the oceanic model, hence a shock can be expected when the oceanic and atmospheric models are coupled. In a truly operational system, presumably a coupled model will be maintained such that it assimilates current observations, thus permitting it to be switched into predictive mode with a minimum of disruption.

The question of soil moisture initialisation also needs to be considered. Presumably reasonably accurate soil moisture values can be obtained from daily numerical weather prediction (NWP) models, if such models routinely calculate soil moisture and transfer the subsequent values to the initialisation scheme used for the long-lead prediction models. Shao *et al.* (1997) have developed an extremely detailed model directed towards such an achievement. They emphasise the sensitivity of the soil moisture content to the specified soil hydraulic parameters. Again, a shock could result if the soil moisture value thus obtained is used in a long-lead prediction model and these values are incompatible with the atmospheric specific humidity in the prediction model. Continuous transfer of NWP soil moisture values on, say, a daily basis to a coupled model being updated in assimilation mode would reduce such shocks. A review of the status of soil moisture specification has been presented by Betts *et al.* (1996).

6.3 SPATIAL AND TEMPORAL RESTRICTIONS

As discussed in the section on analysis of the two-tiered approach, spatial averaging is used to improve signal-to-noise ratios in rainfall predictions. This spatial averaging is needed to obtain a representative sample of rainfall variability as reproduced by the model. Given the inherently small scale of much rainfall, it remains to be seen whether decreasing the spatial scales by using a limited area model driven by GCM outputs will permit smaller areas to be used in such averaging.

Ideally, monthly rather than seasonal mean climatic predictions are desirable, especially for rainfall. Figure 5 shows two detailed temporal variations taken from Figures 3(b) and 3(c). In Figure 5(a) hindcast and observed rainfall time series are compared for northeast Australia for 1982/3. While the 1982 drought conditions were well hindcast, the breaking of the drought occurred two months too early in the hindcast, and represents a major deficiency. In Figure 5(b) the hindcast reproduced the observed rainfall variations over southeast USA/Gulf of Mexico during 1984/85 with remarkable temporal accuracy. At other times poor temporal correspondence occurred for this region as can be seen in Figure 3(c). Thus, it appears that predictions timely to within a month may be *potentially* achievable, but the attainment of this potential may be difficult in many situations.

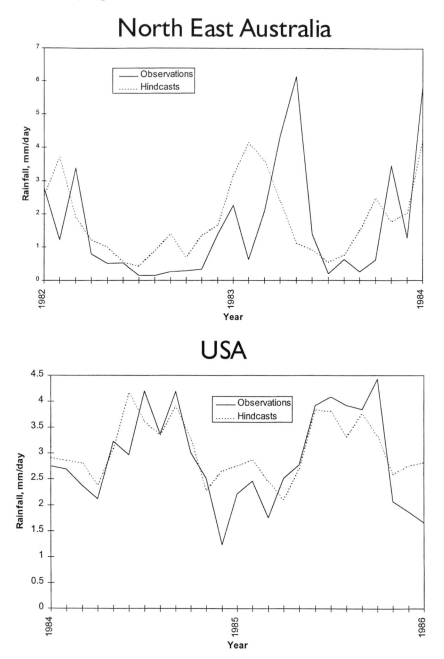

Figure 5. Hindcast and observed rainfall variability for two of the three regions in Figure 3 showing the temporal variability in more detail.

6.4 LIMITED AREA MODELS AND DOWNSCALING

In general, global climatic models lack the necessary horizontal resolution to provide the accuracy needed in many predictive situations. For example, in the Maritime Continent there are numerous regions where complex topographic and orographic features exist. A limited area model, forced by the global climatic model, should be able to improve the accuracy of predictions in this region. Nevertheless, the complexities of the terrain are such that long-lead predictions will always be much more difficult because of the high spatial variability of the rainfall.

Elsewhere, in regions with varying broad-scale orography, the use of limited area models has a demonstrated capacity to improve simulated rainfall distributions. See, for example, the rainfall distributions over eastern Australia in the experiments of Walsh and McGregor (1995). Thus selected combinations of global and limited area models offer the prospects of value adding to the basic global model predictions. Other approaches to downscaling can involve statistical techniques or such devices as neural networks. Statistical downscaling is discussed elsewhere in this volume (Bates et al., 2000) and relies on the use of observed climatic relationships to extend GCM outputs to regional scales.

6.5 ENSEMBLE RUNS

While multiple predictions are needed for a given time period to generate an ensemble because of chaotic influences in the climatic system, considerable information of potential use resides inside such an ensemble. The existence of an ensemble means that probabilistic outcomes for the predictions can be made (Palmer et al., 1990). Examination of the variations between individual members of an ensemble provides guidance concerning the potential skill of any prediction. For example, Dix and Hunt (1995) have shown that in the tropics, and particularly over the Pacific Ocean, a unique outcome in even a difficult climatic variable such as rainfall, is possible with a single realisation. At higher latitudes more runs are necessary to improve the signal-to-noise ratio. Déqué (1997) has provided estimates of how many runs are needed for an ensemble for different latitudinal regions. In addition, an ensemble permits estimates to be made concerning the potential predictability of various regions (Rowell et al., 1995; Anderson and Stern, 1996). As shown by Hunt (1997) there is considerable information associated with the runs constituting an ensemble prediction, and it would appear that this information has still not been fully exploited as an aide to clarifying predictive outcomes.

6.6 EXTENDING THE GEOGRAPHICAL RANGE OF SST PREDICTIONS

Currently SST predictions based on global coupled models are mainly for the low to middle latitude Pacific ocean. This limitation is largely determined by the state of model development, but is also associated with the primary regional interests of the forecasters.

Although ENSO-induced climatic impacts are widely distributed over the globe, predictions for these identified regions cannot be achieved solely with the Pacific Ocean based model. This is because in reality ENSO influences originating from the Pacific produce associated responses from the Indian and Atlantic Oceans, which are manifested as SST variations. These SST variations then produce climatic perturbations.

An indication of the limitations of a Pacific only model is given in Figure 6 where a hindcast of rainfall for South Africa is compared against observations. These results were obtained from the same set of ensemble runs that produced the results shown in Figures 3 and 4. An ENSO signal does exist in observations for this region, but the actual climatic driving force originates in the Indian Ocean (see Jury *et al.*, 1996). The hindcast clearly fails to capture the observed interannual variability.

Thus, to provide predictability for this region and other areas with ENSO signals, such as northeast Brazil, the oceanic model used in predictions needs to be extended to cover at least the whole of the global tropical region.

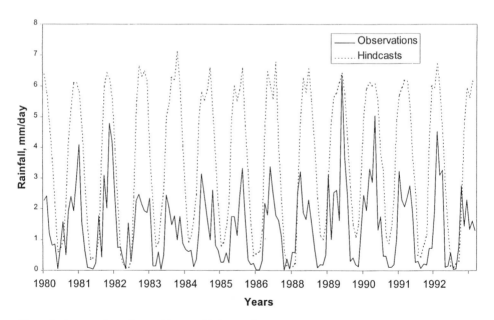

Figure 6. A comparison of ensemble-mean hindcast and observed rainfall variability for South Africa. The results have been averaged over the area 24-30°S, 24-30°E.

6.7 FROM SST PREDICTIONS TO CLIMATIC PREDICTIONS

For obvious reasons the current interest in predictive studies using climatic models has concentrated on producing predictions of Pacific SST anomalies. There has been relatively little interest demonstrated in presentation of long-lead predictions of critical climatic variables such as rainfall as documented in Figure 3 (however, see Kirtman *et al.*, 1997). While there have been numerous simulations with global models forced with observed SST distributions, (AMIP is perhaps the best documented example), these have revealed systematic problems in many models. Given that the SST forcing in predictive studies is limited to the low latitude Pacific Ocean, even the level of achievement documented in the AMIP runs will not be available in predictions.

Hence, in tandem with the SST predictions, there needs to be a consistent effort committed to developing the atmospheric global climatic models, and to forcing them with such SST predictions. Such experiments need to be based on SST predictions for individual years, not composites of El Niño or La Niña years, as such composites distort the actual level of predictability to be expected in a single year. Only by such experiments will it be possible to identify the success and limitations of prediction schemes. The failure of the Zebiak-Cane hindcasts with the current atmospheric model to replicate rainfall over South Africa as shown in Figure 6 is an indication of the issues that need to be addressed. Such experiments will clarify, for example, the necessity or otherwise of initiating soil moisture in prediction experiments, as discussed above.

Much research remains to be done in this area, as well as in the associated area of statistical procedures for assessing the accuracy of results.

Conclusion

Encouraging progress is occurring in the multi-faceted task of developing long-lead climatic prediction schemes. Sea surface temperature anomalies in the low latitude Pacific Ocean associated with ENSO events can now be predicted quantitatively up to 12 months in advance. Considerable improvement is still required as regards both the spatial and temporal characteristics of the SST anomalies, and it remains to be demonstrated that the decadal scale temporal variability of observed ENSO events is predictable. To obtain the maximum advantage of long-lead predictions it will be necessary to extend the SST predictions to at least the global low latitude region, thus requiring the incorporation of oceanic models for the Indian and Atlantic Oceans. An important problem still to be resolved is what is the optimal method of oceanic initialisation? In particular, can such an initialisation scheme be developed to ensure decadal predictability?

The translation of SST predictions into actual climatic predictions of rainfall, frost occurrence etc. using the atmospheric component of the model has still to be fully demonstrated. The hindcasts presented in this paper suggest that rainfall predictions are possible in selected regions, but the present limitation of SST predictions to the Pacific Ocean does not permit potentially predictable regions to be evaluated. In many regions the variance of climate attributable to ENSO events is small or negligible, hence it is important to ensure that unrealistic expectations are not excited for the predictions.

Numerous technical and operational problems remain to be (identified and) resolved, hence considerable research will be necessary. This would suggest that individual research groups will be able to make worthwhile contributions for many years to come. In the meantime, it should be possible to make useful, operational long-lead predictions with existing model techniques, as long as it is appreciated that substantial improvements can be expected over the next decade, so that any short term failures are not used to denigrate present capabilities.

References

Anderson, D.L.T. and McCreary, J.P. (1985) Slowly propagating disturbances in a coupled ocean-atmosphere model. *J. Atmos. Sci.* **42**, 615-629.

Anderson, J.L. and Stern, W.F. (1996) Evaluating the potential predictive utility of ensemble forecasts. *J. Clim.* **9**, 260-269.

Barnett, T.P., Bengtsson, L., Arpe, K., Flugel, M., Graham, N.E., Latif, M., Ritchie, J., Roeckner, E., Schlese, U., Schulzweida, U. and Tyree, M. (1994) Forecasting global ENSO-related climate anomalies. *Tellus* **46A**, 381-397

Bates, B.C., Charles, S.P. and Hughes, J.P. (2000) Stochastic downscaling of general circulation model simulations, in G.L. Hammer, N. Nicholls, and C. Mitchell (eds.), *Applications of Seasonal Climate Forecasting in Agricultural and Natural Ecosystems – The Australian Experience.* Kluwer Academic, The Netherlands. (this volume)

Battisti, D.S. (1988) The dynamics and thermodynamics of a warming event in a coupled tropical atmosphere/ocean model. *J. Atmos. Sci.* **45**, 2889-2919.

Battisti, D.S. and Hirst, A.C. (1989) Interannual variability in a tropical atmosphere-ocean model: influence of basic state, ocean geometry and nonlinearity. *J. Atmos. Sci.* **46**, 1687-1712.

Betts, A.K., Ball, J.H., Beljaars, A.C.M., Miller, M.J. and Viterbo, P.A. (1996) The land-atmosphere interaction: a review based on observational and global modelling perspectives. *J. Geophys. Res.* **101**, 7209-7225.

Cane, M.A. and Zebiak, S.E. (1985) A theory for El Niño and the Southern Oscillation. *Science* **228**, 1085-1087.

Cane, M.A. (2000) Understanding and predicting the world's climate system., in G.L. Hammer, N. Nicholls, and C. Mitchell (eds.), *Applications of Seasonal Climate Forecasting in Agricultural and Natural Ecosystems – The Australian Experience.* Kluwer Academic, The Netherlands. (this volume)

Cane, M.A., Zebiak, S.E. and Dolan, S.C. (1986) Experimental forecasts of El Niño. *Nature* **321**, 827-832.

Chen, D., Zebiak, S.E., Busalacchi, A.J. and Cane, M.A. (1995) An improved procedure for El Niño forecasting. *Science* **22**, 1699-1702.

Déqué, M. (1997) Ensemble size for numerical seasonal forecasts. *Tellus* **49A**, 74-86.

Dix, M.R. and Hunt, B.G. (1995) Chaotic influences and the problem of deterministic seasonal predictions. *Int. J. Climatol.* **15**, 729-752.

Hirst, A.C. (1986) Unstable and damped equatorial modes in simple coupled ocean-atmosphere models. *J. Atmos. Sci.* **43**, 606-630.

Hirst, A.C. (1989) Recent advances in the theory of ENSO. AMOS Newsletter 2, 101-113. Aust. Meteorol. and Oceanogr. Soc., Melbourne.

Hunt, B.G. (1997) Prospects and problems for multi-seasonal predictions: some issues arising from a study of 1992. *Int. J. Clim.* **17**, 137-154.

Hunt, B.G., Zebiak, S.E. and Cane, M.A. (1994) Experimental predictions of climatic variability for lead times of twelve months. *Int. J. Clim.* **14**, 507-526.

Ineson, S. and Davey, M.D. (1997) Interannual climate simulation and predictability in a coupled TOGA GCM. *Mon. Wea. Rev.* **125**, 721-741.

Inoue, M. and O'Brien, J.J. (1984) A forecasting model for the onset of El Niño. *Mon. Wea. Rev.* **112**, 2326-2337.

Ji, M., Kumar, A. and Leetmaa, A. (1994) An experimental coupled forecast system at the National Meteorological Center: some early results. *Tellus* **46A**, 398-418.

Ji, M. and Smith, T.M. (1995) Ocean model responses to temperature data assimilation and varying surface wind stress: intercomparisons and implications for climate forecasts. *Mon. Wea. Rev.* **123**, 1811-1821.

Jin, F.-F. and Neelin, J.D. (1993) Modes of interannual tropical ocean atmosphere-interaction – a unified view. Part III: Analytical results in fully coupled cases. *J. Atmos. Sci.* **50**, 3523-3540.

Jury, M.R., Pathack, B., Rautenbach, C.J. de W. and van Heerden, J. (1996) Drought over South Africa and Indian Ocean SST: statistical and GCM results. *The Global Atmosphere and Ocean System* **4**, 47-63.

Kirtman, B.P., Shukla, J., Huang, B., Zhu, Z. and Schneider, E.K. (1997) Multiseasonal predictions with a coupled tropical ocean - global atmosphere system. *Mon. Wea. Rev.* **125**, 789-808.

McCreary, J.P. (1983) A model of tropical ocean-atmosphere interaction. *Mon. Wea. Rev.* **111**, 370-387.

McCreary, J.P. and Anderson, D.L.T. (1991) An overview of coupled ocean-atmosphere models of El Niño and the Southern oscillation. *J. Geophys. Res.* **96**, 3125-3150.

Munnich, M., Cane, M.A. and Zebiak, S.E. (1991) A study of self-excited oscillations of the tropical ocean-atmosphere system. Part II: Nonlinear cases. *J. Atmos. Sci.* **48**, 1238-1248.

Neelin, J.D. (1989) On the interpretation of the Gill model. *J. Atmos. Sci.* **46**, 2466-2488.

Neelin, J.D. (1990) A hybrid coupled general circulation model for El Niño studies. *J. Atmos. Sci.* **47**, 674-693.

Neelin, J.D. (1991) The slow sea surface temperatures mode and the fast wave limit: analytic theory for tropical interannual oscillations and experiments in a hybrid coupled model. *J. Atmos. Sci.* **48**, 584-606.

Palmer, T.N., Mureau, R. and Motteni, T. (1990) The Monte Carlo forecast. *Weather* **45**, 198-207.

Philander, S.G.H. (1990) El Niño, La Niña and the Southern Oscillation. Academic, San Diego, California, 287 pp.

Rosati, A., Miyakoda, K. and Gudgel, R. (1997) The impact of ocean initial conditions on ENSO forecasting with a coupled model. *Mon. Wea. Rev.* **125**, 754-772.

Rowell, D.P., Folland, C.K., Maskell, K. and Ward, M.N. (1995) Variability of summer rainfall over tropical north Africa (1906-1992): observations and modelling. *Q.J.R. Met. Soc.* **121**, 669-704.

Rowell, D.P., Folland, C.K., Maskell, K., Owen, J.A. and Ward, M.N. (1992) Modelling the influence of global sea surface temperatures on the variability and predictability of seasonal Sahel rainfall. *Geophys. Res. Lett.* **19**, 905-908.

Schneider, E.K., Zhu, Z., DeWitt, D.G., Huang, B. and Kirtman, B.P. (1997) ENSO hindcasts with a coupled GCM. COLA Report No. 39.

Shao, Y., Leslie, L.M., Munro, R.K., Irannejad, P., Lyons, W.F., Morison, R., Short, D. and Wood, M.S. (1997) Soil moisture prediction over the Australian continent. *Met. Atmos. Phys.* **63**, 195-215.

Stern, W. and Miyakoda, K. (1995) Feasibility of seasonal forecasts inferred from multiple GCM simulations. *J. Clim.* **8**, 1071-1085.

Stockdale, T.N., Anderson, D.L.T., Alves, J.O.S., and Balmaseda, M.A. (1998) Global seasonal rainfall forecasts using a coupled ocean-atmosphere model. *Nature* **392**, 370-373.

Suarez, J.J., and Schopf, P.S. (1988) A delayed action oscillator for ENSO. *J. Atmos. Sci.* **45**, 3283-3287.

Walsh, K. and McGregor, J.L. (1995) January and July climate simulations over the Australian region using a limited-area model. *J. Clim.* **8**, 2387-2403.

Xie, P., Rudolf, B., Schneider, U. and Arkin, P.A. (1996) Gauge-based monthly analysis of global land precipitation from 1971 to 1994. *Geophys. Res.* **101**, 19023-19034.

Zebiak, S.E. (1982) A simple atmospheric model of relevance to El Niño. *J. Atmos. Sci.* **39**, 2017-2027.

Zebiak, S.E. and Cane, M.A. (1987) A model El Niño-Southern Oscillation. *Mon. Wea. Rev.* **115**, 2262-2278.

OPERATIONAL EXPERIENCE WITH CLIMATE MODEL PREDICTIONS

NEVILLE NICHOLLS, CARSTEN FREDERIKSEN AND RICHARD KLEEMAN

Bureau of Meteorology Research Centre
PO Box 1289K
Melbourne, Vic 3000, Australia

Abstract

Computer climate models have done a reasonable job in simulating many aspects of the climate variability in the Southern Hemisphere. Based on the success of these models in **simulation**, the Bureau of Meteorology Research Centre has been testing the use of models to **predict** the El Niño - Southern Oscillation and to **predict** Australian rainfall. These models are run every month, in "real time", so examination of the skill of their forecasts provides a good test of the models. We will discuss the accuracy of the model forecasts, especially through the 1997/98 El Niño. Note that the forecasts discussed here are all "operational" forecasts, ie they are prepared and published in "real-time" – this is a more stringent and realistic test of the credibility of the forecasts than the usual research approach of examining "hindcasts".

1. Introduction

In order to effectively simulate climate, computer models need to reproduce atmospheric behaviour over a variety of time scales. The important time-scales that need to be captured by climate models include individual synoptic events (1-5 days), the Madden-Julian Oscillation or MJO (40-50 days), the El Niño - Southern Oscillation (and other interannual phenomena), and decadal and longer trends. As well, the annual cycle should be represented since it represents a major source of variation in rainfall, in at least some areas. Finally, climate models need to be able to represent the diurnal cycle accurately, since much of the variability of precipitation is associated with this time scale. The ability of models to simulate each of these time-scales was reviewed by Nicholls (1996).

Nicholls (1996) concluded that models could reproduce much, but not all, of the variability of the southern hemisphere atmosphere. For instance, it seems that models find it easier to simulate the interannual variability of precipitation in the tropics and sub-tropics, than further polewards. Several studies have demonstrated the ability of

109

G.L. Hammer et al. (eds.), The Australian Experience, 109–119.

models to simulate tropical atmospheric interannual variability, when the model is forced with observed sea surface temperatures (SSTs). The influence of the SSTs on the higher latitudes seems more complex. So far, little work has been done to ensure that the variability simulated in the models is arising from credible representations of physical processes, eg. in the simulations of tropical cyclones. As well, the combined effects of several forcing factors (eg. SSTs, sea-ice, volcanoes) have not been extensively explored as yet. Studies are needed to determine their relative influences, and whether all need to be included to account for most of the observed variability. Significant climate anomalies appear to be produced by changing SSTs in various ocean basins, by changing land surface characteristics, and altering sea ice. As well, the models produce variability even in the absence of external forcings, and other forcings not yet tested in models also probably have an influence. These results suggest that a complete range of forcings needs to be included, if we are to use models to simulate or predict climate variations. The internally generated variability in the models certainly demonstrates that multiple runs (ensembles) are needed to obtain reliable simulations, and that "perfect" simulations or predictions will not be realised.

The performance of atmospheric climate models has improved in recent years, partly because of increased interest and resources devoted to studying their use in simulating and predicting climate variability on a wide range of time and space scales. A recent initiative has been model intercomparison experiments. One such experiment involved many research groups running their models with identical global SST anomalies for June-August 1987 and 1988, in an attempt to simulate the observed differences in the Indian monsoon in these years. There were considerable differences from the various models, especially in regard to monsoon rainfall. An interesting result from this intercomparison was that the model monsoons were mainly affected by SST anomalies in the tropical Pacific. Indian Ocean SST anomalies appeared to have little effect on the Indian Monsoon.

Coupled models designed for the simulation and prediction of the El Niño have developed quickly since the first demonstration that such prediction was feasible (Cane et al., 1986). A variety of coupled models, ranging from simple to complete coupled general circulation models of the ocean and the atmosphere, are now run routinely in various centres. These models generally demonstrate substantial skill in forecasting east equatorial Pacific (NINO3) SSTs when used in hindcast mode.

Model simulations, both stand-alone atmospheric climate models, and coupled ocean-atmosphere models for prediction of the El Niño, seem to have done well enough to justify their operational testing. This has been done in BMRC, with a coupled ocean-atmosphere model being run each month operationally since 1994. Since the start of 1997 a stand-alone atmospheric model has also been run each month, to test its ability to forecast 3-monthly rainfall anomalies.

2. Coupled Model Predictions of the El Niño

BMRC has developed and run an intermediate coupled ocean-atmosphere model (Kleeman *et al.*, 1995) each month, and the results have been widely disseminated, since 1994. The model forecasts are used in the National Climate Centre's *Seasonal Climate Outlook* each month and published every quarter in the *NOAA Experimental Long-lead Forecast Bulletin*. From mid-1996 onwards the model was predicting a 1997 El Niño.

The model forecasts of the 1997 El Niño have been compared with observations and forecasts from other El Niño prediction models published routinely in the *NOAA Experimental Long-lead Forecast Bulletin*. In the December 1996 issue of the *Bulletin*, the following summary of the El Niño forecasts from coupled models was published:

"The improved Scripps/MPI hybrid coupled model predicts mildly cool conditions for winter 1996-97, moderate warming for winter 1997-98. The standard Lamont Doherty Earth Observatory model predicts cool conditions through summer 1997, becoming warmish by winter 1997-98; the new model calls for cool through 1997. The NCEP coupled model calls for warming through the neutral range winter 1996-97, becoming somewhat warm by July 1997. The COLA coupled model forecasts cool Niño 3 SST boreal winter 1996-97, warming to somewhat above normal by October 1997, lasting through winter 1997-98. The Australian BMRC low order coupled model predicts slightly cool conditions for Dec-Jan-Feb 1996-97, becoming slightly warm by Jul-Aug-Sep 1997. The Oxford coupled model calls for moderately negative SST anomalies boreal winter 1996-97, warming to normal by Jul-Aug-Sep 1997."

Within the wide variety of model predictions there were a few (BMRC, NCEP, possibly COLA) models predicting a weak El Niño for 1997. None predicted, before the end of 1996, a strong El Niño for 1997. By early in 1997 the BMRC model was predicting a stronger El Niño. Figure 1 shows the observed east equatorial Pacific (NINO3) SST anomalies, along with the predictions made six months in advance by the BMRC model. The forecast from March 1997, for September 1997, was for a NINO3 anomaly of about +2°C. The observed September NINO3 anomaly was about +3°C. This must count as an excellent result for the model, given that the NINO3 anomaly at the time of forecast was close to zero. Thus the model was NOT simply persisting the observed anomaly. It is impressive, also, that the model during 1997 was predicting, six months in advance, a stronger El Niño signal than had been observed during 1994. The model also did very well in predicting, again six months in advance, the decline in NINO3 after November 1997.

However, the forecasts through 1997 were consistently below the observed strength of the El Niño event. Earlier, the model had difficulty in predicting the decline of the 1994 El Niño episode (Figure 1). At longer lead times the model exhibited little skill over the 1994-97 period. Figure 2 shows the observed and predicted NINO3 anomalies along with the anomalies predicted nine months in advance. Little skill is evident. In particular, the model showed no inclination to predict a strong El Niño episode through 1997. It is not obvious why the model should perform relatively well at six-month lead times, but show little skill at nine months.

The model predictions increased sharply when March 1997 data became available. So the six-month ahead NINO3 anomaly forecast valid for September 1997 was substantially stronger than had been the six-month forecast for August (Figure 1). This appears to have been due to the model reacting to two strong bursts of westerly wind in the west equatorial Pacific around February-March (Moore and Kleeman, 1998). Such synoptic-scale events are unlikely to be predictable months in advance. This would limit the potential skill of any model.

6 MONTH LEAD TIME
FORECAST VERIFICATION

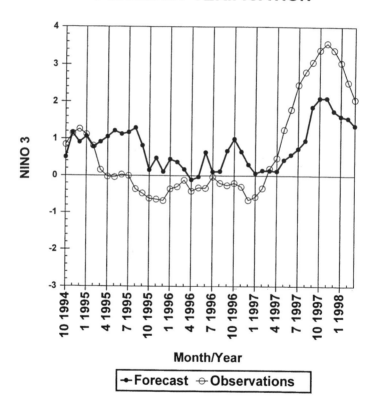

Issued: April 1998

Figure 1. Observed (thin line) NINO3 SST anomalies (°C) and anomalies forecast from the BMRC intermediate coupled model six-month ahead (thick line). All forecasts are real-time operational forecasts issued six-months prior to the time of validity indicated.

9 MONTH LEAD TIME
FORECAST VERIFICATION

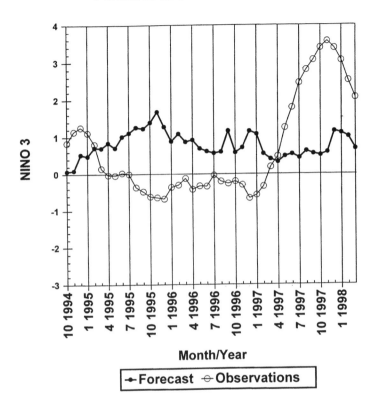

Month/Year

← Forecast ⊖ Observations

Issued: April 1998

Figure 2. Observed (thin line) NINO3 SST anomalies (°C) and anomalies forecast from the BMRC intermediate coupled model six-month ahead (thick line). As in Figure 1, all forecasts are real-time operational forecasts, but in this case they were issued nine months prior to the time of validity indicated.

3. Atmospheric Model Predictions

Frederiksen *et al.* (1995a,b, and 1999) have shown, in a detailed study of simulated Australian rainfall variability over the period 1950-1991, that atmospheric general circulation models (AGCMs), when forced by observed SSTs are capable of capturing much of the observed rainfall variation and its relationship with the Southern Oscillation Index (SOI) and Indian and Pacific Ocean SST anomalies (SSTAs). This is especially true over the northern and eastern portions of the continent. Such studies suggest that, with a sufficiently accurate estimate, or forecast, of the SSTs for the forthcoming months, it might be possible to provide a useful month to seasonal rainfall prediction with a stand-alone AGCM forced by these predicted SSTs. The simplest approach one can take to forecasting the SST, is to add the latest observed SSTAs to the climatological SST for the forthcoming months. Comparisons with other statistical methods (such as, for example, multi-channel singular spectrum analysis) show that, for one to three month forecasts, the use of such *persisted* SSTAs is difficult to improve upon.

Since March 1997, 120-day rainfall forecasts have been conducted routinely each month using the BMRC AGCM forced by persisted SSTAs taken from the observations for the week preceding the forecast period. The model is a global spectral model with rhomboidal 31 horizontal resolution and 17 vertical levels. It includes prognostic equations for the vertical component of relative vorticity, horizontal divergence, temperature, moisture, and surface pressure. The model includes physical parameterizations of convection, radiation, boundary layer effects, diagnostic clouds, gravity wave drag, shallow convection, soil moisture and soil temperature. The main differences between this model and that used in Frederiksen *et al.* (1995a,b, and 1999) is the larger number of vertical levels (17 compared with 9) and the option of using the Tiedtke (1989) mass-flux convective scheme in preference to the Kuo (1974) scheme.

We will concentrate only on the skill of the model in predicting rainfall, surface air temperature, mean sea level pressure (MSLP) and 200hPa geopotential height anomalies over the Australian region. Such anomalies are determined with respect to either a model climatology or a control run using only a SST climatology. To determine the sensitivity of the results to the convective schemes, forecasts were made using both the Kuo and Tiedtke schemes. Two sets of runs were conducted for the model with the Kuo convection scheme; one set uses the climatological weekly SSTs and the other uses the aforementioned method. Thus, the first set of runs constitutes a set of control runs against which the second set can be compared, and differences, or anomalies, can be determined. Only one set of runs was performed for the model with the Tiedtke mass-flux scheme, which only uses the forecast SST. Anomalies for this latter case are determined with respect to a ten year run (using the same model configuration and the observed global SSTs from 1979 to 1988).

An ensemble approach was employed to overcome the model's sensitivity to initial conditions. In each set of experiments, an ensemble of six runs was generated with different initial conditions from the 6-hourly BMRC analysis system near the time when the weekly SSTAs were obtained. Forecast climate anomalies, the average of the six runs, were compared with anomalies derived for the forecast period using observed data for the period and observed climatologies. Observed climatologies were derived from the daily averaged National Center for Environment Prediction (NCEP) reanalysis dataset. Monthly and seasonal anomalies of the NCEP analysis, during the forecast period, were calculated with respect to a 40-year climatology derived from NCEP daily values for the period 1958-1997.

We have used as our principal skill score the linear error in probability space (LEPS) score (Pott *et al.* (1996)). LEPS skill scores for individual forecasts range between −1 and 2, with positive values indicating some skill. A measure of overall skill, taken over all forecasts, was also utilised using a weighted average of individual LEPS scores giving an overall score expressed as a percentage between −100% to 100%. Positive scores are again skilful. A Monte Carlo approach was used to determine a 95% significant level for the overall skill score. Here, we will consider the skill of the forecasts during the period April 1997 − June 1998.

By May 1997, SSTAs in the tropical Pacific Ocean had the characteristic signature of a typical El Niño, with warm and cold SSTAs in the eastern and western Pacific, respectively. Again, the studies of Frederiksen *et al.* (1995a,b, and 1999) suggest that the model, forced by such global SSTAs, should produce generally drier conditions over the eastern half of the Australian continent. Figure 3 shows the LEPS scores for modelled (Kuo) precipitation forecasts of three-month means (April-May-June 1997 (AMJ97) − March-April-May 1998 (MAM98)). While there are areas of skill, for each forecast, it is clear that there is no consistent systematic pattern of skill over Australia throughout the twelve forecasts. Similar conclusions are true for the forecasts based on the Tiedtke convective scheme.

Globally, there are regions (not shown) where our forecasting strategy does show some consistency in skill. Throughout the forecast period, there is appreciable skill in the tropical eastern Pacific, tropical western Indian Ocean, Northeast Brazil and over Indonesia. In particular, the model correctly forecasts the very dry conditions over Indonesia which led to major forest fires and air pollution problems, as well as the failure of crops.

The overall, or total, skill scores for all forecasts and all four climate variables are shown in Figure 4, with only the skill at the 95% significance level plotted. Over Australia, there are some isolated areas of significant skill in rainfall forecasts. Overall, the results are disappointing, given such an intense El Niño event. Interestingly, statistical methods based on SSTs also showed comparably poor skill over this period (Drosdowsky, 1998, personal communication).

In the case of surface air temperature, both sets of forecasts show some significant skill over Australia. In this case, there appears to be much closer qualitative and coherent similarities with the observed anomalies, with the exception of the southern parts of the continent. The Kuo forecasts appear to be more skilful over Australia, and this is reflected in smaller root-mean-square errors (not shown). The skill for the Kuo case is also remarkably similar to that for statistical methods based on SSTs for this same period (Jones, 1998, personal communication). Globally, the major skill is over the tropical/subtropical oceans and landmasses.

MSLP forecasts are also significantly skilful especially over the northern half of the continent. Viewed globally, the overall skill indicates that the Southern Oscillation during the course of the 1997/98 El Niño event is well forecast by the model. While

116

over Australia both sets of forecasts show similar skill, there are significant differences globally. In particular, the Tiedtke forecasts show additional skill over the Indian Ocean but less skill in the eastern Pacific Ocean.

PRECIPITATION LEPS SKILL SCORES (KUO)

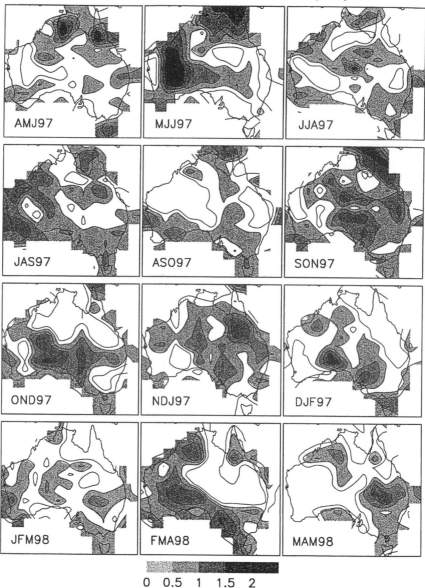

Figure 3. LEPS skill scores for precipitation forecasts using the Kuo convective scheme. The forecasts were prepared using a stand-alone AGCM forced by SST anomalies persisted from the previous month. The forecasts were prepared in real-time. Positive skill scores are shaded.

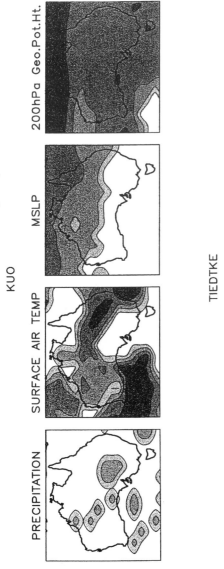

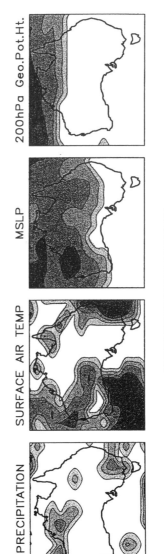

Figure 4. Total LEPS skill score for precipitation, surface air temperature, MSLP and 200hPa geopotential height for Kuo (top panel) and Tiedtke (bottom panel) forecasts. Only skill scores, which are significant at the 95% level, are shown.

STOCHASTIC DOWN-SCALING OF GENERAL CIRCULATION MODEL SIMULATIONS

BRYSON BATES[1], STEPHEN CHARLES[1] AND JAMES HUGHES[2]

[1]CSIRO Land and Water
Private Bag PO
Wembley, WA 6014, Australia

[2]Department of Biostatistics
University of Washington
Seattle, WA 98195, USA

Abstract

Modelling the response of agricultural and natural ecosystems to climate forecasts requires daily data at local and regional scales. General circulation models (GCMs) provide reasonable simulations of atmospheric fields at the synoptic scale. However, they tend to over-estimate the frequency and under-estimate the intensity of daily precipitation. Stochastic downscaling techniques provide a means of linking the synoptic scale with local scales. They can be used to quantify the relation of climate variables at small space scales to the larger scale atmospheric patterns produced by GCMs. This paper reviews downscaling techniques from an applications perspective. It then presents a case study involving the use of a downscaling technique known as the nonhomogeneous hidden Markov model (NHMM). A NHMM fit to a 15-year record of daily atmospheric-precipitation data is used to downscale GCM atmospheric fields for South-West Western Australia. We compare the downscaled and observed 'winter' precipitation statistics at six stations near Perth, Western Australia. The results show that a downscaled GCM simulation provides credible reproductions of observed precipitation probabilities and the frequencies of wet and dry spells at each station.

1. Introduction

Models of the responses of agricultural and natural ecosystems to forecast climatic fluctuations require daily weather data at local and regional scales (Hammer, 2000). The desire for forecasts based on physical rather than purely statistical models is leading to the development of coupled atmospheric-oceanic general circulation models (GCMs) (Hunt and Hurst, 2000). Although existing atmospheric GCMs perform

G.L. Hammer et al. (eds.), The Australian Experience, 121–134.

reasonably well in simulating climate with respect to annual or seasonal averages at sub-continental scales, it is widely acknowledged that they do not provide credible simulations of precipitation at the space and time scales relevant to local and regional impact analyses (Grotch and MacCracken, 1991; Arnell *et al.*, 1996). Differences between GCMs, in terms of simulated precipitation and surface air temperature, also seem to be greater at local and regional scales (Gates *et al.*, 1996).

The above limitations have led to the development of techniques for downscaling GCM simulations. Downscaling methods estimate local and regional scale weather from simulated large-scale atmospheric fields. They are based on the premise that GCMs represent atmospheric conditions at synoptic scales better than at local scales. The concept of downscaling has been used in numerical weather prediction for over 20 years in order to infer local weather elements from global and regional scale models and to remove systematic errors from model simulations (Glahan and Lowry, 1972; Glahan, 1985).

This paper presents a review of downscaling techniques and then focuses on the nonhomogeneous hidden Markov model (NHMM) described by Hughes *et al.* (1998). We describe the application of a NHMM fit to a 15-year record of daily atmospheric and precipitation data for South-West Western Australia to atmospheric fields obtained from a GCM developed by the CSIRO Division of Atmospheric Research. We compare the downscaled GCM winter precipitation statistics with those observed at six stations near Perth. Finally, we discuss on-going research towards the development of algorithms for generating rainfall amounts and downscaled surface temperatures.

2. Review of Downscaling Techniques

Downscaling techniques can comprise either empirical methods or models of the physical dynamics of the climate system (such as limited area or nested regional models). Empirical methods for precipitation modelling typically rely on three key assumptions (Hewitson and Crane, 1996): (1) the dominant rainfall generation mechanism for local and regional scale precipitation is synoptically driven; (2) GCMs provide adequate simulations of atmospheric circulation at the synoptic scale; and (3) the downscaling relationship derived from observational data is time invariant. Examples include statistical downscaling using regression analysis, canonical correlation analysis, statistical analogues, artificial neural networks, and weather classification schemes. To date, precipitation has been the main focus of attention in downscaling studies due to its importance to impact studies and the relative difficulty in obtaining realistic precipitation simulations from GCMs. Correlation techniques are appropriate for continuous variables such as surface air temperature and are less suited to intermittent variables such as daily precipitation. Nevertheless, temperature is also needed for local and regional impact analyses involving environmental systems.

2.1 LIMITED AREA MODELS

Limited area models (LAMs) embed a fine computational grid over a limited domain within the coarse grid of a GCM (Hostetler and Giorgi, 1993; Walsh and McGregor, 1995). Grid meshes for LAMs typically range from 20 to 125 km whereas the meshes for GCMs range from 300 to 600 km (Walsh and McGregor, 1995; McGregor, 1997). The host GCM provides the large-scale synoptic forcing to the LAM through the LAM's boundaries. The use of LAMs allows a more accurate representation of orography, which leads to improvements in the simulation of orographically induced precipitation. They are more economical to run than a GCM with a similar spatial resolution (Walsh and McGregor, 1995). Nevertheless, they tend to over-estimate the frequency and under-estimate the intensity of daily precipitation at the spatial scales of interest in impact analyses (Mearns *et al.*, 1995; Walsh and McGregor, 1995; Bates *et al.*, 1997). This is due to their inability to resolve the full structure of precipitating systems, the use of parameterisations for other sub-grid scale processes and the propagation of any bias from the coarse-resolution GCM into the LAM simulation.

2.2 STATISTICAL ANALOGUES

Statistical analog methods rely on the assumption that the weather during a forecast period will be a replication of weather recorded in the past. Thus long-term observations are required so that an analog can always be found (Stone *et al.*, 2000) demonstrate the use of statistical clustering techniques to identify analog sets based on time series of the Southern Oscillation Index (SOI) and Sea Surface Temperatures (SSTs). Zorita *et al.* (1995) characterised large-scale atmospheric circulation by a vector formed by the coefficients of the first five empirical orthogonal functions (EOFs) of sea level pressure and upper air temperature fields on three consecutive days. The vector of EOF coefficients is called a state. Forecast rainfall is the rainfall corresponding to a previously recorded state that is closest to the state of interest. The method was found to work well in Spain. Martin *et al.* (1997) used an analog method with a daily time-step to compute seasonal snow cover in the French Alps. It produced a systematic under-estimation of seasonal and long-term accumulated precipitation. They attributed this to difficulties in selecting extreme situations associated with high precipitation amounts.

2.3 ARTIFICIAL NEURAL NETWORKS

Although artificial neural networks (ANNs) are similar in application to multiple linear regression, they are inherently nonlinear and well suited to capturing relationships in 'noisy' data. ANNs can represent the behaviour of the atmosphere as a continuously changing function rather than a sequence of a small number of discrete states. However, the relationships obtained from ANNs cannot be subjected to detailed checks for physical realism.

Hewitson and Crane (1996) used artificial neural networks (ANNs) to derive relationships between atmospheric circulation and local precipitation for southern

Africa. Their ANN was successful in capturing the timing of rainfall events and reasonably successful in predicting their magnitude. However, ANNs sometimes fail to reproduce wet-day occurrence statistics (R.L. Wilby, pers. comm., 1997).

2.4 WEATHER CLASSIFICATION SCHEMES

Many empirical downscaling techniques use weather classification schemes to identify dominant large-scale atmospheric circulation patterns, and then model daily precipitation as a stochastic process conditional on the derived patterns (eg. Hay *et al.*, 1991; Bardossy and Plate, 1992; Bogardi *et al.*, 1993; Von Storch *et al.*, 1993; Wilby *et al.*, 1994; Bartholy *et al.*, 1995; Charles *et al.*, 1996). Typically, the patterns are identified by subjective means or by the use of cluster analysis or classification trees. These schemes identify weather patterns that may not be optimal for modelling precipitation occurrence as there is often a temporal discordance between atmospheric data collected at an instant in time and rainfall data accumulated over a 24 hour period. Also, the atmospheric variables used in the schemes often contain little information about the intensity of development of the weather systems involved (Conway *et al.*, 1996). To date, they have had only limited success in reproducing observed wet and dry spell length statistics.

Hughes *et al.* (1998) describe a nonhomogeneous hidden Markov model (NHMM) for downscaling in which the atmospheric circulation is classified into a small number of discrete patterns called 'weather states'. These states are assumed to follow a Markov chain in which the transition probabilities depend on observable atmospheric fields. (A transition probability is the probability of leaving a particular state and entering another state.) Precipitation occurrence is assumed to be conditionally temporally, but not spatially, independent given the weather state. Unlike other schemes, the weather patterns are not determined a priori. The sequence of weather patterns can be recovered from a fitted NHMM and subjected to physical interpretation and checks for physical realism. Many of the downscaling techniques that use the concept of weather states are special cases of the NHMM (Hughes *et al.*, 1998). Applications of the NHMM to observed data from the Pacific North-West region of the United States and the South-West region of Western Australia indicate that the NHMM provides credible reproductions of wet and dry spell length statistics and useful insights into rainfall mechanisms (Hughes and Guttorp, 1994; Charles *et al.*, 1996; Hughes *et al.*, 1998).

For two stations in the United Kingdom with records beginning in 1881, Wilby (1997) found that conditional probabilities for weather classification schemes are not time invariant and are sensitive to circulation type, sample size, geographical location, and regional-scale atmospheric conditions. However, Hughes *et al.* (1998) describe the results of a split sample test in which the first 10 years of data used by Charles *et al.* (1996) were used to fit a NHMM and the remaining 5 years used for validation. They found that the NHMM reproduced observed daily precipitation statistics despite the presence of a small nonstationary shift in the atmospheric data between the fit and test periods.

3. Case Study

We used the NHMM of Hughes *et al.* (1998) to downscale a GCM simulation given its documented performance, generality, and ability to cope with small nonstationary shifts in atmospheric data. The NHMM was fit to historical data first, and this NHMM was then used to downscale the GCM simulation. This permitted an indirect assessment of the realism of the simulated atmospheric circulation patterns at synoptic scales as well as a direct assessment of the realism of downscaled daily precipitation occurrence.

3.1 GENERAL CIRCULATION MODEL

We used a 10-year simulation for present day conditions from the Mk II version of the nine-level atmospheric GCM developed by the CSIRO Division of Atmospheric Research (CSIRO9 GCM). The horizontal resolution of the GCM is about 300 km × 600 km and the computation time required per model day is 30 s on a CRAY Y/MP computer (Walsh and McGregor, 1995). A computational time step of 30 minutes is used.

3.2 NONHOMOGENEOUS HIDDEN MARKOV MODEL

A hidden Markov model involves an underlying unobserved (hidden) stochastic process that can only be observed through another stochastic process that produces the sequence of observed outcomes. The observed process (such as precipitation occurrence at a set of stations) is assumed to be conditionally, temporally independent given the hidden process. The hidden process (such as the temporal evolution of weather states) is assumed to evolve according to a first order Markov chain (Rabiner and Juang, 1986).

Assume that there are a small number of hidden weather states (N) and a fixed number of rainfall stations (M). At each clock time, $t = 1,..., T$, a new state may be entered according to the state transition probability distribution which depends on the state at time $t-1$. After each transition, an observable outcome is produced according to a probability distribution that depends on the current state. The NHMM of Hughes *et al.* (1998) may be defined by a state transition probability matrix (**A**) and a precipitation occurrence probability distribution (**B**):

$$A_{ij}(X) = P(S^t = j | S^{t-1} = i, X^t), (i, j = 1,..., N) \tag{1}$$

$$B_j(r) = P(R^t = r | S^t = j) \tag{2}$$

where $S^t = j$ denotes the jth (unobserved or hidden) weather state at time t, R^t denotes the rainfall process, **r** denotes a vector of observed precipitation occurrences at the M

stations, and X^t denotes the atmospheric data at time t. The NHMM is termed 'nonhomogeneous' as **A** depends on a set of observed covariates **X**.

The parameterisation used herein for **B** is defined by

$$P(R^t = r \big| S^t = s) = \prod_{i=1}^{M} p_{si}^{r_i}(1 - p_{si})^{1-r_i}$$

(3)

where $r_i = 1$ if precipitation occurs at station i and 0 otherwise, s denotes the weather state, $p_{si} = \exp(\alpha_{si})/(1+\exp(\alpha_{si}))$ is the probability of precipitation at station i in weather state s, and the α_{si} are finite and estimated from the data. Although the probabilities of precipitation occurrence are assumed to be spatially independent conditional on the weather state, they are unconditionally correlated due to the influence of the common weather state. Thus (3) is referred to as the 'conditional independence model'.

The parameterisation for **A(X)** is defined by:

$$P(S^t = j \big| S^{t-1} = i, X^t) = \gamma_{ij} \exp\left[-\tfrac{1}{2}(X^t - \mu_{ij})V_x^{-1}(X^t - \mu_{ij})^T\right]$$

(4)

where μ_{ij} is the mean of X^t and V_x is the corresponding covariance matrix, γ_{ij} corresponds to the transition probability matrix for a hidden Markov model, and T in this instance denotes the transpose of a vector or matrix. The constraints $\Sigma_j \gamma_{ij} = 1$ and $\Sigma_j \mu_{ij} = 0$ are imposed to assure the identifiability of the parameters. V_x is used for scaling and numerical stability, and typically is computed directly from the atmospheric variables.

Estimation of $\theta = (A,B)$ for a series of models with different numbers of weather states and different sets of derived atmospheric variables was carried out using an algorithm based on the method of maximum likelihood (Baum *et al.*, 1970). The estimation problem involves the maximisation of $P(R|X,\theta)$ where $R = (R1, R2, ..., R^T)$ and $X=(X^1, X^2, ..., X^T)$. The resulting model fits were evaluated using an approximation to the Bayes factor:

$$B = 2l + \log|V|$$

(5)

where l is the negative likelihood and **V** is the covariance matrix of the parameters (Kass and Raftery, 1995). Although this criterion is useful for comparing models with different combinations of atmospheric variables, the selection of a final NHMM is based the physical realism and interpretability of the weather states. The most probable sequence of weather states can be estimated from the data (Forney, 1978). This permits the assignment of each day to its corresponding weather state.

3.3 DESCRIPTION OF STUDY AREA AND DATA

Six climate stations near Perth, Western Australia, were selected (Table 1). The area encompassed by these stations experiences a 'Mediterranean' climate with mild, wet winters and hot, dry summers. About 80 percent of annual precipitation falls in the period from May to October (hereafter denoted as 'winter'), the wettest months being June and July. Consequently, only 'winter' results will be considered here. The variability of the elevations and the mean 'winter' rainfalls listed in Table 1 are representative of the Perth environs.

TABLE 1. Characteristics of climate stations used in the case study. Data from Bureau of Meteorology (1988).

No.	Station	Latitude (deg)	Longitude (deg)	Elevation (m)	Mean 'Winter' Rainfall (mm)
1	Dalwallinu	30.3	116.7	335	261
2	Moora	30.7	116.0	203	371
3	Wongan Hills	30.8	116.7	305	262
4	Belmont	31.9	116.0	20	686
5	Dandaragan	30.3	115.6	260	496
6	Lancelin	31.0	115.3	4	529

We extracted the 1200 GMT atmospheric fields for the ten 'winter' periods within the CSIRO9 GCM simulation as these offered the closest synchronisation with the observed 1100 GMT atmospheric data used by Charles et al. (1996). The observed data covered the period from 1978 to 1992. The 1100 GMT data were chosen as they are close to the mid-point of the daily precipitation recording period (24 hours to 0100 GMT, 9 am local time). The atmospheric variables included mean sea level pressure (MSLP) and, at the 850 and 500 hPa levels, the geopotential heights (GPH), air temperatures, dew point temperatures, and U- and V- wind components. From these data, a total of 24 atmospheric variables were derived. These included the original variables, the north-south and east-west gradients of MSLP and GPH, north-south and east-west differences in GPH, and MSLP and GPH variables lagged by one day. Both the observed and GCM atmospheric fields for these variables were interpolated to the same $3.75°$ by $2.25°$ rectangular grid.

4. Results

4.1 PARAMETER ESTIMATION

We used the NHMM fit by Charles *et al.* (1996) to the historical data described above. Charles *et al.* (1996) used classification trees and other exploratory analyses to determine which of 24 derived atmospheric variables had the strongest relationship with daily precipitation occurrence over the region. Precipitation was deemed to have occurred at a particular station on a given day if the gauge catch equalled or exceeded 0.3 mm. They adopted a NHMM with six distinct weather states, the selected atmospheric variables for 'winter' being MSLP, north-south MSLP gradient and 850 hPa east-west GPH gradient. Comparison of the spatial patterns in multi-station precipitation probability (hereafter termed 'precipitation occurrence patterns') associated with each state with their corresponding composite MSLP fields suggested that the states were distinct and that the NHMM had a high degree of physical realism (Charles *et al.*, 1996, Figure 2).

4.2 DOWNSCALED GCM SIMULATIONS

The above NHMM was used to downscale the 1200 GMT MSLP, north-south MSLP gradient and 850 hPa east-west GPH gradient fields extracted from the CSIRO9 GCM simulation to produce daily precipitation occurrence sequences at the six stations near Perth. The NHMM was not fit to the atmospheric fields derived from the GCM.

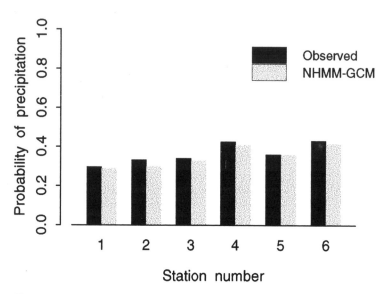

Figure 1. Comparison of observed and downscaled CSIRO9 GCM 'winter' precipitation probabilities for six precipitation stations near Perth, Western Australia.

Figure 1 compares observed and downscaled CSIRO9 GCM 'winter' precipitation probabilities for the six stations near Perth. Although the probabilities obtained from the downscaled GCM simulation under-estimate the observed probabilities, the difference is less than 0.02 for all stations except Station 2 (Moora) where the difference is 0.03.

Figure 2 compares the cumulative distribution function of observed wet spell lengths for each station with the cumulative distribution function derived from the downscaled CSIRO9 GCM simulation. A wet spell is defined as a sequence of consecutive days during which the daily precipitation equalled or exceeded 0.3 mm. Perusal of Figure 2 reveals that the differences between the upper tails of the observed and downscaled cumulative distribution functions are relatively small for Dalwallinu, Wongan Hills, Dandaragan, and Lancelin. The downscaled cumulative distribution function for Belmont has a slightly heavier upper tail than the observed. This contrasts with Moora where the downscaled simulation consistently under-estimates the frequency of the long wet spells. We used the Kolmogorov-Smirnov two-sample test to test the null hypothesis that the observed and downscaled wet spell lengths at each station have identical distribution functions. This is an approximate test as spell length is a discrete rather than continuous variable and consecutive spells may not be independent. The null hypothesis could not be rejected at the 0.10 level in each case.

Figure 3 compares the cumulative distribution function of observed dry spell lengths for each station with the downscaled cumulative distribution function. A dry spell is defined as a sequence of consecutive days during which the daily precipitation remained below 0.3 mm. The downscaled cumulative distribution functions provide good approximations to the observed cumulative distribution functions for all stations and observed spell lengths. We again applied the Kolmogorov-Smirnov two-sample test and found that the null hypothesis that the observed and downscaled dry spell lengths have identical distribution functions could not be rejected at the 0.10 level.

5. Discussion

The success of our downscaling experiment suggests that the CSIRO9 GCM produces reasonable simulations of the synoptic-scale atmospheric circulation over South-West Western Australia. Recall that the NHMM was fit to historical rather than GCM data, and that this NHMM was used to downscale atmospheric fields from the GCM. If the synpotic patterns produced by CSIRO9 GCM were inadequate, the downscaled 'winter' precipitation statistics would have borne less resemblance to the observed.

Our study has shown that downscaled GCM simulations can provide reliable information for studies of climate variability impacts. For seasonal climate forecasting purposes, however, it remains to be shown that downscaling GCM simulations forced by historical sea surface temperatures (SSTs) reproduces the observed at-site interannual variability in 'winter' precipitation. A critical question here is whether the NHMM parameters vary over the length of the SST record. If such an experiment

130

is successful, the NHMM could be used operationally to downscale climate forecasts from dynamical climate models (GCMs and LAMs). This will provide climate information at a higher spatial and temporal resolution than that provided by many other techniques.

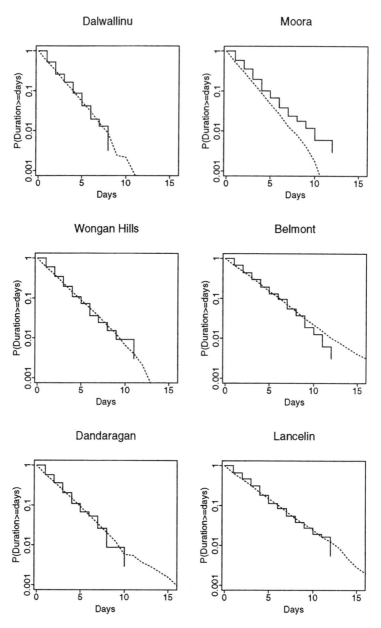

Figure 2. Comparison of wet spell lengths for observed data (solid line) and downscaled CSIRO9 GCM simulation (short-dashed lines).

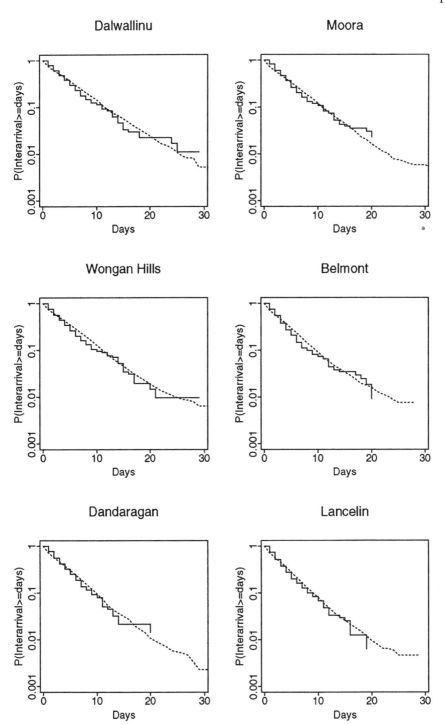

Figure 3. Comparison of dry spell lengths for observed data (solid line) and downscaled CSIRO9 GCM simulation (short-dashed lines).

132

Future work will also involve:

- The development of a precipitation amount model for the NHMM that preserves the spatial correlation structure between stations.

- The determination of the extent to which the NHMM is applicable across southern Australia where precipitation is generated largely by synoptic systems.

- Application of the NHMM to other regions of Australia where convective systems are dominant. This will require the use of atmospheric indices that characterise the vertical as well as the horizontal structure of precipitating systems.

- The development of models based on multivariate observations (eg. precipitation amount, surface temperature). Such models are needed if the NHMM is to reach its full potential for climate forecasting.

Conclusion

A nonhomogeneous hidden Markov model (NHMM) was used to investigate the utility of downscaled GCM simulations. The NHMM was fit to historical atmospheric and precipitation data for South-West Western Australia, and then used to downscale a 10-year simulation from the CSIRO Division of Atmospheric Research atmospheric GCM (CSIRO9 GCM). The downscaled simulation reproduced observed precipitation probabilities and the frequencies of wet and dry spells at six stations located near Perth, Western Australia. This in turn implies that the CSIRO9 GCM produces reasonable simulations of the atmospheric circulation over South-West Western Australia at synoptic scales. Thus it may be inferred that statistical-dynamical approaches to the study of climate variability impacts offer a computationally efficient approach in the short to intermediate term. Further work needs to be done to check whether this finding holds for seasonal climate forecasting.

Acknowledgments

Insights into the synoptic climatology of the study region were provided by P.M. (Mick) Fleming (CSIRO Land and Water, Canberra) and T.J. Lyons (Murdoch University, Perth). Observed atmospheric and precipitation data were provided by the Bureau of Meteorology, and the results of the CSIRO9 GCM simulation by the CSIRO Division of Atmospheric Research. Special thanks are due to Jack Katzfey, John McGregor, and Peter Whetton (CSIRO Division of Atmospheric Research) and Michael Manton, Brian McAvaney, and 'Sri' Srikanthan (Bureau of Meteorology, Australia) for helpful discussions and assistance. This work contributes to the CSIRO Climate Change Research Program and is part funded through Australia's National Greenhouse Research Program.

References

Arnell, N., Bates, B., Lang, H., Magnuson, J.J., and Mulholland, P. (1996) Hydrology and freshwater ecology, in R.T. Watson, M.C. Zinyowera, and R.H. Moss (eds.), *Climate Change1995: Impacts, Adaptations and Mitigation of Climate Change*. Contribution of Working Group II to the Second Assessment Report of the Intergovernmental Panel on Climate Change. Cambridge Univ. Press, Cambridge. pp.325-363.

Bardossy, A. and Plate, E.J. (1992) Space-time model for daily rainfall using atmospheric circulation patterns. *Water Resour. Res.* **28**, 1247-1259.

Bartholy, J., Bogardi, I and Matyasovszky, I. (1995) Effect of climate change on regional precipitation in Lake Balaton watershed. *Theor. Appl. Climatol.* **51**, 237-250.

Bates, B.C., Charles, S.P. and Hughes, J.P. (1997) Stochastic downscaling of numerical climate model simulations. Proceedings International Congress on Modelling and Simulation, MODSIM 97, 8-11 December 1997, Hobart, Australia, Vol. 1 pp.204-209.

Baum, L.E., Petrie, T. Soules, G. and Weiss, N. (1970) A maximization technique occurring in the statistical analysis of probabilistic functions of Markov chains. *Ann. Math. Statist.* **41**, 164-171.

Bogardi, I., Matyasovszky, I., Bardossy, A., and Duckstein, L. (1993) Application of a space-time stochastic model for daily precipitation using atmospheric circulation patterns. *J. Geophys. Res.* **98**(D9), 16653-16667.

Bureau of Meteorology (1988) Climatic Averages Australia. Australian Government Publishing Service, Canberra, Australia. 531 pp.

Charles, S.P., Hughes, J.P., Bates, B.C. and Lyons, T.J. (1996) Assessing downscaling models for atmospheric circulation - local precipitation linkage. Proceedings of the International Conference on Water Resources and Environmental Research: Towards the 21st Century, 29-31 October, Kyoto. Water Resources Research Center, Kyoto University, Japan, Vol. 1, pp. 269-276.

Conway, D., Wilby, R.L. and Jones, P.D. (1996) Precipitation and air flow indices. *Climate Res.* **7**, 169-183.

Forney, Jr., G.D. (1978) *The Viterbi algorithm*. Proc. IEEE **61**, 268-278.

Gates, W.L., Henderson-Sellers, A., Boer, G.J., Folland, C.K., Kitoh, A., McAvaney, B.J., Semazzi, F., Smith, N., Weaver, A.J., and Zeng, Q.-C. (1996) Climate models – evaluation, in J.T. Houghton, L.G. Meira Filho, B.A. Callander, N. Harris, A. Kattenberg, and K. Maskell (eds.), *The Science of Climate Change*, Contribution of Working Group I to the Second Assessment Report of the Intergovernmental Panel on Climate Change, Cambridge Univ. Press, Cambridge. pp.229-284

Glahan, H.R. (1985) Yes, precipitation forecasts have improved. *Bull. Amer. Meteorol. Soc.* **66**, 820-830.

Glahan, H.R. and Lowry, D.A.(1972) The use of model output statistics (MOS) in objective weather forecasting. *J. Appl. Meteorol.* **11**, 1203-1211.

Grotch, S.L. and MacCracken, M.C. (1991) The use of general circulation models to predict regional climatic change. *J. Climate* **4**, 286-303.

Hammer, G.L. (2000) A general systems approach to applying seasonal climate forecasts, in G.L. Hammer, N. Nicholls, and C. Mitchell (eds.), *Applications of Seasonal Climate Forecasting in Agricultural and Natural Ecosystems – The Australian Experience*. Kluwer Academic, The Netherlands. (this volume)

Hay, L.E., McCabe, G.J., Wolock, D.M. and Ayers, M.A. (1991) Simulation of precipitation by weather type analysis. *Water Resour. Res.* **27**, 493-501.

Hewitson, B.C. and Crane, R.G. (1996) Climate downscaling: techniques and application. *Climate Res.* **7**, 85-95.

Hostetler, S.W. and Giorgi, F. (1993) Use of output from high-resolution atmospheric models in landscape-scale hydrologic models: an assessment. *Water Resour. Res.* **29**, 1685-1695.

Hughes, J.P. and Guttorp, P. (1994). A class of stochastic models for relating synoptic atmospheric patterns to regional hydrologic phenomena. *Water Resour. Res.* **30**, 1535-1546.

Hughes, J.P., Guttorp, P. and Charles, S.P. (1998) A nonhomogeneous hidden Markov model for precipitation. *J. R. Statist. Soc., Series C*, **47**, (in press).

Hunt, B.G. and Hirst, A.C. (2000) Global climatic models and their potential for seasonal climatic forecasting, in G.L. Hammer, N. Nicholls, and C. Mitchell (eds.), *Applications of Seasonal Climate Forecasting in Agricultural and Natural Ecosystems – The Australian Experience*. Kluwer Academic, The Netherlands. (this volume)

Kass, R.E. and Raftery, A.E. (1995) Bayes factors. *J. Am. Statist. Ass.* **90**, 773-795.

Martin, E., Timbal, B., and Brun, E. (1997) Downscaling of general circulation model outputs: simulation of the snow climatology of the French Alps and sensitivity to climate change. *Climate Dynamics* **13**, 45-56.

McGregor, J.L. (1997) Regional climate modelling. *Meteorol. Atmos. Phys.* **63**, 105-117.

Mearns, L.O., Giorgi, F., McDaniel, L. and Shields, C. (1995) Analysis of daily variability of precipitation in a nested regional climate model: comparison with observations and doubled CO_2 results. *Glob. Planet. Change* **10**, 55-78.

Rabiner, L.R. and Juang, B.H. (1986) An introduction to hidden Markov models. IEEE Acoustics Speech Signal Processing Mag., pp. 4-16.

Stone, R.C., Smith, I. and McIntosh, P. (2000) Statistical methods for deriving seasonal climate forecasts from GCMs, in G.L. Hammer, N. Nicholls, and C. Mitchell (eds.), *Applications of Seasonal Climate Forecasting in Agricultural and Natural Ecosystems – The Australian Experience.* Kluwer Academic, The Netherlands. (this volume)

Von Storch, H., Zorita, E. and Cubasch, U. (1993) Downscaling of global climate change estimates to regional scales: An application to Iberian rainfall in wintertime. *J. Climate* **6**, 1161-1171.

Walsh, K.J.E and McGregor, J.L. (1995) January and July climate simulations over the Australian region using a limited-area model. *J. Climate* **8**, 2387-2403.

Wilby, R.L. (1997) Non-stationarity in daily precipitation series: Implications for GCM down-scaling using atmospheric circulation indices. *Int. J. Climatol,* **17**, 439-454.

Wilby, R.L., Greenfield, B. and Glenny, C. (1994) A coupled synoptic-hydrological model for climate change impact assessment. *J. Hydrol.* **153**, 265-290.

Zorita, E., Hughes, J.P., Lettenmaier, D.P. and Von Storch, H. (1995) Stochastic characterization of regional circulation patterns for climate model diagnosis and estimation of local precipitation. *J. Climate* **8**, 1023-1042.

STATISTICAL METHODS FOR DERIVING SEASONAL CLIMATE FORECASTS FROM GCM's

ROGER STONE[1], IAN SMITH[2] AND PETER MCINTOSH[3]

[1]*Queensland Centre for Climate Applications*
Queensland Departments Primary Industries and Natural Resources
PO Box 102
Toowoomba, Qld 4350, Australia

[2]*CSIRO Division of Atmospheric Research*
PMB1
Aspendale, Vic 3195, Australia

[3]*CSIRO Division of Oceanography*
GPO Box 1538
Hobart, Tas 7000, Australia

Abstract

A method for developing a 'practical' climate forecast system from the output of a General Circulation Model (GCM) is described. This forecast system is compatible with input needs of agricultural simulation and decision analysis models. The method uses principal components and cluster analysis of GCM-generated forecasts of the time series of the Southern Oscillation Index (SOI) to create SOI 'types' or 'phases'. The SOI predictions were derived from a long-term GCM simulation that was forced with historical sea-surface temperature (SST) data. The GCM-derived 'SOI phases' *(g-phases)* could thus be associated with historical analogue years in a manner similar to SOI phases that have been developed from historical SOI data. Rainfall probability distributions associated with g-phases were calculated from actual rainfall amounts in the analogue year sets associated with each phase. These rainfall probabilities were compared with the currently available distributions that have been derived using *lag* relationships between SOI phases and rainfall. In addition, the historical SST data was analysed in a similar way so that analogue year sets could be formed. Empirical orthogonal function (EOF) analysis and cluster analysis were used to derive SST 'EOF-types' that depended on the temporal dynamics of spatial patterns in the Pacific Ocean and Indian Ocean SST data. Lag relationships between the SST/EOF types and rainfall distributions could then also be derived for comparison. The results show that both the g-phases derived from the GCM forecast of SOI and the SST/EOF-types generally provide larger shifts in rainfall distributions than are currently available, especially at

G.L. Hammer et al. (eds.), The Australian Experience, 135–147.

longer lead times. The methods outlined above may facilitate more practical output from General Circulation Models and analyses of SST data so that connections to agricultural simulation models and software packages are enhanced.

1. Introduction

Crop and pasture simulation models generally require daily inputs of rainfall, evaporation, radiation and other data to function effectively. Certain statistical seasonal climate forecast systems are already able to provide 'forecasts' of these elements through the provision of historical analogue years or seasons corresponding to predefined categories of precursor states. A method used currently is to group similar types of years or seasons based on a classification of patterns of the Southern Oscillation Index (SOI). This type of procedure is referred to as a 'lag-SOI phase' system (Stone and Auliciems, 1992; Stone et al., 1996a). The Southern Oscillation Index referred to here is calculated as the difference in surface pressure anomalies between Tahiti and Darwin divided by the standard deviation of the difference.

The recent development of General Circulation Models (GCMs) has led to new types of forecasts of rainfall, sea-surface temperature information, and the SOI (Hunt and Hirst, 2000). However, these forecast data are not currently in an appropriate form for input into crop and other agricultural simulation systems. In other words, forecast data provided from GCM output are now normally provided either as forecasts of SST anomalies or maps of spatial variability of precipitation probabilities. These forecast data are not normally presented as a typology of months, season, or years, which are needed to form the basis of inputs into crop and pasture simulation models (see Hammer, 2000).

The main aim of this paper is to derive statistical methods that will take the forecast output from GCMs and make that output more appropriate for input into both regional climate forecasting systems and crop simulation models. A second aim is to compare the rainfall probability distributions obtained from GCM produced 'forecasts' with those currently obtainable using lag-SOI phases and those that might be obtained from similar statistical analyses of spatial and temporal patterns in historical SST data.

2. Data Sets and Processing

The 'Global Ice – Sea Surface Temperature' (UK Met Office-GISST 1.1) sea-surface temperature data set was obtained from CSIRO Division of Atmospheric Research. This data set was used to derive both the GCM forecasts of the SOI and Empirical Orthogonal Functions (EOFs) of sea-surface temperatures. The data are on an approximately 2° grid (actually the T63 atmospheric model grid). The original GISST data were on a 1° grid and cover the entire globe for the years 1871-1992. The annual cycle was removed from the data before calculating the EOFs. The annual cycle was computed over the full 121 years. If left in, the annual cycle dominates all other signals, particularly in the Indian Ocean.

3. Methods

3.1 FORECASTS FROM GCM OUTPUTS

Five long-term simulations (1871-1991) were conducted with the CSIRO9 (T63) GCM - a 9-level model with a horizontal resolution of approximately 1.9°. Each simulation was forced by the same GISST data set but was started with slightly different initial conditions (on Jan 1, 1871) that were sufficient to perturb the results and introduce variability. For each simulation the GCM forecast of the SOI was derived. The prediction of SOI by the atmospheric GCM is, in this example, based on a known or given SST value for each month of each year over the entire period of record.

We needed to obtain a classification of season types from the output of the GCM forecasts. To do this we undertook the following:

- Conducted a principal components analysis (PCA) on the time series of *each of the five GCM forecasts* of the SOI derived from the given GISST data set. The PCA was conducted on a matrix consisting of yearly forecasts of the SOI for each month commencing in May of each of the years in the GCM simulation.

- Applied an hierarchical clustering technique to the PCA scores derived from the PCA analyses of the time series of the 'forecasts' of the SOI made by the CSIRO9 GCM, following the method of Stone (1985, 1989) and Kalkstein *et al.* (1987). For a general review of the use of PCA and EOFs in climatological work we suggest Sneyers and Goossens (1985), Richman (1986), Jolliffe (1986), and Yarnal (1993).

- Identified significant break-points in the cluster dendrograms or coefficients in order to isolate GCM SOI-types (hereafter referred to as 'g-phases') of forecast SOI.

- Used the SOI g-phases to identify historical analogue periods, derive associated rainfall probability distributions, and then examine shifts in those probability distributions among g-phases for various locations throughout Australia.

- Tested the significance of the ability of the method(s) to discriminate between rainfall probability distributions associated with the analogue periods derived using the GCM-SOI g-phases. Statistical tests were applied to the various rainfall probability distributions produced to ascertain whether these distributions were significantly different from one another or from climatology.

- Compared the ability of the GCM g-phases to discriminate between rainfall probability distributions with the ability to discriminate using a current system based on PCA/cluster analysis of patterns of the SOI ('lag-SOI phase system') (Stone *et al.*, 1996a,b).

3.2 FORECASTS BASED ON ANALYSIS OF SST DATA

A statistical forecasting system was derived from analysis of the Pacific and Indian Ocean SST data (the GISST 1.1 data set but on a 2° grid produced by CSIRO). In this analysis, cluster analyses were performed on the EOF scores derived from the preliminary EOF analysis of the SST data. The analysis incorporated spatial patterns of SSTs and the trend of these patterns over time. The preliminary EOF analysis identified spatial patterns of SSTs for any given month. The cluster analysis formed groups based on the dynamics of these patterns over time, using the vector of monthly EOF scores as the basis for clustering. In this study, only the periods May-September and October-May were considered, to align with examination of forecasts of summer and winter rainfall, respectively. Again, the aim was to derive analogue years or seasons in much the same manner as that described above for GCM SOI data. However, in this case, lagged relationships between SST EOF clusters and rainfall were identified. The EOF clusters are hereafter referred to as 'EOF-types'. A more detailed description of the procedures employed in the EOF analysis is provided in Appendix A.

Rainfall probability distributions were derived from the analogues associated with each EOF-type. Significance tests were conducted on the resulting rainfall probability distributions associated with both the g-phases and EOF-types. The Kruskal-Wallis (K-W test) one-way analysis of variance is a non-parametric alternative to the analysis of variance or F-ratio. The K-W test is used to compare two or more independent samples to see if they come from identical populations. The test requires ordinal-level data and produces a test statistic - the H-statistic, akin to the chi-square statistic. In this analysis, each rainfall observation was allotted ranking in the data set and then placed within each cluster (in this case g-phase or EOF-type) cell. The mean ranking of rainfall observations associated with each cluster or g-phase was calculated and an H-statistic derived according to the relationship between observed frequency and expected frequency of occurrence. Significance levels associated with the H-statistic were also calculated.

4. Results

4.1 FORECASTS OBTAINED FROM GCM OUTPUT

The PCA analysis on each of the five simulated SOI data sets resulted in a number of factor solutions representing each data set. Only slightly different types of factor solutions were produced when running the data set commencing in May of each year compared to forecast 'runs' commencing in October each year. The PCA analyses accounted for approximately 70 per cent of the variance in the data in each of the analyses.

Rotation of principal components often produces a 'simple structure' (Thurstone, 1947) – ie. the variables generally fall into mutually exclusive groups whose loadings are high on single components, moderate to low on a few components, and of negligible size on the remaining factors or components. Use of the Varimax Rotation method on the PCA analyses (which aimed to simplify the meaning of each component) in this instance resulted in component solutions that were more easily recognisable in terms of histories of the SOI. In the rotated solutions certain months of the year (eg. December) were heavily, positively, loaded on to a certain component while later months in the austral autumn (eg. April/May) were heavily, negatively, loaded onto the component. This result agrees with a general understanding of temporal patterns of the SOI in which there is, generally, an inverse relationship between SOI values during early summer with those of the subsequent autumn. However, this result contrasted with the unrotated solution in which just one main component was produced on which almost all of the months were weakly loaded. It is suggested the above result confirms the view of Richman (1986) in which he suggests that simple structure rotations 'make them good candidates to accurately identify modes of variation in climatological data sets'. The rotated solution, in this instance, more closely conformed to the general understanding of SOI behaviour over time.

Table 1 provides an example of a Varimax Rotated Solution for one of the five forecast runs of the SOI for the October to June period. An interesting result was that Factor 1 in each of the five analyses loaded highly and oppositely on the months of December and May/June (perhaps suggesting a representation of an El Niño/La Niña development/breakdown period), Factor 2 loaded highly on the months of March and April, Factor 3 loaded highly on the months of October and November, and Factor 4 loaded highly on the months of January and February.

TABLE 1. Example of the factor loadings from a rotated PCA of SOI data obtained from the atmospheric GCM run forced using GISST SST data. Note the signs on the factor loadings are arbitrary.

Month	Factor 1	Factor 2	Factor 3	Factor 4
Oct	-.079	.122	.839	-.027
Nov	.046	-.252	.816	.029
Dec	-.724	.295	.142	-.008
Jan	.104	-.106	-.019	.887
Feb	.064	.277	.021	.696
Mar	.026	.581	-.147	.254
Apr	.349	.581	-.147	.254
May	.717	.351	.040	.257
Jun	.709	.298	.062	.031

The coefficients of the components (factors) generated orthogonally independent indices for each SOI entity in the form of component scores. Each of the five PCA analyses provided output of PCA scores. Every month in each of the years loaded to

140

some degree on every component in each of the PCA runs, which provided input into the clustering algorithm. Within the clustering procedure applied (Ward's Hierarchical Method - a method that optimises the minimum variance within clusters) (Ward, 1963; Anderberg, 1973) significant groupings occurred at the five and nine cluster levels. Figure 1 shows the *mean* SOI values associated with clusters at the five-cluster level.

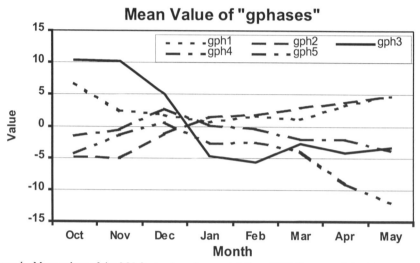

Figure 1. Mean values of the SOI for 'g-phases' generated from GCM forecasts of the SOI. It should be noted these g-phases are produced from an ensemble of GCM output runs. When individual output runs are clustered in the same manner to that above slightly more 'extreme value' SOI means associated with each g-phase were produced.

Figure 2 provides examples of rainfall probability distributions associated with a selection of the five g-phases for a number of locations in Australia. Figure 2a shows the rainfall probability distributions (probability of exceedence) associated with two representative g-phases and two SOI phases for the June to September period for Gunnedah, NSW. The g-phases appear just slightly better able to capture the drier seasons compared to the currently used SOI phases. For example, over the period of rainfall record, the probability of exceeding 170mm at Gunnedah is 35 per cent following a consistently negative SOI phase through April/May compared with 20% associated with (the forecast SOI phase) gphase Type 2. Figure 2b shows the rainfall probability distributions at Charters Towers, Queensland for the October to December period associated with both the g-phases and SOI phases known at the beginning of October. At Charters Towers the g-phases appear slightly better able to capture potentially wetter seasons than the SOI phases.

The K-W test was applied to test whether the rainfall probability distributions associated with each g-phase differed significantly from one another. This test showed that, in general, the rainfall distributions associated with the 'g-phases' were shifted further apart from one another and at a higher level of significance than those associated with the presently employed, lag-SOI phase system. The application of significance tests showed that, where the currently applied forecast method shifted rainfall distributions at, say, the 90% to 95% probability level, the 'g-phases' could shift the distributions at the 99% probability level. It was difficult to identify cases where the forecast 'g-phase' system fared worse than the currently used lag-SOI phase system.

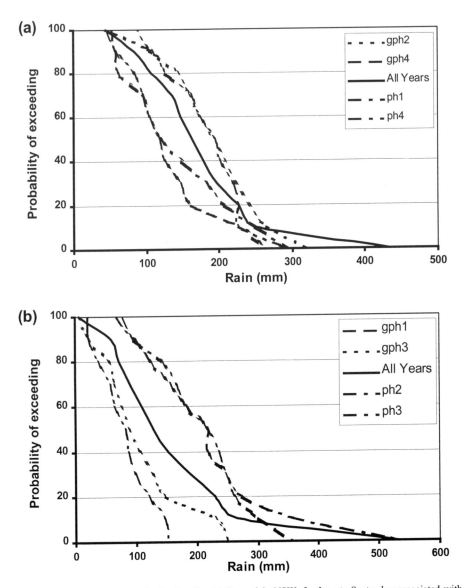

Figure 2. Rainfall probability distributions for (a) Gunnedah, NSW, for June to September associated with two SOI phases (PH1; PH4) and two g-phases (GPH2; GPH4) and for (b) Charters Towers, Qld., for October to December associated with two representative SOI phases (PH2; PH3) and two g-phases (GPH1;GPH3).

It should be emphasised the significance tests applied to these analyses measure only the ability to discriminate between the probability distributions. The system based on the g-phases uses SOI forecasts derived from GCM runs forced with actual SSTs, which may be responsible for some of the improved separation. Further studies are required to determine a true measure of actual forecast 'skill', as may be expected from use in an operational forecasting system, based on forecast SSTs. In addition, in a true forecast mode, measurement of skill would have to be obtained on independent data.

4.2 FORECASTS OBTAINED FROM ANALYSIS OF SST'S

The EOF analysis of the Pacific and Indian Ocean SST data resulted in the first 6 EOFs of each of the data sets accounting for approximately 65 per cent of the variance. The first Pacific Ocean EOF from this analysis is the eastern Pacific El Niño signal. The time series of this EOF has a correlation with the SOI of about 0.6 at zero lag (using the GISST 1.1 data set). The first Indian Ocean EOF is roughly uniform across the basin. Interestingly, the first Indian Ocean EOF time series shows a steady increase from about 1970. The mean SST in the Indian Ocean shows a very similar trend using both the GISST data and the Reynolds' reconstructed SST data set (Reynolds, 1988). The GISST 1.1 Indian Ocean first EOF time series has a correlation with the SOI of about 0.3 and lags the SOI by about three months. A given number of EOF modes explain more of the explained variance in the Indian Ocean than the same number of EOF modes in the Pacific Ocean (data not shown).

When the EOF *scores* for each month for every year for each separate analysis for each ocean were arranged in matrix form a satisfactory cluster analysis could be performed. In this case the data were specifically arranged to commence in May of each year. As the results of this particular analysis were required for input into crop and pasture simulation models only data between May and September or October to May were analysed in this instance. A more generic SST/EOF-based forecast system would require analysis of similar six-month lag periods but for each month of the year. In this way clusters from EOF time series would be created from the entire time series of EOF scores.

Application of Ward's (1963) Hierarchical Clustering algorithm to the output of EOF scores produced 4 clusters at a broad classification level and 9 clusters at a finer level of discrimination. Decision rules that identified an optimal number of clusters (eg. 'Mojena' Rule' that uses the distribution of the within-group error sum of squares to determine the significant changes between each clustering stage (Mojena, 1977)) were applied to derive the final number of clusters to be used in this analysis. The method appears to have classified SST patterns sensibly with Type 1 in the Pacific representing the broad El Niño pattern, and Type 2 representing a reduced or weaker El Niño-type pattern.

In a similar manner to that described above for output from GCM forecasts of the SOI, rainfall probability distributions were derived for a number of locations across Australia from analogues associated with the EOF-types. Figure 3a shows the distributions associated with two of the EOF-types and compares these distributions with those associated with the SOI phases. The EOF-types appear better able, in this example, to capture the drier seasons at Horsham compared with the SOI phases. The K-W test performed on distributions associated with SOI phases shows the rainfall distributions at Horsham are shifted at a level of significance of p=.57 (ie. not statistically significant), while the distributions associated with the EOF-types are shifted at a high level of significance of p=.003.

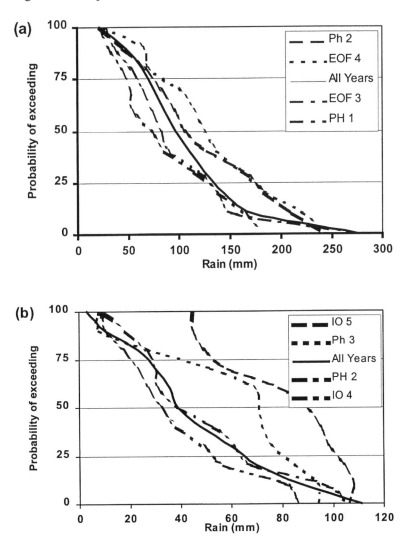

Figure 3. Rainfall probability distributions for (a) Horsham, Victoria, for October to December, associated with two representative EOF types (EOF3; EOF4) and two representative SOI phases (PH1; PH2), and for (b) Merredin, WA, for October to December, associated with two representative EOF-types (IO4; IO5) (Indian Ocean data) and two representative SOI phases (PH2; PH3).

Similarly, the EOF-type analysis provided greater shifts in rainfall probability distributions at Merredin, Western Australia (Figure 3b), than the currently applied SOI phase system *for the seasons analysed*. The Indian Ocean-based EOF-types appear to have greater capacity to capture the relatively wetter seasons (for the October to December period studied) at Merredin than the SOI-phase system.

Finally, when the rainfall probability distributions produced from the 'g-phases' were compared with the rainfall probability distributions produced from the SST-EOFs and the currently used SOI-phase system, it appears the SST EOF-types provided the greatest overall levels of discrimination in rainfall probability distributions out of the three methods compared. Figure 4 illustrates the slightly improved partitioning resulting from use of the entirely statistically-based SST/EOF system. In this example rainfall distributions for Emerald were partitioned according to whether the GCM-based forecast of SOI phases was made or whether a SST-based EOF type system was applied. The probability of obtaining 180mm for the October to December period at Emerald is 28 % applying 'gphase 3', 20% applying 'EOF type 3', 57% following 'gphase 1', and 72% following 'EOF phase 2'. The 'all years' climatological probability of obtaining 180mm at Emerald is 40%. This single site example shows the greatest shift in rainfall distributions occurs when the EOF-SST-type system is applied to the analysis. It should be noted, however, that the SOI forecast used assumes we will have perfect knowledge of future SST in the equatorial Pacific some months in advance.

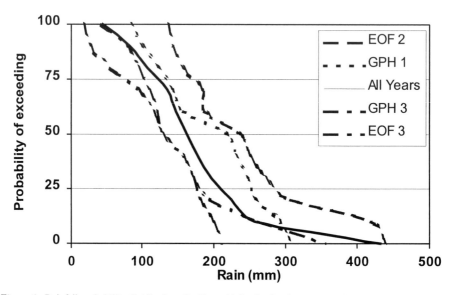

Figure 4. Rainfall probability distributions for Emerald for the October to December period associated with two representative GCM g-phases (GPH1; GPH3) and two representative EOF-types (EOF2; EOF3).

5. Discussion

Significant relationships between rainfall and SOI, where the SOI was 'forecast' by a GCM using given tropical SST anomalies, were identified in the analyses presented. These relationships appeared better than those between the lagged actual SOI phase and rainfall, as currently used in practice. However, the 'forecast' SOI provided by the atmospheric GCM was based on known or given SST for each month for the last 120 years. It may be expected that 'true' forecasts of SOI based on 'true' forecast of SST may provide slightly less significant shifts in rainfall distributions than those shown in this study. However, preliminary results from a fully coupled ocean-atmosphere model (R. Kleeman, personal communication) show forecasts of the SOI at least comparable to those demonstrated above. The results from the fully coupled model were only obtainable for a limited number of years, however, making statistical inferences regarding shifts in rainfall probability distributions difficult to test.

The analytical method presented in this study appears to provide a useful means to convert current output from GCMs to a form immediately accessible by crop and pasture simulation models. In addition, in many instances, the ability to partition the rainfall distributions was similar to or slightly better than that currently available using lag-relationships between an SOI phase and rainfall. Discriminatory ability in this analysis was measured using non-parametric statistical tests applied to the resulting probability distributions associated with each of the (5) GCM 'g-phases'.

A comparison of the ability of the various approaches to partition rainfall probability distributions showed that, in many cases, SST-EOF-generated rainfall distributions were shifted more significantly than were distributions generated from SOI phases. SST EOFs tended to provide greater discrimination between rainfall probability distributions at the rainfall stations analysed than by using either the lag-SOI phase system or the GCM SOI g-phase forecasts, although these results varied somewhat depending on location. This result was especially the case with longer lead times and the SST EOFs appeared to provide better discriminatory ability (as measured by the K-W test) than current methods for rainfall in the spring/summer period. However, for austral winter rainfall, the SST derived EOF phases or GCM produced 'g-phases' did not provide particularly large improvement in ability to discriminate between probability distributions than that currently available using lagged SOI phases at the locations studied.

It should be emphasised that the ability to partition rainfall probability distributions does not necessarily imply the level of *forecasting skill* that would be determined from a test on independent data of forecast model performance. Such tests of forecast skill could be obtainable by applying, for example, the 'LEPS Scoring' method using cross-validation (Potts *et al.*, 1996). Nevertheless, the method outlined here provides a useful means to derive a practical forecasting system from the output of GCMs. Further, the results from the analysis of SST data to derive phases based on both spatial patterns and their temporal dynamics, indicates considerable potential for further development. We hope to be able to conduct similar analyses using the recently developed global SST-based, EOF-derived, seasonal climate forecast system developed by the Bureau of Meteorology Research Centre (following Drosdowsky, 1993a,b).

Acknowledgments

We are grateful to the Land and Water Resources Research and Development Corporation for general funding for this work from the National Climate Variability Programme. We are also grateful to the reviewers for their constructive comments and suggestions of useful reference material.

Appendix A - EOF Analysis

It was possible to compute EOF time series for the whole 121-year period of the data set. Initially, EOFs were computed for the years 1950-1979, because SST data are considered better after the mid-1940s, and because the period from 1980 to the present is sometimes considered atypical. Secondly, EOFs were calculated using just the years 1982-1991 because the SST data include satellite data. Finally, EOFs were computed from the whole 121-year period. The result was that there was a great deal of similarity between the EOFs generated using the three time periods. The 3 time series for the first EOF were essentially the same. For higher EOFs, the time series for the two longer periods EOFs were very similar. The analysis reported here uses EOFs calculated from the entire 121-year period.

The EOFs were computed separately for the Pacific and Indian basins. This seemed the best way to isolate possibly different physical mechanisms in each basin. It also prevented the large Pacific signal from dominating the smaller interannual signal in the Indian Ocean. The other option was to compute EOFs for the combined Pacific and Indian Oceans. For the two shorter time periods, the first two combined EOFs were similar to the sum of the respective individual EOFs. The full 121 year combined EOF was too big to compute easily.

Only data between 30°N and 40°S were used. We were only interested in ocean regions that might directly affect Australia's climate. Also, GISST data in the Southern Ocean are very sparse. The relative scaling of the EOFs and their associated time series is arbitrary. The default scaling gives each EOF a root-mean-square value of unity. Here, each EOF has been multiplied by 10, and each EOF time series divided by 10 so that all numbers are of order 1. The time series multiplied by the EOF gives the contribution of that EOF to the SST anomaly in degrees Celsius.

References

Anderberg, M.R. (1973) *Cluster Analysis for Applications*. Academic Press, New York. pp.359.
Drosdowsky, W. (1993a) An analysis of Australian seasonal rainfall anomalies 1950-1987. I: Spatial patterns. *Int. J. Climatol.* **13**, 1-30.

Drosdowsky, W. (1993b) An analysis of Australian seasonal rainfall anomalies: 1950-1987. II: Temporal variability and teleconnection patterns. *Int. J. Climatol.* **13**, 111-149.

Hammer, G.L. (2000) A general systems approach to applying seasonal climate forecasts, in G.L. Hammer, N. Nicholls, and C. Mitchell (eds.), *Applications of Seasonal Climate Forecasting in Agricultural and Natural Ecosystems – The Australian Experience.* Kluwer Academic, The Netherlands. (this volume)

Hunt, B.G. and Hirst, A.C. (2000) Global climatic models and their potential for seasonal climatic forecasting, in G.L. Hammer, N. Nicholls, and C. Mitchell (eds.), *Applications of Seasonal Climate Forecasting in Agricultural and Natural Ecosystems – The Australian Experience.* Kluwer Academic, The Netherlands. (this volume)

Jolliffe, I.T. (1986) *Principal Component Analysis.* Springer-Verlag, New York. pp.271.

Kalkstein, L.S., Tan, G. and Skindlow, J.A. (1987) An evaluation of three clustering procedures for use in synoptic climatological classification. *J. Clim. Appl. Meteorol.* **26**, 717-730.

Mojena, R. (1977) Hierarchical grouping methods and stopping rules: an evaluation. *Comp. J.* **20**, 359-363.

Potts, J.M., Folland, C.K., Jolliffe, I.T. and Sexton, D. (1996) Revised LEPS Scores for assessing climate model simulations and long-range forecasts. *J Climate* **9**, 34-52.

Reynolds, R.W. (1988) A real-time global sea surface temperature analysis. *J. Climate* **1**, 75-86.

Richman, M.B. (1986) Rotation of principal components. *J. Climatol.* **6**, 293-333.

Sneyers, R. and Goossens, Chr. (1985) The Principal Component Analysis. Application to Climatology and Meteorology. Annex to the Rapporteur on Statistical Methods, Ninth Session of the Commission for Climatology. WMO, Geneva.

Stone, R.C. (1985) Objectively defined weather types at Brisbane, Queensland. Unpublished B.Sc.(Hons) Thesis. University of Queensland. St Lucia, Queensland, Australia. pp.250.

Stone, R.C. (1989) Weather types at Brisbane, Queensland: An example of the use of principal components and cluster analysis. *Int. J. Climatol.* **9**, 3-32.

Stone, R.C. and Auliciems, A. (1992) SOI phase relationships with rainfall in eastern Australia. *Int. J. Climatol.* **12**, (6) 625-636.

Stone, R.C., Hammer, G.L. and Marcussen, T. (1996a) Prediction of global rainfall probabilities using phases of the Southern Oscillation Index. *Nature* **384**, 252-256.

Stone, R.C., Nicholls, N. and Hammer, G.L. (1996b) Frost in north-east Australia: trends and influences of phases of the Southern Oscillation. *J. Climate* **9**, 1896-1909.

Thurstone, L.L. (1947) *Multiple Factor Analysis.* University of Chicago Press, Chicago. pp.535

Ward, J.H. (1963) Hierarchical grouping to optimise an objective function. *J. Amer. Stat. Assoc.* **58**, 236-244.

Yarnal, B. (1993) *Synoptic Climatology in Environmental Analysis.* Bellhaven Press, London. pp.196.

USING SEASONAL CLIMATE FORECASTS TO MANAGE DRYLAND CROPS IN NORTHERN AUSTRALIA – EXPERIENCES FROM THE 1997/98 SEASONS

HOLGER MEINKE AND ZVI HOCHMAN

Agricultural Production Systems Research Unit (APSRU)
Queensland Departments Primary Industries and Natural Resources and
CSIRO Tropical Agriculture
PO Box 102, Toowoomba, Qld 4350, Australia

" ... *os seus ciclos ... abrem-se e encarram-se com um ritmo tão notável que recordam o desdobramento de uma lei natural ainda ignorada.* "
("... the drought cycles ... follow a rhythm in the opening and closing of their periods that is so obvious as to lead one to think that there must be some natural law behind it all, of which we are as yet in ignorance.")

Euclides da Cunha, 'Os Sertões', 1902

Abstract

In Australia, like in many other parts of the world, a significant proportion of rainfall variability is associated with the El Niño / Southern Oscillation phenomenon (ENSO). Significant, physically based lag-relationships exist between an index of the ocean/atmosphere ENSO phenomenon and future rainfall amount and temporal distribution in eastern Australia and many other areas across the globe. A skilful seasonal climate forecast provides an opportunity for farm managers to better tailor crop management decisions to the season. This forms the basis for a probabilistic crop production forecasting system used operationally in Australia where high rainfall variability is the major source of dryland yield fluctuations.

It is a challenge for scientists and farm managers alike to identify decisions that can usefully be aided by climate forecasting. For a forecast to be effective, such decision making ultimately has to improve the long-term performance of the farming enterprise either by increasing profits, by improving sustainability indicators (e.g. erosion, soil organic matter) or by reducing production risks.

Timing and frequency of future rainfall events strongly influences dryland crop growth and yield but the usefulness of rainfall events in terms of their contribution to crop production is difficult to assess. Physiologically based crop simulation models using regional climatic records were used to quantify the relationship between crop performance and phases of the Southern Oscillation Index (SOI). Using the 1997/98 El Niño and the 1998/99 La Niña events as case studies, we demonstrate how a statistical, but physically based seasonal climate forecasting system combined with a crop simulation capability and substantive adviser and producer interaction can aid farm managers in their decision making. Adopting an action learning approach, scientists

G.L. Hammer et al. (eds.), The Australian Experience, 149–165.

working closely with farmers and their advisers identified tactical decisions that can benefit from climate forecasting. The information gained from these interactions was widely disseminated in the rural press not only by scientists and advisers but also by the farmers who are now championing this approach to better farm management.

To be most effective the approach shown here requires an understanding of the probabilistic nature of the information provided, whereby producers must not become disheartened or reckless by any perceived 'failure' or 'win' of the forecast in any given season. In fact, even this terminology should be avoided. The approach must be used consistently for many seasons to truly benefit from it. Further, it must be integrated into the whole decision making process as one of many management tools.

1. Introduction

With European settlement of Australia about 200 years ago, farmers of the 'new' continent were exposed to an environment that differed fundamentally, particularly in terms of climatic variability, from their European experience. Managing for climate variability has only recently become a feature of the Australian farming system. This is in contrast to, for instance, the Nordeste region of Northeastern Brazil, which is also strongly affected by the El Niño weather pattern and where, even last century, farmers and graziers varied their management strategies based on climatological observations (de Cunha, 1995; first published in 1902).

High rainfall variability is the major source of dryland yield fluctuations in northeastern Australia and elsewhere in the semi-arid tropics and sub-tropics (Hammer et al., 1987; Stone et al., 1996a). Although most dramatic at the farm level, the impact of climatic variability is apparent throughout the entire Australian economy and can even affect macroeconomic indicators such as international wheat prices (Chapman et al., 2000), employment or the exchange rate (White, 2000). To remain economically viable in an internationally competitive market, Australian farmers have to devise management options that can produce long-term, sustainable profits in such a variable environment. To assist them in this endeavor is clearly in the interest of the whole nation.

Significant, physically based lag-relationships exist between an index of the ocean/atmosphere El Niño/Southern Oscillation phenomenon (ENSO) and future rainfall amount and temporal distribution in eastern Australia and many other areas across the globe (Stone et al., 1996a). An El Niño event, which generally corresponds to negative Southern Oscillation Index (SOI) values, usually lasts for about one year, beginning its cycle in the austral autumn period of one year and terminating in the autumn period of the following year. During the termination of an El Niño event the SOI may rise sharply. Stone et al. (1996a) have shown how phases of the SOI are related to rainfall variability and are useful for rainfall forecasting for a range of locations in Australia and around the world. For large parts of Eastern Australia, they have shown that a rapid rise in SOI over a two months period is related to a high probability of above long-term average rainfall at certain times of the year. Conversely, a consistently negative or rapidly falling SOI pattern is related to a high probability of below average rainfall for many regions in Australia at certain times of the year. As the SOI pattern tends to be 'phase-locked' into the annual cycle (from autumn to autumn), the SOI phase analysis provides skill in assessing future rainfall probabilities for the season ahead.

A skilful seasonal forecast provides an opportunity for farm managers to better tailor crop management decisions to the season (Hammer *et al.*, 1996). Timing and frequency of future rainfall events strongly influences dryland crop growth and yield but the usefulness of rainfall events in terms of their contribution to crop production is difficult to assess. However, physiologically based crop simulation models can be used as 'filters' to gauge the value of rainfall over a growing season (Meinke and Hammer, 1997; Keating and Meinke, 1998). Using phases of the SOI in conjunction with dynamic simulation models allows better quantification of climatic risk. For instance, for peanuts grown in Northern Australia, Meinke *et al.* (1996) showed that higher yields are generally associated with a consistently positive SOI phase in August/September due to higher and more reliable summer rain coupled with a lower frequency of rain at harvest. Khandekar (1996) showed that El Niño events are usually associated with low grain yields over South Asia and Australia and high grain yield throughout the North American prairies.

For wheat, Meinke and Stone (1997b) showed how the SOI phase system, when combined with a cropping systems simulation capability can be used operationally to assess likely yields at different locations around the world. In northeastern Australia, highest median yields were simulated following a rapidly rising SOI phase in April/May and lowest median yields following a consistently negative phase. Conversely, highest median yields in southeastern Brazil followed a near zero April/May phase and lowest median yields were simulated when the SOI phase was consistently positive.

It is a challenge for scientists and farm managers alike to identify decisions that can usefully be aided by climate forecasting. Such decision making ultimately has to improve the long-term economic performance of the farming enterprise either by increasing profits, improving sustainability indicators (e.g. erosion, soil organic matter) or reducing risk. For wheat, Hammer *et al.* (1996) showed how tactical nitrogen management and cultivar choice based on SOI phases can increase the profitability of cropping in Northern Australia. For peanuts Meinke *et al.* (1996) and Meinke and Hammer (1997) showed how production risks could be reduced using a SOI based forecasting system.

Good farm managers have a rich appreciation of agricultural systems components and their interactions (e.g. the relative importance of stored soil moisture versus in-season rainfall) and are skilled at incorporating new information into the decision making process. The approach demonstrated here can help farmers to replace "gut feeling" about their complex system with (a) hard data about the current state of their system (e.g. stored soil moisture) and (b) with probabilistic information about the way in which the unknown (e.g. future in-season rainfall) will affect the outcome of alternative management decisions.

Using the 1997/98 El Niño and the 1998 La Niña event as case studies, we demonstrate how seasonal forecasting combined with simulation capabilities and substantive adviser and producer interaction can aid farm managers in their decision making processes. Particularly, we

- describe the probabilistic rainfall and crop forecasting system employed,

- outline the climatic conditions prior to and during the winter and spring of 1997 and 1998 in northeastern Australia,
- discuss some of the on-farm management implications of such forecasts,
- and show how the information is currently disseminated and adopted.

The paper is structured into three sections reflecting the three different approaches of using seasonal climate forecasting in tactical decision making. The first section addresses general issues and presents scenario analyses that are not grower specific and can be disseminated via general media outlets. In the second section we present a case study of a dryland grain/cotton grower and document how he currently uses probabilistic climate forecasts operationally. In the third section, issues of growers' and industry awareness of existing forecasting capabilities and their perceived value are addressed.

In this paper, we limit our scope to tactical, single crop issues. Frequently, however, management decisions have to be made that go beyond single seasons. Implications and potential applications of seasonal forecasting on strategic farm management are discussed elsewhere in these proceedings (Hammer *et al.*, 2000; Carberry *et al.*, 2000).

2. General Information and Scenario Analyses

El Niños are usually associated with below-average rain for many parts of Australia, SE Asia, India and South Africa, but above-average rainfall for large areas of South America and parts of the U.SA.. As the SOI pattern tends to be 'phase-locked' into the annual cycle (from austral autumn to autumn), the SOI phase analysis provides skill in assessing rainfall probabilities for the next growing season. A brief summary of this system is given in the introduction of this chapter. A more detailed explanation of the system itself is found in Stone and Auliciems (1992) and Stone *et al.* (1996a). Examples of how this system can be combined with crop simulation models to estimate production risks are given by Meinke *et al.* (1996) and Meinke and Hammer (1997).

2.1 CLIMATIC CONDITIONS IN 1997 AND 1998

During March 1997 SOI values fell sharply. This indicated an eastward shift of the Walker circulation caused by an increase in sea surface temperatures in the Eastern Pacific. This first significant indication that an El Niño was developing (Bate, 1997) was confirmed in April 1997 by further increases in sea surface temperature anomalies in the Eastern Pacific. More importantly, in terms of the climate forecast system used here, the monthly average SOI values dropped drastically in early 1997 from an average value of +12 in February to -14 by the end of April (Table 1). The trend continued and in June 1997, the Australian Bureau of Meteorology officially announced the development of an El Niño. The SOI stayed negative until May 1998 when it rose sharply and was replaced by a consistently positive SOI pattern (typical La Niña pattern).

2.2 WHEAT IN 1997 AND 1998

For such climate information to be useful for producers it is necessary that the information itself together with an outline of likely impacts are disseminated widely using a range of media outlets (rural press, radio, TV, fax-back services, Internet, workshops). Further, the timeliness of the information is vital. The information must be available **before** tactical decisions have to be made. Finally, the information provided must be relevant for a particular region and for the dominant cropping system in this region. This can be achieved only through close interaction among agricultural and climate scientists, agricultural advisers and producers. Such interaction is necessary to ensure that the general information provided remains relevant and useful.

TABLE 1. Average monthly SOI values for 1997 and 1998 and their corresponding 'phases' (Stone *et al.*, 1996a). The SOI phases can be either consistently negative (-), consistently positive (+), rapidly falling (↘), rapidly rising (↗) or near zero (no symbol, did not occur in 1997/98).

1997	J	F	M	A	M	J	J	A	S	O	N	D
Average Monthly SOI	4	14	-9	-14	-19	-24	-9	-19	-14	-17	-14	-11
SOI Phase	+	↗	↘	↘	-	-	-	↘	-	-	-	-

1998	J	F	M	A	M	J	J	A	S	O	N	D
Average Monthly SOI	-22	-22	-26	-23	1	8	14	12	12	12	13	12
SOI Phase	↘	-	-	-	↗	↗	+	+	+	+	+	+

By mid April 1997 there was evidence based on sea surface temperatures and SOI trends that an El Niño event was likely and hence the first alert (El Niño watch) was issued using a range of media outlets. Such timely information was important to producers in the northeastern wheat belt who were getting ready to sow their winter crops. The issuing of the El Niño watch was complimented by a detailed historical rainfall analysis that quantified the chances of receiving planting rain (Meinke and Stone, 1997a). This study concluded that for most locations within the Australian wheat belt the chances of receiving planting rain were only slightly reduced when the SOI phase was negative at the end of May. However, the total amount of winter rain was likely to be lower, but the impact of this possible reduction would differ substantially from region to region and vary according to starting soil moisture conditions.

This study was followed by more detailed estimates of the likely impact of the El Niño weather pattern on wheat yields (Meinke *et al.*, 1997). To evaluate likely wheat yields for the forthcoming season the cropping systems model APSIM configured for wheat was used (McCown *et al.*, 1996; Meinke *et al.*, 1998). The model requires long-term, daily meteorological data (i.e. minimum and maximum temperatures, solar radiation and rainfall) as input. It is responsive to differences in soil water and soil nitrogen status, but does not account for factors such as waterlogging, pests or diseases. Nitrogen limitation can mask the effect of favourable rainfall seasons and, hence, for the simulations it was assumed that nitrogen did not limit production. To allow for different fallow situations, different starting soil moisture conditions were considered, namely a full or a half-full soil profile at sowing. To quantify the impact of negative SOI conditions on wheat yields, each year of simulated yield was categorised according to the corresponding April-May SOI phase. Median simulated yields of years with negative SOI conditions at the end of May were then compared with median yields of the entire 100-year simulation (Table 2).

TABLE 2. Simulated median wheat yields for 6 locations in the northeastern Australian wheat belt. Shown is the median for the 'all years' case and the median for the years that had a consistently negative (-ve) SOI phase by the end of May. Simulations were conducted for a 20 May sowing date on a profile containing either 180mm of stored soil moisture ('full') or 90mm of stored soil moisture ('half full').

Location	Full Profile		Half Full Profile	
	All years	-ve May SOI	All years	-ve May SOI
	$(t\ ha^{-1})$	$(t\ ha^{-1})$	$(t\ ha^{-1})$	$(t\ ha^{-1})$
Dalby	2.85	1.75	1.19	0.58
Emerald	1.97	1.53	0.60	0.58
Goondiwindi	3.48	3.19	1.86	1.65
Moree	3.30	2.73	1.87	0.92
Narrabri	3.32	2.31	1.94	1.43
Roma	2.21	1.87	0.96	0.62

While these simulation results indicate a significant median yield reduction in negative SOI years, they also demonstrate that economically viable yields can be expected even when the climate outlook indicates high changes of below median rainfall, providing adequate soil moisture reserves are available at planting. Through a detailed on-farm monitoring program we were able to confirm that wheat grown on good stored soil moisture reserves in 1997 in north-eastern Australia performed well and although the seasonal rainfall was substantially below average, many crops received timely rain at anthesis resulting in better than expected yields. In contrast, however, only very few wheat crops planted on marginal stored soil water reserves produced economically viable yields.

In May 1998 the SOI started to rise sharply. Most Global Circulation Models predicted a likely break-down in the El Niño pattern for later in 1998, but the sharp rise in the SOI, combined with a marked cooling of sea surface temperatures in the eastern Pacific were the first physically based indicators of such a break-down. Hammer *et al.* (1996) showed that potential profits made from growing wheat in years when the SOI rises sharply during autumn are often much larger than in other seasons. Hence, it is important to alert growers to this potential so that they can ensure that the crop is managed accordingly, for example, adequate nitrogen fertilise is applied and pests and diseases are checked early. Accordingly, in May 1998 a press conference was called and all relevant media outlets were invited. This press conference was used to highlight (a) the probabilistic nature of the information, (b) the high chances of above median rainfall for most wheat-growing regions of Australia, but particularly in the northern wheatbelt, and (c) the potential problems that can be associated with above median rainfall years, such as water logging, problems with planting and harvesting, and increased pressure from diseases favoured by wet conditions. In addition, yield potential estimates derived from simulation models and management recommendations were published in the rural press and in newsletters.

A simulation analysis for the entire Australian wheat belt comparing median yields over the entire climate record with those for years associated with each SOI phase at the end of May showed significant differences with season type (Table 3). The study shows that, providing diseases and the effects of water logging can be avoided, wheat yields can be expected to vary with SOI phase throughout the Australian wheat belt. Under negative SOI conditions (as in 1997), a median yield reduction ranging from 34% at Dalby (Queensland) to 9% at Horsham (Victoria) was simulated. Under rapidly rising SOI conditions (as in 1998) above median yield potentials existed at all locations.

TABLE 3. Simulation analysis for the entire Australian wheatbelt by SOI phases. Results are presented as a percentage of the long-term (100-year) simulated median wheat yield. 'Standard agronomy' with no nitrogen limitation was assumed and results were averaged across a range of starting soil moisture conditions.

Location	Negative	Positive	Falling	Rising	Near '0'	All years
	(eg 1997)			(eg 1998)		
Wongan Hills	84	91	90	120	107	*100*
Minnipa	83	107	71	122	64	*100*
Horsham	91	112	92	119	83	*100*
Wagga Wagga	88	108	75	111	88	*100*
Walgett	73	116	83	115	100	*100*
Dubbo	59	99	76	115	71	*100*
Gunnedah	105	100	75	138	88	*100*
Moree	59	109	77	118	93	*100*
Roma	80	100	86	116	100	*100*
Goondiwindi	61	100	75	138	106	*100*
Dalby	66	94	86	114	96	*100*
St George	83	103	76	116	106	*100*
Emerald	94	85	85	144	121	*100*

While such responses are well documented for northeastern Australia, some corroborating evidence has recently been identified in South Australia (Egan, pers. comm.). Historical district wheat yield data for Minnipa in South Australia for the period 1909 to 1995 show very similar trends to those based on the simulation results shown in Table 3.

By November we were able to confirm that despite the increased potential, the 1998 wheat season was marked by serious problems for many producers in northeastern Australia. Not only were world wheat prices low, but frequent, heavy rainfall either delayed planting or resulted in planted crops being abandoned due to waterlogging and had to be re-sown. Therefore, many crops were planted past their optimal planting date and reached grainfilling at a time when temperatures were already higher than desirable for this developmental stage (Woodruff and Tonks, 1983). By far the biggest problem, however, were leaf diseases favoured by wet conditions. While such diseases are generally not considered a problem in the region, in 1998 speed and intensity of the disease development was such that many producers were unable to take adequate control measures and ended up with yields and grain qualities substantially below their expectations. In spite of this there where large regional differences and some growers did exceptionally well. Our monitoring program showed that producers who avoided paddocks prone to water logging and who sprayed crops early against possible leaf diseases achieved yields considerably above the long-term average.

2.3 THE 1997/98 SORGHUM CROP

By July/August 1997 sea-surface temperature anomalies showed that the El Niño was well established. At the end of September the SOI phase was again consistently negative (Table 1). Hence, based on historical rainfall records, most of Eastern Australia had only a 20 to 40% chance of exceeding the long-term median rainfall for October to December (Stone et al., 1996a). This outlook was used to assess the likely impact of the El Niño weather pattern on sorghum production.

Likely sorghum yields for several locations were estimated using APSIM configured for sorghum (McCown et al., 1996; Hammer and Muchow, 1994). We assumed either a full (180mm) or half-full (90mm) soil water profile at sowing. Based on historical September SOI phase data, the sorghum simulation output was analysed in a similar manner to the wheat simulations presented earlier. This study showed that at the median level, sorghum yields at Dalby on a full soil profile and a half-full profile are likely to be reduced by 25% and 28%, respectively (Figure 1; half-full profile only), when the May SOI phase is consistently negative. This is due to reduced and more erratic summer rain for this season type. For the same location and using a similar approach, Hammer et al. (1997) have shown that the advance knowledge of the season provides an opportunity to adjust crop management to the season by modifying nitrogen fertiliser input.

However, yield distributions at Goondiwindi and Moree did not differ significantly between negative SOI and the 'all years' case at either starting soil moisture conditions (Figure 1, half-full profile only). This reflects the fact that these locations have a good chance of summer storms that are largely unaffected by El Niño. However, the variability at any of the locations is considerable and ranges from 1 to 6 t ha^{-1}. At all locations the median date for receiving a planting rain is one to two weeks later under negative SOI conditions.

The simulation analysis also showed that sowing sorghum later could further reduce the impact of a negative SOI phase on expected yields. Across all locations and soil types, mean yields rose from 2 t ha^{-1} for mid October sowings to 3.4 t ha^{-1} for mid December sowings in seasons where the SOI was consistently negative in September. However, across all season types there was only a slight yield increase with later sowings (2.5 vs 2.8 t ha^{-1}) (Figure 2). This information was published in the rural press on 5 September (prior to the beginning of sorghum plantings in this region), together with a warning about not planting too late in order to avoid possible outbreaks of sorghum ergot, which are likely should the autumn be wet (Meinke and Ryley, 1997). Press releases ensured the widest possible dissemination of this information via radio and TV.

As with the 1997 wheat crop, the conclusion was that with a reasonable amount of stored soil moisture, the chances of growing an economically viable sorghum crop are good, particularly in the more southern parts of the sorghum growing regions. El Niño does not always mean disaster and in fact some regions can still expect good summer storms under such conditions.

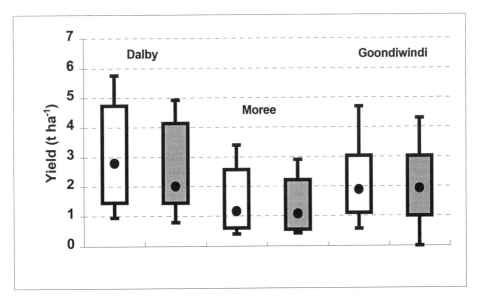

Figure 1. Simulated sorghum yields for Dalby, Moree and Goondiwindi for a soil profile containing 90mm of stored soil moisture at planting. Bars indicate the 90 to 10% yield probability range, boxes show the 75 to 25% probability range, the solid circles show the median. Compared are years when the SOI phase was negative at the end of September (-ve SOI, 25 years, grey boxes) with the entire climate record (all years, 100+ years, white boxes).

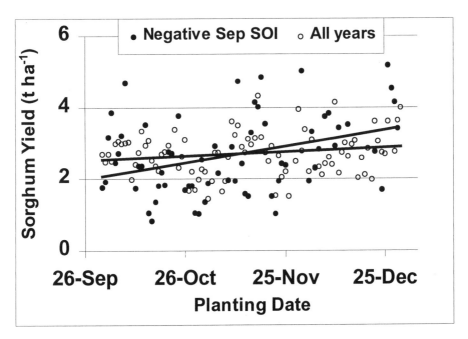

Figure 2. Simulated sorghum yields for locations in southern Queensland and northern NSW for a wide range of starting soil moisture conditions and soil types. The grey and black lines indicate the linear regressions for the all years and the negative SOI years, respectively.

3. Interactions with Growers – A Case Study of Tactical Decision-Making

This is a case study for a dryland grain/cotton farmer (DCF) on the southern Darling Downs. DCF is a farmer-collaborator in an active research program that aims to quantify the value of seasonal climate forecasting on-farm. The research approach requires close interaction with some key collaborators who differ in the cropping systems they employ and/or in their geographical location. This allows the scientists involved to better understand the importance and limitations of using seasonal climate forecasting operationally. Ultimately, insights gained from this interaction feedback into the more general information provided and so increase the relevance and the value of this information. This 'research beyond the comfort zone' requires scientists to re-establish their credibility with a new client group outside the normal, scientific peer-review process. It also requires producers to openly discus their management strategies so that possible advances in their tactical crop management can be identified.

DCF has used climate forecasting in tactical crop management decisions for several years and has recently intensified and streamlined his efforts. He takes the seasonal forecast into account for all his management decisions and aims to increase the proportion of cotton in his rotational system without compromising the long-term sustainability. This approach aims to maximise the profitability of the whole farm operation. To achieve this, several factors are vitally important. Keeping in mind that water is the most limiting resource in this environment, the factors are:

- Avoiding soil fertility decline
- Avoiding erosion and run-off
- Maximising water infiltration
- Developing a surface management system that allows sowing a crop after even minor rainfall events (10 mm) in order to be as close as possible to the optimum sowing date.

Many of the crop management strategies used by DCF are not a direct consequence of a seasonal forecast, but it is difficult to clearly separate the issues. When questioned, DCF pointed out that since he started following and using the seasonal climate outlook his whole thinking about crop and cropping systems management has changed. Although he was always aware of the importance of conserving water and would have implemented management strategies such as no-till regardless of the availability of seasonal outlooks, the speed and thoroughness of the implementation of these strategies was greatly affected by the forecast.

3.1 NITROGEN MANAGEMENT

DCF uses simulated 'target yields' for wheat and sorghum to optimise his nitrogen strategy. These target yields are determined based on the amount of stored soil moisture prior to sowing, historical rainfall records and the long-term rainfall outlook. From this, he estimates the 10 percentile of achievable yield in a given season. Nitrogen requirements are then determined based on the available background nitrogen in the field and the appropriate amount of nitrogen is applied.

Although DCF always aims to reduce evaporative losses by maximising the no-till area and by applying nitrogen in the least soil-disturbing way (e.g. 'knifing' nitrogen into the soil rather than cultivating), this becomes imperative if the outlook indicates a likely dry period. He sees adequate nitrogen nutrition in dry seasons as particularly important to establish a good root system and increase water use efficiency.

3.2 MANAGING FROST RISK IN WHEAT

Frost at or after anthesis can lead to considerable yield reductions in wheat. The anthesis time that corresponds to the highest yield potential is also the period of maximum frost risk. As the season progresses, both frost risk and yield potential are strongly reduced, forcing producers to carefully consider the tradeoffs between risk and production potential (Woodruff and Tonks, 1983). Employing a probabilistic SOI-based frost forecast system allows assessing the likely date of last frost for the coming winter season (Stone et al., 1996b). In February 1997 the SOI rose rapidly and the chances of late damaging frosts for wheat crops appeared low. However, by late February/early March SOI values started to drop sharply. These were the early warning signs that an El Niño was developing in the Pacific. A negative SOI value by the end of May (which now seemed likely) alters the frost outlook significantly with the chances of a late damaging frost for wheat rated at higher than normal (Stone et al., 1996b). Although we cannot be certain about such developments until (usually) the end of May, a carefully worded El Niño alert was issued in April 1997. Based on this information, DCF adjusted his winter crop management. He prepared his wheat cropping area so that he was able to sow some wheat at the earliest planting opportunity in late April. To spread the frost risk, he only planted about half of the area designated to wheat at this time - the rest was planted late May. This 'gamble' paid off and the early-sown wheat yielded close to 5 t ha^{-1}, amongst the highest yielding wheat in the district.

3.3 MARKETING

For DCF grain marketing was one of the most profitable applications of seasonal climate forecasting in 1997. There are two elements to it:

(a) A lag relationship between the SOI phase and futures price of some commodities has been found (Chapman et al., 2000). This reflects that under certain SOI conditions production is likely to deviate from the long-term average. Markets adjust to these changes in supply by adjusting prices. In conjunction with estimates of farm-specific production also based on SOI phases, this information is directly applicable to determine the best marketing strategy by either selling or buying futures and/or options (hedging).

(b) Markets also react directly to the forecast. The trends are foreseeable and will affect all production regions regardless of how strongly they are affected by an SOI signal (in terms of rainfall and its impact on production). Again, producers can utilise this knowledge by buying and selling futures or options and even by anticipating currency movements. DCF has applied this knowledge with great success in the 1997 winter season.

3.4 SPECIFIC MANAGEMENT ACTIONS DURING THE 1997 AND 1998 SEASONS

Following is a detailed summary of DCF's actions/decisions that were affected by the seasonal climate outlook. Using output from dynamic crop simulation models, DCF evaluates alternative management strategies to identify the least risky and/or the most profitable options. In 1997, DCF took the following actions:

- Maximised no-till area;
- Applied nitrogen fertiliser early to allow planting on stored soil moisture at the most appropriate time;
- Planted some wheat on 20 April, the rest in late May to spread frost risk;
- Applied 75 kg ha^{-1} of starter nitrogen to ensure good root establishment;
- Cultivated *early* for 'pupa bust' (required by law for insect control after cotton crops) to save as much soil moisture as possible for a barley double crop (i.e. a winter crop sown straight after a summer crop, often on very little stored soil moisture. This crop yielded between 2.5 to 3 t ha^{-1}, a rare yield level for a double crop even under favourable climatic conditions);
- Conducted early weed control to save moisture;
- Sowed some faba beans and chickpeas as alternative crops to wheat to spread risk (frost risk and different water use pattern);
- Tactically applied additional nitrogen fertiliser at varying amounts based on nitrogen already applied and likely crop performance based on the seasonal outlook;
- Did NOT forward sell grain because of El Niño outlook (yield unknown, but prices usually rise).

During an evaluation session in late 1997, DCF rated all these measures as successful. His yield and grain quality were substantially above the district average and he attributes this positive result largely to his ability to include climate forecasting into his decision making.

Cognisant of the high chances of above median rainfall in 1998 DCF took, among others, the following steps:

- Sowed wheat earlier than usual due to the lower frost risk;
- Applied 100 kg nitrogen to a wheat cover crop grown on a dry profile after cotton. These crops are usually not expected to produce economically viable yields and are only grown to reduce erosion risks. Subsequent rainfall then filled the soil profile, but did not result in any water logging or run-off. Combined with adequate disease control (see point below), this crop yielded 3.8 t/ha, about three times the district average.
- Applied fungicides to wheat crops early and frequently (3 times), to minimise leaf diseases. Fungicide applications in wheat are not common in this region;
- Ensured that machinery was in working order to sow and harvest crops whenever weather conditions allowed;
- Organised grain drying facilities, which allowed him to harvest grain at higher than optimal moisture content, thus minimising harvest losses due to rainfall;
- Did not forward sell chickpeas, assuming those prices would rise due to low volumes of supply.
- Reduced the area sown to chickpeas, which are susceptible to water logging. In addition, he treated chickpeas prior to sowing with a fungicide to reduce disease pressure. In spite of these measures, the chickpea crop suffered from water logging and had to be abandoned.

Based on historic rainfall records in conjunction with dynamic crop simulation models, producers can evaluate alternative management options and quantify the likely effect on farm income and production risk. This was highlighted through interaction with DCF during the 1997 and 1998 seasons. To be most effective, this approach requires an understanding of the probabilistic nature of the information provided, whereby producers must not become disheartened or reckless by any perceived 'failure' or 'win' of the forecast in any given season. The approach must be used consistently for many seasons to truly benefit from it. Further, it must be integrated into the whole decision making process as one of many management tools.

4. Working with Groups of Agricultural Advisers and Farmers

The intensity of scientist/producer interactions discussed in the previous section cannot be maintained beyond specific pilot projects. Methods need to be developed that allow the generalisation of the knowledge gained by scientists and producers through such intense interaction. Regional diversity from a climatic as well as from a farming systems perspective makes such generalisations difficult. One possible avenue might be the closer involvement of farmer groups and their advisers in this process. In the following section we will present two case studies of interaction between scientists, agricultural consultants and farmer groups. The case studies formed part of a much larger APSRU project that aimed to develop suitable methods for agribusiness and/or consultants relating to the generation and dissemination of relevant simulation results and climate forecast information. Jointly with agribusiness firms, this project has established a range of regionally diverse farmer groups.

All meetings were formally evaluated, which served two specific purposes: (a) to allow for reflection about the consequence of research activities in order to improve future interactions, and (b) for researchers to learn about the impact of this work on farm managers. Two types of evaluations were carried out in association with the group activities. The ongoing evaluation activity was an "entry" questionnaire of farmers' intentions before group sessions and an "exit" questionnaire on how they valued various aspects of the group activity. The results of these questionnaires were used to reflect on the group activities and to guide action on improving them. In addition an independent consultant was employed to evaluate the impacts of these activities.

Evaluation results have shown that these group meetings are having a significant impact on the way in which farm managers are thinking about tactical decisions. For example the entry questionnaire asked: "What sort of season do you expect and what is the main influence on your expectation?". At the April 1997 meetings 43% of farmers in four groups included in their answers a reference to seasonal climate forecasting. The same question, asked of the same four groups in August 1997, elicited a 59% reference to seasonal forecasts. In exit questionnaires farmers were asked to respond to the statement: "This presentation and discussion increased my appreciation of how to use SOI information". The average response on a scale of 1 to 5 (where 1 is strongly disagree and five is strongly agree) was 3.9. A similar question on the importance of monitoring soil moisture received an average value of 4.3. Given the relative complexities of the two issues and the probabilistic nature of the SOI information, these results were encouraging.

Case study 1: Wheat or Chickpeas?
Faced with a high probability of below average winter and spring rain in 1997, a number of these farmer groups, all clients of a major Australian agribusiness firm, requested an analysis of the likely impact of this forecast on winter crop production. Calculating the odds of growing an economically successful crop requires access to suitable simulation capabilities but also knowledge of the stored soil moisture reserves, nitrogen requirements, likely price movement, rainfall and frost outlook. Such information needs to be assessed within the context of tactical, crop specific management options (sowing date, area sown to wheat, nitrogen fertilisation strategy etc) and more strategic, rotational options (grow an alternative winter crop, not growing a crop at all).

Chickpeas with their shorter growing season and later planting date are perceived as a potentially less risky option than wheat when rainfall is scarce. However, no quantitative studies have been conducted that objectively compare these options in terms of their long-term profitability and riskiness. Based on measurements that a number of farmers in the group had on their amount of stored soil moisture, wheat and chickpea yields were simulated for the last 100 years and the output segregated into the five possible SOI categories, similar to the wheat and sorghum simulations presented earlier. Under negative May SOI conditions, simulated chickpea yields were, on average 11% higher than those of wheat (Figure 3a). This result differed from the general "rule of thumb" that chickpea yields are up to 40% lower than wheat yields, a perception largely based on limited experiences gained in favourable wheat seasons. In group-sessions, producers 'negotiated' the most appropriate costs and likely prices for wheat and chickpea in order to conduct the necessary gross margin (GM) analyses. The median GM for chickpea was more than double that of wheat for the 1997 scenario. Further, chickpea showed a positive GM in all years, while wheat GM was negative in five of the 15 years (Figure 3b). A monitoring process has now been established to test the validity of these simulation results.

Case study 2: Prospects for the 1997 summer crop
Another example of co-learning resulted from a meeting held in July 1997, after the group's consultant analysed deep soil core samples on two farmers' fields. The group requested that APSRU researchers use this base data as input into APSIM to investigate prospects for summer crops based on these starting conditions and under the current climate outlook.

Learning that is unique to the participatory approach occurred outside the scope of the planned activities when one of the farmers expressed "a hunch". The farmer's hunch was that in El Niño years sorghum sown in late December was likely to out-yield sorghum sown in early October. Using APSIM with soil characterisation data for the farmer's field and the past 25 years local weather data as input, the researchers were able to test and confirm this hypothesis at the meeting (see also Figure 2).

What took place at this meeting was a mutual learning experience. The farmer's hunch was based on very limited experience. Without the availability of reliable simulation techniques (or many years of costly field experimentation) this hunch could not have been "validated". For the researchers the farmer's insight was a testable hypothesis. The product of this collaboration was a timely insight that can impact considerably on farm management decision. Subsequent discussions with agri-business firms provided supporting evidence that the information on time of sowing of sorghum was reflected in the pattern of sale of sorghum seed in the Downs region of southern Queensland.

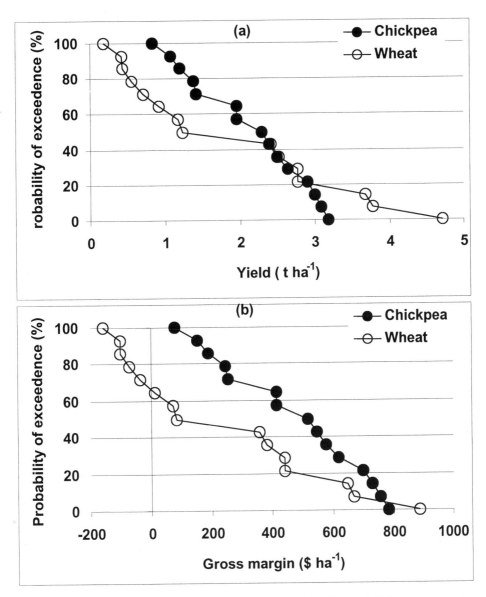

Figure 3. Potential (a) grain yields and (b) gross margins predicted for wheat and chickpeas in years with a consistenly negative SOI phase by the end of May at Jimbour, Queensland. Simulations were conducted for a half-full soil water profile at sowing.

Conclusions

For any forecast to be effective, the information provided must alter a decision (Hammer, 2000). We have shown that seasonal forecasting can be used operationally and that it does influence tactical crop management. However, the range of possible applications of this type of information ranges from very general to highly specific. Further, for producers to apply such knowledge they need to have ownership of the issues. This requires communication processes that are well targeted and range from one-to-one scientist/producer interactions (co-learning) to group interactions to general media releases. The El Niño of 1997/98 followed by a La Niña episode in 1998/99 provided an excellent case study of how producers in northeastern Australia can best benefit from this information and how knowledge gained at one level can be utilised at another.

Acknowledgments

The Queensland Department of Primary Industries (DPI), CSIRO, LWRRDC and GRDC financed this research.

References

Bate, P.W. (1997) The tropical circulation in the Australian/Asian region – November 1996 to April 1997. Aust. Met. Mag. **46**, 237-246.

Carberry, P.S., Hammer, G.L. and Meinke, H. (2000) The potential value of seasonal climate forecasting in managing cropping systems, in G.L. Hammer, N. Nicholls, and C. Mitchell (eds.), *Applications of Seasonal Climate Forecasting in Agricultural and Natural Ecosystems – The Australian Experience.* Kluwer Academic, The Netherlands. (this volume).

Chapman, S.C., Imray, R. and Hammer, G.L. (2000) Relationship between seasonal forecasts and grain prices, in G.L. Hammer, N. Nicholls, and C. Mitchell (eds.), *Applications of Seasonal Climate Forecasting in Agricultural and Natural Ecosystems – The Australian Experience.* Kluwer Academic, The Netherlands. (this volume).

da Cunha, E. (1995) *Rebellion in the backlands.* University of Chicago Press. pp.751.

Hammer, G.L. (2000) A general approach to applying seasonal climate forecasts, in G.L. Hammer, N. Nicholls, and C. Mitchell (eds.), *Applications of Seasonal Climate Forecasting in Agricultural and Natural Ecosystems – The Australian Experience.* Kluwer Academic, The Netherlands. (this volume).

Hammer, G.L. and Muchow, R.C. (1994) Assessing climatic risk to sorghum production in water-limited subtropical environments. I. Development and testing of a simulation model. *Field Crops Res.* **36**, 221-234.

Hammer, G.L., Carberry, P.S. and Stone, R.C. (2000) Comparing the value of seasonal climate forecasting systems in managing cropping systems, in G.L. Hammer, N. Nicholls, and C. Mitchell (eds.), *Applications of Seasonal Climate Forecasting in Agricultural and Natural Ecosystems – The Australian Experience.* Kluwer Academic, The Netherlands. (this volume).

Hammer, G.L., Chapman, S.C. and Muchow, R.C. (1997) Modelling sorghum in Australia: The state of the science and its role in the pursuit of improved practices, in M.A. Foale, R.G. Henzell and J.F. Kneipp (eds.), Proceedings of the Third Australian Sorghum Conference, Tamworth, February 1996. Australian Institute of Agricultural Science, Melbourne. Occasional Publication No. 93.

Hammer, G.L., Holzworth, D.P. and Stone, R.C. (1996) The value of skill in seasonal climate forecasting to wheat crop management in a region with high climatic variability. *Aust.. J. Agric. Res.* **47**, 717-737.

Hammer, G.L., Woodruff, D.R. and Robinson, J.B. (1987) Effects of climatic variability and possible climatic change on reliability of wheat cropping - A modelling approach. *Agric. For. Meteorol.* **41**, 123-142.

Keating, B.A. and Meinke, H. (1998) Assessing exceptional drought with a cropping systems simulator: a case study for grain production in north-east Australia. *Agric. Sys.* **57**, 315-332.

Khandekar, M.L. (1996) El Niño/Southern oscillation, Indian monsoon and world grain yields - a synthesis. *Adv. in Natural and Technological Hazards Res.* **7**, 79-95.

McCown, R.L., Hammer, G.L., Hargreaves, J.N.G., Holzworth, D.P. and Freebairn, D.M. (1996) APSIM: A novel software system for model development, model testing, and simulation in agricultural research. *Agric. Sys.* **50,** 255-271.

Meinke, H. and Hammer, G.L. (1997) Forecasting regional crop production using SOI phases: a case study for the Australian peanut industry. *Aust. J. Agric. Res.* **48,** 789-793.

Meinke, H. and Ryley, M. (1997) Effects of sorghum ergot on grain sorghum production: a preliminary climatic analysis. *Aust. J. Agric. Res.* **48,** 1241-1247.

Meinke, H. and Stone, R.C. (1997a) The chances of receiving planting rain. ProFarmer Newsletter 5(14), 4.

Meinke, H. and Stone, R.C. (1997b) On tactical crop management using seasonal climate forecasts and simulation modelling - a case study for wheat. *Scientia Agricola* **54,** 121-129.

Meinke, H., Stone, R.C. and Hammer, G.L. (1996) Using SOI phases to forecast climatic risk to peanut production: a case study for northern Australia. *Int. J. Climatol.* **16,** 783-789.

Meinke, H., Stone, R.C. and Hammer, G.L. (1997) The climatic outlook and its likely impact on the forthcoming wheat crop. ProFarmer Newsletter 5(17), 4.

Meinke, H., Hammer, G.L., van Keulen, H. and Rabbinge, R. (1998) Improving wheat simulation capabilities in Australia from a cropping systems perspective. III. The integrated wheat model (I_WHEAT*). Europ. J. Agron.* **8,** 101-116.

Stone, R.C. and Auliciems, A. (1992) SOI phase relationships with rainfall in eastern Australia. *Int. J. Climatol.* **12,** 625-636.

Stone, R.C., Hammer G.L., and Marcussen, T. (1996a) Prediction of global rainfall probabilities using phases of the Southern Oscillation Index. *Nature* **384,** 252-55.

Stone, R.C., Nicholls, N. and Hammer, G.L. (1996b) Frost in NE Australia: trends and influences of phases of the Southern Oscillation. *J. Climate* **9,** 1896-1909.

Woodruff, D.R. and Tonks, J. (1983) Relationship between time of anthesis and grain yield of wheat genotypes with differing developmental patterns. *Aust. J. Agric. Res.* **34,** 1-11.

White, B. (2000) The importance of climate variability and seasonal forecasting to the Australian economy, in G.L. Hammer, N. Nicholls, and C. Mitchell (eds.), *Applications of Seasonal Climate Forecasting in Agricultural and Natural Ecosystems – The Australian Experience.* Kluwer Academic, The Netherlands. (this volume).

THE POTENTIAL VALUE OF SEASONAL CLIMATE FORECASTING IN MANAGING CROPPING SYSTEMS

PETER CARBERRY[1,2], GRAEME HAMMER[1,3], HOLGER MEINKE[1,3]
AND MICHAEL BANGE[4]

[1] Agricultural Production Systems Research Unit (APSRU)
CSIRO Tropical Agriculture[2] and Queensland Departments of Primary
Industries[3] and Natural Resources
PO Box 102, Toowoomba, Qld 4350, Australia

[4] Cotton Research Unit
CSIRO Plant Industry
Narrabri, NSW 2390, Australia

Abstract

There is considerable interest in exploring the value of seasonal climate forecasts in assisting farmers to manage cropping systems, not only for short-term decisions on crop management but also for longer-term strategic decisions on crop rotations. This paper reviews a range of applications for climate forecasts, but focuses on cropping systems issues that would benefit from long lead-time forecasts. A specific case study is used to demonstrate the potential for using the Southern Oscillation Index in assisting the incorporation of opportunity cropping into dryland cotton production systems.

In the case study, the standard dryland cropping rotation of long fallowing from sorghum (through a subsequent summer fallow) to cotton is compared to alternative fixed rotations and to a rotation influenced by an SOI forecast. The decision point is the October after sorghum harvest where the manager can choose to proceed with the standard summer fallow or plant sorghum or cotton in that season with the intention in all cases of planting cotton in the following summer. These three fixed rotations (fallow-cotton, sorghum-cotton, cotton-cotton) are compared to an SOI-influenced strategy using a simulation analysis over the long-term climate record for Dalby, Qld.

The simulation case study demonstrated that SOI contributed some skill to improving management decisions over a two-year rotation. By changing between fallow-cotton, sorghum-cotton or cotton-cotton rotations based on the SOI phase in the August-September period preceding the next two summers, average gross margins for the two year period increased by 14% over a standard fallow-cotton rotation. At the same time, soil loss from erosion was reduced by 23% and cash flow was improved in many years

G.L. Hammer et al. (eds.), The Australian Experience, 167–181.
© 2000 Kluwer Academic Publishers. Printed in the Netherlands.

because an extra crop was sown. The SOI-based strategy did however increase the risk of economic loss from 5% of years for the standard fallow-cotton rotation to 9%, but this risk was considerably less than the 15% for sorghum-cotton and 19% for cotton-cotton rotations.

In conclusion, there are many decisions in the management of dryland cropping systems that would greatly benefit from climate forecasts with persistence in skill out to two years in duration. The case study used in this paper demonstrated some skill in using the SOI in choosing a cropping rotation of two-year duration. Such applications are the obvious next frontier both for the development of enhanced forecasting schemes and for their application within the cropping systems of northern Australia and possibly elsewhere.

1. Introduction

In managing cropping systems, farmers make decisions that are influenced by many factors. Certainly economic returns are of primary importance, but decisions will also be made based on perceived risk of economic loss, cash flow, weed and disease control, the risk of soil degradation, and lifestyle among a long list of influences (Blacket, 1996). The fact is that there are not many decisions in farming that are simply based on a single factor nor are they made in line with a purely tactical response to current information. In most cases, management decisions have to fit within a whole farm strategic plan such that many decisions are planned months ahead and their consequences seen months afterwards. This characteristic of managing cropping systems, the requirement for a long lead-time between deciding on a course of action and realising its results, is the subject of this paper. Can seasonal climate forecasting provide a long enough lead-time to contribute to the strategic management of cropping systems in northern Australia?

Just as the general public are becoming increasingly aware of events such as *El Niño*, managers of cropping systems are increasingly utilising climate forecast information in making management decisions (Meinke and Hochman, 2000). The Southern Oscillation Index (SOI) is widely promoted as the basis of a seasonal climate outlook, providing probabilities of achieving above or below average rainfall in upcoming seasons (Stone and de Hoedt, 2000). While the SOI is proving useful to managers, to date its application to cropping systems has been largely for tactical decisions involving short lead-times (Hammer *et al.*, 1996; Meinke and Stone, 1997; Meinke and Hochman, 2000). Even though predictive skill of the SOI declines beyond a 3 to 6 months lead-time (Stone, pers. comm.), its application to management decisions requiring longer lead-times is worth exploring.

The objective of this paper is to explore the potential value of seasonal climate forecasting in the management of dryland cropping systems in northern Australia. While a range of applications for seasonal climate forecasting are reviewed, a specific case study is employed to both demonstrate and assess the potential for using the SOI to aid a management decision, the consequence of which is realised over a two year period.

2. Management of Cropping Systems

Cropping systems, in their simplest sense, are defined by the sequence of crops grown in a rotation and by the agricultural commodities they produce. However, while most farmers see production as a primary goal, the management of cropping systems is done within the context of economic, natural resource, capital and labour, political, marketing and lifestyle influences (Blacket, 1996). How farmers manage their cropping system depends on their own objectives within these varying and at times conflicting factors.

The cropping systems of the northern grain region of Australia are characterised by the opportunity to produce a wide range of cereal, pulse, oilseed, forage and fibre crops. Both summer and winter crops are grown, with yields largely determined by water supply from either in-season rainfall or storage in the soil prior to planting. While the diversity in crop choice and planting time can be seen as advantageous, the high variability in seasonal rainfall means that the prospects for any one crop is often risky (Hammer et al., 1996). Fallowing the soil between crops in order to build up soil moisture storage is a recommended management strategy to offset the risk of low in-season rainfall. However, fallow lengths of up to 18 months result in low cropping frequencies and, in some locations, may be contributing to resource degradation through increased soil erosion or solute leaching (Freebairn et al., 1991; Turpin et al., 1996). As an alternative to rotations of fixed fallow length, opportunity cropping represents the practice of planting a crop whenever a planting opportunity is triggered, based usually on the accumulation of a minimum level of soil moisture storage and occurrence of a planting rain.

A rotational farming program, whether based on long fallowing or opportunity cropping, needs to not only contain a profitable combination of enterprises. It also must be able to maintain soil structure and fertility, facilitate control of insect, weed and disease pests, allow for timely field operations and spread the requirements for labour and machinery. These longer-term strategic considerations need to be balanced with and considered alongside the short-term tactical decisions. For example, the application of some residual herbicides can restrict the opportunity to plant some crops for up to 18 months after application. Likewise, dryland cotton is generally planted following long fallows after preceding sorghum or wheat crops, the stubble of which provides desirable protection from soil erosion during the fallow and cotton crop. Thus, decisions such as applying residual herbicides or planning for dryland cotton generally form part of farmers' strategic management of their cropping system.

For the northern cropping region, Wylie (1996) proposed five important aspects to successful farm business management:
1. Optimum enterprise mix
2. Optimum crop yields

3. Efficiency in resource use
4. Marketing
5. Making it happen (management skill)

There is obvious potential for seasonal climate forecasts to impact on at least the first four of these requirements for managing cropping systems. A number of analyses have already demonstrated value in seasonal climate forecasting for improving tactical agronomic management of crops and optimising yields (Hammer *et al.*, 1996; Meinke *et al.*, 1996; Meinke and Hochman, 2000). Chapman *et al.* (2000) have also explored whether climate forecasting can impact on marketing strategies. However, there has been little attention to date on exploring the potential of seasonal climate forecasts in the more strategic management decisions regarding crop rotations and resource sustainability.

The potential to utilise climate forecasts to improve enterprise mix or resource use requires forecasts of sufficient duration to extend beyond a single crop into the subsequent fallow and even as far as the next crop. This will require forecasting skill out beyond 12 to 18 months from the time of forecast. If this was the case, one may be able to tradeoff length of fallow, and consequent soil moisture storage, for improved prospects of future rainfall. Reduced fallow lengths from opportunity cropping have been suggested as advantageous not only in terms of profitability but also for resource use efficiency and sustainability (Keating *et al.*, 1995; Turpin *et al.*, 1996; Wylie, 1996). However, opportunity cropping is riskier - in some seasons one may be sacrificing an assured good crop after long fallowing for two mediocre crops. The prospect of climate forecasting being able to reduce this risk is an appealing one.

3. Analysis of Cropping Systems

Assessing the value of a climate forecast in a cropping system is difficult to achieve through experience alone. Experiments just cannot be run for sufficient duration to sufficiently sample the distribution of seasons experienced in the northern cropping region. Alternatively, crop simulation models combined with the historical climate record have been used to analyse the value of seasonal climate forecasting for tactical decision making (Hammer *et al.*, 1996; Meinke *et al.*, 1996; Meinke and Hochman, 2000).

The ability to assess the value of a climate forecast in a cropping system requires an ability to simulate the key components of the relevant system (Hammer, 2000). The Agricultural Production Systems Simulator (APSIM) is a simulation framework designed to simulate the production and resource consequences of agricultural systems, including fallowing and cropping sequence (McCown *et al.*, 1996). APSIM has been specified for, and its simulations tested against, a range of crop and farming systems (Carberry *et al.*, 1996a,b; Turpin *et al.*, 1996; Keating *et al.*, 1995; Probert *et al.*, 1998). Such testing has also demonstrated that APSIM is suitable for simulating commercial yields in the northern cropping region (Foale and Carberry, 1996).

In the following case study, APSIM is used to explore the potential for using a climate forecast in influencing management that has consequences beyond the yield of a single crop.

4. Case Study: Cotton in an Opportunity Cropping System?

Dryland cotton production in northern Australia is a high return, high risk cropping option. For this reason, farmers who decide to grow cotton generally become dedicated cotton growers, with grain production often a secondary priority. Other crops are grown as rotation crops with cotton, usually to provide stubble cover to protect against soil erosion or as a disease break. Because of its high production costs and high risk, the recommendation for dryland cotton is for planting after long fallowing from either sorghum or winter cereal. Thus, the decision to produce dryland cotton is usually made 6-12 months ahead of planting time and income received 6-8 months thereafter. In many years long fallowing will improve yields and reduce risk of crop failure by allowing sufficient time for the soil profile to recharge with moisture. But in other years, the soil water profile may have been full long before planting time or in-crop rainfall may have been sufficient to discount the value of pre-plant moisture storage. The degree to which the efficiency of dryland cotton systems could be improved if they became more flexible through an opportunity cropping strategy has not previously been explored.

In this case study, three set crop rotations (Figure 1) are compared with an opportunity cropping rotation. The recommended sorghum-long fallow-cotton rotation (SFC) is compared with the option of planting sorghum (SSC) or cotton (SCC) in the second year of the rotation and short-fallowing through to cotton in the third year. All three rotations are committed to sorghum in year 1 and cotton in year 3 based on the rationale that the farm is geared for cotton production in terms of infrastructure (machinery, labour), markets (futures trading) and rotation of farm strips. While this degree of rigidity may not be completely realistic, it is assumed for the purpose of this case study.

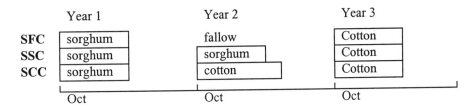

Figure 1. Schematic representation of the three crop rotations used in the simulation study.

In order to simplify the comparison of the three rotations, subsequent analyses assumed that the decision point is the 1st October in year 2. That is, the farm manager has come out of sorghum in year 1 and is faced with the question of which of three rotations to select given known soil resources. At this point, the manager has the option of planting a sorghum or cotton crop or of fallowing through to the set cotton crop in year 3. This case study is thus considering a tactical decision for a given decision point and with soil water and nitrogen values set at the beginning of a two year period. A more strategic consideration of this decision would need to accommodate a range of decision points and soil resource conditions, and the full three year rotation.

APSIM version 1.40 was configured to simulate crop rotations grown on a Brigalow soil type at Dalby (27°S, 153°E) over the historical climate record between 1887 and 1997. Over the 100 years, the simulations were run continuously with the system's status reset on 1st October every two years - soil water was reset to 122mm available soil water, representing a 47% full profile to 1.8m depth. The three rotations, fallow-cotton, sorghum-cotton or cotton-cotton, were simulated with the initial sorghum or cotton crops planted on 1st October and the final cotton crop planted based on a sowing criterion of receiving 25mm rainfall over a 5 day period between the 1st October and 25th November. For sorghum, the cultivar Buster was sown at 100,000 plants/ha. For cotton, a Siokra-type variety was planted in single-skip configuration (two rows planted followed by one missed row) at 12 plants/m row. Both sorghum and cotton crops were fertilised with 150kg N /ha. Each simulation run was repeated a second time with the starting year offset by one in order to enable the crops in rotation to be represented in all years.

As can be seen in the example output of simulated yields for the three rotations (Figure 2), fallowing in year 2 resulted in some guarantee in cotton yields in year 3. Planting sorghum or cotton in year 2 dramatically affected subsequent cotton yields in some years, whereas there are other years where there was little or no effect. Lower year 3 cotton yields were a consequence of the previous crop reducing the soil water available to the following cotton crop. This effect of a preceding crop was reduced in those years where pre-season or within-season rainfall was adequate to produce good yields for the cotton crop in year 3. Interestingly, there are a number of years when a preceding crop resulted in increased cotton yields due to a reduced incidence of waterlogging.

At the decision point of 1st October in year 2 of the rotation, managers can either select the same rotation every year or choose to opportunity crop, ie. select a different option (SFC, SSC or SCC) each year based on knowledge of their system status at that point in time. Given in this case study that soil resources are reset to the same nominated value at this point in time, this decision can be based solely on a forecast of future climate. In this case, the five-phase SOI system (Stone and Auliciems, 1992; Stone et al., 1996) is employed whereby the value of the August-September SOI phase determines which of the three rotations will be chosen in an opportunity cropping situation. This approach is analogous to the studies by Hammer et al. (1996) and Meinke et al. (1996) on tactical management of wheat and peanut crops.

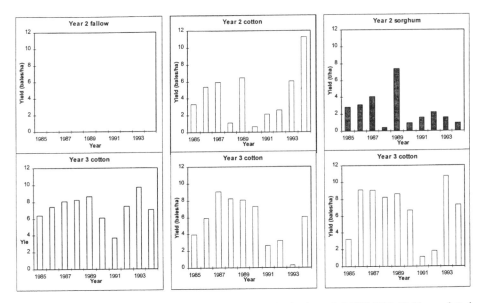

Figure 2. Simulated yields for the three rotations (vertical panels) for the period 1985-1994. Yields are plotted against the year in which the decision point was made (ie. year 2 of the three year rotation).

The selection of a rotation for each SOI phase is not straight forward as there is no clear advantage of any one rotation in the analogue years consistent with each SOI phase. Therefore, the choice between any rotational strategy would depend on a manager's attitude to returns and risk. In order to select a preferred rotation for each SOI phase, average gross margin is plotted against a measure of economic risk for each of the three rotations to create a return-risk tradeoff space over all years and for each SOI phase (Figure 3). The assumption is that risk-efficient strategies dominate others by higher mean gross margin return at a given level of risk. Choice between rotations in this return-risk space depends on attitude to risk of individual decision-makers, which can be expressed as a utility function. Within this return-risk space, iso-utility or indifference lines can be drawn tangential to the points of highest return or lowest risk. A horizontal indifference line tangential to the strategy of highest mean return would represent the choice of a risk-neutral farmer. Indifference lines of increasing slope depict increasing aversion to risk, culminating in a vertical indifference line, for a highly risk-adverse farmer, tangential to the strategy of lowest risk. This depiction of return-risk tradeoff and the application of iso-utility or indifference lines has been successfully employed in a number of previous studies (Barah *et al.*, 1981; Hammer *et al.*, 1991; McCown *et al.*, 1991; Carberry *et al.*, 1993; Muchow and Carberry, 1993; Keating *et al.*, 1994; Parton and Carberry, 1995; Hammer *et al.*, 1996).

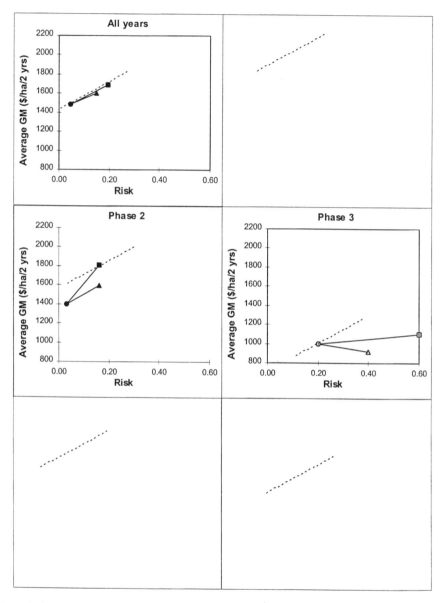

Figure 3. Average gross margin ($/ha/2 years of rotation) for the SFC (●), SSC (▲) and SCC (■) rotations plotted against a measure of risk (defined as the proportion of years where the accumulated 2-year gross margin was less than $500/ha) for all years and for subsets of years associated with each SOI phase. The dashed line represents the nominal farmer utility function adopted for this analysis.

In dryland cotton production regions, the recommended strategy is to long fallow into cotton, analogous to the SFC rotation. Accordingly, the slope of the indifference line depicted in Figure 3 was determined as that required to select the SFC strategy marginally ahead of the next best rotation over all years. The rationale for choosing this slope for the indifference line was to depict a decision-maker whose current preference is for the recommended SFC rotation but who would also not be far from choosing a higher return, higher risk strategy (eg. SCC). In fact, over all years, the three fixed rotations are close to forming a linear efficiency frontier (SSC is slightly below this frontier) whereby the chosen indifference line would find it difficult to discern between the three rotations. Nevertheless, by using an indifference line with this same slope within the return-risk tradeoff space for years associated with each SOI phase (Figure 3), a preferred rotation can be selected in each case, namely:

Phase 1	(SOI negative)	= SSC
Phase 2	(SOI positive)	= SCC
Phase 3	(SOI rapidly falling)	= SFC
Phase 4	(SOI rapidly rising)	= SCC
Phase 5	(SOI near zero)	= SFC

By selecting the rotation corresponding to the SOI phase at the time of the decision point in year 2 of the rotation, an SOI-responsive tactic can be determined for each year of the simulation analysis. Gross margins calculated for the 100-year climate record based on this SOI-responsive strategy are presented in Figure 4, where the annual gross margin advantage/disadvantage for adopting this responsive strategy compared to the standard SFC rotation is also provided.

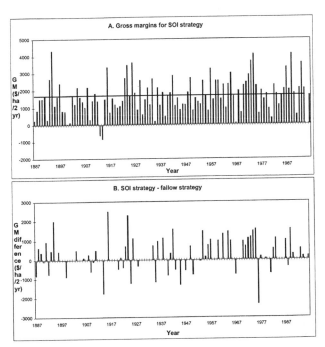

Figure 4. (a) Gross margins ($/ha/2 years of rotation) calculated from simulated yields for the SOI responsive strategy and (b) the difference in gross margin each year between the SOI responsive strategy and the SFC rotation.

The performance of the three set rotations and the SOI-responsive tactic can be compared using a number of criteria. In this case, performance criteria included:

(i)   gross margin accumulated for the simulated two year period (years 2 and 3 of the rotation);

(ii) the risk of economic loss, quantified as the percentage of years when the accumulated gross margin for the two-year simulation period was less than $500/ha/2 years (an estimate of the fixed costs required to maintain a typical cotton farm at Dalby);

(iii) cash flow, quantified as the gross margin attained at the end of year 2 in a three year rotation; and

(iv) relative soil loss, quantified as the simulated soil erosion loss for each rotation relative to the simulated soil loss for the recommended SFC rotation (set to 1.0).

A final assessment of the four rotational systems is to compare their performance against a rotation where the choice at the decision point in year 2 is determined by future knowledge of the rotation in each year that produces the highest gross margin (criterion (i) above). This rotation is termed "perfect knowledge" (PK) and is an indicator of the potential economic performance of the farming system.

TABLE 1.   Average performance of five 2-year crop rotation systems, based on a 100-year simulation analysis at Dalby, Qld. The abbreviations for the rotation systems are defined in the text.

		SFC	SSC	Crop Rotation System SCC	SOI	PK
Yield						
	Year 2 (bale or t/ha)	0.0	3.2	4.2	-	-
	Year 3 (bale/ha)	6.0	5.2	4.3	-	-
Gross margin ($/ha2/yr)		1482	1605	1691	1683	2226
Risk (%yrs GM<$500)		5	15	19	9	3
Cash flow (year 2)		-56	380	820	405	578
Soil loss (relative to SFC)		1	0.49	0.72	0.77	0.65

A summary of the average performance of the three set rotations, the SOI-responsive rotation and the rotation based on perfect knowledge is given in Table 1. Long fallowing into cotton (SFC) clearly produced the highest average cotton yields in year 3 of the rotation. It produced a high expected gross margin with the lowest risk of economic loss. However, long fallowing performed poorly in other performance criteria in having negative cash flow in year 2 of the rotation and the highest risk of soil erosion. Rotations where sorghum or cotton replaces the summer fallow (SSC or SCC) reduced final cotton yields, on average by 13% and 28% respectively. However, the compensation for lower final cotton yields was the increased overall productivity of the SSC and SCC rotations due to the additional crops planted in year 2 of these rotations. The SCC rotation increased average gross margin by 14% and SSC by 8% relative to the recommended SFC rotation. While cash flow and erosion risk were also positively affected by these more intensive rotations, a significant downside was the large increase in risk of economic loss. Economic risk increased from 5% of years with SFC to 19% and 15% of years with SCC and SSC, respectively. So, while each of the three set rotations performed best in at least one performance criterion, no one rotation was best overall. In fact, using the farmer indifference line nominated in Figure 3, the lower return but lower risk outcome for the recommended SFC rotation would result in its preference over the other two set rotations (Figure 5).

The SOI-responsive rotation performed well against the three set rotations yet it was not best in any one performance criterion (Table 1). It was significantly better than SFC in terms of average gross margin, cash flow and erosion risk, but it was also almost twice as risky on average. The SOI-responsive rotation produced very close to the average gross margin and soil loss values of SCC yet was considerably less risky and it was better than the SSC rotation in all but the soil loss criterion. Perfect knowledge produced significantly greater average gross margins at lower risk than any other of the management strategies. However, perfect knowledge, selected on a basis of maximisation of gross margins, did not produce the best outcomes for cash flow nor erosion. No one rotation, even including perfect knowledge, provided a best option for all selection criteria.

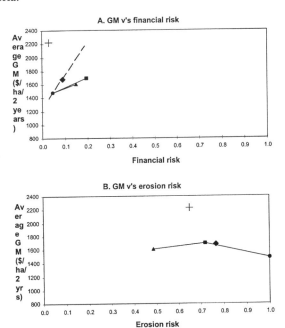

Figure 5. Average gross margin ($/ha/2 years of rotation) for the SFC (●), SSC (▲), SCC (■), SOI (◆) and PK (+) rotations plotted against (a) financial risk, defined as the proportion of years where the accumulated 2-year gross margin was less than for $500/ha (the dashed line represents the minimum slope of the indifference line that would be required for preference of SFC over the SOI-based rotation), and (b) erosion risk, defined as the simulated soil erosion loss for each rotation relative to that for the SFC rotation.

For a decision-maker with an indifference line similar to that used in Figure 3, the SOI-responsive rotation would be the preferred rotation based on a trade-off between gross margin return and risk (Figure 5a). For a decision-maker to not select the SOI-responsive rotation, the slope of the indifference line would have to increase by a factor of 3.3. It is important to note, however, that a number of decision-makers could remain indifferent to the SOI-responsive rotation if their attitude to risk is represented by an indifference line of slope greater than this value.

The previous analyses traded off gross margin return against risk of financial loss. However, other criteria may be of greater importance to some decision-makers. Figure 5b presents a trade-off space for gross margin return versus erosion risk over all years for the three fixed rotations and the rotations based on SOI and perfect knowledge. Such a trade-off space is relevant to a decision-maker wishing to maximise gross margin while minimising soil erosion. In this situation, there are only two risk-efficient strategies, SSC and SCC, which have equal or higher average gross margins at lower levels of erosion risk than the other rotations. Thus, a rotational system using an SOI-based seasonal climate forecast provided no advantage to such a decision-maker - the SCC rotation provided equivalent returns at a lower risk of soil loss. While the SCC rotation would be selected by a decision-maker who was indifferent to erosion, selection of the SSC rotation would only require a small increase in the slope of the indifference line away from the horizontal. Hence, the SSC rotation would be a preferred rotation for decision-makers concerned with soil loss from erosion. In fact, some decision-makers who were extremely adverse to erosion risk would even select SSC in preference to a rotation that used perfect knowledge to maximise gross margin.

This case study examined the use of an SOI-based seasonal climate forecast in incorporating cotton into an opportunity cropping system. The example dealt with only one decision point and a given set of soil water and nitrogen values. Further analyses are obviously warranted for alternative decision points and soil conditions – values for soil water set closer to 0% or 100% would undoubtedly favour the SFC and SCC rotations respectively. And in reality, the assumed obligation to lock into cotton in year 3 could be reassessed prior to its planting, based on soil water and the SOI outlook at that point in time. Nevertheless, faced with the situation of a 47% full soil water profile and a planting opportunity in early October, this case study was able to demonstrate that an SOI-based forecast at this time could be advantageous to making a decision with consequences seen 18 months thereafter. As the persistence in forecast skill beyond 3-6 months is low, it is likely that the SOI-based forecast was providing an indicator of soil moisture storage 6 months hence, which in turn strongly influenced the summer cropping potential 12-18 months after the original decision point.

Conclusions

Seasonal climate forecasts can undoubtedly assist farmers in managing cropping systems, either in short-term tactical decisions or long-term strategic decisions. This paper reviewed a range of applications for climate forecasts and concluded that there are a number of systems issues that would benefit from long lead-time forecasts. A specific case study demonstrated considerable potential for using the Southern Oscillation Index in assisting the incorporation of opportunity cropping into dryland cotton production systems. While this example dealt with only a limited situation, the suggested benefits derived from using SOI-based forecasts in selecting crop rotations clearly warrant further exploration.

The value of SOI-based forecasts has been assessed previously in relation to crop performance within 3-6 months of the forecast. The results from the case study presented here demonstrated useful impacts on crop rotations ending 18 months after the initial forecast. Considerable value could be attributed to the forecast via responsive management of the crop rotation. Such value would have been partly associated with the effects on soil water storage at the end of the initial 3-6 month period. This telegraphing of an SOI-based forecast beyond 6 months via soil water storage provides opportunities, beyond any direct effects on seasonal rainfall, for use of the SOI in longer-lead time forecasting in cropping systems.

This paper has demonstrated that the value of even a good climate forecast depends on the risk preferences of the decision-maker. The decision-analysis framework adopted in the case study indicated that the SOI-based rotation would not have been selected by very risk averse decision-makers (Figure 5a). Similarly, a decision-maker who would place soil erosion as a higher concern than financial risk would also not adopt the SOI-based strategy developed in this paper (Figure 5b). All decision-makers, including farmers, make decisions for a multitude of reasons and so, while a seasonal climate forecast may be of use to some, it may not necessarily be useful to all.

Finally, this paper introduced a decision-analysis framework to assess the value of a seasonal climate forecast against multiple criteria. This analysis framework proved useful in exploring trade-offs between conflicting objectives in assessing the value of a climate forecast. This approach is subsequently used by Hammer *et al.* (2000) to assess the relative merits of several alternative forecasting systems in assisting management of cropping systems.

Acknowledgments

The authors wish to thank James Gaffney for providing assistance in the economic analyses undertaken in this paper. The provision of the OZCOT cotton model by the CSIRO Cotton Research Unit, Narrabri, for inclusion within the APSIM framework is gratefully appreciated.

References

Barah, B.C., Binswanger, H.P., Rana, B.S. and Rao, N.G.P. (1981) The use of a risk aversion in plant breeding: Concepts and application. *Euphytica* **30**, 451-458.

180

Blacket, D. (1996) From teaching to learning: social systems research into mixed farming. Queensland Dept. Primary Ind., Publication QO96010. pp.118.

Carberry, P.S., Adiku, S.G.K., McCown, R.L. and Keating, B.A. (1996b) Application of the APSIM cropping systems model to intercropping systems, in O. Ito, C. Johansen, J.J. Adu-Gyamfi, K. Katayama, J.V.D.K. Kumar Rao and T.J. Rego (eds.), *Dynamics of Roots and Nitrogen in Cropping Systems of the Semi-Arid Tropics*. Japan International Research Center for Agricultural Sciences, International Agricultural Series No. 3., pp.637-648.

Carberry, P.S., McCown, R.L., Muchow, R.C., Dimes, J.P., Probert, M.E., Poulton, P.L. and Dalgliesh, N.P. (1996a) Simulation of a legume ley farming system in northern Australia using the Agricultural Production Systems Simulator. *Aust. J. Exp. Agric.* **36**, 1037-48

Carberry, P.S., Muchow, R.C. and McCown, R.L. (1993) A simulation model of kenaf for assisting fibre industry planning in northern Australia: 4. Analysis of climatic risk. *Aust. J. Agric. Res.* **44** ,713-30.

Chapman, S.C., Imray, R.J. and Hammer, G.L. (2000) Can seasonal climate forecasts predict movements in grain prices? in G.L. Hammer, N. Nicholls, and C. Mitchell (eds.), *Applications of Seasonal Climate Forecasting in Agricultural and Natural Ecosystems – The Australian Experience*. Kluwer Academic, The Netherlands. (this volume)

Foale, M.A. and Carberry, P.S. (1996) Sorghum in the farming system: Reviewing performance, and identifying opportunities by doing on-farm research, in M.A. Foale, R.G. Henzell and J.F. Kneipp (eds). Proceedings Third Australian Sorghum Conference, Tamworth, 20 to 22 February 1996. Australian Institute of Agricultural Science, Melbourne, Occasional Publication No. 93. pp.63-74.

Freebairn, D. M. Littleboy M. Smith G. D. and Coughlan K. J. (1991) Optimising soil surface management in response to climatic risk, in R.C. Muchow and J.A. Bellamy (eds.), *Climatic Risk in Crop Production: Models and Management in the Semiarid Tropics and Subtropics*. CAB International, Wallingford. pp.283-305.

Hammer, G.L. (2000) A general systems approach to applying seasonal climate forecasts, in G.L. Hammer, N. Nicholls, and C. Mitchell (eds.), *Applications of Seasonal Climate Forecasting in Agricultural and Natural Ecosystems – The Australian Experience*. Kluwer Academic, The Netherlands. (this volume)

Hammer, G.L. Carberry, P.S., and Stone, R. (2000) Comparing the value of seasonal climate forecasting systems in managing cropping systems, in G.L. Hammer, N. Nicholls, and C. Mitchell (eds.), *Applications of Seasonal Climate Forecasting in Agricultural and Natural Ecosystems – The Australian Experience*. Kluwer Academic, The Netherlands. (this volume)

Hammer, G.L., Holzworth, D.P., and Stone, R. (1996) The value of skill in seasonal climate forecasting to wheat crop management in a region with high climatic variability. *Aust. J. Agric. Res.* **47** , 717-737.

Hammer, G.L., McKeon, G.M., Clewett, J.F. and Woodruff, D.R. (1991) Usefulness of seasonal climate forecasts in crop and pasture management. Proc. First Aust. Conf. Agric. Meteorol., Melbourne. Bureau of Meteorology. pp. 15-23.

Keating, B.A., McCown, R.L. and Cresswell, H.P. (1995) Paddock-scale models and catchment-scale problems: The role for APSIM in the Liverpool Plains. In: Binning, P., Bridgman, H. and Williams, B. (eds.), MODSIM - Proceedings, International Congress on Modelling and Simulation, Nov 27-30, 1995, University of Newcastle.

Keating, B.A., Wafula, B.M., Watiki, J.M. and Karanja, D.R. (1994) Dealing with climatic risk in agricultural research - a case study with modelling in semi-arid Kenya, in E.T. Crasswell and J. Simpson (eds.), *Soil Fertility and Climatic Constraints in Dryland Agriculture*. Australian Centre for International Agricultural Research, Proceedings No. 54. pp.105-114.

McCown, R.L., Hammer, G.L., Hargreaves, J.N.G., Holzworth, D.P. and Freebairn, D.M. (1996) APSIM: A novel software system for model development, model testing, and simulation in agricultural research. *Agric. Sys.* **50**, 255-271.

McCown, R.L., Wafula, B.M., Mohammed, L., Ryan, J.G. and Hargreaves, J.N.G. (1991) Assessing the value of a seasonal rainfall predictor to agronomic decisions: The case of response farming in Kenya, in: R.C. Muchow and J.A. Bellamy (eds.), *Climatic Risk in Crop Production: Models and Management in the Semi-arid Tropics and Subtropics*. CAB International, Wallingford. pp.383-409

Meinke, H. and Hochman, Z. (2000) Using seasonal climate forecasts to manage dryland crops in northern Australia, in G.L. Hammer, N. Nicholls, and C. Mitchell (eds.), *Applications of Seasonal Climate Forecasting in Agricultural and Natural Ecosystems – The Australian Experience*. Kluwer Academic, The Netherlands. (this volume)

Meinke, H. and Stone, R.C. (1997) On tactical crop management using seasonal climate forecasts and simulation modelling - a case study for wheat. *Sci. Agric., Piracicaba* **54**,121-129.

Meinke, H., Stone, R.C. and Hammer, G.L. (1996) Using SOI phases to forecast climatic risk to peanut production: a case study for northern Australia. *Int. J. Climatol.* **16**, 783-789.

Muchow, R.C. and Carberry, P.S. (1993) Designing improved plant types for the semiarid tropics: agronomists' viewpoints, in F.W.T. Penning de Vries *et al.* (eds.), *Systems Approaches for Agricultural Development.* Kluwer Academic Publishers, The Netherlands. pp.37-61.

Parton, K.A. and Carberry, P.S. (1995) Stochastic efficiency and mean-standard deviation analysis: some critical issues. *Aust. J. Agric. Res.* **46**, 1487-91.

Probert, M.E., Carberry, P.S., McCown, R.L. and Turpin., J.E. (1998) Simulation of legume-cereal systems using APSIM. *Aust. J. Agric. Res.* **49**,317-27.

Stone, R.C. and Auliciems, A. (1992) SOI phase relationships with rainfall in eastern Australia. *Int. J. Climatol.* **12**, 625-636.

Stone, R.C. and de Hoedt, G. (2000) The development and delivery of current seasonal forecasting capabilities in Australia, in G.L. Hammer, N. Nicholls, and C. Mitchell (eds.*), Applications of Seasonal Climate Forecasting in Agricultural and Natural Ecosystems – The Australian Experience.* Kluwer Academic, The Netherlands. (this volume)

Stone, R. C., Hammer G. L., and Marcussen, T. (1996) Prediction of global rainfall probabilities using phases of the Southern Oscillation Index. *Nature* **384**, 252-55.

Turpin, J. E. Huth N. I. Keating B. A. and Thompson J. P. (1996) Computer simulation of the effects of cropping rotations and fallow management on solute movement. Proceedings 8th Australian Agronomy Conference, Toowoomba 1996. pp.558-61.

Wylie, P.B. (1996) Farming systems – the need for economics and integration. Proceedings 8th Australian Agronomy Conference, Toowoomba 1996. pp.598-601.

COMPARING THE VALUE OF SEASONAL CLIMATE FORECASTING SYSTEMS IN MANAGING CROPPING SYSTEMS

GRAEME HAMMER[1,2], PETER CARBERRY[1,3] AND ROGER STONE[1,2]

[1] *Agricultural Production Systems Research Unit (APSRU)*
Queensland Departments of Primary Industries[2] and Natural Resources and CSIRO Tropical Agriculture[3]
PO Box 102, Toowoomba, Qld 4350, Australia

Abstract

Cropping systems involve sequences of crops and management actions designed for profitable production and maintenance of resource integrity. Decisions made on crop choice and management in one season have ramifications on the sequencing of crops in subsequent seasons and on potential for resource degradation. In the previous paper, Carberry *et al.* (2000) found that a seasonal forecast based on 2-month SOI phases could be used to improve management in the subsequent two years of a dryland cotton cropping system at Dalby, Qld. Their study showed that the intensity of cropping could be increased by planting either sorghum or cotton in place of the fallow in specific season types identified by the forecasting system. The objective of this paper was to compare a range of forecasting systems in relation to their value in managing a cropping system in northern Australia. The same dryland cotton cropping system was used to compare the potential value of four seasonal climate forecasting systems. The four systems were -
- the SOI phase system, which utilises SOI patterns over a 2-month period
- a system based on SOI patterns over a 9-month period
- a system based on Pacific Ocean sea surface temperatures over a 5-month period
- a system based on projected SOI patterns from GCM runs for a 7-month period

All forecasting systems showed skill in shifting the median rainfall for the 6-month summer cropping period of October-March. For the 12- and 18-month rainfall totals commencing in October, the separation of medians among groups defined by each system was greatest for the SST-based system and least for the GCM-based system.

All forecasting systems showed some value in improving management decisions over the 2-year period examined in the dryland cropping system. In all cases, this was associated with increasing the intensity of cropping in specific forecast year types, either by introducing a sorghum or a cotton crop to replace the fallow in the second year of the rotation. The best outcome was associated with the forecasting systems based on

G.L. Hammer et al. (eds.), The Australian Experience, 183–195.

either the 5-month SST patterns or the 9-month SOI patterns, which gave a 20% increase in gross margin, a 30% decrease in soil erosion, but an increased risk of making a financial loss. The other two forecasting systems gave a gross margin increase of about 13%. There were some differences in relation to financial risk and erosion outcomes among forecasting systems.

The superior performance associated with tactical use of the forecasting systems reflected their ability to select years to increase cropping intensity. This was likely related to both the magnitude of predicted shifts in climate and the associated variability. The analysis highlighted the potential for seasonal forecasts to have value at long lead times. While the results suggested that systems based on SST patterns or longer term SOI patterns may be more useful, it was not possible to reach a general conclusion. A far wider range of locations and management scenarios need to be tested. This paper presents an appropriate framework for approaching this task. It also provides a means to integrate design of a forecasting system with its effective application.

1. Introduction

Cropping systems involve sequences of crops and management actions designed for profitable production and maintenance of resource integrity. Decisions made on crop choice and management in one season have ramifications on the sequencing of crops in subsequent seasons and on potential for resource degradation, such as through soil erosion. To have value in decision-making in relation to cropping sequences, a climate forecast with skill up to 12-18 months ahead is desirable. This would enable planning of management responses over a number of seasons, particularly in relation to crop sequence decisions. These decisions often require a long lead-time as a consequence of their major influence on the logistics of whole farm operation.

In the previous chapter, Carberry *et al.* (2000) examined whether seasonal climate forecasting could provide a long enough lead-time to contribute to the strategic management of cropping systems in northern Australia. They examined the potential value of seasonal forecasting in the strategic management of a dryland cropping system at Dalby, Queensland. In that system, the conventional practice is to long fallow from a sorghum crop in one summer, through to a cotton crop two years later. This long fallow system allows build up of soil moisture to ensure a reliable cotton crop. The moisture reserve minimises the risk of low yield and economic loss from the cotton crop, which, while having potential for high returns, also has high growing costs. Their case study showed that using a simple climate forecast based on the SOI phase (Stone *et al.*, 1996) in August-September preceding the planned summer fallow, the intensity of cropping could be increased by planting either sorghum or cotton in October in place of the fallow in specific season types. The key decision point is in early October at the beginning of the planned fallow. The adjusted rotations, based on the seasonal forecast, gave increased profit and reduced soil erosion risk, but slightly increased risk of economic loss over the following two summer cropping seasons, that is, over the next

18 months. The fact that the forecast generated value over such a long period was perhaps a surprising result given a forecasting system based on analogue years derived using patterns of movement of the SOI over only the preceding 2-month period.

Can we do better with other seasonal climate forecasting systems? A number of studies have suggested that improved forecasts with longer lead times are possible using analyses of sea surface temperature (SST) data or dynamic ocean and atmosphere models (GCM's) (Drosdowsky and Allan, 2000; Hunt and Hirst, 2000; Stone *et al.*, 2000). There has, however, been a paucity of comparative studies of forecasting approaches. This is partly due to the different nature of the forecast products and the differing lengths of historical data available for testing. Stone *et al.* (2000), however, have provided methods for generating analogue years from SST data and GCM output that enable comparison with SOI-based methods. The ability to derive analogue years also facilitates comparative analyses of the value of this range of forecasting schemes using the systems approach detailed earlier (Hammer, 2000).

The objective of this paper is to compare a range of forecasting systems in relation to their economic and resource conservation value in the management of a dryland cropping system in northern Australia. This is achieved by examining a specific case study that builds on the study reported by Carberry *et al.* (2000) in the previous chapter.

2. Cropping System Analysis

The simulation analysis for the dryland sorghum-cotton cropping system at Dalby, Queensland, as detailed in the previous chapter (Carberry *et al.*, 2000) was used in this study. Briefly, dryland cotton is usually grown in a sorghum-long fallow-cotton (SFC) rotation over three years, to minimise financial and erosion risks associated with growing cotton. The fallow allows the soil profile to recharge, thus increasing the likelihood of a successful cotton crop in the subsequent year. This standard rotation was compared with the option of planting sorghum (SSC) or cotton (SCC) in the second year of the rotation and short-fallowing through to cotton in the third year (see Figure 1, Carberry *et al.*, 2000). All three rotations were committed to sorghum in year 1 and cotton in year 3 assuming restrictions associated with the overall farm plan.

As all three rotations are the same until the beginning of the second crop, the key decision point is 1st October in year 2 of the rotation. At this point one of the three possible rotations must be chosen. As the intent was to examine the influence of various seasonal forecasting systems on a tactical decision at this time, soil water was set to 122mm (48% of soil water holding capacity) and the simulation started at this time. A more strategic consideration of this decision would need to accommodate a more comprehensive range of decision points and soil resource conditions. In this study, these factors were excluded to simplify the analysis and focus on the comparative value of the forecasting systems derived from implications on climate and the cropping system response after the decision point. The last two years of each of the three rotations were simulated using the cropping system simulator APSIM (McCown *et al.*,

1996) for the length of the historical climate record (1887-1997). Each simulation run was repeated a second time with the starting year offset by one in order to enable the crops in rotation to be represented in all years. The soil and crop management details and costs and prices used are detailed in Carberry *et al.* (2000). In this study, only the period from 1887 to 1991 could be used to facilitate comparison across forecasting systems.

The simulation results were used to derive three performance criteria -
(i)   gross margin ($/ha) accumulated for the simulated two year period (years 2 and 3 of the rotation);
(ii)   the risk of economic loss, quantified as the percentage of years when the accumulated gross margin for the two year simulation period was less than $500/ha (an estimate of the fixed costs required to maintain a typical cotton farm at Dalby);
(iii)   relative soil loss, quantified as the simulated soil erosion loss for each rotation relative to the simulated soil loss for the recommended SFC rotation (which was set to 100).

The performance of each of the three set rotations was quantified using these criteria and compared with the performance when the rotation was tactically adjusted in the second year using a number of climate forecasting systems. The methods for determining tactical rotation decisions in response to a forecast involved a trade-off between return and risk. The method to examine options in the return-risk space, detailed by Carberry *et al.* (2000) for the forecasting system based on SOI phases, was used for all forecasting systems examined in this study. This method involved defining the indifference line that quantified the trade-off between return (criterion (i)) and economic risk (criterion (ii)) among competing decision options. The indifference line was defined by plotting the average simulated return against economic risk for each of the three fixed rotation systems simulated over all years. A line was drawn through the point for the SFC rotation, which is accepted standard practice, and its slope was adjusted until the points for the other options fell just below the line in the return-risk space. This indifference slope was then employed to determine decision options in other situations, such as in subsets of years associated with each forecasting system.

3. Climate Forecasting Systems

Four seasonal climate forecasting systems were used in this study. The forecasting systems are -
- the SOI phase system (Stone *et al.*, 1996) as used in the previous chapter. This system utilises SOI patterns over a 2-month period preceding the October decision (August-September)
- a system based on SOI patterns over the 9-month period preceding the October decision (January-September) (after Drosdowsky, 1994)
- a system based on Pacific Ocean sea surface temperature (SST) patterns over the 5-month period preceding the October decision (May-September) (Stone *et al.*, 2000)

- a system based on projected SOI patterns from GCM runs for the 7-month period following the October decision (October-April) (Stone *et al.*, 2000)

To facilitate ease of subsequent decision analysis and comparison across the four systems, five groups of analogue years were identified for each forecasting system.

SOI phase system

This system is defined by Stone and Auliciems (1992) and Stone *et al.* (1996) and has been used extensively in application studies (Hammer, 2000). Five groups (phases) are defined from a cluster analysis of SOI trends over a 2-month period. There is a significant shift in rainfall and temperature likelihood in the season following each SOI phase at many times of year and in many parts of the world. The five phases are described as SOI consistently negative, SOI consistently positive, SOI rapidly falling, SOI rapidly rising, and SOI near zero.

9-month SOI patterns

This forecasting system is based on the approach reported by Drosdowsky (1994). The monthly values of the SOI from January to September for each year from 1878-1991were used as vectors in a hierarchical classification procedure (Ward, 1963) to identify distinct groups of years based on similarity of their SOI patterns. In this way, season types and distinct groups of analogue years could be identified and used in a manner similar to that of the 2-month SOI phase system. The five SOI patterns found (Figure 1) show a range of trends including rapid change from negative to positive over this period (Group I), gradually increasing positive values (Group II), and change from near zero to negative (Group IV).

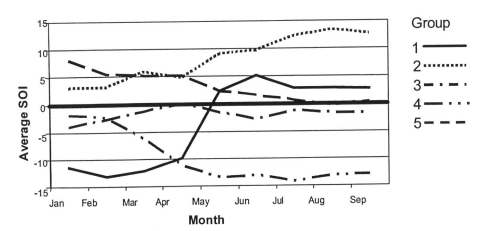

Figure 1. The average SOI values for January to September for the five groups of years identified by hierarchical classification of patterns using historical SOI data from 1878 to 1991.

SST patterns

This system is based on the approach reported by Stone *et al.* (2000), as applied to Pacific Ocean SST data (derived from GISST 1.1 data set) for the 5-month period May-September. Briefly, the seasonal pattern is removed from the raw SST data, and the remaining anomalies are subjected to EOF analysis. The EOF scores for each month from May to September in each year were then used as vectors in a hierarchical classification procedure (Ward, 1963) to identify distinct groups of years based on similarity of the dynamics of their SST patterns over the May to September period. The analysis of Stone *et al.* (2000) identified four main groups. The first group represented a strong El Nino development pattern and the second group represented a weaker form of the same pattern. The fifth group used in this study was a subset of the first group identified by Stone *et al.* (2000).

GCM SOI forecast patterns

This system is also based on the approach reported by Stone *et al.* (2000), as applied to SOI predictions from the GCM for the 9-month period October-June. Briefly, the SOI predictions were derived from the ensemble run of the CSIRO9 (T63) GCM, which was conducted by forcing with actual SST data and slight perturbation of starting conditions to generate the five runs of the ensemble. The SOI predictions (9 monthly values for each ensemble run for each year) were subjected to principal component analysis (PCA). The resulting PCA scores were combined across ensemble runs for each year and used in a hierarchical classification to identify groups of years based on similarity of the SOI predictions over the October to May period. The five groups identified reflected a wide range of predicted SOI patterns (see Figure 1, Stone *et al.*, 2000).

4. Skill in Rainfall Forecasting

The skill in rainfall forecasting associated with each forecasting system was gauged by the degree of shift in rainfall likelihood associated with each group defined in each of the forecasting systems. The rainfall probability distribution for each group was derived from the rainfall amounts received in each of the analogue years that were classified in that group. This was done only for Dalby, Queensland – the site chosen for the cropping system decision analysis case study.

All four forecasting systems showed skill in shifting the median rainfall for the 6, 12, and 18-month periods commencing in October (Figure 2). After 6 months (Figure 2a) the median forecast rainfall amounts for October to March had diverged considerably in relation to the all years median of 440mm. The greatest spread in the medians (about 120mm) occurred for groups identified by SST patterns, although this was largely due to the outcome for one group (Group 4), which isolated a group of much wetter than normal years. The spread in median values among groups within the other three forecasting systems was similar (about 80mm). This first summer period after the forecast in October is the time that an extra crop may be grown in place of the fallow if a sufficiently reliable forecast was available.

(a)

6 Months

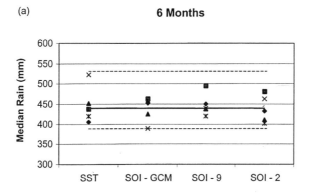

(b)

12 Months

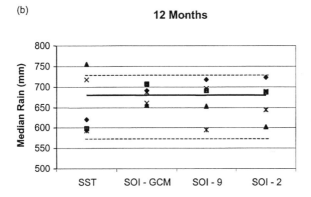

(c)

18 Months

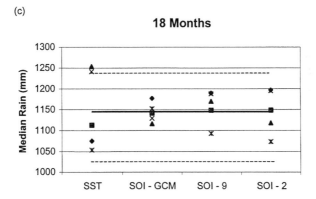

Figure 2. Median rainfall predicted at the end of September by each of four climate forecasting systems (SST patterns from May to September (SST); SOI patterns predicted by GCM for October to May (SOI-GCM); 9-month SOI patterns from January to September (SOI-9); 2-month SOI phases for August-September (SOI-2)) for –

(a) the following 6-month period from October to March the following year,
(b) the following 12-month period from October to September the following year, and
(c) the following 18-month period from October to March two years hence.

The solid line is the long term median rainfall for the period and the dashed lines are the 30 and 70 percentiles of the rainfall distribution for the period.

The spread in forecast medians for groups identified by SST patterns maintains a level equivalent to a +/-20% probability shift from the long-term median throughout the 18-month forecast period (Figure 2). This indicates maintenance of forecasting skill over this range of lead times, giving an effect through into the cotton crop, which would be grown in the 12-18 month outlook interval. The forecasts based on the SOI phases and 9-month SOI patterns maintain a slightly lower shift in total median rainfall. The shift in median increases to near the +/-20% probability level at the 12-month outlook, but then diminishes. This suggests reduction in skill at the long lead times. However, the movement of individual groups within this range over time (e.g. Group 1 in the SOI phase system) can be equally important as the overall range of all the groups. Significant shifts in relative median rainfall for individual groups across time periods indicates specific periods where that forecasting system may still have useful skill. The forecast system based on SOI forecast patterns from the GCM has least spread in median rainfall across groups and the spread has diminished by the 12-month outlook.

5. Analysis of Comparative Value of Forecasting Systems

The simulation analysis of the three cropping systems over the entire period used in this study (1887-1991) gave similar results (Table 1) to that of Carberry *et al.* (2000) as expected. The small differences from the former study were caused by the slight difference in time period required for this study to allow use of the range of forecasting systems. The greatest gross margin resulted from the sorghum-cotton-cotton (SCC) rotation but this was associated with the greatest financial risk. The least soil erosion was associated with the sorghum-sorghum-cotton (SSC) rotation but this was also associated with relatively high financial risk. The standard sorghum-fallow-cotton rotation (SFC) is likely adopted as standard practice because the low associated financial risk more than compensates for the reduced gross margin and the increased chance of soil erosion. The magnitude of this trade-off quantifies managers' attitudes to decision-making under risk and was outlined earlier and discussed in detail in the companion paper (Carberry *et al.*, 2000).

Table 1. Average simulated outcomes over the period 1887-1991 for three performance criteria for three rotational cropping systems at Dalby, Queensland.

Performance Criterion	Rotational Cropping System		
	SFC	SSC	SCC
Gross margin ($/ha/2 yr)	1457	1596	**1676**
Risk (% yrs GM < $500)	**5**	14	20
Soil loss (relative to SFC)	100	**49**	72

(a)

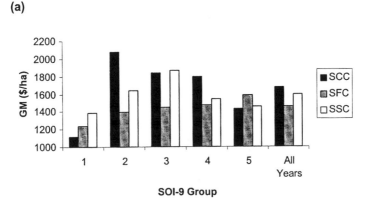

(b)

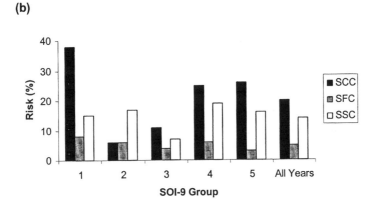

(c)

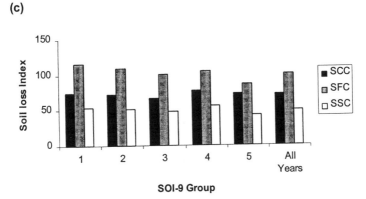

Figure 3. Simulated performance of three cropping systems (sorghum-fallow-cotton (SFC), sorghum-sorghum-cotton (SSC), and sorghum-cotton-cotton (SCC)) at Dalby, Queensland, for each year type group defined by the climate forecasting system based on 9-month SOI patterns. Performance was evaluated by -
(a)   average gross margin ($/ha/2yr),
(b)   risk of making a loss (% years with GM<$500), and
(c)   relative soil erosion (scaled to a value of 100 for SFC rotation over all years)
The performance of the three cropping systems over all years (1878-1991) of the simulation is included for comparison.

When the simulation analysis was examined by groups of analogue years associated with each forecasting system, the gross margin, financial risk, and erosion risk all varied. The outcome for the 9-month SOI pattern (SOI-9) forecasting system (Figure 3) demonstrated a typical outcome. In all groups of years, the erosion risk was least with SSC and greatest with the standard SFC. If rotation choice was based on minimising erosion risk alone, then the forecast would have no value as the decision would always be SSC. There were, however, changes in ranking in gross margin and financial risk among groups. In Group II the decision option of SCC and in Group V the option SFC, both gave highest gross margin and least risk and so would be the preferred options based on return-risk considerations. For the other groups, no single option could be identified in this way, so the choice must be determined by trade-off between gross margin and risk. For example, in Group I, SSC gave higher return than SFC but at considerably higher financial risk. Is the extra return sufficient to trade-off the extra risk? Using the level of indifference between return and risk defined by the outcomes over all years (Carberry *et al.*, 2000) the decision options preferred in each group in this case were SSC, SCC, SSC, SCC, and SFC.

Tactical changes in crop rotation decisions were generated for all four forecasting systems when the same approach used for the SOI-9 system was applied (Table 2). For each system, except the predicted SOI patterns from the GCM (SOI-GCM), all possible options were represented among groups of analogue years. This was associated with increasing the intensity of cropping in specific forecast year types, by introducing either a sorghum or cotton crop to replace the fallow. This occurred most often with the SOI-9 forecasting system and least with the SOI-GCM system. The greatest frequency of introducing additional cotton crops occurred with the 2-month SOI phase system (SOI-2) and was least with the SST patterns system (SST).

Table 2. Rotation decision option and number of years for each of the five groups of analogue years defined for each of four climate forecasting systems. The preferred decision is based on the trade-off between gross margin and financial risk. Group numbers are the same as in Figure 2.

	Climate Forecasting System			
	SOI-2	SOI-9	SOI-GCM	SST
Group I	SSC	SSC	SCC	SFC
No of years	19	13	24	28
Group II	SCC	SCC	SCC	SFC
No of years	31	18	13	24
Group III	SFC	SSC	SFC	SSC
No of years	10	27	36	19
Group IV	SCC	SCC	SFC	SCC
No of years	15	16	19	21
Group V	SFC	SFC	SFC	SSC
No of years	30	31	13	13

The cropping system performance criteria differed substantially when the tactical rotation management options associated with each forecasting system were evaluated using hindcasting (Table 3). All forecasting systems resulted in improved returns and decreased erosion risk when compared with the standard SFC rotation. However, with the exception of the SOI-GCM forecasting system, all systems also resulted in increased financial risk. Employing the same level of indifference between return and risk as used earlier, the outcomes for SOI-9 and SST systems were equivalent and best. After that the preference would be SOI-GCM, followed by SOI-2. Tactical use of the SOI-9 or SST systems resulted in a gross margin increase of about 20%, a soil erosion decrease of about 30%, but a doubling of risk of making a loss. Use of the SOI-2 or SOI-GCM systems resulted in a gross margin increase of about 13%, but differences in the other criteria. Financial risk was unaffected by tactical use of the SOI-GCM system whereas it was doubled using SOI-2. In contrast, erosion risk was reduced 23% with SOI-2, but only 10% with SOI-GCM.

A hypothetical perfect forecast, in which the best option would be chosen every year, gave a 50% increase in gross margin, with reduced risk and soil erosion (Table 3). This was associated with replacing the fallow by crop in 79 of the 105 years (40 as SSC and 39 as SCC). On the basis of change in gross margin alone, this gives the impression that the forecast has captured about 40% of the increase possible, which is considerably greater than the proportion reported in single crop studies (Hammer, 2000). However, this must be adjusted for the concurrent differences in risk. When this is done using the indifference lines defined earlier, the best forecasting systems in this example (SOI-9 and SST) account for about 30% of the increase possible, which remains considerably higher than previous reports for gross margin increases.

Table 3. Average simulated outcomes over the period 1887-1991 for three performance criteria for tactical rotational cropping systems based on a range of climate forecasting systems at Dalby, Queensland. Outcomes without use of a forecasting system (none) and those associated with a perfect forecast are included for comparison.

Forecasting System	Performance Criterion		
	Gross Margin ($/ha/2 yr)	Risk (% years GM<$500)	Soil Loss Index (Relative to SFC)
None (SFC)	1457	5	100
SOI-2	1658	10	77
SOI-9	1761	10	68
SOI-GCM	1642	5	90
SST	1733	9	74
Perfect	2186	3	65

The superior performance associated with tactical use of the forecasting systems reflects their ability to select years to increase cropping intensity, without jeopardising the high returns from growing cotton nor increasing the financial risks too much. The differences among the systems reflect their precision in this regard. The SOI-GCM system was most conservative, retaining the SFC system in 68 of the 105 years simulated, but opting for SCC in the other 37. This generated increased return without changing risk of making a loss. The SOI-2 system identified 28 more years to replace the fallow with crop (19 SSC and 46 SCC) but did not result in substantially increased return while doubling the financial risk. The additional cropping did, however, reduce erosion risk. There is a similar contrast between the SST and SOI-9 systems. The SST system replaced the fallow with crop in 53 years (32 SSC and 21 SCC) whereas for the SOI-9 system this was 74 years (40 SSC and 34 SCC). Yet both systems gave similar performance. This may be explained by the nature of the distribution of years in each group of the forecasting systems.

The value of the forecasting system likely relates not only to the predicted shifts in climate but also to the associated precision. That is, the confidence interval on the likely climate scenario shift, determined from the spread of the analogue years, may be as important as the magnitude of the shift. This has been demonstrated in studies with hypothetical changes to forecasting systems (Hammer et al., 1993). For example, the shifts in median rainfall associated with the SOI-GCM system are less than the other systems (Figure 2), yet Groups I and II in that system identify analogue years for reliable introduction of an extra cotton crop. In contrast, the greater shift in median rainfall for groups of the SST system than that for the SOI-9 system allows the same level of improvement with 30 fewer years of replacing fallow with crop. Clearly, both the shift and distribution around that shift are key aspects of a forecast in relation to its value in decision-making. Further analysis and targeted study of this point is required to identify most valuable improvements that could be made to seasonal climate forecasting systems.

This analysis highlights the potential for seasonal forecasts to have value beyond the 3-6 month horizon relevant to tactical management of individual crops. The analysis suggests that systems based on SST patterns or longer term SOI patterns may be more useful in this regard. However, it is not possible to reach a general conclusion until a far wider range of locations and management scenarios are tested. This paper presents an appropriate framework for approaching this task. It also provides a means to integrate design of a forecasting system with its effective application.

Acknowledgments

The assistance of Mr Dean Holzworth and Mr Greg McLean with data analysis is gratefully acknowledged.

References

Carberry, P.S., Hammer, G.L., Meinke, H., and Bange, M. (2000) The potential value of seasonal climate forecasting in managing cropping systems. in G.L. Hammer, N. Nicholls, and C. Mitchell (eds.), *Applications of Seasonal Climate Forecasting in Agricultural and Natural Ecosystems – The Australian Experience.* Kluwer Academic, The Netherlands. (this volume).

Drosdowsky, W. (1994) Analog (nonlinear) forecasts of the Southern Oscillation Index time series. Weather and Forecasting **9,**78-84.

Drosdowsky, W. and Allan, R. (2000) The potential for improved statistical seasonal climate forecasts. in G.L. Hammer, N. Nicholls, and C. Mitchell (eds.), *Applications of Seasonal Climate Forecasting in Agricultural and Natural Ecosystems – The Australian Experience.* Kluwer Academic, The Netherlands. (this volume)

Hammer, G.L. (1993) Farming strategies to address climate variability. in, W. Kinninmonth (eds.), Proc. Western Pacific Workshop on Seasonal to Interannual Climate Variability: Strategies for Improved Forecasting and Applications for Social and Economic Sectors, June, 1993. Bureau of Meteorology, Melbourne.

Hammer, G.L. (2000) A general systems approach to applying seasonal climate forecasts. in G.L. Hammer, N. Nicholls, and C. Mitchell (eds.), *Applications of Seasonal Climate Forecasting in Agricultural and Natural Ecosystems – The Australian Experience.* Kluwer Academic, The Netherlands. (this volume)

Hunt, B.G and Hirst, A.C. (2000) Global climatic models and their potential for seasonal climatic forecasting. in G.L. Hammer, N. Nicholls, and C. Mitchell (eds.), *Applications of Seasonal Climate Forecasting in Agricultural and Natural Ecosystems – The Australian Experience.* Kluwer Academic, The Netherlands. (this volume)

McCown, R.L., Hammer, G.L., Hargreaves, J.N.G., Holzworth, D.P., and Freebairn, D.M. (1996) APSIM: A novel software system for model development, model testing, and simulation in agricultural research. *Agric. Sys.* **50**, 255-271.

Stone, R.C., and Auliciems, A. (1992). SOI phase relationships with rainfall in eastern Australia. *Int. J. Climatol.* **12,** 625-636.

Stone, R.C., Hammer, G.L, and Marcussen, T. (1996). Prediction of global rainfall probabilities using phases of the Southern Oscillation Index. Nature **384,** .252-256.

Stone, R.C., Smith, I., and McIntosh, P. (2000) Statistical methods for deriving seasonal climate forecasts from GCMs. in G.L. Hammer, N. Nicholls, and C. Mitchell (eds.), *Applications of Seasonal Climate Forecasting in Agricultural and Natural Ecosystems – The Australian Experience.* Kluwer Academic, The Netherlands. (this volume)

Ward, J.H. (1963). Hierarchical grouping to optimise an objective function. *J. Amer. Stat. Assoc.* **58**, 236-244.

MANAGING CLIMATIC VARIABILITY IN QUEENSLAND'S GRAZING LANDS - NEW APPROACHES

PETER JOHNSTON[1], GREG MCKEON[2], ROSEMARY BUXTON[3], DAVID COBON[4], KEN DAY[2], WAYNE HALL[2], JOE SCANLAN[5].

[1]Sheep and Wool Institute
Queensland Department of Primary Industries
GPO Box 3129, Brisbane, Qld 4001, Australia

[2]Climate Impacts and Natural Resource Systems
Queensland Department of Natural Resources
80 Meiers Rd, Indooroopilly, Qld 4068, Australia

[3]CSIRO Division of Wildlife and Ecology
PO Box 2111, Alice Springs, NT 0871, Australia

[4]Queensland Department of Primary Industries
PO Box 102, Toowoomba, Qld 4350, Australia

[5]Robert Wicks Research Centre
Queensland Department of Natural Resources
PO Box 318, Toowoomba, Qld 4350, Australia

Abstract

The grazing industries based on beef cattle and sheep are the major land use by area in Queensland, occupying greater than 80% of the State and contributing over 30% of total value of agricultural products in terms of meat, live animals and wool. Animals feed mostly on native and sown perennial grass pastures growing across a range of climates, soils and vegetation types. At both a location and regional scale, year-to-year variability in rainfall and other climatic variables is high, with the El Niño/Southern Oscillation phenomenon as a major cause.

Preliminary studies on the use of seasonal climate forecasting indicate potential for better management decisions such as –
(a) herd management by forecasting rates of reproduction and mortalities, and
(b) pasture management by improved use of pasture burning, improved legume establishment and assessing risks of overgrazing.

G.L. Hammer et al. (eds.), The Australian Experience, 197–226.

The management of stocking rate in Queensland's variable climate is very important for the profitability and sustainability of grazing enterprises. Current approaches to achieve both goals include using -

- 'safe' carrying capacity, which is based on low levels of utilisation of pasture growth on average,
- flexible grazing management, which involves changing animal numbers each year at the end of the growing season,
- tactical grazing management, which involves rapid rotation of stock through a large number of paddocks as in 'time controlled grazing', and
- 'tactical rest', which involves rules for seasonal changes in stocking rate based on condition of perennial grasses.

Current research and experience with these approaches is described and the possible role of seasonal climate forecasting discussed.

1. Introduction

The extensive grazing lands of Queensland are currently (1997) grazed by 10.5 million cattle and 10.1 million sheep (Australian Bureau of Statistics). Peak cattle and sheep numbers occurred in 1976-78 (11 million) and 1964/5 (24 million) respectively. The gross value of Queensland's rural production from beef cattle and sheep for 1997/8 was estimated at $1882 million being 33% of the total gross value of Queensland's agricultural products ($5499 million). In 1997/8, Queensland meat and wool exports were worth approximately $1800 million or 11% of the total value of exports ($16295 million). Thus the products of Queensland's grazing industries are a major source of income for the State. Native pastures account for 85% of Queensland's 173 million ha with the majority of beef cattle and nearly all sheep in Queensland located on these pastures. This is in contrast to most other Australian States where stock are concentrated on introduced pastures in the high rainfall and wheat-sheep zones (Mott *et al.*, 1981).

For the major grazing areas of Queensland (O'Rourke *et al.*, 1992), spatial variability in rainfall and distribution within the year is large. For example, average annual rainfall varies from <250 to >1000 mm (O'Rourke *et al.*, 1992) with 10% of rain falling in winter (April to September) in the north of the State increasing to 30% in the south (Weston, 1988). Other major sources of short and long-term variability affecting grazing enterprises in Queensland are year-to-year variability in rainfall (Figures 1 and 2) and other climate elements, which is associated with the El Niño - Southern Oscillation (Figure 1) and sea surface temperature patterns (Power *et al.*, 1998). Long-term trends in climate variables (McKeon *et al.*, 1998) and in atmospheric CO_2 (IPCC, 1996), as well as trends and variability in markets, prices (Figure 3) and costs (Anon. 1998) also have significant effects on grazing enterprises.

Major vegetation changes have been occurring in native pastures and include woodland thickening (Burrows, 1995), spreading of buffel grass (O'Rourke *et al.*, 1992), loss of desirable perennial grasses and weed invasion (O'Rourke *et al.*, 1992; Tothill and Gillies, 1992). Managing Queensland's grazing lands therefore requires balancing socio-economic goals of production with goals of sustainability and protecting biodiversity in an environment characterised by ecological complexity and high climatic variability.

In this paper we examine -
(i) the role of climate variability in Queensland's grazing lands, including historical trends and fluctuations, and
(ii) options for better grazing management in relation to climate variability with particular attention to the potential use of seasonal climate forecasting.

Of the grazing management options available to graziers, we concentrate on stocking rate decisions, reviewing new approaches and the experience of graziers and scientists.

2. Climate Variability in Queensland's Grazing Lands

The El Niño/Southern Oscillation (ENSO) phenomenon is the major source of year-to-year variability in rainfall and other climatic variables in Queensland (Pittock, 1975; McBride and Nicholls, 1983; McKeon *et al.*, 1990; Clewett *et al.*, 1991; Hammer *et al.*, 1991; Stone *et al.*, 1996a).

The Southern Oscillation Index (SOI) is a commonly used measure of ENSO behaviour. Year-to-year variation in rainfall (Figure 1) has been strongly correlated with SOI for extended periods in the last hundred years. Generally, below-average rainfall is likely when the SOI is strongly negative as in El Niño years, and above average rainfall is likely when the SOI is strongly positive as in La Niña years (McBride and Nicholls 1983; Clewett *et al.*, 1991; Stone *et al.*, 1996a). However, the impact of ENSO varies with time of year, location, and rate of change in the SOI (Stone *et al.*, 1996a).

In the grazing lands of Queensland, high year-to-year variability in rainfall occurs not only at individual locations, but also at a regional level (Figure 2). For example, annual (April-March) rainfall has a coefficient of variation of 29% even when averaged across the major grazing areas of Queensland (McKeon *et al.*, 1998). As a consequence, large areas have experienced climatic extremes at the same time, thus amplifying the effects on state-wide production and the need for government support. The 5-year drought of 1991 to 1995 was estimated to have cost Queensland $3 billion directly through lost production, which represents 14% of the total gross value of Queensland's 1997/8 agricultural production (expressed as an average annual loss), and a further $3 billion through flow-on effects (Anon, 1996). The effect of this extended drought on resilience of the grazing resource can only be assessed once years with above-average rainfall have occurred (eg. 1998/9) (Walker, 1988). A major resource issue will be the extent to which herbaceous and woody weeds replace desirable perennial grasses given that graziers have already identified weeds as a problem in many regions (O'Rourke *et al.*, 1992).

200

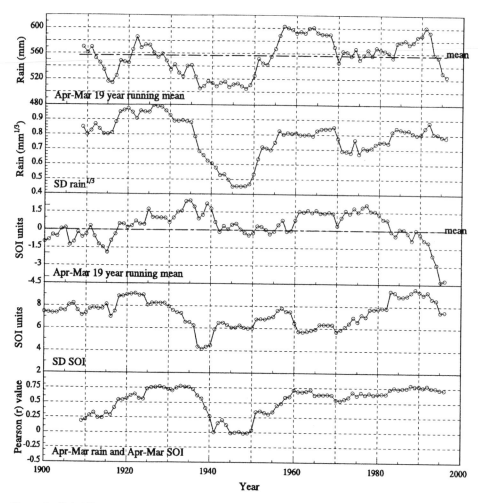

Figure 1. Rainfall and SOI statistics calculated over 19-year periods for Queensland's pastoral/cropping zone. Starting at the top the panels show the moving average of April - March rainfall; the moving standard deviation of April - March rainfall (cube root transformation); the moving average of April - March SOI; the moving standard deviation of April - March SOI; and the moving Pearson correlation of April - March SOI and April - March rainfall (bottom panel) (after McKeon *et al.,* 1998).

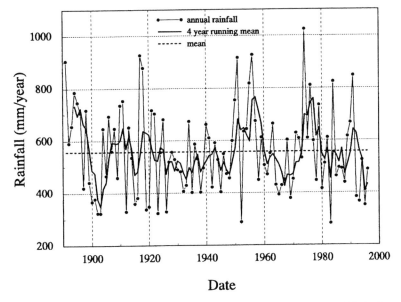

Figure 2. Annual rainfall (1April – 31 March) and four-year running mean of rainfall for Queensland's pastoral/cropping zone (after McKeon *et al.,* 1998).

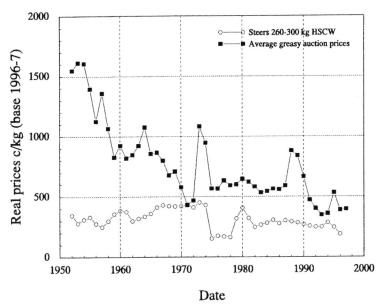

Figure 3. Time series of Australian wool and cattle prices. Wool prices are average annual greasy prices as reported by the Australian Bureau of Agricultural Resource Economics. Cattle price time series is an amalgamation of data from Daly (1983), who reported prices for 300-320 kg dressed weight (DW) steers between 1952-1974, and Australian Meat and Livestock Corporation who reported prices for 251-300 kg DW steers for 1975-80, then in 1988 for 260.1-300 kg hot standard carcass weight.

2.1 DROUGHTS AND FLOODS

Droughts have had a major impact on grazing enterprises throughout Queensland's history. The historical rainfall record includes major drought periods, such as 1898/9-1902/3 (Gibbs and Maher, 1967) and extremely dry single years, such as 1901/2 and 1911/2. Regional and property stock records indicate substantial reduction in stock numbers occur during droughts (Heathcote, 1965; Lilley, 1973; Daly, 1994). For example, Daly (1994) reported that sheep numbers, which he regarded as more reliable than cattle numbers, declined by 63% from a peak in 1897 to a low in 1903.

The period 1991/2-1995/6 had a similar four-year mean rainfall to the 1898/9-1902/3 drought when averaged across the grazing lands (Figure 2). In the 1902 calendar year, 70% of Queensland and NSW received less than decile 1 rainfall (Gibbs and Maher, 1967) and remains the worst on record at many locations. The hundred year instrumental rainfall record has been extended back to 1736 using reconstructions from fluorescence bands in coral collected at the mouth of a major river system in Queensland (Lough, 1991). This reconstruction indicates that 1901/2 (October-September) was the driest year in the 245 years of coral record (1736-1980). Although rare, the low rainfall in the periods 1898/9-1902/3 and 1991/2-1994/5 highlights the potential for extreme droughts that have major impacts on natural resources and human enterprises. We believe that some research resources should be allocated to the examination of the historical extremes to evaluate to what extent they can be forecast. Although 1902 was associated with an El Niño episode, the unheralded repeat of such a widespread and extreme event would be damaging not only to agricultural industries but also to the credibility of the science of climate forecasting.

Periods of high rainfall generally have a favourable impact on grazing enterprises through growth of perennial grasses providing pasture feed, fuel for pasture burning, soil surface cover and soil organic matter. However, there are also increased risks of stock losses with low quality feeds due to nutrient dilution, floods (Gillard and Monypenny, 1990), and increased disease and pests (Sutherst, 1990).

2.2 RESPONSES TO DROUGHT IN QUEENSLAND

Daly (1994) has reviewed the history of drought in Queensland with particular reference to the operation of the Queensland government drought relief schemes. He argued that a government supported system, which results in drought declaration of more than 30% of the State in 15 out of 26 years (1965-89), indicated both an inadequate appreciation of the actual climate variability, and a political process that was too willing to respond to moderate rainfall deficits. Daly's analysis is supported by previous historical reviews of drought by geographical researchers (Heathcote, 1994), graziers (eg. Mann, 1993), and sociologists (West and Smith, 1996).

Government drought policies have been recently reviewed and now include support for self reliance and the objective assessment of exceptional circumstances using historical climatic records (O'Meagher et al., 1998). However, 'drought' remains a highly emotive concept in the community (West and Smith, 1996). It is unclear to what extent the availability of seasonal forecasting is likely to change public and political perception of drought. Knowledge that El Niño and La Niña episodes change the year-to-year probabilities of experiencing drought or flood may change the view of drought as a random unforseen and damaging occurrence to that of an accepted risk in managing grazing enterprises (McKeon and White, 1992).

Simulation of drought declaration schemes has provided an alternative view. Stafford Smith and McKeon (1998) examined the criteria required for declaration and revocation of drought in two Queensland locations, such that the proportion of the hundred years in drought was only 5%. Under such strict criteria only three periods in the 1900s, 1930s and 1990s qualified. In another exploratory study, criteria of declaring on a low percentile annual pasture growth (percentile 5) and revocation on the occurrence of median annual pasture growth (percentile 50) resulted in approximately 20% of the State being drought declared when averaged across years - similar to that which has occurred in the operational scheme since 1964 (K.A. Day, pers. comm.). These findings highlight the importance of discriminating between the frequency of events at which droughts are declared (trigger points), and the duration for which monetary support is provided. Given the importance of drought to the performance of grazing enterprises and use of public money (Daly, 1994), the evaluation of the implication of declaration/revocation rules should be a high priority if policy is to be based on objective criteria.

3. Trends and Fluctuations in the Climate of Queensland's Grazing Lands

Major fluctuations and changes have occurred in climatic phenomena affecting the climate of Queensland's grazing lands. Ward and Russell (1980) examined the behaviour of winds in south-east Queensland since 1887 to understand changes in coastal dune movement. They found that the period 1922-52 had very few calm periods compared to previous and subsequent periods. This period of high winds was associated with below-average rainfall both in Queensland (Figures 1 and 2) and other rangeland regions (Gibbs and Maher, 1967), and also with wind-driven soil erosion and frequent dust storms (Ratcliffe, 1937). This period also had low correlations between measures of ENSO (ie. SOI) and rainfall (Figure 1) (Allan *et al.,* 1996; McKeon *et al.,* 1998). Given the apparent damaging impact on grazing lands, and the finances of graziers and governments (Payne and McLean, 1939; Gibbs and Maher, 1967), the development of a forecasting capability that explains this period and provides warning of its repetition would be clearly advantageous.

Since the 1970s night-time temperatures have been increasing across Queensland's grazing lands (0.25^0C per decade) with the most rapid change in May night-temperature (0.7^0C per decade) (Figure 4). Low temperatures in autumn, winter and spring limit the growth of C_4 tropical grasses if moisture and nutrients are available (Ivory and Whiteman, 1978). Similarly, frost has a damaging impact on nutritional value of tropical grasses (Wilson and Mannetje, 1978). Thus the night-time temperature increase and associated reduced frequency of frosts (Stone *et al.*, 1996b) are likely to be beneficial in terms of animal nutrition.

Carbon dioxide concentration has increased from 300 to 355 ppm over the last 40 years (Gifford, 1997). Physiological studies indicate that the water use efficiency (growth per mm) of C_4 tropical grasses is likely to increase by about 40% for a doubling of CO_2 to 700 ppm if nutrients (eg. nitrogen) are available (Howden *et al.*, 1999). There is also likely to be increased growth of C_3 species including herbaceous weeds (Patterson *et al.*, 1984) and trees (Duff *et al.*, 1994), and hence the likely outcome in terms of C_3 tree/C_4 grass competition is uncertain. Anthropogenic emissions of greenhouse gases are expected to continue and a doubling of CO_2 is likely to occur in the next 100 years (IPCC, 1996) unless policies to reduce emissions are implemented. The lack of knowledge on likely responses to increased CO_2 of the many plant species that occur in native pastures (Tothill and Gillies, 1992) limits our capability to forecast future effects to making conservative generalisations (Hall *et al.*, 1998).

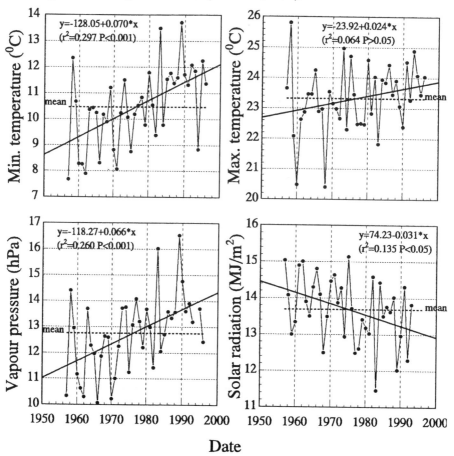

Figure 4. Trends in climatic variables (minimum and maximum temperature, vapour pressure, and solar radiation) in May for Queensland's pastoral/cropping zone (after McKeon *et al.*, 1998).

The behaviour of ENSO since 1976 is regarded as 'unusual' based on the available historical record (IPCC, 1996). The role of increasing greenhouse gas concentrations, and current and expected global warming in changing the behaviour of this global phenomena is the subject of much debate (Nicholls *et al.*, 1996a). The recent 1991-5 El Niño episode (Trenberth and Hoar, 1997) and the reduced frequency of La Niña events between 1976-96 has been associated with below average rainfall in Queensland (Figure 1). The attribution of causes between natural variability and global warming has important implications for the future productivity of Queensland grazing lands and public attitudes towards the seriousness of reducing national and global greenhouse gas emissions (McKeon *et al.*, 1998). The greenhouse gas emissions from the management of Queensland grazing lands are major components of the national greenhouse inventory as both sources through tree clearing and methane emissions, and as sinks through woodland thickening (Burrows, 1995) and soil organic matter (Ash *et al.*, 1996). Thus the assessment of likely changes in ENSO due to increased greenhouse gas emissions (eg. Timmermann *et al.*, 1999) will have impacts on both projected productivity of Queensland's grazing lands and international political agreements on resource management (Burrows, 1995; Sampson *et al.*, 2000).

Because of the importance of ENSO to the climate and production of Queensland's grazing lands we quote at length from a recent evaluation of the impact of future greenhouse emissions on ENSO (Timmermann *et al.*, 1999) using a General Circulation Model (GCM) -

'The tropical Pacific climate system is thus predicted to undergo strong changes if emissions of greenhouse gases continue to increase. The climatic effects will be threefold. First, the mean climate in the tropical Pacific region will change towards a state corresponding to present-day El Niño conditions. It is therefore likely that events typical of El Niño will also become more frequent. Second, a stronger interannual variability will be superimposed on the changes in the mean state, so year-to-year variations may become more extreme under enhanced greenhouse conditions. Third, the interannual variability will be more strongly skewed, with strong cold events (relative to the warmer mean state) becoming more frequent. Although the model was successful in simulating and predicting ENSO, the ENSO response to greenhouse warming may depend on processes that are not well understood, such as cloud feedback, so we cannot exclude the possibility that the results are sensitive to the model formulation.'

The future stability in skill of statistical forecasting systems based on the SOI has been evaluated for Queensland using GCM simulations under increasing greenhouse gas emissions (Suppiah *et al.*, 1998). The results suggest that the correlations between SOI and rainfall are likely to continue, although the relationships will change as may have been already occurring in the last 20 years (Nicholls *et al.*, 1996b). Given the uncertainty associated with future climate change, especially rainfall (Suppiah *et al.*, 1998), the continued monitoring of the performance of forecasting systems should have a high priority (McKeon *et al.*, 1998).

4. Options for Use of Seasonal Climate Forecasting in Grazing Enterprises

There are few historical records of the outcomes of the many management decisions that graziers make. Where they are available, it is not necessarily possible to separate the effects of climate variability from management changes (eg. change in cattle breed). In the case of pasture improvement, an individual grazier's experience may be limited to a few sowings of new species. As a result it is rarely possible to assess the value of climate forecasting by directly comparing historical SOI time series with grazier records. We have adopted the systems approach, similar to that outlined by Hammer (2000), of developing and utilising simulation models of components of the grazing system. The models combine both grazier experience and research trials, to allow the long term historical climate records to be transformed into simulated records of physical, biological and financial variables, such as runoff, plant growth, and cash flow. These links between simulated variables and the forecasting system can then be evaluated and better decision rules derived.

4.1 STOCKING RATE

Most grazing enterprises (wool and cattle) are based on a self-replacing nucleus of breeding animals, which also generate the surplus animals for sale (eg. wethers, cull ewes, steers, cull cows). The choice of target carrying capacity for the property and the year-to-year management of stocking rate (head/ha) have the major impacts on both financial performance of grazing properties (Clark et al., 1992; Buxton and Stafford Smith, 1996) and their resource condition (Tothill and Gillies, 1992). Average stocking rates, or pasture utilisation rates, often vary more than two-fold on properties in the same region (Beale, 1985; Beale et al., 1986; Scanlan et al., 1994; Johnston, 1996). Stocking rate affects pasture availability, which in turn impacts –

- availability of fuel for burning to control weeds,
- level of surface soil cover, which influences runoff, and
- pasture composition and condition

All of these factors combine to a major influence on resource productivity. Because of the major impacts of stocking rate decisions on the performance of grazing enterprises, subsequent papers in this volume (McKeon et al., 2000; Ash et al., 2000; Stafford Smith et al., 2000) concentrate on the use of seasonal forecasting in stocking rate management.

4.2 RISK OF OVERGRAZING

Analysis of grazing trials indicates that the consumption of more that 30% of summer pasture growth results in large decreases in grass basal area of desirable perennial grasses (Orr et al., 1986; McKeon et al., 1990; Orr et al., 1993). In eastern Queensland, especially the north-eastern region, some skill exists in forecasting summer pasture growth (November to March) using SOI from August to October (McKeon et al., 1990). Hence the risk of overgrazing can be forecast for a given stocking rate (McKeon et al., 1990; Hammer et al., 1991; Willcocks et al., 1991). For example, a simulation of pasture growth over 105 years at Charters Towers (1890 to 1995) was used to examine the risks of exceeding 30% utilisation of summer growth when stocking at a constant 'safe' stocking rate (McKeon *et al.*, 1990; Hammer *et al.,* 1991). Over all years the chance of exceeding 30% utilisation was 23%, but this chance nearly doubled in El Niño years (SOI August to October <-5) and was zero in La Niña years (SOI August to October >+5). Hence, spring SOI <-5 provides warning of years when resource damage is more likely to occur so that actions such as early destocking in reaction to drought may be implemented with confidence. Although the SOI-based forecast gives a warning, it has indicated only about half of the years when there have been high risks of pasture degradation. Resource management would be greatly aided by forecasts of major non-ENSO droughts (eg. 1960/1, 1968/9, 1984/5, 1985/6), which have been as damaging economically and ecologically as ENSO-related droughts.

4.3 FORECASTING PRODUCT PRICES

Sensitivity studies (Hall, 1996; Stone and Hall, 1999) with grazing enterprise models confirm the importance of product price for wool and beef on enterprise income. Cattle prices are influenced by state-wide climatic variability on a weekly timescale and changes in export markets, which may also be driven by climatic variability affecting both global and regional crop yields and prices (eg. Chapman *et al.*, 2000). A preliminary analysis of historical beef prices in relation to ENSO indicated some effect of La Niña years on prices (W. Hall, pers. comm.). However, such analyses have to be treated with caution given the major market perturbations that have occurred in export markets since 1974 (Figure 4).

4.4 FORECASTING ANIMAL PRODUCTION

There are few published time series of actual property reproduction and mortality rates, steer growth or wool cut. Nevertheless, the few studies reviewed by Hall *et al.* (1998) support the strong link between climatic indices and animal production.

Simulation of climatic indices showed that annual steer growth and wool cut per head were highly correlated with the proportion of the year that plant growth or soil moisture indices exceeded a given threshold, rather than the total amount of plant growth (Reid and Thomas, 1973; McCown, 1980-1; McCaskill, 1991; McCaskill and McIvor, 1993; Hall *et al.*, 1998). Thus climate forecasting systems that have skill in those seasons with high year-to-year variability in temperature and moisture (eg. spring) are likely to be suitable for forecasting seasonal and annual animal production. For example, SOI based forecasting systems have skill in forecasting variability in winter and spring temperature and moisture variation in many regions of Queensland and hence, not unexpectedly, can discriminate between years of high or low production in wool or simulated steer liveweight gain (Hammer *et al.*, 1991; McKeon *et al.*, 2000). Similarly, O'Rourke *et al.* (1991) found that the pregnancy rate of mature lactating cows in north-eastern Queensland measured in June-July was correlated with the previous August to October SOI (r^2=0.68). Such forecasts are likely to be beneficial in the planning of a large number of animal husbandry (eg. diet supplementation, drought feeding, disease and pest prevention) and marketing decisions.

4.5 PASTURE MANAGEMENT DECISIONS

The success of a large number of pasture management decisions (eg. pasture burning, establishment of improved pastures, fertiliser application, tree killing, native pasture seed collection, rangeland revegetation, and forage cropping) is dependent on favourable seasonal conditions. In the past, these practices have been carried out without the benefit of climate forecasting and hence the chance of success has often been regarded as unreliable, resulting in limited adoption (O'Rourke *et al.*, 1992).

Spring pasture burning in much of Queensland is regarded as a controversial practice (Scattini *et al.*, 1988). There are beneficial effects on animal nutrition and weed control if reasonable rain follows burning. However, if rain does not eventuate, burning may result in problems of feed availability and poor surface soil protection (McIvor *et al.*, 1995). Over much of Queensland, spring rainfall and simulated pasture growth are correlated with the SOI in the previous winter (McBride and Nicholls, 1983; Clewett *et al.*, 1994) and hence some forecasting skill is available in predicting pasture growth following spring burning (McKeon *et al.*, 1990; Willcocks *et al.*, 1991). For example, spring native pasture growth was simulated for 119 years at Gayndah, south-east Queensland (Hammer *et al.*, 1991). At a conservative stocking rate (0.2 beast/ha), 600 kg DM/ha of pasture growth is required for a safe level of utilisation (ie. <30%). Over the 119 years simulated, this level of pasture growth was exceeded in 54% of years. In years when the winter SOI > +5, the probability of exceeding requirements was 81%. When the winter SOI < -5, pasture growth was greater than 600 kg DM/ha in only 36% of years. Thus avoiding burning when the winter SOI < -5 would reduce the risks of overgrazing and low surface cover leading to soil erosion in the following spring.

The success of sowing improved pasture species is regarded as unreliable by graziers (O'Rourke *et al.*, 1992) and limits adoption (Gramshaw *et al.*, 1993). Menke *et al.* (1999) developed a simulation model of legume (*Stylosanthes*) establishment from research data collected across northern Australia (Gramshaw *et al.*, 1993) and validated the model with data from repeated sowings over six years at Brian Pastures, south-east Queensland. Analysis of simulations of repeated sowing over longer time periods using historical rainfall (1958-94) showed the chance of successful establishment was related to phases of the SOI (Table 1). The results suggest that substantial improvement in reliability could be achieved by limiting sowing decisions to years with SOI phase (Stone *et al.*, 1996a) consistently positive or rising.

These examples suggest that there are several aspects of grazing systems where climate forecasting could contribute to better management of climate variability and therefore increase the reliability of outcomes from risky management options. However, the major challenge is to evaluate the use of seasonal climate forecasting in decision making in a whole property context.

Table 1. Establishment of seed sown in November at Brian Pastures, south-east Queensland, classified according to the SOI phase in October (after Menke *et al.*, 1999). Seedling establishment was simulated for Brian Pastures for the period 1958 to 1994 assuming sowing every day in November, giving a total of 1110 simulated sowings. SOI phases are as defined by Stone *et al.* (1996a): I – consistently negative, II – consistently positive, III – rapidly falling, IV – rapidly rising, V – near zero.

Variable	SOI Phase					
	I	II	III	IV	V	All
Number of sowings	270	300	90	60	390	1110
Average establishment (%)	17	26	15	30	22	22
Years with successful establishment (%)	11	34	1	70	26	25

5. Stocking Rate Management

Of the large number of management decisions that graziers make, we regard stocking rate as the most important because of its major impact on financial performance (Gramshaw, 1995) and long-term resource condition (Tothill and Gillies, 1992). Through its effects on pasture availability, stocking rate influences a wide range of other decisions, such as frequency of burning, individual animal performance, impact of drought, weed invasion and the need for supplementation. Thus we review current approaches to stocking rate management in Queensland and other rangeland regions.

There have been major changes in management and animal husbandry over the last century including the use of better adapted animals and improved drought management resulting in a buffering of the impact of climate variability on wool production and cattle turnoff (Gramshaw and Lloyd, 1993). Livestock numbers have increased in the 1970s and 1980s leading to increased pasture consumption, reduced burning, woodland thickening, woody weed invasion and soil loss (Tothill and Gillies, 1992; Burrows, 1995).

Sheep and cattle have physiological attributes that dampen climatic-induced variability in animal nutrition, for example wool growth continues during periods of weight loss (Hall, 1996). Both sheep and cattle have lower metabolic rates during periods of low nutrition (SCA, 1990) and are able to exhibit compensatory growth once nutrition improves (Kidd and McLennan, 1998). Thus sheep and cattle enterprises have the capacity to achieve high utilisation rates of pasture growth even in areas with highly variable climates. Indeed, most short term (<10 years) grazing trials and simulation studies derived from the data show that maximum animal production is achieved at 40-60% utilisation of average annual pasture growth (Day *et al.*, 1997a; McKeon *et al.*, 2000), well above recommended rates of utilisation. However, the interaction of rainfall deficits and 'high' utilisation rates may lead to –

- increased frequency of intervention (drought feeding)
- major perturbations to flock/herd dynamics (increased mortalities, reduced reproduction and destocking), and most importantly

- deterioration and degradation of the grazing resource.

Thus the determination of appropriate livestock numbers and grazing management strategies is regarded as the major issue affecting the long-term use of Queensland's grazing lands (Tothill and Gillies, 1992).

5.1 DEGRADATION EPISODES IN AUSTRALIA'S GRAZING LANDS AND CLIMATE VARIABILITY

Australia's native grazing lands have experienced episodes of obvious degradation including soil loss caused by wind and water, and expansion of woody and herbaceous weeds (Condon *et al.*, 1969). These episodes are characterised by periods of favourable rainfall resulting in a build up of animal numbers including pests such as the rabbit; followed by periods of low rainfall (Beadle, 1948; Williams and Oxley, 1979). Graziers attempted to maintain stock numbers resulting in high levels of utilisation of desirable perennial species (Beadle, 1948). The loss of productivity coupled with financial difficulties has regularly resulted in submissions for government assistance through drought relief, and in extreme conditions, Royal Commissions into the pastoral industries (eg. NSW 1900, South Australia 1927, Western Australia 1940), or government supported property amalgamations (eg. the current South-West Strategy in Queensland – Anon, 1993).

Despite these major perturbations in Australia's native grazing lands, there has been little attempt to enforce stocking rates and penalise resource damage (Gardener *et al.*, 1990). An exception is South Australia where the degradation episodes of the 1930s (Ratcliffe, 1937) resulted in legislation controlling stock numbers and subsequent comprehensive monitoring of every property in the rangelands (R. Tynan, pers. comm.).

An alternative approach (described later) has been developed in south-west Queensland (Johnston *et al.*, 1996a,b) in which the experience of successful graziers is being extrapolated to other properties in the region through models that calculate 'safe' carrying capacity as a function of climate and resource information.

Climate variability plays a major role in the perception of resource deterioration and irreversible degradation. For example, Tothill and Gillies (1992) found approximately 40% of northern Australia was in a deteriorating condition that could be reversed by above-average rainfall. Similarly, Condon (1986a) stated that a substantial proportion of the degradation that occurred in NSW from 1890 to 1945 was halted or repaired by the period of well above average rainfall in the 1950s and 1970s. However, such periods have also contributed to substantial increases in woody weeds where pasture burning was not practised because of preference or over-grazing (Anon, 1969; Condon, 1986b). An important test of the value of seasonal climate forecasting will be to examine to what extent these past degradation episodes could have been avoided by the drought and pasture-degradation alert systems such as that implemented by Carter *et al.*, (2000).

5.2 GAINING THE INSIGHTS TO ACHIEVE A BALANCE

Grazing management science has developed a range of tools that can contribute to understanding and better management of grazing systems. These tools include formal experimental grazing trials (eg. Jones and Sandland, 1974), producer demonstration sites, and simulation models of pastures, animals (Hall, 1996) and grazing enterprises (White, 1978; Holmes, 1995; Stafford Smith *et al.*, 2000). To determine the sensitivity of extensive grazing industries to spatial and temporal variation in climate, a systems approach has been adopted (eg. Campbell *et al.*, 1996, 1997). In this approach the grazing system is represented as a flow of plant dry matter (carbon) and nutrients (nitrogen and phosphorus) that determines both carrying capacity (animals per ha) and production per head (including wool, meat, births and mortalities). The biophysical components of the grazing system are climate, soils, and plant species, which interact to produce the plant dry matter consumed by sheep and cattle for maintenance and production.

Grazing trials have been a key experimental tool used by researchers to measure animal production and derive recommendations for grazing land management (predominantly stocking rate recommendations). However, a recent review (Ash and Stafford Smith, 1996) of stocking rate trials in rangelands has questioned the extent to which data collected on small paddocks with a limited number of animals can be applied to 'real' paddocks or properties.

An alternative to formal grazing trials is the examination of grazing practices on commercial properties. Individual graziers are beginning to document how they have learnt from past difficulties in managing this complex climatic and economic environment (Mann, 1993; Fahey, 1997; Landsberg *et al.*, 1998; Anon, 1999). Surveys of graziers (eg. O'Rourke *et al.*, 1992) have also provided regional data on actual management of grazing enterprises. Grazier groups are also forming to develop property profiles (Clark *et al.*, 1992) and benchmark analyses (Smith, 1998; Kernot, 1999). Similarly, Buxton *et al.* (1995) and Buxton and Stafford Smith (1996) combined grazier data and simulation models in order to allow individual graziers to conduct simulation experiments for their own property.

As the ability to quantify the variability between properties in terms of biophysical resources increases, there will be a greater opportunity for graziers to rapidly evaluate a range of strategies for managing climatic and economic variability. Fundamental to this capability has been the development of computer models to quantify the effects of historical climatic variability on plant and animal production from agronomic and grazing trial data (Fitzpatrick and Nix, 1970; Rose *et al.*, 1972; McCown *et al.*, 1974; McCown, 1980-81; McCaskill and McIvor, 1993; Hall, 1996; Day *et al.*, 1997a). However, limitations to this approach are the lack of suitable property records or an understandable unwillingness of producers to share their property data for fear of the information being used to their disadvantage.

5.3 STRATEGIC MANAGEMENT OF CLIMATIC VARIABILITY WITH 'SAFE' CARRYING CAPACITY ASSESSMENT

The high year-to-year variability in climate and pasture production of Queensland's and Australia's grazing lands has long been recognised as a major source of risk in terms of herd/flock management, financial performance and resource degradation (White, 1978; Gardener *et al.*, 1990; Stafford Smith and Foran, 1992; Tothill and Gillies, 1992). The importance of these issues to graziers is supported by current rural and city media attention given to reporting alternative views of graziers on pasture utilisation (eg. McCosker, 1998; Anon, 1999; Quirk, 1999). We regard this public debate as a useful contribution to what is an extremely sensitive issue.

Some graziers have effectively managed these risks by adopting conservative or 'safe' annual/seasonal stocking rates, or by managing around a 'safe' 'long-term' carrying capacity (Johnston, 1996; Johnston *et al.*, 1996a). The demonstrable evidence that these stocking rates or carrying capacities are 'safe' is provided by the long-term survival of these enterprises and the subjective assessment of property condition relative to more degraded properties (Johnston *et al.*, 1996a,b).

The adoption of 'safe' stocking is likely to result in a range of synergistic benefits that are difficult to separate. The benefits include better individual animal production, higher reproductive and lower mortality rates, better soil conditions in terms of infiltration and organic matter, more opportunities for pasture burning to control woody weeds, higher proportion of desirable perennial grasses and less impact of rainfall deficits (drought) on pasture feed availability (Day *et al.*, 1997a).

The objective calculation of 'safe' long-term carrying capacity has been a major theme of recent work (McKeon *et al.*, 1994; Scanlan *et al.*, 1994; Johnston *et al.*, 1996a,b; Day *et al.*, 1997b). This follows the pioneering work of Condon (1968) and Condon *et al.* (1969) who developed quantitative systems for calculating 'safe' carrying capacities for both western NSW and central Australia (NT). The fundamental component of the method was the relationship between average rainfall and the carrying capacities estimated from regional ratings or benchmark properties. In this way, both successful community experience and climatic and ecological knowledge were combined. The system also included the effects of varying tree density, land systems and resource condition.

This approach has been modified in Queensland by formally simulating these biophysical effects on average pasture growth (Scanlan *et al.*, 1994; Johnston *et al.*, 1996a,b; Day *et al.*, 1997b) with equations developed from the GRASP model (Day *et al.*, 1997a). For three regions of Queensland (south-west, south-east, north-east), similar relationships exist between 'safe' carrying capacity and simulated average annual pasture growth. In these case studies 'safe' carrying capacity was estimated by graziers for the different land systems within their properties (Figure 5). Analysis of the pooled data from these regions (Day *et al.*, 1997b; Hall *et al.*, 1998) showed that simulated pasture growth accounted for 77% of the variation in carrying capacity across a wide range of soil types, land systems, pasture communities, grazing enterprises and climatic zones. When estimates of 'safe' carrying capacity were expressed as the ratio of animal intake per ha to average annual pasture growth (ie. pasture utilisation), 47% of estimated 'safe' carrying capacities fell between 15 and 25% pasture utilisation, and 72% of estimated 'safe' carrying capacities were equivalent to less than 25% utilisation of average annual pasture growth (Figure 6).

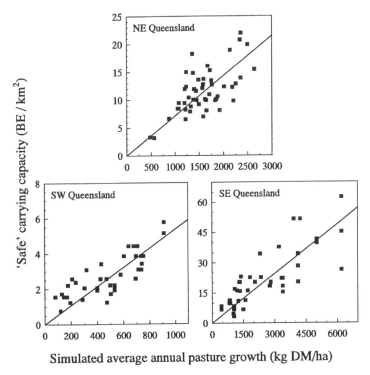

Figure 5. Relationships between estimated 'safe' carrying capacity (400 kg beast equivalents/ha) and simulated average annual pasture growth (kg DM/ha) for land systems in north-east Queensland (Scanlan *et al.* 1994), south-west Queensland (Johnston *et al.* 1996a), and south-east Queensland (Day *et al.* 1997b) (after Hall *et al.,* 1998).

For each region, the regression of pasture growth and dry matter intake was used to calculate their 'safe' utilisation rates: south-west Queensland, 14%; north-east Queensland, 19%; and south-east Queensland, 22%. Hall *et al.* (1998) suggested that this low variation in 'safe' utilisation across the three regions was surprising given that they differ substantially in –

- year-to-year variability in rainfall,
- disappearance rates of dry matter,
- nitrogen dilution by plants,
- frequency of burning to control woody plants and improve animal nutrition,
- animal type (sheep versus cattle), and

214

- length of the plant growing season, expressed as the percentage of days in the period when the calculated pasture growth index exceeded a specific threshold (after McCown, 1980-1 and McCaskill and McIvor, 1993).

To some extent these differences compensate for each other. For example, in the drier south-west region the higher temporal variation in plant growth can be buffered by lower disappearance rates, higher nutrient content of dead plant material and low utilisation rates. The low utilisation rates in the more climatically 'benign' eastern environments provide more opportunities for pasture burning and hence benefits in animal nutrition and pasture management.

Hall *et al.* (1998) expanded the analysis for the 12 cattle producing regions of Queensland identified by O'Rourke *et al.* (1992). Regional safe stocking rates were calculated from estimates of Tothill and Gillies (1992) and pasture growth calculated from spatial data on climate, soils, and vegetation (including tree density) (Carter *et al.*, 2000). Calculated safe utilisation rates ranged from 6% in north Queensland to 30% in coastal areas and were strongly correlated with length of the plant growing season ($r^2 = 0.91$, n = 12), ie. a climatic index of animal nutrition.

These studies suggest that wide variation in 'safe' stocking rates derived from successful grazier experience and expert opinion can be quantified in terms of climate, vegetation and soil resources, and hence successful experience in managing for climatic variability can be extrapolated to other properties.

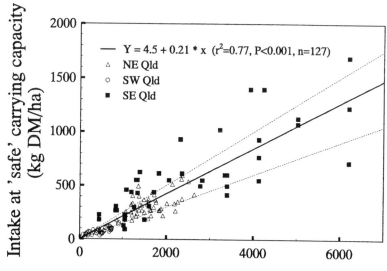

Figure 6. Intake at 'safe' carrying capacity for three regions in Queensland covering a range of annual pasture growth rates. The lower and upper dashed lines represent utilisation rates of 15 and 25% respectively (after Hall *et al.*, 1998).

5.4 APPLICATION IN SOUTH-WEST QUEENSLAND

The approach presented above has been applied in south-west Queensland in response to grazier (Warrego Graziers Association, 1988) and government initiatives towards a more sustainable grazing industry (Anon, 1993). A total of 217 properties have participated in objective carrying capacity assessments from March 1994 to August 1998 representing 37% of the properties in the Murweh, Paroo, Quilpie, Bulloo and Barcoo shires covering a total of 137,439 km^2 (45% of the area of these shires).

Properties ranged greatly in size from 2,376 ha to 866,949 ha. Estimated carrying capacities ranged from 9 Dry Sheep Equivalents per km^2 (DSE/km^2) to 305 DSE/km^2 reflecting the wide range in resource productivity. Where owner livestock figures were available, the range in owner carrying capacities was 9.4 DSE/km^2 to 334 DSE/km^2. On 165 properties the owner/manager provided their own estimates of long-term carrying capacities, either derived from their records or experience. On 66% of properties where owner figures were available, the assessed carrying capacity was within $\pm$ 10% of the owner estimated carrying capacity.

On 23% of properties, the assessed carrying capacity was 10% or more below the owner nominated carrying capacity. Conversely, on 11% of properties the assessed carrying capacity was 10% or more above the owner nominated capacity. The ratio of owner carrying capacity to assessed carrying capacity (range 0.3 – 1.7) was not significantly related to property size, flock size or proportion of 'downs' country on the 165 properties where owner figures were available.

An evaluation of this work was conducted from February to June 1998 to assess the impact of the project within the community and to determine whether any change had occurred. To achieve this a telephone survey of 60 producers was conducted. One third of producers surveyed, who had participated in the project, indicated they had made some change to their management practices following assessment. Two thirds of those making changes believed they had seen some change in pasture condition since implementing the changes. Producers reported that where they agreed with the results, they used the information as a benchmark or reference for making stocking decisions or as a check of what they were presently doing.

There are several major limitations to the adoption of 'safe' carrying capacities:
(i)   Conservation of pasture biomass is unlikely to be successful unless there is control of total grazing pressure, including native and introduced herbivores. Similarly, studies in south-west Queensland suggest that non-domestic herbivores contribute between 30-40% of total grazing pressure (L. Pahl, pers. comm.). Thus graziers see little advantage in reducing stocking rates when native and feral herbivores cannot be controlled. However, where these herbivores have been removed, substantial benefits in production and resource condition have been reported.
(ii)   The decline in prices received for wool and beef over the last 30 years (Figure 4) has resulted in marginal economic viability for many grazing enterprises

(Gramshaw, 1995). Small property sizes are likely to exacerbate the problems of marginal economics preventing the use of lower stocking rates.

Despite these limitations, where graziers have greatly reduced stock numbers (by 30-50%), substantial benefits in resource condition and cash flow have been reported (Landsberg et al., 1998; Fahey, 1997; Anon, 1999). These graziers reported benefits in terms of reducing the impact of droughts on their production systems once they adopted lower stocking rates as well as more opportunities for pasture burning to control weeds.

Because of variation between properties in terms of land, vegetation and climate resources, it is difficult to extrapolate actual stocking rates (beasts/ha) from property to property. The approach of Johnston et al. (1996 a,b) described above, provides a guide as to where an individual grazier may be operating relative to objective assessment of safe carrying capacity. The use of enterprise simulation models (Stafford Smith and Foran, 1992; Holmes, 1995) provides an approach whereby graziers can assess the financial impact of adopting lower stocking rates. For example, Buxton and Stafford Smith (1996) report a study in which a model of a cattle property in the Kimberley region of Western Australia was derived from the grazier's records. A simulation experiment examining the impact of a low (10,000 cows) and high stocking rate (15,000 cows), combined with grazier estimates of the effects on herd reproduction and mortality rates, found the lower stocking rate resulted in increased annual cash flow by 42% over a 40 year period.

5.5 TACTICAL GRAZING MANAGEMENT

The adoption of safe stocking rates involving low levels of pasture utilisation may be limited as some graziers wish to maximise use of available pasture with higher stocking rates in an attempt to increase profitability (Beale, 1985; Beale et al., 1986; Johnston, 1996; McCosker, 1998).

In regions of high climatic variability, graziers using high stocking rates are likely to experience greater impacts of drought when further amplified by resource damage, eg. loss of desirable perennial grass species and increased woody vegetation. Given that such enterprises are likely to require frequent changes in stock numbers, it is possible that management will benefit from the use of flexible grazing strategies including the use of seasonal climate forecasting.

5.5.1 Responding to Pasture Condition – The GrazeOn Example
Flexible grazing strategies in which stock numbers are adjusted each year at the end of the growing season have been tested in grazing trials in highly variable climates such as western Queensland (Beale, 1985; Beale et al., 1986; Orr et al., 1993). Results have shown that the matching of stocking rate to feed supply at low levels of utilisation (20-30%) can maintain pastures in good condition (Orr et al., 1986). Cobon and Buxton (1999) have reported simulations of the application of this approach at a property level. Several properties at Longreach in central west Queensland were studied. The current owner/managers of these properties were interviewed and their management plans,

animal reproduction rates, animal production and financial positions were incorporated into a herd dynamics and property economics model, Herd-Econ (Stafford Smith and Foran, 1992). To include the climate variability in the models, pastoralists were asked to interpret the historical monthly rainfall for nearby rainfall stations. The modelling approach is described in more detail in Buxton *et al.* (1995) and Cobon and Buxton (1999). The current management of the properties was simulated over a 40 (1957-96) or 105 (1893-1997) year period of seasonal rainfall variability and the resulting production and cash flow determined.

A responsive strategy (GrazeOn) was simulated for each property as well as current management. On each property current management closely resembled a conservative or low stocking rate strategy. A GrazeOn strategy adjusts stock numbers relative to total standing dry matter (TSDM) at the end of the pasture growing season. In these case studies the stocking rate adjustments for the GrazeOn strategy were made on the 1st June each year. Furthermore, stocking rates were calculated with in-built safety mechanisms to provide more certainty of the strategy being sustainable in terms of soil and pasture stability. These safety mechanisms included a non-utilised 500 kg/ha pasture threshold and the equivalent of the wet season pasture growth not utilised until the 1st June. Each year in June animals were either purchased or sold according to the stocking rate calculated from the GrazeOn strategy and the property size.

Table 2. Simulated comparison of current management and responsive management (GrazeOn strategy) for two properties at or near Longreach. Entries show the annual mean and co-efficient of variation (in parenthesis) for rainfall and simulated variables. Simulations were conducted for the period 1893-1997.

Variable	Current Management		Responsive Management	
	Property A	*Property B*	*Property A*	*Property B*
Rainfall (mm/yr)	431 (46%)	431 (46%)	431 (46%)	431 (46%)
Pasture growth (kg/yr)	1791 (67%)	1791 (67%)	1791 (67%)	1791 (67%)
GFW[1] (kg/hd)	3.8 (7%)	4.1 (12%)	3.5 (11%)	3.3 (16%)
Total wool (kg)	25200 (19%)	20700 (17%)	30200 (34%)	23800 (53%)
Stocking rate (hd/ha)	0.63 (14%)	0.61 (12%)	0.7 (27%)	0.83 (47%)
ACS[2] ($)	38300 (48%)	39500 (43%)	60800 (69%)	54700 (116%)

[1] Greasy fleece weight
[2] Accumulated cash surplus

Table 2 shows the coefficient of variation for wool growth on the two Longreach properties to be relatively low (7-19%) compared to annual rainfall (46%), pasture growth (67%) and accumulated cash surplus (ACS) (43-48%). The responsive (GrazeOn) management strategy on these same properties improves production and profit by making better use of the variable pasture supply. As a result, the variability of production parameters increases, mean wool production per head decreases, and mean stocking rate and total wool production increase. The responsive stocking strategy significantly increased the average profit of grazing enterprises compared to a conservative stocking strategy (by 24 and 73% on these two western Queensland properties). This is achieved at 22% utilisation of total standing pasture dry matter over all seasons, which converts to a stocking rate that is higher than the conservative rate (44 and 74%) following good seasons but lower than it following bad seasons (36 and 53%). Taking advantage of the good seasons generates most of the extra profit.

5.5.2 *Time Controlled Grazing*

An alternative form of flexible stocking rate management is that proposed by McCosker (1993) in which animals are rapidly rotated through small paddocks, often called 'cell grazing' (Norton, 1998) or 'time controlled grazing' (TCG, McCosker, 1993). Stock numbers are adjusted by monitoring resource and vegetation attributes and comparing grazing days with rainfall (McArthur, 1998). Advocates for these intensive management systems list benefits in terms of increased profitability, animal production (per unit area), better plant production and resource condition (McCosker, 1993).

Experienced pasture/animal researchers (Jones, 1993; Morley, 1995; Quirk, 1999) who have reviewed comparisons of cell grazing with continuous grazing are less certain that there is evidence of these benefits. McCosker (1994) draws attention to the disparity between research and practical experience and provides preliminary results from a property in central Queensland suggesting a 40% increase in live weight gain per ha following the adoption of TCG with only a marginal increase in stocking rate (5%). McCosker (1994) highlights the difficulty of researching TCG because of the managerial components involved in using the system at a property scale and the rapid evolution of the approach. Similar research difficulties exist for the application of other approaches to grazing management at a property scale. Property simulation models (Holmes, 1995; Buxton and Stafford Smith, 1996) appear to be acceptable to graziers as a way of comparing systems but considerable improvement in modelling nutrient cycling, degradation processes and diet selection will be required to accurately represent the principles underlying TCG (McKeon *et al.*, 2000).

5.5.3 *Tactical Rest*

Several studies have shown that desirable perennial grasses die more quickly when the stresses of drought and grazing are combined (Scattini, 1973; Orr *et al.*, 1986; Orr *et al.*, 1993; Hodgkinson, 1995). Hodgkinson (1996) has proposed tactical management decision rules in which paddocks are deliberately spelled when certain criteria are reached, i.e. rainfall less than critical thresholds and average grazed height of desired perennial grasses less than 10 cm. The effect of adopting these rules is currently being tested in 10 sites in western Queensland and western New South Wales.

5.5.4 *Matching Stocking Rate to Feed Availability*

One of the important features of TCG and responsive management ('GrazeOn') is the matching of stocking rate to carrying capacity (McArthur, 1998; Cobon and Buxton, 1999). These tactical approaches differ in the frequency of stocking rate decision, subdivision of property and frequency of stock movements. Theoretical simulations of matching stocking rate to pasture growth suggest considerable advantages above 'safe' constant stocking rates can be achieved in animal production per unit area (McKeon *et al.*, 2000). However, it is problematic whether this potential advantage can be translated into cash flow for conventional beef cattle herds because of constraints in buying and selling (Stafford Smith *et al.*, 2000). McKeon *et al.* (2000) point out that the more plant material that is consumed for animal production, the less there is available for maintaining soil organic matter, soil surface protection, or fuel for burning, which increases the likelihood of death of perennial plants (Scattini, 1973; Orr *et al.*, 1986; Gardener *et al.*, 1990; McKeon *et al.*, 1990; Orr *et al.*, 1993). Hence McKeon *et al.* (2000) suggest that their own theoretical simulations of potential animal production benefits are of limited value because the deleterious feedbacks of increased soil loss and reduced frequency of burning on pasture production have not been included. The history of Australia's rangelands suggests that many of these degradation and resilient processes are episodic in nature (e.g. Gardener *et al.*, 1990; Walker, 1993) and short-term experimental trials or experience may not cover the extremes of climatic variation that lead to rapid changes. They warn that unless these long term effects on resource productivity are included in simulation studies, experimental grazing trials, or in the analysis of producer innovations, incorrect conclusions on the production advantages of high pasture utilisation are likely to occur.

The examples described above show that there is a range of tactical grazing management approaches being trialed on properties in Queensland. A key component of these approaches is the matching of stock numbers to rainfall, pasture availability and condition of the resource (Quirk, 1999). Simulations of property performance (Cobon and Buxton, 1999) and property demonstration (McCosker, 1994) suggest financial as well as resource benefits. These approaches would be amenable to including seasonal climatic forecasting as part of tactical stocking rate management rules providing there is skill in forecasting at the time of decision making.

6. Adoption of Seasonal Climate Forecasting in Grazing Management

Given the marginal economic environment in which much of the grazing industry operates, it is not unexpected that graziers report that the perceived unreliability of many improved practices prevents their rapid adoption. High climate variability contributes to this cautious judgement (O'Rourke *et al.*, 1992). It might be reasoned that the use of seasonal forecasting in decision making increases the chance of success and should contribute to greater confidence in adopting new approaches to management. However, seasonal climate forecasting could also be judged as an unreliable innovation especially given the media's past record of giving high profile coverage to graziers who have made decisions that were regarded as unsuccessful because expected seasonal conditions did not occur. Such occurrences are inevitable with probabilistic forecasts and hence it is not surprising that Ash *et al.* (2000) report cautious attitudes towards the use of seasonal forecasting within the grazing community.

The attitudes of graziers to the use of seasonal forecasting are likely to be changing as they build up experience in accessing information and seeing the results of their own decisions. Paull *et al.* (1999) report the results of a small survey of 65 Queensland farmers/graziers conducted in 1998. Respondents were mostly involved with beef cattle (86%) and reported that:

- the most important decisions were selling/agisting stock (83%), buying stock (44%), sowing crops/pastures (47%), burning pastures (34%) and weed/disease/pest control (34%);
- judgement of future climatic condition was moderately (29%) to very (56%) important in decision making;
- 58% agreed that climate forecasts are better expressed in terms of probabilities and that probability information was regarded as moderately (37%) to very (12%) useful;
- 45% currently use seasonal climate forecasts; and
- 38% said they would use seasonal climate forecasts to decide what number of stock to carry through periods of feed shortage.

Individual comments highlighted the lack of accuracy of recent forecasts, confirming that necessarily probabilistic forecasts, such as those based on SOI, may not satisfy the needs of 40 to 50% of the grazing community. Given that graziers have only had access to seasonal forecasting information since 1991, the reported rate of adoption (35-45%) is relatively high. Other papers in this volume (e.g. Clewett *et al.* 2000; Meinke and Hochman, 2000) describe extension and action research programs designed to address the issues of education about, and application of, probabilistic seasonal climate forecasting information.

Conclusion

Grazing industries based on sheep and beef cattle have been subjected to decline in commodity prices directly affecting profitability of grazing enterprises (Gramshaw, 1995; Hall, 1996; Anon, 1998). Graziers have had to manage for high climatic variability against this background. The adoption of high stocking rates to increase short-term profitability has resulted in long-term increases in risks of resource degradation (Tothill and Gillies, 1992). Seasonal climate forecasting appears to have a role in reducing these risks and hence in contributing to more sustainable grazing systems (McKeon *et al.*, 1990). However, the value of using probabilistic forecasts (i.e. not 100% accurate) has to be compared with other flexible grazing management strategies, such as responsive management ('GrazeOn'), 'time controlled grazing' and 'tactical rest', which respond to known conditions of pasture availability and resource condition.

There are a large number of enterprise decisions that may benefit from the use of seasonal climate forecasting. Simulation modelling would appear to be the only alternative to evaluate the impact of these decisions on the whole property. Models of properties are being developed from grazier records and simulation results are becoming accepted by graziers, even for such sensitive issues as the objective assessment of property carrying capacity. This latter success indicated the benefit of combining grazier experience and ecological models developed from fifty years of pasture and grazing experiments. The further combination of grazier knowledge and grazing science can only accelerate this process of providing the best information for better grazing management.

Acknowledgments

We would like to thank John McIvor, Mark Howden and Ken Rickert for their comments on the manuscript.

References

Allan, R., Lindesay, J. and Parker, D. (1996) *El Niño Southern Oscillation and Climatic Variability*. CSIRO, Melbourne.

Anon (1969) Report of the interdepartmental committee on scrub and timber regrowth in the Cobar-Byrock district and other areas of the Western Division of New South Wales. Government Printer, Sydney.

Anon (1993) Mulga Region-A study of the inter-dependence of the environment, pastoral production and the economy. Queensland Department of Lands, Brisbane.

Anon (1996) Surviving the Drought - 'Returning to Profitability'. Attachment to Submission No. 00585. Queensland Department of Primary Industries, Brisbane.

Anon (1998) Australian Farm Surveys Report 1998. Australian Bureau of Agricultural and Resource Economica, Canberra.

Anon (1999) Stock cuts beat debt cycle. Queensland Country Life, 1 March 1999.

Ash, A.J. and Stafford Smith, M. (1996) Evaluating stocking rate impacts in rangelands: Animals don't practice what we preach. *Rangeland Journal* **18**, 216-43.

Ash, A.J., Howden, S.M. and McIvor, J.G. (1996) Improved rangeland management and its implications for carbon sequestration. Proceedings of the Fifth International Rangeland Congress, Salt Lake City, Utah, USA, 23-28 July, 1995, Volume 1, pp. 19-20.

Ash, A.J., O'Reagain, P.J., McKeon, G.M. and Stafford Smith, M. (2000) Managing climate variability in grazing enterprises: A case study of Dalrymple Shire, in G.L. Hammer, N. Nicholls, and C. Mitchell (eds.), *Applications of Seasonal Climate Forecasting in Agricultural and Natural Ecosystems – The Australian Experience*. Kluwer Academic, The Netherlands. (this volume)

Beadle, N.C.W. (1948) The vegetation and pastures of western New South Wales with special reference to soil erosion. Department of Conservation of New South Wales, Sydney.

Beale, I.F. (1985) Animal plant interactions in native pastures of western Queensland. Final Report to the Australian Wool Corporation for Project W.R.T.F K/2/900C. Queensland Department of Primary Industries, Brisbane.

Beale, I.F., Orr, D.M., Holmes, W.E., Palmer, N., Evenson, C.J. and Bowly, P.S. (1986) The effect of forage utilisation levels on sheep production in the semi arid south west of Queensland, in 'Rangelands: A Resource Under Siege'. Proceedings of the Second International Rangeland Congress, Adelaide. p.30.

Burrows, W.H. (1995) Greenhouse revisited - land-use change from a Queensland perspective. Climate Change Newsletter **7**, 6-7.

Buxton, R. and Stafford Smith, M. (1996) Managing drought in Australia's rangelands: four weddings and a funeral. *Rangeland Journal* **18**, 2292-308.

Buxton, R., Cobon, D.H. and Stafford Smith, M.D. (1995) Drought Plan: Developing with graziers, profitable and sustainable strategies to manage for rainfall variability. Regional Report No 1: Longreach/Richmond, western Queensland. CSIRO, Alice Springs.

Campbell, B.D., McKeon, G.M., Gifford, R.M., Clark, H., Stafford Smith, D.M., Newton, P.C.D. and Lutze, J.L. (1996) Impacts of atmospheric composition and climate change on temperate and tropical pastoral agriculture, in. W.J. Bouma, G.I. Pearman, and M.R. Manning (eds.), *Greenhouse: Coping with Climate Change*. CSIRO, Australia. pp.171-89.

Campbell, B.D., Stafford Smith, D.M. and McKeon, G.M. (1997) Elevated CO_2 and water supply interactions in grasslands: A pastures and rangelands management perspective. *Global Change Biology* **3**, 177-87.

Carter, J.O., Hall, W.B., Brook, K.D., McKeon, G.M., Day, K.A. and Paull, C.J. (2000) Aussie GRASS: Australian grassland and rangeland assessment by spatial simulation, in G.L. Hammer, N. Nicholls, and C. Mitchell (eds.), *Applications of Seasonal Climate Forecasting in Agricultural and Natural Ecosystems – The Australian Experience.* Kluwer Academic, The Netherlands. (this volume)

Chapman, S.C., Imray, R.J. and Hammer, G.L. (2000) Can seasonal climate forecasts predict movements in grain prices? in G.L. Hammer, N. Nicholls, and C. Mitchell (eds.), *Applications of Seasonal Climate Forecasting in Agricultural and Natural Ecosystems – The Australian Experience.* Kluwer Academic, The Netherlands. (this volume)

Clark, R.A., Davis, G.P., Cheffms, R.C. and Esdale, C.R. (1992) Integrating cattle breeding technologies into beef property management. Proceedings of the Australian Association of Animal Breeding and Genetics **10**, 345-8.

Clewett, J.F., Clarkson, N.M., Owens, D.T. and Abrecht, D.G. (1994) Australian Rainman: Rainfall information for better management. Queensland Department of Primary Industries, Brisbane.

Clewett, J.F., Cliffe, N.O., Drosdowsky, L.M., George, D.A., O'Sullivan, D.B., Paull, C.J., Partridge, I.J. and Saal, R.L. (2000) Building knowledge and skills to use seasonal climate forecasts in property management planning, in G.L. Hammer, N. Nicholls, and C. Mitchell (eds.), *Applications of Seasonal Climate Forecasting in Agricultural and Natural Ecosystems – The Australian Experience.* Kluwer Academic, The Netherlands. (this volume)

Clewett, J.F., Howden, S.M., McKeon, G.M and Rose, C.W. (1991) Use of systems analysis and the Southern Oscillation Index to optimise management of grain sorghum production from a farm dam irrigation system, in R.C. Muchow and J.A. Bellamy (eds.), *Climatic Risk in Crop Production: Models and Management for the Semiarid Tropics and Subtropics.* CAB International. pp. 307-328.

Cobon, D.H. and Buxton, R. (1999) Safe stocking rates based on pasture supply and demand for the Mitchell grasslands in central and north-west Queensland. I. Development of a model and likely economic implications. *Rangeland Journal* (in press).

Condon, R.W. (1968) Estimation of grazing capacity on arid grazing lands, in G.A. Stewart (ed.), Land Evaluation: Papers of a CSIRO symposium organised in cooperation with UNESCO. CSIRO Division of Land Research, Canberra. pp. 112-124.

Condon, R.W. (1986a) Recovery from catastrophic erosion in western New South Wales, in 'Rangelands: A Resource Under Siege'. Proceedings of the Second International Rangeland Congress, Adelaide. p.39.

Condon, R.W. (1986b) Scrub invasion on semi-arid grazing lands - causes and effects, in 'Rangelands: A Resource Under Siege'. Proceedings of the Second International Rangeland Congress, Adelaide. p.40.

Condon, R.W., Newman, J.C. and Cunningham, G.M. (1969) Soil erosion and pasture degeneration in Central Australia. Part III -The assessment of grazing capacity. *Journal of the Soil Conservation Service of New South Wales* **25**, 225-50.

Daly, J.J. (1983) The Queensland beef industry from 1930 to 1980: Lessons from the past. *Queensland Agricultural Journal* **109**, 61-97.

Daly, J.J. (1994) *Wet as a Shag, Dry as a Bo*ne. Queensland Department of Primary Industries, Brisbane.

Day, K.A., McKeon, G.M. and Carter, J.O. (1997a) Evaluating the risks of pasture and land degradation in native pasture in Queensland. Final report for Rural Industries and Research Development Corporation - Project DAQ124A. Queensland Department of Natural Resources, Brisbane.

Day, K.A., Scattini W.J. and Osborne, J.C. (1997b) Extending carrying capacity calculations to the central Burnett region of Queensland, in D.M. Stafford Smith, J.F. Clewett, A.D. Moore, G.M. McKeon and R. Clark (eds.), DroughtPlan Working Paper No. 10. CSIRO, Alice Springs.

Duff, G.A., Berryman, C.A. and Eamus, D. (1994) Growth, biomass allocation and foliar nutrient content of two *Eucalyptus* species of the wet-dry tropics of Australia grown under CO_2 enrichment. *Functional Ecology* **8**, 502-08.

Fahey, D. (1997) Changing grazing management on "Keen-Gea" – the strategies and findings – a producers experience. An Introduction to Management Practices in the Dalrymple Shire. Grazing Land Management Unit, Queensland Department of Primary Industries, Charters Towers. pp.16-17.

Fitzpatrick, E.A. and Nix, H.A. (1970) The climatic factor in Australian grassland ecology, in R.M. Moore (ed.), *Australian Grasslands.* Australian National University Press, Canberra. pp. 3-26.

Gardener, C.J., McIvor, J.G. and Williams, J. (1990) Dry tropical rangelands: Solving one problem and creating another. *Proc. Ecol. Soc. Aust.* **16**, 279-90.

Gibbs, W.J. and Maher, J.V. (1967) Rainfall deciles as drought indicators. Bulletin 48, Bureau of Meteorology, Melbourne. pp.118.

Gifford, R.M. (1997) Effects of vegetation on climate and atmospheric CO_2 and climate change, in *Australians and Our Changing Climate.* Australian Academy of Science, Canberra. pp. 42-47.

Gillard, P. and Monypenny, R. (1990) A decision support model to evaluate the effects of drought and stocking rate on beef cattle properties in northern Australia. *Agricultural Systems* **34**, 37-52.

Gramshaw, D. (ed.) (1995) *Integrated Management for Sustainable Forage-based Livestock Systems in the Tropics.* Queensland Department of Primary Industries, Brisbane.

Gramshaw, D. and Lloyd, D. (1993) *Grazing the North: Creating Wealth and Sustaining the Land.* Queensland Department of Primary Industries, Brisbane.

Gramshaw, D., McKeon, G.M. and Clem, R.L. (1993) Tropical pasture establishment. 1. A systems perspective of establishment illustrated by legume oversowing in the subtropics. *Tropical Grasslands* **27**, 261-75.

Hall, W.B. (1996) Near-real time financial assessment of the Queensland wool industry on a regional basis. Unpub. Ph.D. Thesis, University of Queensland, Brisbane.

Hall, W.B., McKeon, G.M., Carter, J.O., Day, K.A., Howden, S.M., Scanlan, J.C., Johnston, P.W. and Burrows, W.H. (1998) Climate change in Queensland's grazing lands: II. An assessment of the impact on animal production from native pastures. *Rangeland Journal* **20**, 177-205.

Hammer, G.L. (2000) A general systems approach to applying seasonal climate forecasts, in G.L. Hammer, N. Nicholls, and C. Mitchell (eds.), *Applications of Seasonal Climate Forecasting in Agricultural and Natural Ecosystems – The Australian Experience.* Kluwer Academic, The Netherlands. (this volume)

Hammer, G.L., McKeon, G.M., Clewett, J.F. and Woodruff, D.R. (1991) Usefulness of seasonal climate forecasts in crop and pasture management. Proc. First Australian Conference on Agricultural Meteorology. Bureau of Meteorology, Melbourne. pp.15-23.

Heathcote, R.L. (1965) *Back of Bourke - Study of Land Appraisal and Settlement in Semi-arid Australia.* Melbourne University Press, Melbourne.

Heathcote, R.L. (1994) Australia, in M.H. Glantz (ed.), *Drought Follows the Plow: Cultivating Marginal Areas.* Cambridge University Press, Cambridge, U.K. pp.91-102.

Hodgkinson, K. (1995) A model for perennial grass mortality under grazing. Proceedings of the Fifth International Rangeland Congress, Salt Lake City, Utah, USA. 23-28 July, 1995. pp.240-1.

Hodgkinson, K. (1996) Tactical rest for rangeland management in a scientist/grazier project. CSIRO Division of Wildlife and Ecology, Canberra.

Holmes, W.E. (1995) Breedcow and Dynama herd budgeting software package: Version 4.00 for DOS. Queensland Department of Primary Industries, Townsville, Queensland, October 1995.

Howden, S.M., McKeon, G.M., Walker, L., Carter, J.O., Conroy, J.P., Day, K.A., Hall, W.B., Ash, A.J. and Ghannoum, O. (1999) Global change impacts on native pastures in south-east Queensland, Australia. *Environmental Modelling and Software* **14**, 307-16.

IPCC (1996) Technical Summary, in J.T. Houghton, L.G. Meira Filho, B.A. Callander, N. Harris, A. Kattenberg and K. Maskell (eds*.*), *Climate Change 1995: The Science of Climate Change.* Contribution of Working Group 1 to the Second Scientific Assessment Report of the Intergovernmental Panel on Climate Change. Cambridge University Press, Cambridge. pp.13-49.

Ivory, D.A. and Whiteman, P.C. (1978) Effect of temperature on growth of five subtropical grasses. I. Effect of day and night temperature on growth and morphological development. *Australian Journal of Plant Physiology* **5**, 86-99.

Johnston, P.W. (1996) Grazing capacity of native pastures in the Mulga Lands of south western Queensland: A modelling approach. Unpub. Ph.D. Thesis, University of Queensland, Brisbane.

Johnston, P.W., McKeon, G.M. and Day, K.A. (1996a) Objective "safe" grazing capacities for south-west Queensland Australia: development of a model for individual properties. *Rangeland Journal* **18**, 244-58.

Johnston, P.W., Tannock, P.R. and Beale, I.F. (1996b) Objective "safe" grazing capacities for south-west Queensland Australia: a model application and evaluation. *Rangeland Journal* **18**, 259-69.

Jones, R.J. and Sandland, R.L. (1974) The relation between animal gain and stocking rate. *Journal of Agricultural Science* **83**, 335-42.

Jones, R.M. (1993) A review of time – controlled grazing, in 'Will cells sell?', Proceedings of Grazing Systems Seminar. Queensland Branch of Soil and Water Conservation Association of Australia. Brisbane. pp.47-55.

Kernot, J. (1999) Benchmarking in north Queensland. Queensland Country Life, 1 April, pp.39.

Kidd, J.F. and McLennan, S.R. (1998) Relationship between liveweight change of cattle in the dry season in northern Australia and growth rate in the following wet season. *Animal Production in Australia* **22**, 363.

Landsberg, R.G., Ash, A.J., Shepherd, R.K. and McKeon, G.M. (1998) Learning from history to survive in the future: Management evolution on Trafalgar Station, north-east Queensland. *Rangeland Journal* 20, 104-18.

Lilley, G.W. (1973) *The Story of Lansdowne: The History of a Western Queensland Sheep Station.* Lansdowne Pastoral Co. Ltd., Melbourne.

Lough, J.M. (1991) Rainfall variation in Queensland, Australia 1891-1986. *International Journal of Climatology* 11, 745-68.

Mann, T.H. (1993) Flexibility – the key to managing a northern beef property. Proceedings of the XVII International Grassland Congress. pp.1961-4.

McArthur, S. (1998) Practical evidence supports cell grazing benefits. Australian Farm Journal: Beef, 8-9.

McBride, J.L. and Nicholls, N. (1983) Seasonal relationships between Australian rainfall and the Southern Oscillation. *Monthly Weather Review* 111, 1998-2004.

McCaskill, M.R. (1991) Prediction of cattle growth rates in northern Australia from climate information. Proc. First Australian Conference on Agricultural Meteorology, Bureau of Meteorology, Melbourne. pp. 85-8.

McCaskill, M.R. and McIvor, J.G. (1993) Herbage and animal production from native pastures and pastures oversown with *Stylosanthes hamata* 2. Modelling studies. *Australian Journal of Experimental Agriculture* 33, 571-9.

McCosker, M. (1998) Greens wrong on grazing pasture value. Queensland Country Life, 6 August, pp1.

McCosker, T. (1993) The principles of time control grazing. The 3rd National Beef Improvement Association Conference. Australia. pp.87-95.

McCosker, T. (1994) The dichotomy between research results and practical experience with time control grazing. Australian Rural Science Annual 1994. pp.26-31.

McCown, R.L. (1980-81) The climatic potential for beef cattle production in tropical Australia: Part 1 - Simulating the annual cycle of liveweight change. *Agricultural Systems* 6, 303-17.

McCown, R.L., Gillard, P. and Edye, L.A. (1974) The annual variation in yield of pastures in the seasonally dry tropics of Queensland. *Australian Journal of Experimental Agriculture and Animal Husbandry* 14, 328-33.

McIvor, J.G., Williams, J. and Gardener, C.J. (1995) Pasture management influences run-off and soil movement in the semi-arid tropics. *Australian Journal of Experimental Agriculture* 35, 55-65.

McKeon, G.M. and White, D.H. (1992) El Niño and better land management. *Search* 23, 197-200.

McKeon, G.M., Ash, A.J., Hall, W.B. and Stafford Smith, D.M. (2000) Simulation of grazing strategies for beef production in north-east Queensland, in G.L. Hammer, N. Nicholls, and C. Mitchell (eds.), *Applications of Seasonal Climate Forecasting in Agricultural and Natural Ecosystems – The Australian Experience.* Kluwer Academic, The Netherlands. (this volume)

McKeon, G.M., Brook, K.D., Carter, J.O., Day, K.A., Howden, S.M., Johnston, P.W., Scanlan, J.C. and Scattini, W.J. (1994) Modelling utilisation rates in the black speargrass zone of Queensland. Proceeding of 8th Australian Rangelands Conference, Katherine, Australia. pp.128-32.

McKeon, G.M., Day, K.A., Howden, S.M., Mott, J.J., Orr, D.M., Scattini, W.J. and Weston, E.J. (1990) Northern Australian savannas: management for pastoral production. *Journal of Biogeography* 17, 355-72.

McKeon, G.M., Hall, W.B., Crimp, S.J., Howden, S.M., Stone, R.C., and Jones, D.A. (1998) Climate change in Queensland's grazing lands. I. Approaches and climatic trends. *Rangeland Journal* 20, 151-76.

Meinke, H. and Hochman, Z. (2000) Using seasonal climate forecasts to manage dryland crops in northern Australia, in G.L. Hammer, N. Nicholls, and C. Mitchell (eds.), *Applications of Seasonal Climate Forecasting in Agricultural and Natural Ecosystems – The Australian Experience.* Kluwer Academic, The Netherlands. (this volume)

Menke, N., McKeon. G.M., Hansen, V., Quirk, M.F. and Wilson, B. (1999) Assessing the climatic risk for *Stylosanthes* establishment in south-east Queensland. Proceedings of the VI International Rangelands Congress, Townsville, Australia. July 1999.

Morley, F. (1995). Comparing grazing systems. *Australian Society of Animal Production Federal Newsletter* 72, 4-6.

Mott, J.J., Tothill, J.C. and Weston, E.J. (1981) Animal production for the native woodlands and grasslands of northern Australia. *Journal of the Australian Institute of Agricultural Science* 47, 132-41.

Nicholls, N., Gruza, G.V., Jouzel, J., Karl, T.R., Ogallo, L.A. and Parker, D.E. (1996a) Observed climate variability and change, in J.T. Houghton, L.G. Meira Filho, B.A. Callander, N. Harris, A. Kattenberg and K. Maskell (eds.), *Climate Change 1995: The Science of Climate Change.* Contribution of

Working Group 1 to the Second Scientific Assessment Report of the Intergovernmental Panel on Climate Change. Cambridge University Press, Cambridge, U.K. pp.137-92.

Nicholls, N., Lavery, B., Fredericksen, C., Drosdowsky, W. and Torok, S. (1996b) Recent apparent changes in relationships between the El Niño-Southern Oscillation and Australian rainfall and temperature. *Geophysical Research Letters* **23**, 3357-60.

Norton, B.E. (1998) The application of grazing management to increase sustainable livestock production. *Animal Production in Australia* **22**, 15-26.

O'Meagher, B., Du Pisani, L.G. and White, D.H. (1998) Evaluation of drought policy and related science in Australia and South Africa. *Agricultural Systems* **57**, 231-58.

O'Rourke, P.K., Doogan, V.J., Entwistle, K.W., Fordyce, G. and Holroyd, R.G. (1991) Early seasonal indicators to aid management of cattle properties in north Australia. Proc. First Australian Conference on Agricultural Meteorology. Bureau of Meteorology, Melbourne. pp..81-4.

O'Rourke, P.K., Winks, L. and Kelly, A.M. (1992) North Australia Beef Producer Survey 1990. Queensland Department of Primary Industries and the Meat Research Corporation, Brisbane.

Orr, D.M., Bowly, P.S. and Evenson, C.J. (1986) Effects of grazing management on the basal area of perennial grasses in *Astrebla* grassland, in 'Rangelands: A Resource Under Siege'. Proceedings of the Second International Rangeland Congress, Adelaide, pp.56-7.

Orr, D.M., Evenson, C.J., Lehane, J.K., Bowly, P.S. and Cowan, D.C. (1993) Dynamics of perennial grasses in *Acacia aneura* woodlands in south-west Queensland. *Tropical Grasslands* **27**, 87-93.

Patterson, D.T., Flint, E.P. and Beyers, J.L. (1984) Effects of CO_2 enrichment on competition between a C_4 weed and a C_3 crop. *Weed Science* **32**, 101-5.

Paull, C.J., O'Sullivan, D.B. and Cliffe, N.O. (1999) Using spatial seasonal information products in sustainable rangeland management. Proceedings of the VI International Rangelands Congress, Townsville, Australia. July, 1999.

Payne, W.L. and McLean, W.M. (1939) Report of the wool advisory commission appointed to inquire into the economic condition of the wool industry in Queensland. Queensland Government Printer, Brisbane.

Pittock, A.B. (1975) Climate change and the patterns of variation in Australian rainfall. *Search* **6**, 498-504.

Power, S., Tsietkin, F., Mehta, V., Lavery, B., Torok, S. and Holbrook, N. (1998) Decadal climate variability in Australia during the 20th century. BMRC Research Report No. 67. Bureau of Meteorology, Melbourne.

Quirk, M. (1999) Cell grazing debate: Producers need full story on grazing management. Queensland Country Life, 29 April 1999.

Ratcliffe, F.N. (1937) Further observations on soil erosion and sand drift, with special reference to south-western Queensland. Council for Scientific and Industrial Research, Pamphlet No. 70, Melbourne.

Reid, G.K.R. and Thomas, D.A. (1973) Pastoral production, stocking rate and seasonal conditions. *Quarterly Review of Agricultural Economics* **26**, 217-27.

Rose, C.W., Begg, J.E., Byrne, G.F., Torssell, B.W.R. and Goncz, J.H. (1972). A simulation model of growth: field environment relationships for Townsville Stylo (*Stylosanthes humilis* H.B.) pasture. *Agricultural Meteorology* **10**, 161-83.

SCA (1990) *Feeding Standards for Australian Livestock: Ruminants*. CSIRO, Melbourne.

Scanlan, J.C., Hinton, A.W., McKeon, G.M., Day, K.A. and Mott, J.J. (1994) Estimating safe carrying capacities of extensive cattle-grazing properties within tropical, semi-arid woodlands of north-eastern Australia. *Rangeland Journal* **16**, 64-76.

Scattini, W.J. (1973) A model for beef cattle production from rangeland and sown pasture in south eastern Queensland, Australia. Unpub. Ph.D. thesis, University of California, Berkley.

Scattini, W.J., Orr, D.M., Miller, C.P., Holmes, W.E. and Hall, T.J. (1988) Managing native pastures, in W.H. Burrows, J.C. Scanlan, and M.T. Rutherford (eds.), *Native Pastures in Queensland, The Resources and Their Management*. Queensland Department of Primary Industries, Brisbane. pp. 52-71.

Smith, P.C. (1998) Benchmarking – the buzz word – Handle with care. *Northern Muster* **62**, 11-12.

Stafford Smith, D.M. and Foran, B.D. (1992) An approach to assessing the economic risk of different drought management tactics on a South Australian pastoral sheep station. *Agricultural Systems* **39**, 83-105.

Stafford Smith, D.M. and McKeon, G.M. (1998) Assessing the historical frequency of drought events on grazing properties in Australian rangelands. *Agricultural Systems* **57**, 271-99.

Stafford Smith, M., Buxton, R., McKeon, G.M. and Ash, A. (2000) Seasonal climate forecasting and the management of rangelands: Do production benefits translate into enterprise profits? in G.L. Hammer, N. Nicholls, and C. Mitchell (eds.), *Applications of Seasonal Climate Forecasting in Agricultural and Natural Ecosystems – The Australian Experience*. Kluwer Academic, The Netherlands. (this volume)

Stone, G.S. and Hall, W.B. (1999) Extra calves: are they worth it? Proceedings of the VI International Rangelands Congress, Townsville, Australia. July, 1999.

Stone, R.C., Hammer, G.L. and Marcussen, T. (1996a) Prediction of global rainfall probabilities using phases of the Southern Oscillation Index. *Nature* **384**, 252-5.

Stone, R.C., Nicholls, N. and Hammer, G.L. (1996b) Frost in NE Australia: trends and influence of phases of the Southern Oscillation. *Journal of Climate* **9**, 1896-1909.

Suppiah, R., Hennessy, K., Hirst, T., Jones, R., Katzfey, J., Pittock, B., Walsh, K., Whetton, P. and Wilson, S. (1998) Climate change under enhanced greenhouse conditions in northern Australia. Third Annual Report, 1996-1997. CSIRO Division of Atmospheric Research, Melbourne.

Sutherst, R.W. (1990) Impact of climate change on pests and diseases in Australia. *Search* **21**, 230-2.

Timmermann, A., Oberhuber, J., Bacher, A., Esch, M., Latif, M. and Roeckner, E. (1999) Increased El Niño frequency in a climate model forced by future greenhouse warming. *Nature* **398**, 694-6.

Tothill, J.C. and Gillies, C. (1992) The pasture lands of northern Australia: Their condition, productivity and sustainability. Tropical Grasslands Society of Australia Occasional Publication No. 5.

Trenberth, K.E. and Hoar, T.J. (1997) El Niño and climate change, *Geophysical Research Letters* **24**, 3057-60.

Walker, B.H. (1988) Autecology, synecology, climate and livestock as agents of rangeland dynamics. *Australian Rangeland Journal* **10**, 69-75.

Walker, B.H. (1993) Stability in rangelands: Ecology and economics. Proceedings of the XVII International Grassland Congress. pp.1885-90.

Ward, W.T. and Russell, J.S. (1980) Winds in south east Queensland and rain in Australia and their relationship with sunspot number. *Climate change* **3**, 89-104.

Warrego Graziers Association (1988) Submission to the United Graziers Association on the degradation of south-west Queensland. Unpublished report. UGA, Queensland.

West, B. and Smith, P. (1996) Drought, discourse, and Durkheim: a research note. *The Australian and New Zealand Journal of Sociology* **32**, 93-102.

Weston, E.J. (1988) The Queensland Environment, in W.H. Burrows, J.C. Scanlan and M.T. Rutherford, (eds.), *Native Pastures in Queensland: The resource and their Management.* QDPI Information Series QI870323, Queensland Department of Primary Industries, Brisbane. pp.21-33.

White, B.J. (1978) *A Simulation Based Evaluation of Queensland's Northern Sheep Industry.* James Cook University, Department of Geography, Townsville. Monograph Series No. 10.

Willcocks, J.R., McKeon, G.M. and Day, K.A. (1991) Using Southern Oscillation Index to predict the growth of *Heteropogon contortus* pasture in south-east Queensland. Proc. First Australian Conference on Agricultural Meteorology. Bureau of Meteorology, Melbourne. pp.36-39.

Williams, O.B. and Oxley, R.E. (1979) Historical aspects of the use of chenopod shrublands, in *Studies of the Australian Arid Zone IV: Chenopod Shrubs.* CSIRO, Canberra. pp.5-16.

Wilson, J.R. and Mannetje, L.'t (1978) Senescence, digestibility and carbohydrate content of buffel grass and green panic leaves in swards. *Australian Journal of Agricultural Research* **29**, 503-16.

SIMULATION OF GRAZING STRATEGIES FOR BEEF PRODUCTION IN NORTH-EAST QUEENSLAND

GREG MCKEON[1], ANDREW ASH[2], WAYNE HALL[1] AND MARK STAFFORD SMITH[3]

[1] *Climate Impacts and Grazing Systems*
Queensland Department of Natural Resources
80 Meiers Rd, Indooroopilly, Qld 4068, Australia

[2] *CSIRO Tropical Agriculture*
Private Mail Bag
Aitkenvale, Qld 4814, Australia

[3] *CSIRO National Rangelands Program*
PO Box 2111
Alice Springs, NT 0871, Australia

Abstract

A simulation study was conducted to compare diverse grazing strategies for steers grazing open woodlands in northeast Queensland. Simulations included a wide range of possible stocking rates and pasture utilisation levels using 108 years (1889-1996) of daily climate data for Charters Towers. Five strategies were compared in terms of steer liveweight gain per ha, risk of weight loss, pasture availability, frequency of burning and soil loss. The strategies included constant stocking, stocking in response to available feed, and stocking in response to predicted future feed availability based on a climate forecast. For strategies achieving an average annual liveweight gain per head of about 100 kg, the simulation studies indicated that a responsive stocking rate strategy in June using a forecast of the next year's pasture growth would increase liveweight gain per ha by about 10%, reduce the risk of liveweight loss by 57%, reduce risk of low pasture yield, but would slightly increase the risk of soil loss (4%). Maximum LWG/ha was achieved at high utilisation rates (> 35%). However, at such high levels of utilisation burning was achieved in less than 10% of years and soil loss was 30-40% more than at levels of utilisation regarded as safe ($\approx$20%). The simulations highlighted the potential value of achieving in June, the skill from seasonal forecasting that is now available in November using average SOI in the Aug-Oct period as the indicator of season type. Assumptions in the model development are outlined and future work required is discussed. Despite the complexity of the simulation analysis, it is concluded that there is a trade-off between production and environmental damage, and that improved forecasting may improve production and/or reduce damage.

G.L. Hammer et al. (eds.), The Australian Experience, 227–252.
© *2000 Kluwer Academic Publishers. Printed in the Netherlands.*

1. Introduction

Year-to-year variation in rainfall is a major problem for beef enterprises in the semi-arid tropics of northeast Queensland (Scanlan *et al.,* 1994). The degree of variability is shown by annual (July to June) rainfall at Charters Towers, which has a coefficient of variation of 40% for the period 1883 to 1992. Partly as a result of this variability, Dalrymple Shire (the local government area containing Charters Towers) was drought declared 12 times between 1964 and 1989, with a frequency of declaration of 1 in 2.2 years, and total time declared of 34% (Daly, 1994). One cause of climatic variability in this region is the effect of the El Niño/Southern Oscillation (ENSO) phenomenon (McBride and Nicholls, 1983; Stone and Auliciems, 1992). The Southern Oscillation Index (SOI) in spring has been shown to explain 25% of the variation in subsequent summer (November to March) rainfall recorded from 1958 to 1988 (McKeon *et al.,* 1990).

Decade-to-decade and generation-to-generation variations in rainfall are also high (Lough, 1991; McKeon *et al.,* 1998b). Reconstruction of Burdekin river flow from fluorescence bands in coral has suggested that both drier and wetter decades and 30-year periods occurred before the current century (Lough, 1991). The grazing management decisions that are made against this background of climatic uncertainty include stocking rate, frequency of burning, control of woody weeds, use of animal supplements and herd management. Stocking rate or the management of grazing pressure has a major impact on resource condition (Gardener *et al.,* 1990). Stocking rate affects pasture yield (ie. standing crop) not only through the processes of consumption and trampling, but also through the effects of grazing on pasture composition (McIvor *et al.,* 1995a). Increased grazing pressure can result in less pasture yield, reduced pasture growth, increased run-off and soil loss (McIvor *et al.,* 1995b; Scanlan *et al.,* 1996a), reduced burning opportunities and increased risk of woody weed growth (Burrows *et al.,* 1990).

Many possible strategies (Stafford Smith, 1992) are available to manage grazing pressure. These include -
* constant year-to-year stocking rate
* stocking rate adjusted each year to eat a fixed proportion of existing pasture yield over the next year, and
* stocking rates adjusted each year to eat a constant proportion of expected future or forecast pasture growth.

Graziers use some of these options (Scanlan *et al.,* 1994) but no data are available to formally compare the effects of these options on production and resource status. Not only are different strategies available but the optimal level of pasture utilisation is also uncertain judging by the large variation in pasture utilisation among properties (Scanlan *et al.,* 1994). In eastern Queensland, historical grazing trials with beef cattle have concentrated on comparing only varying levels of constant stocking rate. Grazing trials comparing different stocking rate 'decision rules' have only just commenced (P.C. Smith, pers. comm; P. O'Reagain, pers. comm.). Simulation modelling provides one approach to extrapolate existing research findings and producer experience to formally compare a wide range of grazing strategies over a wider range of climatic conditions (using historical climate data) than can be experienced in a typical field trial (White, 1978; McKeon *et al.*, 1990). The use of simulation modelling in a general systems approach to applying seasonal climate forecasts is detailed elsewhere in this volume (Hammer, 2000). This paper describes the modification of a simulation model (GRASP) to evaluate the impact of grazing strategies on steer growth and risk of resource damage at one location (Charters Towers). Studies at other locations and using a range of seasonal climate forecasting systems are described by Ash *et al.* (2000). Economic evaluation of grazing management strategies was achieved by coupling GRASP to a dynamic herd and financial model and the results are reported by Stafford Smith *et al.* (2000).

2. Simulation Methodology

2.1 WORD MODEL OF PROCESSES IN SOIL-PASTURE-ANIMAL SYSTEM

The semi-arid tropical grazing lands, when in good condition, are mainly an open woodland with an understorey of perennial tussock grasses growing on a wide variety of soils of low slopes (2-3%). The major hydrological and erosion processes are -

- run-off, which is a function of surface cover, rainfall intensity and soil-water deficit (Scanlan *et al.*, 1996a)
- evapotranspiration made up of soil evaporation, grass transpiration and tree transpiration
- drainage (Rickert and McKeon, 1982; Scanlan and McKeon, 1993), and
- soil loss, which is a function of cover, run-off and slope (McIvor *et al.*, 1995b,; Scanlan and McIvor, 1993).

Standing pasture yield is the net result of the processes of pasture growth, death, detachment, consumption and trampling (McKeon *et al.*, 1982; McKeon *et al.*, 1990). Pasture growth in this summer rainfall environment is either limited by moisture or soil fertility, primarily nitrogen (McCown, 1973; McCown *et al.*, 1974; Mott *et al.*, 1985). Potential perennial pasture growth when green cover or light interception is low is proportional to grass basal area (McKeon *et al.*, 1990). Steer liveweight gain (LWG) is a function of the length of the growing season and pasture utilisation, ie. the proportion of growth which has been eaten (McCown *et al.*, 1980-81; McKeon and Rickert, 1984; McIvor and Monypenny, 1995). Pasture burning increases potential LWG by about 15 kg/hd/yr (Ash *et al.*, 1982).

Intake restrictions are likely to occur when pasture yield is below 300 kg DM/ha (Ash *et al.*, 1982). However, it has been observed (A. Ash, pers. comm.) that cattle with high *Bos indicus* content in large scale commercial paddocks are able to graze even lower yields without an obvious reduction in animal performance.

Increased grazing pressure has several effects on pasture production -

- grass basal area is reduced by the effect of grazing on plant death and the reduction of basal area of surviving plants, especially during periods of drought (Scattini, 1973; Ash et al., 1996; D.M. Orr, pers. comm.)

- photosynthetic area is reduced by defoliation which reduces future growth (McKeon et al., 1982)

- pasture composition (eg. percentage perennials) is affected by heavy grazing of live tissue resulting in loss of perennial species and replacement by annuals with less production or faster senescence (McIvor et al., 1995a; Ash et al., 1996; D. Cooksley, unpublished data)

- increased pasture utilisation increases run-off because of reduced surface cover (McIvor et al., 1995b) and fewer preferred infiltration sites such as areas occupied by perennial tussocks (K.A. Day, pers. comm.)

- increased soil loss reduces nutrients available for plant growth, infiltration and available water range (Williams and Chartres, 1991), and

- reduced pasture yields result in fewer fires and increased survival of woody weed and tree seedlings (Burrows et al., 1990; Burrows, 1995).

In the model development, we have addressed the first five of these feedbacks. However, a dynamic model of woody plant growth (Scanlan and McKeon, 1990) has not yet been implemented and thus the effects of burning and heavy grazing (Scanlan et al., 1996b) on woody plants are not included. However, because of the importance of fire management, we examine the frequency of pasture burning as one indicator of resource condition.

2.2 THE GRASS AND ANIMAL PRODUCTION MODEL – GRASP

GRASP (GRASs Production) is a deterministic, point-based model of soil-water, grass growth and animal production (sheep and cattle), developed and validated on tropical grasslands (McKeon et al., 1990; Day et al., 1993; Day et al., 1997a). A full description of equations and assumptions in the pasture model is included in Day et al. (1997a). The model has been calibrated for over 40 pasture communities in Queensland. Soil-water is simulated for given soil attributes (texture and depth) from daily inputs of rainfall, temperature, humidity, pan evaporation and solar radiation. Plant growth is calculated from transpiration, but includes the effects of vapour pressure deficit, temperature, radiation interception, nitrogen availability and grass basal area. The competitive effect of variable tree density is simulated via the effects of trees on water use and nitrogen uptake. Animal production responses are calculated by multiple regression equations for live weight change (and wool growth for sheep (Hall, 1996)) as a function of utilisation and length of growing season. These equations have been developed from grazing trials.

The soil-water model in GRASP is a simple one-dimensional, multi-layered tipping bucket type model. Infiltration occurs from layer to layer only when the water content of each layer has reached a user-defined field capacity. Four layers are simulated (0-10cm, 10-50cm, 50-100cm and 100cm down to an estimate of the bottom of the profile). Soil water in the deepest layer (>100cm) is available only for tree transpiration. Before infiltration, run-off is calculated as a function of cover, rainfall intensity and soil-water deficit using an equation derived from measured run-off data (Scanlan *et al.*, 1996a). Deep drainage is also calculated although this component of the soil-water budget is yet to be validated.

For the purposes of this analysis, the most important hydrological processes are run-off due to low surface cover, soil evaporation simulated for the 0-10 cm and 10-50 cm soil layers as a function of soil type and surface cover, and transpiration from grass and trees. Tree density was assumed to be constant in the simulations reported here. Transpiration from grass is calculated from green cover, soil-water availability and Class A pan evaporation (or estimates where recorded data were not available). Green cover is recalculated each day from the simulated green biomass pool, itself the net result of gain by simulated growth and loss by animal intake and death due to aging, frost or soil-water deficit. Thus the simulations of soil-water and plant growth respond dynamically to the interaction of daily climate inputs (rainfall, pan evaporation, temperature, solar radiation and vapour pressure deficit) and management (stocking rate, fire, tree density).

No feedbacks of soil loss on nutrient availability or infiltration characteristics have been modelled. Thus GRASP effectively allows for the simulation of the risks of production losses but not the long-term consequences of taking those risks. It thus provides a framework for estimating animal production and short to medium-term pasture sustainability as a result of different sequences of climate and stocking strategies.

2.3 CHANGES IN PASTURE COMPOSITION

The decline in the percentage of desirable perennial grasses under heavy grazing was simulated by linking the percentage of desirable perennials to a yearly time scale (Figure 1) derived by Ash *et al.* (1996) from observed data. Grazing pressure is expressed as the percentage of total growth that is eaten as green material. We assume, based on rotational grazing studies on black speargrass (*Heteropogon contortus*) pastures (D. Cooksley, unpublished data), that the heavy consumption of dead material (eg. during the dry season) does not influence pasture composition.

The GRASP model was modified to include the following pasture composition rules based on utilisation of green material by the end of growing season (30th April) -
- < 22.5% utilisation - percentage perennials increase by one perennial-index year equivalent as in Figure 1
- > 22.5 but < 35% utilisation - no change, and
- > 35% utilisation - percentage perennials decrease by one perennial-index year equivalent.

Thus continued grazing with more than 35% of total growth consumed as green material results in a decline from 90% perennials to zero over 11 years. Thus the variable 'perennial-index' year represents the history of grazing impacts on perennial grasses.

The loss of desirable perennial grasses associated with heavy utilisation results in reduced pasture production (McIvor *et al.*, 1995a; Day *et al.* 1997a). This consequence was simulated by changing the model parameters as described below. Field measurements and observations (eg. Ash and McIvor, 1995; Day *et al.* 1997a) indicate that the following changes occur as the percentage of perennials decline from 90 to 0% -

- the soil-water index threshold at which growth declines increases from 0.4 to 0.99
- maximum nitrogen uptake declines from 12 to 6 kg N/ha/yr, and
- minimum nitrogen content to which plants can dilute nitrogen and continue growth increases from 0.5% to 1%, thus reducing overall nitrogen use efficiency.

These changes in parameters result in a reduction in average annual pasture growth from 1500 kg DM/ha to 600 kg DM/ha as percentage perennials decline from 90% to 0%.

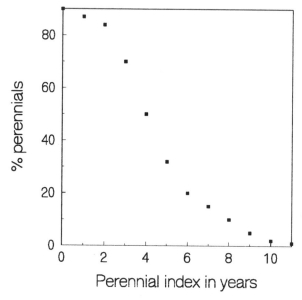

Figure 1. The time course in perennial composition of pasture (%) with continued heavy utilisation (after Ash *et al.*, 1996). The X-axis has the units 'years' and represents the memory of the perennial grass system.

2.4 BEEF CATTLE PRODUCTION MODEL IN GRASP

Hall *et al.* (1998) developed a general annual beef production model for the coastal black speargrass zone of Queensland from grazing trials at Kangaroo Hills (Gillard, 1979), Galloway Plains (Burrows, 1997) and Brian Pastures (R. Roberton, unpublished data) in which annual liveweight gain (LWG; kg/hd/day) was given by -

$$\text{LWG} = 0.060 + 0.00483 * \%\text{gidays} - 0.00206 * \%\text{utilisation} \quad (r^2 = 0.71, n = 76) \quad [1]$$

where %gidays is the percentage of days in the year with growth index ≥ 0.05, and
%utilisation is pasture eaten expressed as a percentage of pasture growth for
the year

Using the same form of relationship, annual LWG for treatments with trees (Kangaroo
Hills data; Figure 2) was found to be -

LWG = 0.0239 + 0.005586 * %gidays - 0.00229 * %utilisation ($r^2 = 0.80$, n=20)   [2]

The stocking rate effect on LWG in these relationships is similar to that found by Ash
and Stafford Smith (1996) in a review of stocking rate relationships in rangelands,
which showed a 20% decline in LWG per head for a 100% increase in stocking rate.

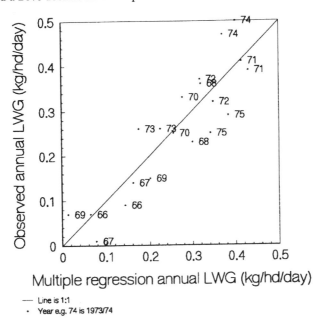

Figure 2. Relationship between predicted and observed LWG (kg/hd/day)
for Kangaroo Hills (equation [2]). Note that 74 is the observation for
1973-74.

The terms in the regression are also affected by grazing effects on pasture. Increasing
utilisation reduces green cover and increases soil-water availability and hence the
percentage of days when the growth index is above 0.05. Other modelling analyses of
grazing trials (Day *et al.,* 1997a) have shown that the soil-water threshold at which
growth stops increases with increasing pasture utilisation either due to reduced root
activity or a change in species composition to early flowering species. In this model,
the critical threshold for accumulating growth index days is changed from 0.05 to 0.20
as the percentage of desirable perennials decline to zero.

A risk in modelling annual LWG using a multiple regression is that basic energy conversion limitations in converting pasture dry matter intake to LWG will be overridden. Hence in this model, we calculate an upper limit to annual LWG (LWG_{max} kg/hd/yr) for steers (200 kg initial liveweight) based on average daily dry matter intake per 200 kg weaner equivalent (DMI, kg/hd/day) for the year as -

$$LWG_{max} = 365 * (DMI * 0.304 - 0.800)$$
[3]

This relationship was estimated from model output and differs substantially from the relationship between intake and LWG given by Minson and McDonald (1987). These differences between intake-LWG relationships (eg. McKeon and Rickert, 1984; Minson and McDonald 1987; Coates, 1995) can only be resolved by further experimentation measuring animal intake (Hall *et al.*, 1999).

Although this form of beef production model (equations [1] and [2]) was established over a wide range of locations and rainfall variability, its application in the simulation of grazing strategies requires further attention to the effects on LWG of -
- intake restrictions due to low pasture yield
- effects of pasture burning, and
- loss of desirable perennial grasses, ie. poor pasture condition.

2.5 INTAKE RESTRICTIONS

Intake restrictions occur when low pasture yield reduces bite size and animals are unable to compensate by increasing biting rate and/or grazing time. In tussock grasslands, restrictions probably occur below 200-500 kg DM/ha (Hendricksen *et al.*, 1982) although no definitive data are available. In GRASP we have included a restriction on intake as part of modelling pasture yield under heavy grazing. We proposed the following relationship between relative intake (RI_1) and total standing dry matter (TSDM kg DM/ha) -

$$RI_1 = min (1.0, TSDM / P_{144})$$
[4]

where P_{144} is pasture yield above which intake is not restricted by pasture availability.

By calibrating the model for heavily grazed pasture we have found that P_{144} is approximately 300 kg DM/ha for a range of black speargrass pastures grazed by British breeds of cattle (data from Scattini, 1973).

GRASP also calculates a second restriction on intake (RI_2) due to the declining availability of desirable pasture components such as leaf (Ash *et al.*, 1982; McKeon and Rickert, 1984). This is represented mathematically by accumulating utilisation since the start of the growing season (December 1) -

$$RI_2 = 1.0 - 0.30 * \frac{\%utilisation}{100}$$
[5]

The coefficient (-0.30) has been derived from an analysis of average stocking rate relationships for a wide range of recent grazing trials (M. Quirk, pers. comm.). This slope coefficient is flatter than that found by McKeon and Rickert (1984) and that also derived from data for short term or rotationally grazed trials (Ash et al., 1995). A coefficient of -0.4586 (McKeon and Rickert, 1984) was used in GRASSMAN (Scanlan and McKeon, 1990) to simulate LWG and hence produced a greater decline in LWG with increasing stocking rate than simulated here by GRASP.

The actual restriction on intake is the minimum of RI_1 or RI_2. This approach of reducing simulated intake by either restrictions on pasture yield or declining availability of desirable pasture components allows reasonable simulation of pasture yield under heavy grazing and/or drought.

2.6 LIVEWEIGHT LOSS ASSOCIATED WITH LOW PASTURE YIELDS

In the 10 years of the Kangaroo Hills grazing trial, there were a few months where simulated yields were less than 300 kg DM/ha. For these pastures, weight loss had a linear relationship with pasture yield when the proportion of green in the pasture was low (< 10%). Thus, in the annual LWG model we also calculate weight loss (expressed as a negative LWG) when these conditions occur using the relationship -

$$\text{daily liveweight loss (kg)} = \text{TSDM} / 300.0 - 1.0 \qquad [6]$$

This daily liveweight loss is summed over the year and deducted from the annual LWG calculated using equation [2]. However, it should be noted that the LWG data from Kangaroo Hills may under-estimate liveweight loss on native pastures during the dry season because of the presence of high quality dead material from the annual legume Townsville stylo (*Stylosanthes humilis*) (Norman and Begg, 1973).

Annual weight losses greater than 25% of the steer starting weight (300 kg) are likely to result in death. This has not been simulated in this analysis. Nevertheless, the effects of these extreme weight losses are included to allow a plausible comparison of the grazing management strategies.

2.7 EFFECT OF PASTURE BURNING ON LWG

Burning is a common practice in black speargrass areas to control woody weeds, provide access to green feed following spring rains, and re-distribute grazing pressure. Of the trials used to develop equation [1] for LWG only Mt Bambling (Brian Pastures Research Station) was regularly burnt. The impact of burning on annual liveweight has been estimated as +15 kg/hd from the experiments of Ash et al. (1982) at Brian Pastures and from data of McLennan et al. (1986) at Swans Lagoon. In the multiple regression analysis the coefficient for pasture burning effect on LWG derived for the Mt Bambling site (dummy variable for burning = 1) compared to the other sites (dummy variable = 0) was 7 kg/hd but was not significant (P>0.05). This lower than expected value may result from other site differences (eg. phosphorus, legume presence, winter feed quality). In this paper we have added 15 kg to the annual LWG calculated from equation [2] in those years when fires were simulated.

2.8 EFFECT OF LOSS OF DESIRABLE PERENNIAL GRASSES ON LWG

The loss of desirable perennial grasses (DPGs) can result in reduced pasture production (McIvor *et al.*, 1995a; Wandera, 1993). However, where DPGs are replaced by unpalatable perennials such as wiregrass *(Aristida* spp.) or rats tail *(Sporobolus* spp.), primary production may be maintained whilst still affecting secondary production. Simulation of pastures where DPGs have been replaced by unpalatable species would require modifications to the LWG multiple regression such that only the palatable production was included.

Ash and McIvor (1995) found that pasture quality may increase with a decline in pasture condition. In short term (8 weeks duration) grazing trials, poor-condition pastures had higher LWG/hd at low utilisation rates but had a greater decline in LWG/hd with increasing utilisation rate (Ash *et al.*, 1995).

In the current simulation analysis we consider only the effects of changing pasture condition on pasture production in the LWG multiple regression. Thus, for pastures in poor 'condition' but of high quality, such as where perennial species have been replaced by annual species, the model is likely to underestimate LWG/hd at low utilisation.

2.9 EFFECT OF ANNUAL PERIOD ON LWG

The annual LWG model was developed from LWG of drafts of steers grazed from the early dry season (June) or winter over the following 10-13 months. In the present study the model is also used to simulate LWG from November to October although no validation data of steer growth for this period were available. It should be noted that measurement of liveweight gain from June to May includes compensatory growth in the summer growing season associated with the preceding winter/dry season. However measurement from November to October includes compensatory gain associated with the winter/dry season prior to the 12-month liveweight gain being simulated. We assume, not unreasonably, that the contribution of compensatory growth to the annual liveweight gain does not vary markedly between years.

3. Grazing Strategies

There is possible confusion about the appropriate use of the words 'strategy' and 'tactics' when discussing grazing management. 'Strategies' are concerned with longer time frames than 'tactics'. In these simulation studies we are evaluating long term strategies involving annual stocking rate decisions. When compared to the constant stocking rate strategy, varying stocking rates each year based on a rule could be described as 'tactical' because each year is treated as a separate decision. However, we prefer the term 'strategy' as we are comparing decision rules to vary stocking rate each year based on information available at the time. These 'varying stocking rate strategies' represent rational approaches to deal with the full spectrum of rainfall variability. They are not really 'tactical' when compared to real world decisions that involve responses to short term (monthly to seasonal) fluctuations in markets, rainfall and government regulations. In this paper, we evaluate different decision rules as if they were inflexible recipes. No constraints were placed on the year-to-year change in stock numbers. This assumption is supported by the findings of Scanlan *et al.* (1994) who reported that herd numbers for 45 properties changed by up to ±90% per year.

Various forecasting systems suitable for use in land management have been developed based on the sound understanding of climatic mechanisms and the uncertainties attached to the use of imperfect forecasts (Pittock, 1975; Nicholls, 1983; Allan, 1988; Clewett *et al.*, 1988; Stone and Auliciems, 1992; Stone *et al.*, 1996). The development of climate forecasting systems in Australia is reviewed by Stone and de Hoedt (2000) elsewhere in this volume. Appreciation of the potential value of seasonal rainfall forecasting based on ENSO has increased in parallel with the forecasting systems (McKeon *et al.*, 1990; Hammer *et al.*, 1991; McKeon and White, 1992; Stone and McKeon, 1992; Hammer *et al.*, 1996).

In this study we examine only the 'lag' SOI system (McBride and Nicholls, 1983; Clewett *et al.*, 1988) and the theoretical potential of current climate models under development to forecast El Niño or La Niña years (Cane, 2000). The 'lag' SOI system allows years to be classified based on the average SOI for a 3-monthly period (eg. August to October) and its application in this study is described in more detail later. Further studies are required to examine the use of more recent climate forecasting systems, such as the SOI phase system (Stone and Auliciems, 1992; Stone *et al.*, 1996) and systems based on sea surface temperatures and climate modelling (Hunt, 1991; Hunt *et al.*, 1994; Smith, 1994; Kleeman *et al.*, 1995; Hunt and Davies, 1997; Drosdowsky and Chambers, 1998).

Five general grazing strategies were tested for a range of utilisation rates using 108 years of rainfall for Charters Towers, north Queensland. A base simulation record of pasture growth, as required to simulate some of the strategies, was generated in accordance with the approach detailed by McKeon *et al.* (1990). In this approach an ungrazed exclosure is simulated with yields reset to 100 kg DM/ha green and 200 kg DM/ha dead material on November 1st each year (as if the pasture was mown). Thus this base simulation could be compared with field measurements. Simulated pasture growth in individual years was stored and used to provide the basis for analysis of the grazing strategies. Average annual growth was calculated from the base simulation record.

3.1 STRATEGY 1: 'CONSTANT' STOCKING RATE

The stocking rate of yearling steers (300 kg LW) was calculated to eat a constant proportion of long-term average annual growth as calculated from the base simulation of exclosure pasture growth. Thus stocking rate does not change from year-to-year. In GRASP, stocking rates are converted to 200 kg weaner equivalents for the purposes of calculating feed intake (McKeon and Rickert, 1984).

3.2 STRATEGY 2: 'RESPONSIVE' BASED ON JUNE PASTURE YIELD

Each year's stocking rate was changed in June to eat a fixed proportion of existing TSDM over the next 12 months. No estimate of future growth was included in the calculation.

3.3 STRATEGY 3: 'PERFECT KNOWLEDGE'

Each year's stocking rate was changed in either June or November to eat a constant proportion of a perfectly known pasture growth for the next 12 months. The 'perfect forecast' pasture growth was derived in the initial base simulation of an exclosure. Stocking rate affects pasture growth during the growing season and hence the approach adopted here was to establish a multiple regression to estimate intake (kg DM/ha) and actual pasture growth as a function of both stocking rate and 'perfectly known' pasture growth in an exclosure. Equations [7] and [8] were solved iteratively to find the stocking rate resulting in the desired constant utilisation of predicted pasture growth -

$$\text{predicted growth (t DM/ha)} = 0.744 + 0.804 * \text{growth} + 0.01429 * \%\text{perennials}$$
$$+ 0.0030565 * \text{TSDM} - 0.002397 * \text{SR} \qquad [7]$$

$$\text{predicted eaten (kg DM/ha)} = 1.4725 * \text{SR} * (0.0235 * \text{TSDM} + 1) *$$
$$(0.2943 * \text{growth} + 1) * (0.06996 * \%\text{perennials} + 1) \qquad [8]$$

where   growth is simulated pasture growth in exclosures (t DM/ha),
         TSDM is total standing dry matter (t/ha),
         %perennials is the percentage of perennials in TSDM,
         SR is stocking rate (200kg weaner equivalents /100 ha).

By iteratively solving these equations for different stocking rates, the stocking rate at which predicted utilisation was equal to the target utilisation was determined. Predicted utilisation was calculated as -

$$\text{Predicted utilisation} = \frac{\text{predicted eaten}}{\text{predicted growth} * 1000} \qquad [9]$$

This approach was adopted to speed up simulation. However it did not result in the exact target utilisation being achieved every year. The target utilisation was reached on average and the coefficient of variation of year-to-year utilisation (CV=17%) was substantially less than that for the constant (CV=55%) and responsive (CV=64%) strategies.

3.4 STRATEGY 4: 'CONSTANT + FORECAST' IN NOVEMBER

The constant stocking rate calculated in strategy 1 (on the basis of consuming a proportion of long-term average growth) was adjusted each year in November based on the average value of the SOI recorded during the previous three months (August - October). Three types of year were identified (after Clewett *et al.,* 1988) –

- SOI ≤ -5, referred to here as El Niño years,
- SOI between -5 and +5, and
- SOI ≥ +5, referred to here as La Niña years.

Within this broad strategy there were three methods of modifying stocking rates –

- decrease stocking rate by 50% in El Niño years
- increase stocking rate by 50% in La Niña years, and
- do both of the above

3.5 STRATEGY 5: 'RESPONSIVE + LONG-LEAD FORECAST' IN JUNE

Recent developments in the modelling of ENSO behaviour suggest that forecasts of El Niño or La Niña years can be made six to 12 months beforehand (Kleeman *et al.,* 1995; Cane, 2000). For example, some successful forecasts of the 1998/9 La Niña were made in May 1998. Although there remains some uncertainties with skill of various dynamic climate models (Cane, 2000; Nicholls *et al.,* 2000) the more recent statistical forecasting systems (eg. Stone *et al.* (1996)) show predictive skill by June.   In strategy 5 we evaluate the impact of combining this potential forecast capability with the responsive strategy (strategy 2) described above.   There are no available hindcasts of this 'long-lead' type for the last 100 years so we have used the same classification of years into El Niño and La Niña years as given in strategy 4 above, ie. using the SOI from August to October. In effect we are evaluating the benefit of forecasting the August to October SOI on the 1st June with 100% accuracy. The actual long-term accuracy of these developments in long-lead forecasts of ENSO behaviour is yet to be assessed.

Stocking rate was changed each year in June, as in strategy 2, to eat a varying proportion of TSDM. The calculated stocking rate was then adjusted for El Niño and La Niña year types using the three methods identified in strategy 4. For example, a stocking rate would be calculated each June to eat 40% of current TSDM.   This calculated stocking rate would, using the third method in strategy 4, be increased by 50% in La Niña and decreased by 50% in El Niño years. No change to the calculated stocking rate would occur for other years.

3.6 IMPACT OF STOCKING RATE ADJUSTMENT FOR EL NIÑO YEARS ONLY

The impact of magnitude of adjustment of stocking rate for El Niño years only was examined further by modification of the two strategies (4 and 5) involving climate forecasts as follows -

- *'Responsive + long-lead forecast' in June*– the stocking rate calculated in June of each year to eat 40% of TSDM was adjusted by a constant value ranging from 0.25 to 1.25 in El Niño years only, and
- *'Constant + forecast' in November* - the constant stocking rate calculated to eat 30% of long term average annual growth was adjusted in November by a constant value ranging from 0.25 to 1.25 in El Niño years only.

4. Simulation Results

Table 1 reports the results from 19 simulations selected from over 100 simulation treatments. The selected treatments include combinations of grazing strategies (*perfect-knowledge, responsive, constant, constant + forecast, responsive + long-lead forecast*), pasture utilisation rate (0-60%), and month of stocking rate change (June or November). Detailed results are given in McKeon *et al.* (1998a).

The output variables were averaged over the period of simulation and included -
- components of soil-water balance, ie. run-off, drainage, tree water use, soil evaporation and grass transpiration (not shown)
- pasture variables, ie. pasture growth, TSDM at time of stocking rate change (June 1st or November 1st), grass basal area, frequency of pasture burning, and minimum TSDM at any time in the year
- animal intake and numbers, ie. amount eaten, annual utilisation, stocking rate and coefficient of variation of stocking rate, and
- animal production variables, ie. LWG/hd, LWG/ha, liveweight loss, and percentage of years with annual LWG less than 100 kg/hd.

The variable LWG/ha has been calculated each year and then averaged over the 108 years. Hence in Tables 1 and 2, average LWG/ha does not equal average LWG/hd multiplied by average stocking rate (hd/ha). In contrast average annual utilisation in Tables 1 and 2 was calculated as the ratio of pasture eaten (kg DM/ha) averaged over 108 year to pasture growth averaged over the same time period. This approach avoids bias due to the limited number of years with very high utilisation and is directly compatible with the approaches used to calculate safe stocking rates (Johnston *et al.*, 1996).

Results will be evaluated separately in terms of liveweight gain per ha and stocking rate; risk of liveweight loss and low pasture yield; and resource condition.

4.1 LIVEWEIGHT GAIN PER HA AND STOCKING RATE

The highest LWG/ha of 29.2 kg (treatment #1) was achieved at an intended utilisation of 40% of predicted annual (June-May) growth ie. *'perfect knowledge'*. Above this utilisation level the composition of desired perennials declined and production of both pasture and animals was greatly reduced (results not shown).

TABLE 1. Results of simulation study for a range of grazing strategies using 108 years of historical daily climate for Charters Towers, Qld. Details of grazing strategies are given in the text. The treatment number given here (#) is used in discussing results for these strategies.

#	Grazing Strategy	Mean Annual LWG/ha (kg)	Stocking Rate (hd/km²)	Stocking Rate CV (%)	Utilisation (%)	Mean Annual LWG/hd (kg)	Years LWG/hd <100 kg (%)	Mean Annual Weight Loss (kg/hd)	Worst Weight Loss in any Year (kg/hd)	Worst TSDM (kg/ha)	Decile 1 TSDM (kg/ha)	Years Able to Burn (%)	Annual Soil Loss (kg/ha)
June change in stocking rate													
1	Perfect 40%	29.2	27.8	36	38	99	48	-4.0	-94	77	150	9	1333
2	Perfect 30%	24.7	21.0	36	29	112	33	-2.5	-63	109	247	31	1064
3	Responsive 60%	24.3	31.2	47	42	79	57	-8.3	-116	79	146	1	1349
4	Responsive 40%	23.1	24.3	41	33	96	43	-3.9	-145	83	231	12	1022
5	Responsive 30%	20.7	19.3	40	27	107	33	-2.4	-111	106	242	36	975
6	Constant 40%	20.3	26.7	0	35	76	47	-11.0	-169	6	110	18	1270
7	Constant 30%	20.2	20.0	0	27	101	36	-7.1	-132	48	167	44	1091
8	Constant 20%	15.6	13.4	0	18	117	28	-3.8	-88	88	286	60	908
9	Responsive -50%	23.0	21.4	49	30	105	32	-1.4	-69	135	282	22	955
10	Responsive +50%	25.0	26.5	46	36	93	46	-4.1	-145	83	220	6	1128
11	Responsive ±50%	25.2	23.9	58	33	102	37	-1.7	-69	135	271	16	1062
November change in stocking rate													
12	Perfect 35%	27.3	25.1	33	37	102	45	-2.2	-81	107	224	6	1199
13	Perfect 30%	24.8	21.6	33	32	109	36	-1.4	-60	139	294	14	1009
14	Constant 40%	19.9	26.9	0	37	74	50	-12.4	-277	6	82	8	1320
15	Constant 30%	19.9	20.2	0	29	99	33	-7.4	-121	50	134	33	1112
16	Constant 20%	15.6	13.4	0	20	116	22	-4.3	-97	75	213	55	920
17	Constant + forecast -50%	18.7	16.9	28	25	110	26	-4.0	-79	72	172	39	992
18	Constant + forecast +50%	22.4	23.0	20	33	95	39	-7.6	-124	49	121	27	1236
19	Constant + forecast ±50%	21.3	19.7	40	29	106	32	-3.9	-79	71	166	29	1096

TABLE 2. Results of simulation study examining a range of stocking rate adjustments in El Niño years only for the relevant subset of grazing strategies from those considered in the main simulation study (see Table 1). The simulations were conducted using 108 years of historical daily climate for Charters Towers, Qld. The treatment number given here (#) is as defined in Table 1. Details of grazing strategies are given in the text.

#	Stocking Rate Multiplier (for El Niño years only)	Mean Annual LWG/ha (kg)	Stocking Rate (hd/km²)	Stocking Rate CV (%)	Utilisation (%)	Mean Annual LWG/hd (kg)	Years LWG/hd <100 kg (%)	Mean Annual Weight Loss (kg/hd)	Worst Weight Loss in any Year (kg/hd)	Worst TSDM (kg/ha)	Decile 1 TSDM (kg/ha)	Years Able to Burn (%)	Annual Soil Loss (kg/ha)
June change in stocking rate[1]													
1	0.25	22.2	19.8	61	28	110	31	-0.7	-25	180	295	27	924
2	0.50	23.0	21.4	49	30	105	32	-1.4	-69	135	283	22	955
3	0.75	23.3	22.9	42	31	100	38	-2.7	-111	106	244	18	990
4	1.00	23.1	24.3	41	33	96	43	-3.9	-145	83	231	12	1022
5	1.25	22.4	25.6	44	34	92	45	-4.2	-103	102	213	10	1073
November change in stocking rate[2]													
6	0.25	17.4	15.2	47	22	115	24	-2.6	-46	72	208	43	921
7	0.50	18.7	16.9	28	25	110	26	-4.0	-79	72	172	39	992
8	0.75	19.5	18.5	13	27	104	30	-5.8	-102	72	152	37	1073
9	1.00	19.9	20.2	0	29	99	33	-7.4	-121	50	134	35	1112
10	1.25	20.2	21.8	11	30	95	38	-7.7	-138	34	117	29	1161

[1] June base stocking rate set to eat 40% of standing dry matter.
[2] November base stocking rate set to eat 30% of long-term average pasture growth.

There was 1.9 kg/ha difference in maximum LWG/ha associated with stocking rate change in June or November (treatment #1 compared to treatment #12). For stocking rate change in November an intended utilisation of 40% and above resulted in detrimental pasture composition changes but intended utilisation of 40% of annual growth starting in June could be achieved without such effects. This is likely to be an artefact of the model resulting from the fact that potential nitrogen availability is reset each year at the end of the dry season (November 1st). Thus annual pasture growth (Nov-Oct) has a maximum possible growth of 2068 kg DM/ha/yr due to nitrogen limitations whilst the period June-May can have higher values of annual growth. The development of continuous functions to simulate nitrogen mineralisation to replace the resetting date of nitrogen availability in the model is likely to eliminate this artefact. At lower intended utilisation, eg. 30% (treatments #2 and #13) there was less difference (0.1 kg LWG/ha) between June and November stocking rate change.

Maximum LWG/ha for *perfect knowledge* (29.2 kg LWG/ha) exceeded the best production from the *responsive* strategies (treatment #3) by 4.9 kg/ha and the best *constant* stocking rate (treatment #6) by 8.9 kg/ha. These latter two treatments (#3, #6) had average LWG/hd of 79 and 76 kg respectively, which would be too low to ensure turn-off of steers at a 'reasonable' age, eg. to turn-off 450 kg steer at 42 months from a weaning weight of 150 kg would require an average of 100 kg LWG/hd/yr.

The 40% utilisation *responsive* treatment (#4) averaged 96 kg LWG/hd whilst the *constant* stocking rate treatment intended to utilise 30% of average growth (#7) averaged 101 kg/hd/yr. Thus for three treatments (#1, #4, and #7) averaging approximately the same LWG/hd (96-101 kg), LWG/ha was 29.2, 23.1 and 20.2 kg for the *perfect knowledge*, *responsive* and *constant* stocking rates respectively. Interpolating between stocking rates to give a LWG/hd of approximately 100 kg, LWG/ha would be in the order of 28.8, 22.3 and 20.2 kg for the *perfect knowledge*, *responsive* and *constant* stocking rates respectively.

Six treatments involving climate forecasts were simulated. For the *constant + forecast* strategy, and modifying stocking rate for both El Niño and La Niña year types, LWG/ha (21.3 kg/ha, #19) exceeded the best *constant* rate treatment (#14) by 1.4 kg LWG/ha or 7%. Greater advantages were achieved when stocking rates were increased only in La Niña years (#18). However, *constant + forecast* strategies in November (#17-19) had lower LWG/ha than *responsive* strategies in June (#3 and #4).

The value of achieving the same forecasting skill in June, as is now available in November (ie. strategy 5 - *responsive + long-lead forecast*), was evaluated in treatment #11. Stocking rate was changed in June to eat 40% of TSDM over the next 12 months as in treatment #4 with a 50% reduction in El Niño years and a 50% increase in La Niña years. Compared to the *responsive* strategy the *responsive + long-lead forecast strategy* improved LWG/ ha from 23.1 to 25.2 kg or 9% and increased the proportion of years in which a LWG of 100 kg/hd was achieved from 57% to 63%.

The advantages of variable stocking rates in terms of LWG/ha were achieved by greatly varying year-to-year stock numbers. For example, the *responsive* and *responsive + long-lead forecast* strategies had very high CV for stocking rate (40-58%). The issues of how such flexibility in stock numbers may be achieved in reality is discussed in Johnston *et al.* (2000) and Stafford Smith *et al.* (2000).

4.2 RISK OF LIVEWEIGHT LOSS AND LOW PASTURE YIELD

The risk of liveweight loss and pasture deficits was evaluated via a range of variables - average annual liveweight loss over the 100 years; lowest pasture yield; and decile one pasture yield. Compared to *perfect knowledge*, the *responsive* and *constant* strategies that maximised LWG/ha (#3 and #6) resulted in at least twice the average weight loss (-4 compared to -8 and -11 respectively). Reducing the intended utilisation rate for the *responsive* strategy from 40% to 30% of TSDM halved the average liveweight loss and substantially reduced the risks when expressed in terms of lowest TSDM. Reducing the intended utilisation in the *constant* stocking rate treatment similarly reduced risks of liveweight loss but not as much as the *responsive* treatment.

Both treatments involving climate forecasting (#11 and #19) reduced average liveweight loss compared to their comparative constant treatments (#4 and #15 respectively). For strategies other than *perfect knowledge*, the *responsive + long-lead forecast strategy* produced the lowest average and lowest extreme liveweight loss, the highest minimum TSDM, and highest average LWG/ha achieving 86% of the potential LWG/ha, ie. that achieved with *perfect knowledge* of annual pasture growth (#1).

4.3 RESOURCE CONDITION

The three indicator variables of resource condition were grass basal area, percentage years with a fire and average annual soil loss. The impact of changing grass basal area has been modelled through the effects on pasture growth. The effects of varying fire frequency on woody weed regrowth and of soil loss on infiltration, nutrient availability and available water range have not been simulated in these studies.

In these simulations, 33% of the pasture was burnt if yield exceeded 1200 kg DM/ha. Thus a fire frequency of over 60% of years is required to ensure each location in the pasture is burnt at least once in five years. Only the safe *constant* stocking (#7) achieved this frequency of burning. As would be expected, treatments that maximised LWG/ha by consuming more pasture minimised the frequency of burning (Figure 3a).

Similarly, soil loss and LWG/ha were related, with soil loss generally increasing with LWG/ha (Figure 3b). However, for the treatments which maximised LWG/ha (#1, #3 and #6), soil loss was similar, 1270 to 1333 kg/ha/yr. The strategies involving climate forecasting (#11 and #19) had similar soil loss (1022 to 1096 kg/ha/yr) but still exceeded soil loss calculated for conservative utilisation treatments - #8, *constant* 20% utilisation; and #9, responsive + *long-lead forecast* in El Niño years only.

Thus the *responsive + long-lead forecast strategy* (#11), when compared to the safe *responsive* treatment (#4), had increased LWG per head and reduced liveweight loss but only allowed a frequency of burning of one in twenty years, and increased soil loss slightly (< 4%).

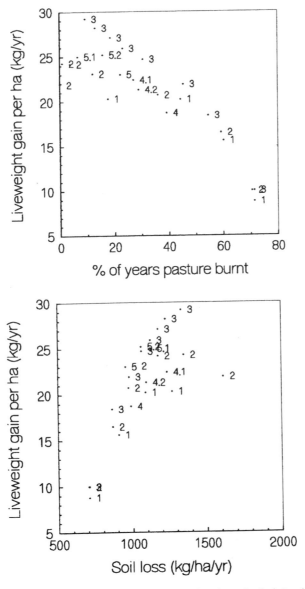

Figure 3. Relationship between average liveweight gain per ha (kg/yr) and (a) % of years in which pasture was burnt, and (b) average soil loss (kg/ha/year) as simulated for all strategies. The numbers represent the grazing strategies as described in the text.

4.4 IMPACT OF STOCKING RATE ADJUSTMENT FOR EL NIÑO YEARS

Table 2 shows the impacts of differing stocking rate adjustments in El Niño years only. For *responsive* stocking rate changes in June, a reduction in stocking rate in El Niño years only did not substantially reduce LWG/ha. In fact a 25% reduction in stocking rate (ie. 0.75 multiplier) slightly increased LWG/ha as well as increasing LWG/hd, reduced the risk of weight loss (by 31% from -3.9 to -2.7), increased the minimum TSDM (28%) and reduced soil loss (3%).

For changes in stocking rate in November, a 25% reduction in stocking rate in El Niño years only resulted in a 2% reduction in LWG/ha, a 22% reduction in average liveweight loss, and smaller reductions in minimum TSDM and soil loss. To achieve substantial (>10%) reductions in soil loss, greater reductions in stocking rate in El Niño years (50%) were required, but this also reduced LWG/ha by 6%.

5. Discussion

5.1 FORECASTING STRATEGIES

The simulation analysis presented highlights the trade-off between increased production and increased risk of feed shortage, reduced fire frequency and increased risk of soil loss. Adjusting strategies by including climate forecast information (ie. #19 vs #14 and #11 vs #4) resulted in increased production (7-9% respectively) without substantial increase in risk of resource damage and, in the case of pasture availability, a substantially reduced risk of running out of feed.

We regard the simulation of the irreversible soil losses as conservative as no feedbacks on soil physical (eg. infiltration) or biological (eg. loss of soil organic matter) attributes have been incorporated as yet. For example, exposure of surface soil to direct rainfall impact is likely to lead to surface sealing in some soils. Similarly the effects of heavy grazing can lead to surface structure changes resulting in reduced infiltration (Williams and Chartres, 1991). If these effects were included then the advantage of reducing stocking before, rather than after, seasons of low pasture growth would be greater than simulated in these studies.

5.2 STOCKING RATES IN FORECAST YEAR TYPES

Table 3 shows the stocking rates used in each year type for the different strategies. The *perfect knowledge* strategy resulted in a 19% reduction in stocking in the 35 El Niño years and a 24% increase in the 30 La Niña years. The *responsive* strategy did not utilise a forecast at all and hence average stocking rates in each type were the same. The two strategies involving climate forecasting used the forecast year type to determine stocking rate and the results reflect this link. Thus, the *responsive* strategy is likely to be one year out in responding after a drought or good year. However, where poor or good years occur in a sequence, the *responsive* strategy is likely to have benefits over a strategy using a forecast with imperfect skill. Long sequences of good and poor pasture growth years have occurred at Charters Towers (eg. good runs in 1953 to 1960 and 1970 to 1982, poor in 1991 to 1996), although it is not clear whether these are chance occurrences or structured non-random sequences.

Future simulation studies will concentrate on evaluating to what extent the *constant + forecast* strategy can be improved by using other variables (such as TSDM, grass basal area) known at the time of the stocking rate decision (ie. November).

TABLE 3. Stocking rates associated with selected grazing strategies by year type (average Aug-Sep SOI < -5, referred to as 'El Niño'; average Aug-Sep SOI > +5, referred to as 'La Niña'; and average Aug-Sep SOI between –5 and +5, referred to as 'other') for the simulation study conducted using 108 years of historical daily climate for Charters Towers, Qld. The treatment number given here (#) is as defined in Table 1. Details of grazing strategies are given in the text.

#	Grazing Strategy	Stocking Rate (hd/km^2)	Stocking Rate CV (%)	Stocking Rate in El Niño Years (hd/km^2)	Stocking Rate in La Niña Years (hd/km^2)	Stocking Rate in Other Years (hd/km^2)
June change in stocking rate						
2	Perfect 30%	21.0	36	17	26	21
5	Responsive 30%	19.3	40	19	19	20
7	Constant 30%	20.0	0	20	20	20
9	Responsive + forecast -50%	21.4	49	12	26	26
November change in stocking rate						
19	Constant + forecast +50%	19.7	40	10	30	20

5.3 ISSUES FOR FURTHER MODELLING

An annual multiple regression LWG model was developed for this simulation study. The data included in the model were from steers grazed from June to May and hence included any potential compensatory growth as a result of weight loss during winter months. The application of this equation to an annual period from November to October, as used in this paper, has not been validated, as this period does not correspond to any available experimental data.

The annual LWG model is also limited in not allowing steer growth to be simulated for shorter time periods, eg. wet and dry seasons. Hence in this study it was not possible to simulate the effects of several stocking rate changes in one year as is likely to occur in actual enterprises. Thus the development of an accurate shorter time step model of LWG is a priority. Such a model will have to include known sources of variation such as management, supplementation, climate (more correctly weather), pasture, genetics and previous grazing history. Of particular importance to the above studies is the modelling of the following factors related to LWG -

- supplementation to prevent animal weight loss and death
- compensatory gain after severe weight loss
- reduced metabolic demand in response to low nutrition
- contribution of browse, forbs and legumes to diet
- effect of pasture phenology and composition on diet quality including effects of drought and pasture burning; and
- effect of extreme wet weather on animal grazing behaviour, metabolic demand, pasture quality, pests and diseases.

The LWG model used in these studies does not include explicitly any of the negative effects of high rainfall listed above. Hence the model could over-estimate production in these years and hence over-estimate the advantage of stocking up in response to or anticipation of these high rainfall seasons. However, the Kangaroo Hills LWG data did include 1973-74 which had an extremely wet summer. The observed LWG for this year was adequately simulated or even under-estimated by the multiple regression model (Figure 2). Nevertheless, greater attention should be given to producer experience in regard to the impact of extreme years (Gillard and Moneypenny, 1990) so that the risks of extreme events are correctly evaluated. The impact of these events is rarely measured in formal grazing trial experiments.

5.4 SAFE CARRYING CAPACITY

Scanlan et al. (1994) surveyed 45 properties in the Dalrymple shire, north Queensland, and compared grazier estimates of 'safe carrying capacity' with simulated property pasture growth. On average, 'safe carrying capacities' represented 20% utilisation of average annual growth (Hall et al, 1998), a value consistent with similar studies in south-west Queensland (Johnston et al., 1996) and south-east Queensland (Day et al., 1997b). The simulations conducted in this study (Table 1) indicated that the 20% constant strategy (#8 and #16) had lower LWG/ha (15.6 vs 23.1 kg) than the responsive strategy (#4) but had increased LWG/hd (119 vs 96 kg), least risk of not achieving a target annual LWG of 100 kg/hd (28 vs 43%), greater frequency of burning (60 vs 12% of years) and reduced soil loss (908 vs 1022 kg/ha/yr). Thus the simulation results highlight the dilemma for graziers in which the long-term benefits of resource condition have to be balanced against the short-term costs of reduced production per ha. Some graziers, however, have reported resource, production, marketing and other benefits after adopting low stocking rate strategies (Lansberg et al., 1998; Fahey, 1997).

The simulation results presented in this paper highlight the need to model the trade-off between production (LWG/ha and LWG/hd) and resource condition (frequency of burning and soil loss). Generally the treatments that increased LWG/ha decreased resource condition as a result of increased utilisation of grass production, which reduced fuel for burning and soil surface cover.

Strategies using current climate forecasting skill in November (#19) to adjust a constant stocking strategy were not superior to responsive strategies in June. However, modifying the responsive stocking rate strategy in June according to a climate forecast (*responsive* + *long-lead forecast*) increased LWG/ha by 9% (25.2/23.1), reduced the risk of liveweight loss by 57% (from -3.9 to -1.7 kg/hd), reduced risk of low pasture yield, but increased the risk of soil loss (+4%). Alternative strategies (eg. #9) that could reduce annual soil loss (7%) without reducing production (LWG/ha) were also found. There were small advantages for the best *responsive* strategy (#4) over strategies using current forecasting skill in November (#19). Thus the development of long-lead forecasts has considerable potential to contribute to better management of climate variability in these grazing lands.

Further evaluation of grazing management strategies using the model described above are given in Ash *et al.* (2000) and Stafford Smith *et al.* (2000). These evaluations incorporate spatial variation and a financial analysis considering the limitations imposed by managing a herd.

Acknowledgments

We would like to thank Joe Scanlan and Peter O'Reagain for their comments on the manuscript; and acknowledge the financial support of the Climate Variability in Agriculture Program (administered by the Land and Water Resources Research and Development Corporation).

References

Allan, R.J. (1988) El Niño Southern Oscillation influences in the Australasian region. *Progress in Physical Geography*, 12, 313-48.

Ash, A.J. and McIvor, J.G. (1995) Land condition in the tropical tallgrass pasture lands. 2. Effects on herbage quality and nutrient uptake. *Rangeland Journal* 17, 86-98.

Ash, A.J. and Stafford Smith, M. (1996) Evaluating stocking rate impacts in rangelands: Animals don't practice what we preach. *Rangeland Journal* 18, 216-43.

Ash, A.J., Brown, J.R. and Cowan, D.C. (1996) Transitions between vegetation states in tropical tallgrass: Falling over cliffs and slowly climbing back. Proc. 9th Australian Rangelands Conference, Port Augusta, September 1996. pp.219-20.

Ash, A.J., McIvor, J.G., Corfield, J.P. and Winter, W.H. (1995) How land condition alters plant-animal relationships in Australia's tropical rangelands. *Agriculture, Ecosystems and Environment* 56, 77-92.

Ash, A.J., O'Reagain, P.J., McKeon G.M. and Stafford Smith, D.M. (2000) Managing Climate Variability In Grazing Enterprises: A Case Study Of Dalrymple Shire, North-Eastern Australia, in G.L. Hammer, N. Nicholls, and C. Mitchell (eds.), Applications *of Seasonal Climate Forecasting in Agricultural and Natural Ecosystems – The Australian Experience.* Kluwer Academic, The Netherlands. (this volume)

Ash, A.J., Prinsen, J.H., Myles, D.J. and Hendricksen, R.E. (1982) Short-term effects of burning native pasture in spring on herbage and animal production in south-east Queensland. *Proc. Australian Society of Animal Production* 14, 377-80.

Burrows, W.H. (1995) Greenhouse revisited - land-use change from a Queensland perspective. *Climate Change Newsletter* 7, 6-7.

Burrows, W.H. (1997) Effects of stocking rate, legume augmentation, supplements and fire on animal production and stability of native pastures, in B. Walker (ed.), *The North Australia Program: 1997 review of improving resource management projects.* Meat Research Corporation, Sydney. pp.13-18.

Burrows, W.H., Carter, J.O., Scanlan, J.C. and Anderson, E.R. (1990) Management of savannas for livestock production in northeast Australia: Contrasts across the tree grass continuum. *Journal of Biogeography* 17, 503-12.

Cane, M.A. (2000) Understanding and predicting the world's climate system, in G.L. Hammer, N. Nicholls, and C. Mitchell (eds.), *Applications of Seasonal Climate Forecasting in Agricultural and Natural Ecosystems – The Australian Experience.* Kluwer Academic, The Netherlands. (this volume)

Clewett, J.F., Young, P.D. and Willcocks, J.R. (1988) Effect of climate change on agriculture in Central Queensland: Rainfall variability analysis, in E.R. Anderson (ed.), *The changing climate and central Queensland agriculture.* Australian Institute of Agricultural Science, Rockhampton. pp.43-52.

Coates, D.B. (1995) Effect of phosphorus as fertiliser or supplement on the forage intake of heifers grazing stylo-based pastures. *Australian Journal of Experimental Agriculture* 35, 181-8.

Daly, J.J. (1994) *Wet as a shag, dry as a bone.* Queensland Department of Primary Industries, Brisbane.

Day, K.A., McKeon, G.M. and Carter, J.O. (1997a) Evaluating the risks of pasture and land degradation in native pasture in Queensland. Final report for Rural Industries and Research Development Corporation project DAQ124A. Queensland Dept. Natural Resources, Brisbane.

Day, K.A., Scattini W.J. and Osborne, J.C. (1997b) Extending carrying capacity calculations to the central Burnett region of Queensland. DroughtPlan Working Paper No. 10. CSIRO, Alice Springs.

Day, K.A., McKeon, G.M. and Orr, D.M. (1993) Comparison of methods for assessing productivity of native pastures in Queensland. Proc. XVII International Grassland Congress, Palmerston North, New Zealand. pp. 784-5.

Drosdowsky, W. and Chambers, L. (1998) Near global sea surface temperature anomalies as predictors of Australian seasonal rainfall. BMRC Research Report No. 65. Bureau of Meteorology, Melbourne.

Fahey, D. (1997) Changing grazing management on "Keen-Gea" – the strategies and findings – a producers experience. An Introduction to Management Practices in the Dalrymple Shire. Grazing Land Management Unit, Queensland Department of Primary Industries, Charters Towers.

Gardener, C.J., McIvor, J.G. and Williams, J. (1990) Dry tropical rangelands: Solving one problem and creating another. *Proc. Ecological Society of Australia* 16, 279-86.

Gillard, P. (1979) Improvement of native pasture with Townsville stylo in the dry tropics of sub-coastal northern Queensland. *Aust. J. Exp. Agric. and Anim. Husb.* 19, 325-36.

Gillard, P. and Moneypenny, R. (1990) A decision support model to evaluate the effects of drought and stocking rate on beef cattle properties in northern Australia. *Agricultural Systems* 34, 37-52.

Hall, W.B. (1996) Near-real time financial assessment of the Queensland wool industry on a regional basis. Unpub. Ph.D. thesis, University of Queensland, Brisbane.

Hall, W.B., McKeon, G.M., Carter, J.O., Day, K.A., Howden, S.M., Scanlan, J.C., Johnston, P.W. and Burrows, W.H. (1998) Climate change and Queensland's grazing lands: II. An assessment of impact on animal production from native pastures. *Rangeland Journal* 20, 174-202.

Hall, W.B., Rickert, K.G., McKeon, G.M. and Carter, J.O. (1999) Simulation studies of diet selection by sheep in western Queensland. *Australian Journal of Agricultural Research (in press).*

Hammer, G.L. (2000) A general systems approach to applying seasonal climate forecasts, in G.L. Hammer, N. Nicholls, and C. Mitchell (eds.), *Applications of Seasonal Climate Forecasting in Agricultural and Natural Ecosystems – The Australian Experience.* Kluwer Academic, The Netherlands. (this volume)

Hammer, G.L., McKeon, G.M., Clewett, J.F. and Woodruff, D.R. (1991) Usefulness of seasonal climate forecasts in crop and pasture management. Proc. First Australian Conference on Agricultural Meteorology. Bureau of Meteorology, Commonwealth of Australia, Melbourne. pp.15-23.

Hammer, G.L., Holzworth, D.P. and Stone, R. (1996) The value of skill in seasonal climate forecasting to wheat crop management in a region with high climatic variability. *Aust. J. Agric. Res.* 47, 717-737.

Hendricksen, R.E., Rickert, K.G., Ash, A.J. and McKeon, G.M. (1982) Beef production model. *Proc. Australian Society of Animal Production* 14, 204-8.

Hunt, B. (1991) The simulation and prediction of drought, in A. Henderson-Sellers and A.J. Pitman (eds.), *Vegetation and Climate Interactions in Semi-arid Regions.* Kluwer Academic Publishers, The Netherlands. pp.89-103.

Hunt, B. and Davies, H.L (1997) Mechanism of multi-decadal climatic variability in a global climatic model. *International Journal of Climatology* 17, 565-80.

Hunt, B., Zebiak, S.E. and Cane, M.A (1994) Experimental predictions of climatic variability for lead times of twelve months. *International Journal of Climatology* 14, 507-26.

Johnston, P.W., McKeon, G.M., Buxton, R., Cobon, D.H., Day, K.A., Hall, W.B. and Scanlan, J.C. (2000) Managing climatic variability in Queensland's grazing lands - current status, in G.L. Hammer, N. Nicholls, and C. Mitchell (eds.), Applications *of Seasonal Climate Forecasting in Agricultural and Natural Ecosystems – The Australian Experience.* Kluwer Academic, The Netherlands. (this volume)

Johnston, P.W., McKeon, G.M. and Day, K.A. (1996) Objective "safe" grazing capacities for south-west Queensland Australia: Development of a model for individual properties. *Rangeland Journal* **18**, 244-258.

Kleeman, R., Moore, A.M. and Smith, N.R. (1995) Simulation of sub-surface thermal data into an intermediate tropical coupled ocean atmosphere model. *Monthly Weather Review* **123**, 3103-13.

Landsberg, R.G., Ash, A.J., Shepherd, R.K. and McKeon, G.M. (1998) Learning from history to survive in the future: Management evolution on Trafalgar Station, north-east Queensland. *Rangeland Journal* **20**, 104-118.

Lough, J.M. (1991) Rainfall variation in Queensland, Australia 1891-1986. *Int. J. Climatol.* **11**, 745-68.

McBride, J.L. and Nicholls, N. (1983) Seasonal relationships between Australian rainfall and the Southern Oscillation. *Monthly Weather Revue* **111**, 1998-2004.

McCown, R.L. (1973) An evaluation of the influence of available soil-water storage capacity on growing season length and yield of tropical pastures using simple water balance models. *Agricultural Meteorology* **11**, 53-63.

McCown, R.L., Gillard, P. and Edye, L.A. (1974) The annual variation in yield of pastures in the seasonally dry tropics of Queensland. *Aust. J. Exp. Agric. and Anim. Husb.* **14**, 328-33.

McCown, R.L., Gillard, P., Winks, L. and Williams, W.T. (1980-81) The climatic potential for beef cattle production in tropical Australia. 2. Liveweight change in relation to agroclimatic variables. *Agricultural Systems* **7**, 1-10.

McIvor, J.G. and Monypenny, R. (1995) Evaluation of pasture management systems for beef production in the semi-arid tropics: Model development. *Agricultural Systems* **49**, 45-67.

McIvor, J.G., Ash, A.J. and Cook, G.D. (1995a) Land condition in the tropical tallgrass pasture lands. 1. Effects on herbage production. *Rangeland Journal* **17**, 69-85.

McIvor, J.G., Williams, J. and Gardener, C.J. (1995b) Pasture management influences run-off and soil movement in the semi-arid tropics. *Aust. J. Exp. Agric.* **35**, 55-65.

McKeon, G.M. and Rickert, K.G. (1984) A computer model of the integration of forage options for beef production. *Proc. Aust. Soc. Anim. Prod.* **15**, 15-19.

McKeon, G.M. and White, D.H. (1992) El Niño and better land management. *Search* **23**, 197-200.

McKeon, G.M., Ash, A.J., Hall, W.B. and Stafford Smith, D.M. (1998a) Simulation of grazing strategies for beef production in north-east Queensland. DroughtPlan Working Paper No. 8. Queensland Department of Natural Resources, Brisbane.

McKeon, G.M., Day, K.A., Howden, S.M., Mott, J.J., Orr, D.M., Scattini, W.J. and Weston, E.J. (1990) Management of pastoral production in northern Australian savannas. *J. Biogeog.* **17**, 355-72.

McKeon, G.M., Hall, W.B., Crimp, S.J., Howden, S.M., Stone, R.C. and Jones, D.A. (1998b) Climate change in Queensland's grazing lands. I. Approaches and climatic trends. *Rangeland Journal* **20**, 147-173.

McKeon, G.M., Rickert, K.G., Ash, A.J., Cooksley, D.G. and Scattini, W.J. (1982) Pasture production model *Proc. Aust. Soc. Anim. Prod.* **14**, 201-4.

McLennan, S., Clem, R.L., Mullins, T.J. and Shepherd, R.K. (1986) The effect of burning on pasture yield and animal performance in the sub-coastal speargrass region of north Queensland. Proceedings of the Second International Rangeland Congress, Adelaide, May 1984. pp.605-6.

Minson, D.J. and McDonald, C.K. (1987) Estimating forage intake from the growth of beef cattle. *Tropical Grasslands* **21**, 116-22.

Mott, J.J., Williams, J. Andrew, M.H. and Gillison, A.N. (1985) Australian savanna ecosystems, in J.C. Tohill and J.J. Mott (eds.), *Ecology and Management of the Worlds Savannas*. Australian Academy of Science, Canberra. pp.56-82.

Nicholls, N. (1983) Predictability of 1982 Australian drought. *Search* **14**, 154-5.

Nicholls, N., Fredericksen, C., and Kleeman, R. (2000) Operational experience with climate model predictions, in G.L. Hammer, N. Nicholls, and C. Mitchell (eds.), *Applications of Seasonal Climate Forecasting in Agricultural and Natural Ecosystems – The Australian Experience*. Kluwer Academic, The Netherlands. (this volume)

Norman, M.J.T. and Begg, J.E. (1973) Katherine Research Station: A review of published work 1965-72. CSIRO Division of Land Research Technical Paper No. 33. CSIRO, Canberra.

Pittock, A.B. (1975) Climatic change and the patterns of variation in Australia rainfall. *Search* **6**, 498-504.

Rickert, K.G. and McKeon, G.M. (1982) Soil water balance model: WATSUP. *Proc. Aust. Soc. Anim. Prod.* **14**, 198-200.

Scanlan, J.C. and McIvor, J.G. (1993) Pasture composition influences soil erosion in *Eucalyptus* woodlands of northern Queensland. Proc. XVII International Grassland Congress, Palmerston North, New Zealand. pp.65-6.

Scanlan, J.C. and McKeon, G.M. (1990) Grassman. A computer program for managing native pastures in eucalypt woodlands. Version 1. Queensland Department of Primary Industries, Brisbane.

Scanlan, J.C. and McKeon, G.M. (1993) Competitive effects of trees on pasture are a function of rainfall distribution and soil depth. Proc. XVII International Grassland Congress, Palmerston North, New Zealand. pp.2231-2.

Scanlan, J.C., McKeon, G.M., Day, K.A., Mott, J.J. and Hinton, A.W. (1994) Estimating safe carrying capacities in extensive cattle grazing properties within tropical semi-arid woodlands of north-eastern Australia. *Rangeland Journal* **16**, 64-76.

Scanlan, J.C., Pressland, A.J. and Myles, D.J. (1996a) Runoff and soil movement on mid-slopes in north-east Queensland grazed woodlands. *Rangeland Journal* **18**, 33-46.

Scanlan, J.C., Pressland, A.J. and Myles, D.J. (1996b) Grazing modifies woody and herbaceous components of north Queensland woodlands. *Rangeland Journal* **18**, 47-57.

Scattini, W.J. (1973) A model for beef cattle production from rangeland and sown pasture in south eastern Queensland, Australia. Unpub. Ph. D. thesis, University of California, Berkeley.

Smith, I. (1994) Indian ocean sea-surface temperature patterns and Australian winter rainfall. *Int. J. Climatol.* **14**, 287-305.

Stafford Smith, D.M. (1992) Stocking rate strategies across Australia. Range Management Newsletter 92, 1-3.

Stafford Smith, M., Buxton, R., McKeon, G.M. and Ash, A. (2000) Seasonal climate forecasting and the management of rangelands: do production benefits translate into enterprise profits? in G.L. Hammer, N. Nicholls, and C. Mitchell (eds.), *Applications of Seasonal Climate Forecasting in Agricultural and Natural Ecosystems – The Australian Experience.* Kluwer Academic, The Netherlands. (this volume)

Stone, R.C. and Auliciems, A. (1992) SOI phase relationships with rainfall in eastern Australia. *Int. J. Climatol.* **12**, 625-636.

Stone, R.C. and de Hoedt, G. (2000) The development and delivery of current seasonal climate forecasting capabilities in Australia, in G.L. Hammer, N. Nicholls, and C. Mitchell (eds.), *Applications of Seasonal Climate Forecasting in Agricultural and Natural Ecosystems – The Australian Experience.* Kluwer Academic, The Netherlands. (this volume).

Stone, R.C. and McKeon, G.M. (1992) Tropical pasture establishment. 17. Prospects for using weather predictions to reduce pasture establishment risk. *Tropical Grasslands* **27**, 406-13.

Stone, R.C., Hammer, G.L. and Marcussen, T. (1996) Prediction of global rainfall probabilities using phases of the Southern Oscillation Index. *Nature* **384**, 252-5.

Wandera, P.F. (1993) Patches in a *Heteropogon contortus* dominated grassland in southeast Queensland - their characteristics, probable causes, implications and potential for rehabilitation. Unpub. Ph. D. thesis, University of Queensland, Brisbane.

White, B.J. (1978) *A Simulation Based Evaluation of Queensland's Northern Sheep Industry.* Monograph Series, Department of Geography, James Cook University, Townsville.

Williams, J. and Chartres, C.J. (1991) Sustaining productive pastures in the tropics. 1. Managing the soil resource. *Tropical Grasslands* **25**, 73-84.

MANAGING CLIMATE VARIABILITY IN GRAZING ENTERPRISES: A CASE STUDY OF DALRYMPLE SHIRE, NORTH-EASTERN AUSTRALIA

ANDREW ASH[1], PETER O'REAGAIN[2], GREG MCKEON[3] AND MARK STAFFORD SMITH[4]

[1] CSIRO Tropical Agriculture
PMB PO
Aitkenvale, Qld 4814, Australia

[2] Queensland Department of Primary Industries
PO Box 976, Charters Towers, Qld 4820, Australia

[3] Queensland Department of Natural Resources
Climate Impacts and Spatial Systems
80 Meiers Rd, Indooroopilly, Qld 4068, Australia

[4] CSIRO Wildlife and Ecology
PO Box 2111
Alice Springs, NT 0871, Australia

Abstract

In this paper we examine approaches to managing climate variability in the Dalrymple Shire of north-east Queensland. We use the forage-animal production model GRASP to evaluate the production and resource implications of grazing management and seasonal climate forecasting strategies. In this study, five forecasting strategies were assessed at each of nine test rainfall stations in Dalrymple Shire. Forecasting strategies used were: (a) spring SOI, (b) spring SOI phases, (c) an SOI phase system "tuned" to Charters Towers rainfall, (d) winter Pacific Ocean sea surface temperatures (EOF analyses), and (e) winter Pacific and Indian Ocean sea surface temperatures (EOF analyses). Stocking rates were adjusted annually according to analogue year types provided by the forecasts. In forecast "dry" years stock numbers were reduced by 50% and in "wet" years they were increased by 30%. Stocking rate changes were made in either November, when all forecasts were available, or in June, which assumes some improvement in forecasting lead time, particularly for the SOI.

Results from these analyses show that:
- seasonal climate forecasting provides more benefit to animal production when the forecasting information is available in June rather than November

G.L. Hammer et al. (eds.), The Australian Experience, 253–270.
© 2000 *Kluwer Academic Publishers. Printed in the Netherlands.*

- the relative value of forecasting is greater for constant grazing strategies than for flexible ones
- increased animal production derived from applying a forecast is not at the expense of the resource base and if increased animal production is not the desired aim of the forecast then significant reductions in soil loss can be achieved
- using localised forecasts does not provide any extra skill in production simulations.

Despite the demonstrated benefits of using a forecasting strategy based on the SOI, there is considerable reluctance amongst producers to adopt such forecasts. A producer survey indicated that even if more reliable forecasts were developed, most would wait until extreme events had an impact on their enterprise before making critical stocking decisions. A grazing trial has been established that will compare grazing management strategies, including a seasonal forecasting strategy, which should assist in demonstrating the potential benefits of seasonal forecasting. Even in the absence of seasonal forecasts, grazing management in the rangelands of Australia could be vastly improved through better incorporating existing understanding of climate variability into stocking decisions. In this paper we show that further value maybe added through the application of seasonal forecasting but that perhaps this added sophistication should only be contemplated after basic grazing management principles are incorporated into whole property planning.

1. Introduction

"Perhaps the most tantalizing phenomenon of the Central Division of Queensland is its rainfall. The whole continent suffers in turn from droughts and floods, but in few parts, except the far interior, are the annual rainfalls so spasmodic and unreliable..... Innumerable theories have been advanced to explain the constantly recurring droughts and floods, but no solution has apparently yet been hit upon" (Bird 1904).

The early settlers of northern Australia were clearly perplexed by climate variability and quickly recognised the severe constraints that it placed on the development of a viable grazing industry. The damage and loss caused by droughts and floods featured regularly in the diaries of the early pastoralists. Despite many early setbacks the sheep and cattle industries survived and grew and by the turn of the century occupied more than 90% of the semi-arid tropical zone of northern Australia.

Today, pastoralists in northern Australia have to cope with the same climate variability but they have the advantage of a much better climatological understanding and much improved on-farm technologies to help buffer some of the climatic extremes. Rainfall in northern Australia is summer dominant with over 80% of rain being recorded between November and April. Seasonality of this rainfall is greatest in the Top End of the Northern Territory and in the Kimberley region of Western Australia while rainfall is most variable close to the east and west coasts. A combination of unreliable rainfall, nutrient poor soils and tropical grasses that are low in nutritive value contribute to grazing lands that are relatively unproductive and have a low carrying capacity.

Consequently, the grazing industry in northern Australia is characterised by properties that are large in size and have low inputs and only modest outputs.

In this study, we focus on the Dalrymple Shire in north-east Queensland. This shire covers 67,687 km^2 and measures about 370 km north to south and 260 km east to west. Open eucalypt woodlands dominate the shire though there are significant areas of spinifex desert in the south-west and pockets of rainforest in the north-east. Rainfall ranges from 500 to 700 mm. Inter-annual rainfall has coefficients of variation that range from 37% to 49% (Figure 1). The correlation between the Southern Oscillation Index (SOI) and summer rainfall in this region is quite strong. This, combined with the relatively good understanding of climate impacts on the forage-animal system makes the Dalrymple region suitable for a case study. The nine rainfall stations shown in Figure 1 were used in simulation studies described later.

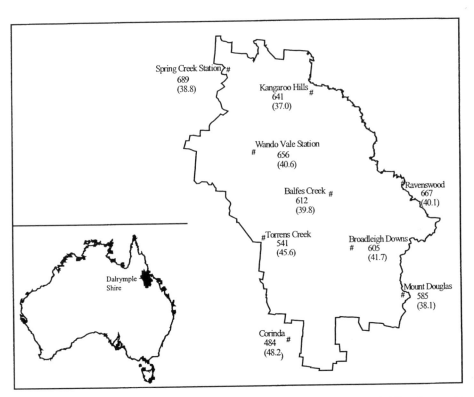

Figure 1. Map of Dalrymple Shire, its location within Australia (inset), and the nine rainfall stations used in subsequent analyses. Figures beneath station names are mean annual rainfall (mm) and coefficient of variation (%) of annual rainfall (shown in brackets).

A survey of Dalrymple Shire has revealed that enterprises are relatively unproductive, with gross margins only a little more than $5 per ha (Hinton 1993). With declining terms of trade and recurrent droughts, most properties in the Shire have been returning substantial losses in recent years. Therefore, any improvement in production or prevention of losses would be most welcome by the industry in the extensively used regions. In this paper we explore the potential role of seasonal forecasting in adding value to production in favourable years, preventing losses in production in drought years, and in minimising resource damage. We first examine how the grazing industry currently copes with climate variability through various grazing strategies. This provides the context, both historical and contemporary, in which we can assess the likely success of new grazing management strategies. We then assess the value of seasonal forecasting using the forage-animal production model, GRASP (McKeon *et al.* 1990) to expand on the companion study of McKeon *et al.* (2000).

2. Coping With Rainfall Variability

2.1 HISTORICAL GRAZING MANAGEMENT

When pastoralists moved into northern Australia they brought with them British (*Bos taurus*) breeds of cattle that were poorly adapted to the climate and the very low forage quality of tropical grasses. Their response to these constraints was to stock at very light rates so that cattle had the best opportunity to select a diet of reasonable quality from the low nutrient pastures. Low stocking rates also ensured that feed shortages did not occur during droughts. Because of the nutritional constraints, calving percentages were also low so that herd sizes did not expand rapidly. In the mid-1890s ticks and associated tick fever spread through northern Queensland. Ticks were most prolific in wetter years so even years that were favourable in terms of pasture growth and forage quality carried with them the constraints of pests and disease. Together, these constraints acted to keep cattle populations in check as the cattle numbers at Hillgrove Station (Tom Mann, personal communication) in the Dalrymple Shire clearly show (Figure 2).

It is clear that during this period cattle enterprises were very much at the mercy of the climate and to a great degree the climate "managed" the cattle herds for producers. The climatic constraints on stock numbers meant that there were few resource management problems as utilisation rates were rarely and consistently high enough to influence vegetation dynamics or reduce cover levels to allow increased erosion. Fire was regularly used to promote a more nutritious "green pick" for the cattle (Ash *et al.* 1982) so a fairly stable open eucalypt woodland was maintained throughout the region.

2.2 POST-1960S GRAZING MANAGEMENT - NEW TECHNOLOGIES AND NEW PROBLEMS

In the 1960s, tick and drought resistant Brahman (*Bos indicus*) cattle were being widely adopted by graziers. Brahman cattle have an enhanced ability to forage and digest low quality pastures compared with British breeds (Frisch and Vercoe, 1977), so this advantage, combined with their tick resistance, ensured their success in tropical Australia. Supplementary feeding technologies, largely based on phosphorus and urea-molasses mixes, greatly reduced mortality.

These new technologies gave producers the ability to control much better their herd size and make decisions on stocking strategies without as much reliance on prevailing climatic and forage supply conditions. Herd sizes began to increase rapidly through the 1960s and this rate of increase was accelerated by a beef slump in the 1970s, which saw most producers in the Dalrymple Shire retain cattle rather than sell them for what they considered unrealistically low prices. Consequently, cattle numbers doubled during this period as the stock records for Trafalgar Station show (Figure 2). A run of drier than average years during the 1980s saw many producers initially retain their high stock numbers and attempt to survive the drought by embarking on extensive drought feeding and supplementation programs. However, an economic survey of Dalrymple Shire properties indicated that greater than 90% of the direct costs of production are incurred through purchasing supplements, fodder or agisting cattle and that these costs have a major impact on enterprise profitability (Hinton 1993). As these drought mitigation programs became too expensive to sustain, producers were forced to sell off their herds. The controlling influence on herd numbers was in the end unfavourable economic conditions rather than direct biophysical constraints such as high mortality rates.

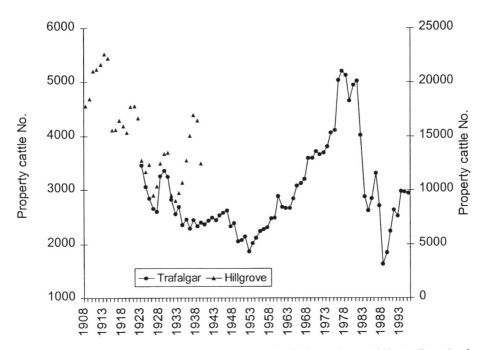

Figure 2. Historical record of cattle numbers on Hillgrove Station (80 km north-west of Charters Towers) and Trafalgar Station (60 km south-west of Charters Towers).

The last three decades have shown that producers can manage their herds much more proactively than in the previous 80 to 100 years. While climate variability is still a strong *influencing* factor on animal production and herd numbers it is no longer the *controlling* factor. However, while the new breeding and feeding technologies have given producers the tools to better cope with climate variability in terms of production they have also induced resource management problems previously of no great concern to the grazing industry. Increased utilisation rates brought about by increased stocking rates have resulted in loss of valuable perennial grasses and associated biodiversity, increased soil erosion and run-off, and increased density of eucalypts, shrubs and exotic woody weeds as a consequence of reduced fire frequency (Gardener *et al.*, 1990). Current grazing management strategies, in trying to cope with climate variability, must therefore take into account both production and resource management implications.

3. Contemporary Grazing Management Strategies and Their Production and Resource Consequences

In the Dalrymple Shire three approaches or philosophies to grazing management and coping with climate variability are currently practised (described in 3.1, 3.2, and 3.3). Most grazing trials are usually too short to assess properly the longer term production and resource consequences of various grazing management strategies, particularly where issues such as vegetation change and soil loss are involved. We have used the currently operational forage and animal production model, GRASP, to simulate the longer term production and resource implications of these grazing strategies. A brief description of GRASP is given in Box 1.

3.1 STOCK WITH A RELATIVELY HIGH NUMBER OF ANIMALS AT A RELATIVELY CONSTANT LEVEL

This stocking strategy is widely practised throughout the region. In times of drought the herd is fed supplements and/or parts of the herd are agisted in areas where forage supply is more plentiful. Despite the economic constraints this sort of strategy would appear to impose through the costs of supplementation and agistment, as indicated above and in the study of Hinton (1993), it is still the preferred method of grazing management by many producers in the region. Grazing trials throughout Australia and overseas (Ash and Stafford Smith, 1996; Stafford Smith *et al.,* 1997) show that production per ha is maximised at high stocking rates in the short to medium term (3-10 years). By adopting such a strategy many producers perceive that maintaining high stock numbers is also financially rational, even taking into account the costs of supplementation and agistment.

However, the longer term production, financial and resource management implications of a high stocking strategy are less clear, at least from empirical studies. Certainly, the experience of past overstocking events in southern, central and western Australia (Anon 1901; Ratcliffe, 1936; Wilcox 1963), and more recently in Dalrymple Shire, leaves little doubt as to the serious resource implications of pursuing a high stocking strategy in the longer term. Although there may be some diet quality benefits from a changed vegetation state as a result of overstocking, the loss of pasture productivity far outweighs these benefits (Ash *et al.*, 1995). The GRASP simulations highlight the longer term implications of pursuing this strategy. Eventually a series of dry years forces a decline in perennial grasses from which the system does not recover. Production also is greatly reduced and soil loss increases dramatically (Figure 3).

3.2 STOCK CONSERVATIVELY TO MINIMISE DROUGHT IMPACTS

This strategy aims to utilise some safe amount (eg. 25%) of the average forage growth by stocking with a conservative number of cattle. This strategy accepts that overutilisation and feed shortages will occur in extreme years but it is assumed that such seasons will be sufficiently infrequent to keep pasture degradation and/or economic loss to a minimum. Although production per unit area may be low relative to the heavier stocking strategy (see 3.1), conservative stocking has the advantage of maintaining or improving the resource base, minimising production costs and maximising individual animal performance (Foran and Stafford Smith, 1991; Pratchett and Gardiner, 1993). Consequently, while the strategy may have relatively lower returns in a good year due to lost opportunity cost, in a poor season it is perceived by its practitioners to outperform other strategies due to lower costs and better animal performance. An added advantage of this strategy is that there will be increased opportunities for fire to control unwanted increase in woody plants due to increased fuel loads and management flexibility in being able to rest and burn paddocks. The simulations show that while production per ha is lower at lower levels of utilisation (Figure 3) fire is an option in about 60% of years. The simulations also show that there is a sharp decline in production once the optimum stocking rate is passed. Given the relatively small increases in production between conservative stocking and the optimum with a constant grazing strategy (Figure 3) there would appear to be little benefit in attempting to get close to the optimum because of the risk of a dramatic decline. As rainfall variability increases this safety margin between conservative and the decline threshold would need to be increased to take account of the greater variability in forage production.

3.3 STOCK IN A FLEXIBLE WAY BY RESPONDING TO CHANGES IN FORAGE SUPPLY

This stocking strategy typically involves assessing the availability of forage at some fixed point in the season when further rainfall is unlikely, for example at the end of the wet, and adjusting stock numbers according to feed availability. This strategy requires considerably better herd management and marketing skills as it is inherently riskier. An error in judgement with this strategy could have more sudden and serious consequences for farm finances and the resource base. Flexible stocking should theoretically outperform conservative stocking as it capitalises on good years but avoids the economic and environmental costs of overstocking by reducing stock numbers during forage deficits. Simulation studies show this to be the case with substantially higher production with a flexible grazing strategy compared with a constant strategy. At the production optimum for the Broadleigh Downs example in Figure 3 there was an 18%

production advantage with flexible stocking. Interestingly, the relative benefit of adopting the flexible grazing strategy increases as rainfall variability increases compared with a constant strategy (Figure 4). This is logical as the constant stocking rate required to maintain the resource base in a more variable environment will be lower as there will be a higher proportion of years with low forage production.

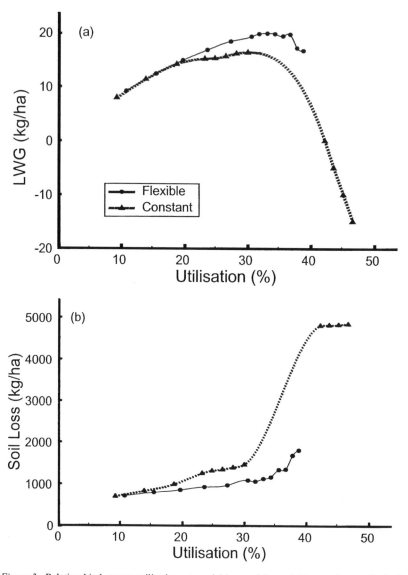

Figure 3. Relationship between utilisation rate and (a) annual liveweight gain of steers (kg/ha/) and (b) annual soil loss (kg/ha) for constant and flexible grazing strategies. Relationships are derived from GRASP simulations using approximately 100 years of historical rainfall records for the Broadleigh Downs site in Dalrymple Shire.

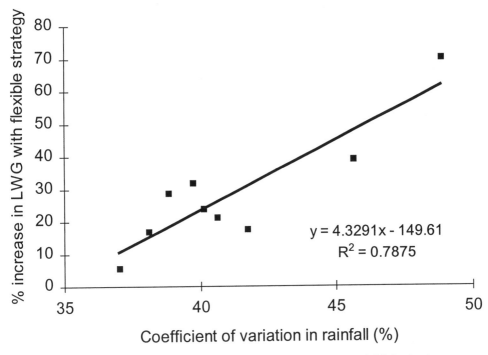

Figure 4. Relationship between coefficient of variation in interannual rainfall for the nine rainfall stations in the study and the relative benefit of the flexible stocking rate over the constant stocking rate for maximum liveweight gain (kg/ha/year).

GRASP is a model simulating the above-ground growth and biomass of a sward dominated by perennial native grasses in the tropics and sub-tropics. The model grows grass as a function of rainfall, infiltration, soil moisture balance, evapo-transpiration, radiation interception and nitrogen uptake. Management effects such as stocking rate and fire influence grass growth, basal area of perennial grasses, perennial grass/annual grass composition, ground cover and soil loss. Animal production (liveweight gain of growing steers) is simulated as a function of the length of the growing season and level of utilisation. The model and its application are more fully described in McKeon *et al.* (1990) and a full description of each equation is given in Littleboy and McKeon (1997). In our study we carried out simulations for nine rainfall stations in Dalrymple Shire. Historical daily rainfall records of these stations were used to predict the implications of the various grazing management and seasonal forecasting strategies. For other daily climate indices (pan evaporation, minimum and maximum temperatures, radiation) we used the records (actual and generated) for Charters Towers for all nine daily rainfall stations. These rainfall stations are shown in Figure 1.

Box 1. Description of the forage-animal production model GRASP.

4. Can Seasonal Climate Forecasting Add Value to Production and/or Minimise Resource Damage?

We now want to assess whether such stocking strategies can be improved by the use of seasonal forecasting technologies.

A variety of seasonal climate forecasts are in operation and most of these can be applied to crop and pasture-animal production models to assess the value of seasonal climate forecasting (Hammer, 2000). Seasonal forecasts range from simple statistical procedures based on linear lagged regression between summer rainfall and the spring Southern Oscillation Index (SOI), simple cluster analysis of SOI trends over time to identify analogue seasons, sophisticated time-series analyses which predict the SOI in the future based on analogue years, more complex statistical analysis of sea surface temperature (SST) patterns in the Indian and Pacific Oceans, to more dynamical forecasts linking the ocean and atmosphere system to generate predictions of sea surface temperature (SST) in the eastern Pacific (Stone and de Hoedt, 2000; Nicholls, 1996). Currently, statistical forecasts based on the SOI (Stone and Auliciems, 1992; Stone et al., 1996) are largely used with crop and pasture-animal simulation models, with the application of such forecasts being more advanced in the cropping industry (Hammer et al., 1996; Stone, 1996).

In Dalrymple Shire the lag regression of summer rainfall on spring SOI is significant though the amount of variance explained is not high ($R^2=0.15$). When modelled pasture growth (using the same summer rainfall) is regressed against spring SOI the relationship is improved ($R^2=0.20$). However, stocking decisions based on seasonal climate forecasting in the extensive grazing industry are not likely to be made on such a continuous or sliding scale but may be simpler and more categorical ie. to reduce stock numbers by a certain proportion in forecast dry years with the reverse applying in forecast wet years. Categorising the SOI into year types (based on the approach of Clewett et al., 1991) and relating these year types to modelled pasture growth, which is similarly divided into categories, shows that there is some potential promise in applying an SOI based forecast in the extensive grazing industry (Table 1).

TABLE 1. Frequency of occurrence of simulated pasture growth categories relative to preceding spring SOI (average August-October) categories for the 106 years from 1890 to 1996 at Charters Towers, north-east Queensland. Average simulated pasture growth is 1500 kg/ha.

Simulated pasture growth (kg/ha)	SOI Category		
	< -5	-5 → +5	> +5
< 1000 kg/ha	22	15	4
1000 → 2000 kg/ha	9	11	5
>2000 kg/ha	5	16	19

4.1 ASSESSING THE VALUE OF DIFFERENT FORECASTS FOR GRAZING MANAGEMENT IN DALRYMPLE SHIRE

The forage-animal production model GRASP was used to assess a number of seasonal climate forecasts at the nine rainfall stations in Dalrymple Shire shown in Figure 1. All forecasting systems used were lag forecasts based on observed values of either the SOI or SST. SOI and SST indices were available as far back as the historical rainfall records of the nine stations. Forecasts were simplified into three year types; above average, average and below average. Stocking rates were adjusted each year in the model to reflect the forecast year type and remained at that stocking rate for the year until the next forecast was available. In forecast "dry" years stock numbers were reduced by 50% and in "wet" years they were increased by 30%. The different multiplier for increasing and decreasing stock numbers was in recognition that it is easier for producers to reduce numbers either by sale or agistment than it is to quickly increase animal numbers. The optimum amount to decrease and increase herd size in response to seasonal climate forecasts is assessed by Stafford Smith *et al.* (2000) elsewhere in this volume. Stocking rates were adjusted in either November, when all forecasts were available, or in June, which assumes some improvement in forecasting lead time, particularly for the SOI indices. The availability of a forecast in May/June is likely to be important for the extensive grazing industry as this is just after completion of the main mustering and weaning round and saleable animals are in their best condition after the wet season. Decisions on how many stock to carry through the next year, which animals are most appropriate for sale, culling of unwanted heifers and old cows, and the general structure of the herd are often made at this time so a climate forecast for the next wet season would be most beneficial at this time. The forecasts were assessed with two stocking strategies; either constant or flexible, as described earlier.

The five forecasts used were:

(i)   Spring SOI (August to October average, <-5 = dry, -5→+5 = average, >+5 = wet)

(ii) SOI Phases (5 phases (falling, negative, neutral, positive, rising) of Stone and Auliciems (1992) reduced to three groups (negative/falling, neutral, positive/rising)

(iii) Spring SOI "locally" optimised into the three year type categories by relating spring SOI average and SOI phases to Charters Towers historical rainfall

(iv) Winter Pacific Ocean SST (EOF time series based on 2° grid with pattern analysis to generate three clustered year types)

(v)   Winter Pacific and Indian Ocean SST (EOF time series based on 2° grid with pattern analysis to generate three clustered year types)

The five forecasting systems were compared with a control simulation of each management strategy conducted without the use of a seasonal forecast.

5. Results and Discussion of the Analysis of Strategies Using Seasonal Forecasting

The simulation studies showed there was a significant impact of forecasting but the magnitude of this effect varied according to forecasting system, lead time of forecast, the stocking strategy used and the rainfall location in the shire. Overall the benefits to gross production from forecasting are modest but the net benefits may be significantly greater as the costs of introducing a forecasting system into management of the enterprise may not be great. These economic implications are explored in more detail in Stafford Smith *et al.* (2000).

The comparisons of stocking strategies outlined below are based on the stocking rate that gave the maximum liveweight gain per unit area (ha). There was no single forecasting system that gave consistently better results across all stocking strategy and forecast lead time combinations. The Spring SOI forecast gave the best results for a constant stocking strategy applied in June while the Local SOI forecast gave the best production when a constant stocking strategy was applied in November. For the flexible stocking strategies the SOI Phase forecast gave the most benefit in both June and November. Figure 5(a) illustrates the relative benefit to production of the "best" forecasting system for each of the four stocking strategy x lead time combinations while Figure 5(b) shows the implications of these same forecasting systems for soil loss. The SST forecasting systems consistently gave less benefit than the SOI based forecasts. These forecasts have a longer lead time than the SOI forecasts but, at least for the SST forecasts used in this analysis, had less skill. The SOI forecast locally "optimised " to Charters Towers rainfall generally provided no extra skill compared with the more general Spring SOI or SOI Phase forecast systems. This highlights the inherent dangers in developing local forecasts that have little additional skill (ie. false skill) when used in other locations, even within the same region.

Applying a seasonal forecast in June was of more benefit than when applied in November. This indicates that improvement in lead time in current statistical forecasts could be expected to increase the benefit of that forecast. This production benefit may translate to even greater benefits in profitability when livestock prices are taken into consideration (see Stafford Smith *et al.*, 2000).

Even without forecasts the flexible grazing strategy gives significant improvement in production and at the same time reduces soil loss compared with a constant grazing strategy (Figure 3). Indeed the simulated benefit derived from adopting a flexible grazing strategy is considerably greater than from any seasonal climate forecasts. When a forecast is applied to the flexible grazing strategy, the additional benefit is small and is about half that of the same forecast applied to a constant grazing strategy (Figure 5). Because of the much greater level of management and marketing skill required to successfully operate a flexible grazing strategy, there may be greater acceptance amongst pastoralists of a constant grazing strategy enhanced with a seasonal climate forecast. This approach would still require a greater level of management input than most producers currently provide.

(a)

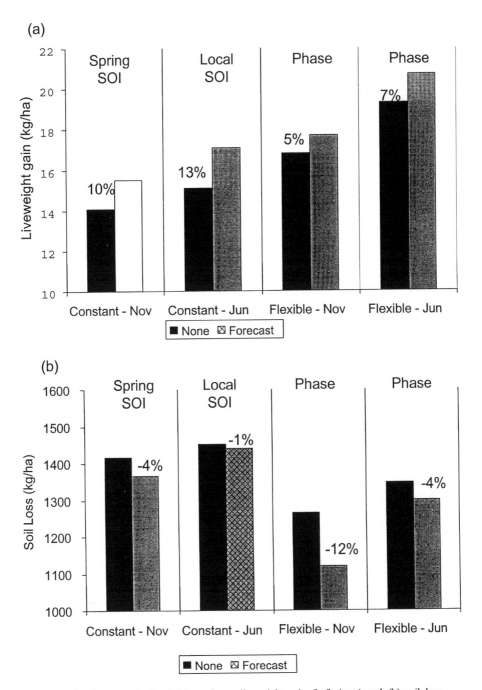

Figure 5. Average simulated (a) maximum liveweight gain (kg/ha/year) and (b) soil loss (kg/ha/year) at maximum liveweight gain (kg/ha/year) for constant and flexible grazing strategies with June or November stocking rate changes either with or without use of a seasonal forecast. For each comparison only the forecasting method that gave most benefit over the control (no forecast) is shown. This forecasting method is indicated at the top of each panel and the % change associated with its use is noted next to the bars.

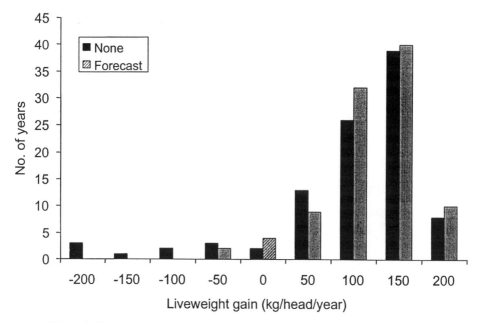

Figure 6. Frequency distributions of annual liveweight gain categories for a Spring SOI forecast system applied to a constant stocking strategy in November and the associated control (no forecast). The GRASP simulations are for the Broadleigh Downs site and use 95 years (1902-1996) of historical rainfall records.

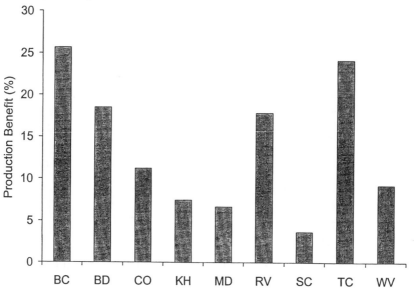

Figure 7. Spatial variation in the value of forecasting across the nine rainfall study sites in Dalrymple Shire. The relative forecasting value for each rainfall site is based on a comparison between the Spring SOI forecasting system applied in June and the control (no forecast) applied in June with the "constant" grazing strategy. BC = Balfes Creek, BD = Broadleigh Downs, CO = Corinda, KH = Kangaroo Hills, MD = Mt Douglas, RV = Ravenswood, SC = Sprong Creek, TC = Torrens Creek, WV = WandoVale.

One of the benefits of a forecasting strategy not evident from the mean values of a 100 year simulation, is the change in variability in production. Figure 6 shows a frequency distribution of liveweight performance of steers from one of the simulation runs (constant stocking strategy, forecast applied in November) comparing a Spring SOI forecast with a control no-forecast system. Without a forecast there are a significant number of years where a large liveweight loss was predicted. In reality, such a loss would generally be prevented by drought feeding or agisting cattle and the costs of these measures can be substantial. In contrast, with the seasonal climate forecast there were only a small number of years where a liveweight loss was predicted to occur. The decrease in production variability with the application of a seasonal climate forecast may be just as important as improvements in overall production, particularly where producers may have forward contracts which stipulate minimum liveweights.

Improvements in production through applying a seasonal climate forecast were not at the expense of the soil resource (Figure 5) or opportunities to use fire as a management tool (data not shown). Further, soil loss for strategies using a forecast was considerably less (between 10 and 20%) than the control simulations when the stocking rate in forecast simulations was reduced to achieve the same level of animal production as the optimum in non-forecast simulations i.e. if the aim of the forecast is to minimise the impact of grazing on the resource rather than maximise animal production then the reduction in soil loss can be significant. Although this reduction in soil loss may not have any benefits in the short term, in the longer term retention of top soil may have significant production implications through improved water infiltration and reduced scalding and run-off.

There was considerable spatial variation in the benefit of forecasting as the results in Figure 7 demonstrate for one of the simulation runs (Spring SOI, Constant grazing, June stocking change). Within a forecasting method the spatial differences in the value of the forecast were reasonably consistent across stocking strategies and timing of stocking rate change. However, there were large differences in the spatial patterning of benefits among the forecasting methods. Inter-annual rainfall variability, which explained much of the variation in the spatial differences between constant and flexible grazing strategies (Figure 4), was not related at all to the spatial differences due to seasonal forecasting.

In summary, seasonal climate forecasts can add modest but significant value to animal production even though the responses are vary with forecasting method and across the region. Specific outcomes from these simulation analyses are:

- adopting a flexible grazing strategy alone provides more benefit than a strategy employing seasonal forecasting ie. there is potential for considerable improvement in production and resource management via improved grazing management strategies even without forecasting strategies
- seasonal climate forecasting provides more benefit to animal production when the forecasting information is available in June rather than November

- the relative value of forecasting is greater for constant grazing strategies than for flexible ones, largely because constant stocking strategies are starting from a lower production base
- increased animal production derived from applying a forecast is not at the expense of the resource base and if increased animal production is not the desired aim of the forecast then significant reductions in soil loss can be achieved
- using localised forecasts do not provide any extra skill in production simulations

6. Improving the Use of Seasonal Climate Forecasts by the Grazing Industry

Seasonal climate forecasts will only be of benefit if they are applied and used as a management tool to aid grazing management decisions. A survey of forty-one producers in the Dalrymple Shire found that while nearly 60% watch and take some interest in the SOI, only 8% are using seasonal climate forecasting as a management tool (Table 2). Even if more reliable forecasts were available, 30% of respondents still wouldn't make stocking decisions based on such forecasts, instead preferring to react to opportunities or hazards as they arose. Of the 70% who said they would consider using seasonal forecasts as a management tool if more reliable forecasts were forthcoming, most nominated a reduction in stock numbers in forecast dry years and retention of store cattle in forecast wet years as their preferred response. Some also said they would borrow money for improvements or for improved pasture seed if reliable forecasts of wet years were available (Table 2).

TABLE 2. Results of a survey of producers in Dalrymple Shire as to their use of seasonal climate forecasting.

Issue	Response
Percentage of producers who take an interest in the SOI	59%
Percentage of producers who use the SOI as a management tool	8%
Reliability of forecasts required before greater acceptance and use	Reliable 7-8 years out of 10
If greater reliability available percentage of producers who would use forecasts	70%
Management response in predicted dry years	Reduce stocking rates, agist cattle, wean earlier, live export stores
Management response in predicted wet years	Buy cattle esp. store cattle, borrow money for improvements, burn paddocks for weed control, sow improved pasture species, retain stores for ox market

Increased adoption of seasonal climate forecasts as a management tool is more likely if there is an on-the-ground demonstration of its value rather than relying solely on the simulation modelling approach outlined in this paper. A grazing trial has been initiated in Dalrymple Shire with the aim of assessing and demonstrating the ability of different stocking strategies to cope with rainfall variability in terms of animal production, economic performance and pasture condition. Grazing management strategies in the trial include constant stocking, flexible stocking and constant stocking with a seasonal climate forecast. Operational grazing management decisions will be made in conjunction with a producer advisory committee, which will assist in the acceptance and adoption of the results. This grazing trial will, as one of its aims, provide field validation of the modelling simulations and will hopefully demonstrate the value of seasonal climate forecasting in the extensive grazing industries of northern Australia.

References

Anon. (1901) Report of the Royal Commission to Inquire into the Conditions of the Crown Tenants, Western Division of New South Wales. Government Printer: Sydney.

Ash, A.J. and Stafford Smith, D.M. (1996) Evaluating stocking rate impacts in rangelands: animals don't practice what we preach. *The Rangeland Journal* 18, 216-43.

Ash, A.J., McIvor, J.G., Corfield, J.P. and Winter, W.H. (1995) How land condition alters plant-animal relationships in Australia's tropical rangelands. *Agriculture, Ecosystems and Environment* 56, 77-92.

Ash, A.J., Prinsen, J.H., Myles, D.J. and Hendricksen, R.E. (1982) Short-term effects of burning native pasture in spring on herbage and animal production in south-east Queensland. Proceedings of the Australian Society of Animal Production, 14, pp.377-80.

Bird, J.T.S. (1904) The early history of Rockhampton, dealing chiefly with events up till 1870. The Morning Bulletin, Rockhampton: Queensland.

Clewett, J.F., Howden, S.M., McKeon, G.M., Rose, C.W. (1991) Optimising farm dam irrigation in response to climatic risk, in R.C. Muchow and J.A. Bellamy (eds.), *Climatic risk in crop production: models and management for the semi-arid tropics and sub-tropics.* CAB International: Wallingford, UK. pp. 307-328

Foran, B.D. and Stafford Smith, D.M. (1991) Risk, biology and drought management strategies for cattle in central Australia. *Journal of Environmental Management* 33, 17-33.

Frisch, J.E. and Vercoe, J.E. (1977) Food intake, eating rate, weight gains, metabolic rate and efficiency of feed utilization in *Bos taurus* and *Bos indicus* crossbred cattle. *Animal Production* 25, 343-58.

Gardener, C.J., McIvor, J.G. and Williams, J. (1990) Dry tropical rangelands: solving one problem and creating another. Proceedings of the Ecological Society of Australia, 16. pp279-86.

Hammer, G.L. (2000) A general systems approach to applying seasonal climate forecasts, in G.L. Hammer, N. Nicholls, and C. Mitchell (eds.), *Applications of Seasonal Climate Forecasting in Agricultural and Natural Ecosystems – The Australian Experience.* Kluwer Academic, The Netherlands. (this volume)

Hammer, G.L., Holzworth, D.P. and Stone, R.C. (1996) The value of skill in seasonal climate forecasting to wheat crop management in a region with high climatic variability. *Australian Journal of Agricultural Research* 47, 717-37.

Hinton, A.W. (1993) Economics of beef production in the Dalrymple Shire. Queensland Department of Primary Industries: Brisbane, pp.100.

Littleboy, M. and McKeon, G.M. (1997) Subroutine GRASP: Grass Production Model. Documentation of the Marcoola Version of Subroutine Grasp. Appendix 2 of "Evaluating the risks of pasture and land degradation in native pasture in Queensland". Final Project report for Rural Industries Research and Development Corporation DAQ124A, pp.76.

McKeon, G.M., Ash, A.J., Hall, W.B. and Stafford Smith, D.M. (2000) Simulation of grazing strategies for beef production in north-east Queensland, in G.L. Hammer, N. Nicholls, and C. Mitchell (eds.), *Applications of Seasonal Climate Forecasting in Agricultural and Natural Ecosystems – The Australian Experience.* Kluwer Academic, The Netherlands. (this volume)

McKeon, G.M., Day, K.A., Howden, S.M., Mott, J.J., Orr, D.M., Scattini, W.J. and Weston, E.J. (1990) Northern Australian savannas: management for pastoral production. *Journal of Biogeography* **17**, 355-72.

Nicholls, N. (1996) Improved statistical and model approaches for seasonal climate forecasting, Proceedings of Managing with Climate Variability Conference, LWRRDC Occasional Paper CV03/95, Land and Water Resources Research and Development Corporation: Canberra. pp.6-9.

Pratchett, D. and Gardiner, G. (1993). Does reducing stocking rate necessarily mean reducing income? in A. Gaston, M. Kernick and H. Le Houerou (eds.), *Proceedings of the Fourth International Rangeland Congress*, Association Francaise De Pastoralisme: Montpellier. pp.714-716.

Ratcliffe, F.N. (1936). Soil drift in the arid pastoral areas of South Australia. CSIR Pamphlet No. 64, Government Printer: Melbourne.

Stafford Smith, M., Morton, S. and Ash, A. (1997) On the future of pastoralism in Australia's rangelands, in N. Klomp and I. Lunt (eds.), *Frontiers in Ecology: Building the Links*, Elsevier Science Ltd: Oxford, UK. pp7-16.

Stafford Smith, M., Buxton, R., McKeon, G.M. and Ash, A. (2000) Seasonal climate forecasting and the management of rangelands: do production benefits translate into enterprise profits?, in G.L. Hammer, N. Nicholls, and C. Mitchell (eds.), *Applications of Seasonal Climate Forecasting in Agricultural and Natural Ecosystems – The Australian Experience.* Kluwer Academic, The Netherlands. (this volume)

Stone, R. (1996) Meeting farmers' needs for seasonal forecasting, in Proceedings of Managing with Climate Variability Conference. LWRRDC Occasional Paper CV03/95, Land and Water Resources Research and Development Corporation: Canberra. pp.30-34

Stone, R.C. and Auliciems, A. (1992) SOI phase relationships with rainfall in eastern Australia. *International Journal of Climatology* **12**, 625-636.

Stone, R.C. and de Hoedt, G.C. (2000) The development and delivery of current seasonal climate forecasting capabilities in Australia, in G.L. Hammer, N. Nicholls, and C. Mitchell (eds.), *Applications of Seasonal Climate Forecasting in Agricultural and Natural Ecosystems – The Australian Experience.* Kluwer Academic, The Netherlands. (this volume)

Stone, R.C., Hammer, G.L., and Marcussen, T. (1996) Prediction of global rainfall probabilities using phases of the southern oscillation index. *Nature* **384**, 252-255.

Wilcox, D.G. (1963) The pastoral industry of the Wiluna-Meekatharra area in: Lands of the Wiluna-Meekatharra Area, Western Australia. CSIRO Australian Land Research Series No. 7.

SEASONAL CLIMATE FORECASTING AND THE MANAGEMENT OF RANGELANDS: DO PRODUCTION BENEFITS TRANSLATE INTO ENTERPRISE PROFITS?

MARK STAFFORD SMITH[1], ROSEMARY BUXTON[1], GREG MCKEON[2] AND ANDREW ASH[3]

[1]*CSIRO Wildlife and Ecology*
PO Box 2111
Alice Springs, NT 0871, Australia

[2]*Queensland Department of Natural Resources*
PO Box 631
Indooroopilly, Qld 4068, Australia

[3]*CSIRO Tropical Agriculture*
PMB PO
Aitkenvale, Qld 4814, Australia

Abstract

There are increasing opportunities for farmers to use climate forecasts to assist with their management of grazing enterprises, but the value of the forecasts needs to be assessed in the whole enterprise context, with the realistic consequences of management decision-making. We used a linked model of pasture growth (GRASP) and herd dynamics and property economics (Herd-Econ) to simulate whole enterprise management on a cattle station in north-east Queensland. The simulations realistically included the costs and benefits of buying and selling stock, and the impacts of differing stocking rates on resource condition and animal production rates (growth, death and birth rates). We examined a Constant stocking rate and a trading Reactor strategy over the 104-year weather record, then added 6 or 12 month forecast information to the farmers' decision-making. We found that -
(i) In general the optimal level of response to a forecast (increasing stock numbers in good years and decreasing them in bad) **increased** with **increasing** certainty. An inappropriate response to the forecast could be worse than no response, highlighting the sensitivity of forecasting advice to this factor. The differences in cash flow between different levels of response to a forecast could be considerably larger than the difference between strategies.
(ii) At the optimal response levels, forecasts provided a modest benefit in cash flow over baseline strategies, with more benefit to be gained from a 12 month forecast than a 6-month forecast. The same level of cash flow could be achieved for a much lower risk

G.L. Hammer et al. (eds.), The Australian Experience, 271–289.

of environmental degradation (as measured by a soil loss index) with forecasting; however, if advantage was taken of an increased cash flow, then the risk of soil loss changed little. The benefits found in production per unit area (see companion papers by McKeon *et al.* (2000) and Ash *et al.* (2000)) do not translate to economic output at the whole enterprise level in a simple way.

(iii) The outcomes were sensitive to changes in market prices. In general, trading strategies (Reactor, and 6 or 12-month forecasts) were relatively more favoured over a Constant stocking rate strategy as sale prices rose, and especially as the margins between sale and purchase prices increased. Higher stocking rates were also favoured at higher prices in Reactor strategies.

The study highlights the importance of assessing the value of forecasts in the context of the whole management system. There are many other aspects of risk management from which producers could benefit before forecasting becomes the limiting factor in management. However, a seasonal forecast with current skill has some value for production and resource protection when used to trigger appropriate responses; future developments should deliver more.

1. Introduction

Recent years have seen a rapid evolution in our understanding of climatic variability and the effects of the El Niño/Southern Oscillation (ENSO) phenomenon on Australia's weather (eg. Clewett *et al.*, 1994; Glantz, 1996; Kane, 1997). The potential for using the Southern Oscillation Index (SOI) to forecast total rainfall in the coming summer growth season has been extensively explored, and new indices based on sea surface temperature (SST) fields are evolving (Drosdowsky and Allan, 2000; Stone *et al.*, 2000). Dynamic models of the climate and ocean system show promise for increasing both the lead time and reliability of forecasts in the near future (Hunt and Hirst, 2000; Nicholls *et al.*, 2000). Consequently, there are increasing opportunities for farmers to use climate forecasts to assist with their management of grazing enterprises.

However, a forecast can only be used if it fits with a manager's decision-making processes. That use is only converted into value if the net benefits of using the forecast outweigh any costs arising from implementing the actions (eg. losses from selling animals early) and from errors caused by incorrect forecasts. The latter include both missed opportunities for action, and actions taken inappropriately. In southern Australian production systems, Bowman *et al.* (1995) showed that the financial benefits of even accurate seasonal forecasts for sheep producers might be modest indeed. It is therefore important to assess the economic benefits of using forecasts for real decision-making at the whole enterprise level.

Two companion papers have explored this issue in terms of animal production and resource condition in a cattle grazing system in north-eastern Australia (McKeon *et al.*, 2000; Ash *et al.*, 2000). Here we extend that analysis to examine the economic implications when the whole integrated process of herd dynamics and marketing is added to liveweight gain (Figure 1). Specifically, we ask what level of response to a forecast is most beneficial in this example system, whether this results in a net gain over not using the forecasts, and how these gains are affected by market prices.

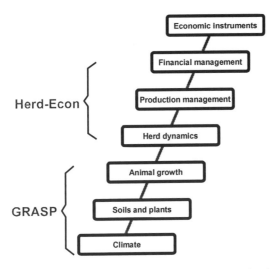

Figure 1. Basic components of the whole grazing system showing the scope of each constituent model.

2. Modelling Approach

We used a linked model (Figure 1) of pasture growth (GRASP) and herd dynamics and property economics (Herd-Econ) to simulate whole enterprise management on a cattle station in north-east Queensland over a century of weather data. The simulations included the costs and benefits of buying and selling stock, and the impacts of differing stocking rates on resource condition and animal production rates (growth, death and birth rates). The GRASP component very closely follows the model as described in McKeon *et al.* (2000); only significant differences and factors relevant to whole enterprise modelling are discussed here.

2.1 GRASP

GRASP is a deterministic, point-based model of soil water, grass growth and animal (sheep and cattle) production, devised and validated on tropical grasslands (McKeon *et al.*, 1990; Day *et al.*, 1997). The model has been calibrated for 40 pasture communities in Queensland (Day *et al.*, 1997). From daily inputs of rainfall, temperature, humidity, pan evaporation and solar radiation, soil water is simulated for given soil attributes (texture and depth). Plant growth is calculated from transpiration, but includes the effects of vapour pressure deficit, temperature, radiation interception, nitrogen availability and grass basal cover. The effect of variable tree density is simulated through the effects of trees on water use and nitrogen uptake. Animal production responses are captured through empirical equations for live weight change as a function of utilisation and length of growing season, developed using data from experiments with grazing cattle.

The feedbacks of increased grazing pressure on plant production are simulated through effects on runoff, perennial grass basal cover, sensitivity to soil water deficit and nitrogen uptake. As yet, no feedbacks of soil loss on nutrient availability or infiltration characteristics have been modelled. Effectively, GRASP allows the simulation of the risks of production losses but not the long-term consequences of taking those risks. Thus it is possible that it under-predicts the chronic impacts of grazing on production in the long term, and consequently somewhat over-predicts the optimal stocking levels on pastures. However, the simulations presented here identify economically optimal stocking rates at a level considerably below the point at which major pasture damage is predicted, so that in practice this problem is probably minimal for the conclusions of the current study. GRASP thus provides a well-validated framework for estimating animal production levels and short to medium-term pasture sustainability as a result of different sequences of climate and stocking levels.

2.2 RANGEPACK HERD-ECON

The emergent characteristics of the whole enterprise are handled by RANGEPACK Herd-Econ, a deterministic herd dynamics and property economics model (Stafford Smith and Foran, 1992). Herd-Econ allows users to define a grazing enterprise comprising any number of animal classes by age groups, each of which has specified biological (mortality, growth and reproduction) rates. The model then tracks animal numbers on a monthly time-step, subject to different buying and selling strategies and transfers between animal classes. The economics module uses animal numbers to adjust variable costs, and combines trading results with other expense and receipt items to generate a cash flow.

To link Herd-Econ to the outputs of GRASP, relationships are also needed between the condition and growth rate of animals and their reproduction and mortality rates. These relationships are the subject of ongoing research (Pepper et al., 1996), so relationships based on the data from the study region (from Gillard and Monypenny (1988) and collected by us during property surveys in 1996) are used in the simulations described here. The value of Herd-Econ is that it provides a flexible and context-independent method for describing the tactical details and dynamics of alternative stock management strategies.

It is worth noting that large integrated models have proved useful for improving our understanding of systems in the past, but few have contributed to useable decision-support tools. There is a quandary here. On the one hand, plenty of evidence suggests that large numbers of parameters do not necessarily add resolution or precision to information that is useful to decision makers (eg. Zeide, 1991). With increasing numbers of parameters, validation and interpretation become more difficult, optimisation surfaces are often relatively flat, and day-to-day use is cumbersome. There is also an increasing risk of propagating undetected errors between models. On the other hand, models must incorporate the processes to which they are intended to be sensitive, and understanding the interaction between ecological and economic sustainability in grazing systems means encompassing the many processes linking grass growth to net financial performance. Here, GRASP, Herd-Econ and a number of *post hoc* economic analyses are linked hierarchically to minimise these problems. This enables an appropriate level of process to be incorporated at each level whilst allowing the information passed between levels to be limited to a minimum necessary data set. We do not describe the validation of the models here, but Day *et al.* (1997) present extensive validation of GRASP, Stafford Smith and McKeon (1998) discuss validation under dry conditions, and various publications (eg. Buxton and Stafford Smith, 1996) have assessed the use of Herd-Econ in many case studies. Weak points in the linked model remain; in particular these are our limited knowledge of the relationships affecting reproduction and mortality rates, and the lack of modelling the long-term feedback of grazing on pasture production through soil and plant compositional changes. These issues are discussed in the above publications and are the subject of on-going research.

3. Study System and Simulations

3.1 STUDY REGION AND ENTERPRISES

Our study region in the Dalrymple Shire of north-eastern Queensland is described in detail by Ash *et al.* (2000). Dalrymple Shire covers 67,687 km^2 and is dominated by open eucalypt woodlands though there are significant areas of spinifex desert in the south-west and pockets of rainforest in the north-east. The simulations are based on the woodlands. The region is subject to strongly summer-dominant rainfall; mean annual rainfall for the 104 simulated years was 665 mm (June-May), of which 80% fell during Nov-May. The majority of pasture growth for the year is therefore complete by about May and the next wet season does not usually start before Nov. The majority of stock management is carried out between these dates. Supplementary feeding is often used towards the end of the dry season to maintain body condition, and phosphorus may be fed year round on P-deficient soils. GRASP parameters are as outlined by Ash *et al.* (2000); additional data for reproduction and mortality rates, and for the economics and management decision-making, have been drawn from Hinton (1993) and Buxton (1997).

The simulated enterprise is a self-replacing cattle herd in the Charters Towers region of north-eastern Queensland. The aim is to sell bullocks at around 600 kg live weight to the Japanese market (live exports are also common in the region, but these are not considered here). Cows in the herd have their first calf at 3 years old, and breed to a maximum of 10 years old; they are then speyed once empty and sold at 10 or 11 years old in good condition. In addition to sales of old cows, 5% of all age groups are culled every year as normal management. 4% of heifers are also culled each year at 2 years old, irrespective of herd size. After this, sufficient heifers are retained to keep cow numbers at a target level and the surplus sold. Calves are assumed to be born year round, but with 77% in the October to February period; weaning occurs twice, in April and October. Bulls are maintained at 4% of cows, with replacements bought at 2 and sold at 6 years old.

In general 20% of bullocks are assumed to reach the target sale weight at 3 years old, with the remainder sold a year later; however, growth rates are tracked in response to seasonal conditions, so this proportion varies from year to year. All normal sales take place between March and May. Bullock weights are tested in March; if they are heavy enough, they are sold. The process is repeated in May.

Steer growth rates were constrained not to fall below 50 kg y^{-1}, with increasing supplementation costs included if pasture conditions were too poor to support this; 90% of growth was assumed to occur between Dec-April. Branding and mortality rates were based on data of Gillard and Monypenny (1988); supplementation of growth rates effectively constrains reproduction rates to a maximum of 80% and cow mortalities to 40%.

Results are very sensitive to marketing and price assumptions. All cattle are sold direct to abattoirs on the coasts at an assumed distance of 300 km from the station; freight and levy costs are included. Baseline prices for cows are $420/head, heifers $350/head, bullocks $1.10/kg and bulls $600/head. Additional marketing may be required to change stocking rates under reactive and forecasting strategies. When reducing numbers, the order of selling until the target is met is: sell any remaining old cows (at $400), then cull up to 15% of all age groups of cows (at $400), then sell up to 50% of 2 year old steers to the live export market ($1.00/kg), then up to 80% of heifers ($320), then any steers purchased in previous years ($1.05/kg), then a further 50% of 4 year old bullocks ($1.05/kg) finally a further 50% of 3 year old bullocks ($1.05/kg). The remaining core herd is assumed to be preserved (with appropriate supplement or agistment costs) under any circumstances. When increasing numbers, if breeder numbers are low, the enterprise first buys in 2 and 3 year old heifers ($350 & $400/head), and then trades in steers by buying up to a maximum of 1000 head of 2 year old steers ($385/head). A maximum is assumed on the basis of the limited availability of both capital and stock in good years.

The sale and purchase prices have been subjected to sensitivity analysis. It is worth noting that one might expect prices to be related to season. In fact there is a great deal of noise in any such relationship (Milham et al., 1995), so that the effect may not be great in the relatively inelastic markets of the rangelands; thoroughly examining its implications is another study in itself (currently underway). In the simulations here, prices received on a per weight basis will be somewhat lower in poor years (especially if sales are late) because growth rates will be depressed. The sensitivity analysis provides an indication of how important further price fluctuations could be.

3.2 SIMULATIONS

Two underlying approaches to managing stocking rates are (i) to aim at a more-or-less constant stocking rate, or (ii) to adjust numbers every year to suit the conditions. In practice, producers can adopt an infinite range of approaches between these two extremes by varying the degree to which they perform this adjustment. Here we simulated some baseline strategies, then added forecasting to each in a number of ways (Table 1); these strategies followed those of Ash *et al.* (2000) in principle, but the details were implemented differently, as follows, to allow for decision-making on a real whole enterprise.

The baseline strategies were:

- *Constant*, adjusting numbers every May with the aim of maintaining a constant number of breeders; and,

- *Reactor*, adjusting numbers every May with the aim of maintaining a constant utilisation rate of the total standing dry matter then on offer in the pasture at the end of the wet season.

We also included a double Reactor strategy, which was essentially as for the baseline *Reactor* except that the numbers were then re-adjusted in Nov in relation to expected summer growth, to assess whether making two decisions a year had any intrinsic value.

In north-east Queensland, the Aug-Oct average SOI provides one of the best relationships available in Australia for estimating forthcoming summer grass growth (Clewett *et al.*, 1994). The forecast strategy that producers can most easily implement today is therefore a response to this information at the end of October. This 6-month forecast was added to the baseline strategies by a round of buying or selling at the start of November; for Constant strategies, this took the form of buying or selling some number of head, and by increasing or decreasing the target utilisation rate for the Reactor-based strategies. Since SOI forecasts are rapidly being augmented by information from SSTs in ways that are likely to provide the same level of reliability earlier in the season, we also assessed the value of knowing the Aug-Oct average SOI by early June. This was implemented as a 12-month forecast response added to Constant and Reactor by acting in May in a similar fashion to the response in Nov.

Finally, 'perfect knowledge' was simulated to assess what the best Forecast ever obtainable could return to the enterprise; 'Perfect' was expressed here as a reliable knowledge of when seasons would be in the best or worst tercile. We counted how many years had SOI >+5 or <-5 (30 and 31 of the 104 years respectively), then simulated the system at a medium stocking rate, from which we identified the 30 best and 31 worst years for annual growth. The Perfect strategies reacted in the same ways as the Forecasts but correctly identified these best and worst years.

TABLE 1. Management strategies simulated in this study. All strategies sell key marketable steers and cull cows in May each year before the following amendments are made. Asterisks mark the baseline strategies that require no forecasting.

Basis	Description	Name used
Constant *	Targets a constant number of cows each May	Constant
Constant +6 month Forecast	Sets a constant number of cows each May then buys or sells some animals in Nov if the Aug-Oct mean SOI is outside ±5	Const+Fore6m
Constant + 6 month Perfect	Sets a constant number of cows each May then buys or sells some animals in Nov if the summer season is going to be in the top or bottom tercile	Const+Perf6m
Constant + 12 month Forecast	Sets a constant number of cows each May then buys or sells some animals in May if the Aug-Oct mean SOI is outside ±5	Const+Fore12m
Constant + 12 month Perfect	Sets a constant number of cows each May then buys or sells some animals in May if the summer season is going to be in the top or bottom tercile	Const+Perf12m
Reactor *	Buys or sells animals in May to target a constant utilisation rate (U) of the currently standing dry matter	Reactor
Reactor + 12 month Forecast	Buys or sells animals in May to target a constant U of the currently standing dry matter, where U is adjusted up or down if the Aug-Oct mean SOI is outside ±5	React+Fore12
Reactor + 12 month Perfect	Buys or sells animals in May to target a constant U of the currently standing dry matter, where U is adjusted up or down if the summer season is going to be in the top or bottom tercile	React+Perf12
Reactor + Nov Reactor *	Buys or sells animals in May to target a constant U of the currently standing dry matter, then buys or sells animals in Nov to target the same U of the median expected growth over summer	React+React
Reactor + 6 month Forecast	Buys or sells animals in May to target a constant U of the currently standing dry matter, then buys or sells animals in Nov to target the same U of the median expected growth over summer, except this U is adjusted up or down if the Aug-Oct mean SOI is outside ±5	React+Fore6
Reactor + 6 month Perfect	Buys or sells animals in May to target a constant U of the currently standing dry matter, then buys or sells animals in Nov to target the same U of the median expected growth over summer except this U is adjusted up or down if the summer season is going to be in the top or bottom tercile	React+Perf6

All strategies were assessed at a range of different stocking rate targets and resulting utilisation rates, to identify an optimal level. All Forecast and Perfect knowledge simulations were tested for a wide range of possible management responses to the above and below average years. The final comparisons are made with the best responses for the optimal stocking level for each strategy, applied consistently across other stocking rates.

A large number of outputs were recorded from Herd-Econ and checked; here only key economic and environmental indicators are reported. The mean utilisation rate (total dry matter eaten divided by total growth) provides a measure of the long-term stocking rate actually achieved by each strategy. Annual cash flow was recorded, and was made the subject of a sensitivity analysis with respect to the prices of cattle; purchase and sale prices were altered separately, but all purchase or all sale prices were assumed to be altered by the same proportion across age classes and cattle type. A simple post-tax assessment of profit was made by allowing for changes in stock value and a net depreciation, which balanced average expenditure on capital items, and then applying current (1997) Australian tax averaging rules. This was not intended to be accurate, but rather to give a sense of whether tax averaging might alter the relative benefits of strategies that resulted in different levels of variability of cash flow. It also accounts for any changes in herd capital value that might arise as a result of some strategies systematically selling or buying stock near the end of the climatic sequence.

As an integrated measure of environmental impact of different strategies, the mean predicted soil loss for a fixed slope was also recorded; this should be treated as an indicator of soil loss rather than an absolute measure, since real landscapes vary greatly. However, it integrates the effects of plant cover and rainfall intensity into the likelihood of soil loss (Scanlan and McIvor, 1993). All averaging, profit and sensitivity analysis calculations were performed after the model runs by transferring relevant data into a spreadsheet.

4. Results

Some 300 simulations were run and examined. Here we summarise the key points that emerged.

4.1 VARYING STOCKING RATES ON BASELINE STRATEGIES

The model produced reasonable patterns of economic returns when either baseline strategy was simulated through 104 years (Figure 2). Increasing target stocking rates resulted in greater variability in many measures between years (Figure 3), as the higher pressure on the pasture caused bigger fluctuations in birth and death rates for *Constant* and a greater requirement on buying and selling in *Reactor*. As a consequence, mean annual cash flow and profit increased with increasing stock numbers up to a point with either strategy, and then declined again. This decline is associated with loss of pasture condition in occasional dry years, until the loss is so severe that recovery is impossible within a realistic time frame. As a result, an optimum stocking target can be identified for any management strategy.

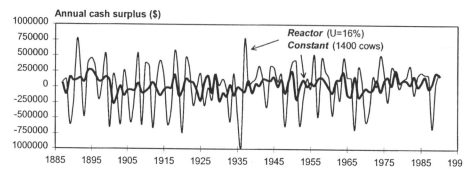

Figure 2. Annual cash flow ($) for the 104-years run for each baseline strategy at an intermediate stocking level (as shown as optima in Table 3). Note that the *Reactor* strategy is intrinsically more variable than the *Constant* strategy, due to deliberate selling and buying every year.

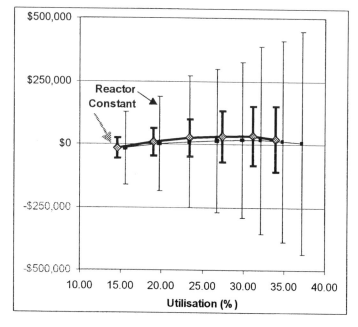

Figure 3. Mean annual cash flow ($) and standard deviation by year plotted against realised utilisation rate (%) for each simulated stocking target for the two baseline strategies. Note that the variability of both increases with increasing utilisation rate.

4.2 VARYING REACTION LEVEL FOR FORECASTS

In all cases, running a series of different levels of reaction to a forecast (or perfect knowledge) eventually provided an optimum combination of reactions in terms of the benefit in mean annual cash flow. Sometimes the surface was very flat so that there was little to choose between several similar combinations. We illustrate this for *Constant +*
6 month Forecast (Table 2), where it is apparent that (i) perfect knowledge conveys a small advantage over a forecast, (ii) the response surface for perfect knowledge is steeper than for forecast, (iii) perfect knowledge tends to have a larger optimal reaction than the less certain forecast, and, most importantly, (iv) there are combinations of reactions which are actually worse than no reaction at all. These four points usually held true for the other combinations, although perfect knowledge often conveyed more relative advantage than this. On the basis of examining the surfaces for all forecasting approaches, a compromise level of reaction was chosen for each combination of strategies and used for all subsequent runs (Table 3). However, it should be noted that the optimal reaction level is price dependent; for example, if the differential between sale and purchase prices rises by 10%, the optimum reaction to a 6 month forecast (Table 2) rises to +200/-200 head. However, the marginal benefit over no action only increases to $10,720, so the effect of using a slightly non-optimal reaction is not large in this case.

TABLE 2. Benefit in mean annual cash flow (relative to no reaction at all, at baseline prices) produced by various combinations of reaction to a 6 month forecast and perfect knowledge, with respect to the *Constant* strategy with a target of 1400 cows. Note the higher optimum but greater number of negative values (*ie.* more concave optimising surface) for perfect knowledge. A compromise optimum of +100/-100 was chosen for subsequent simulations of this forecast, but this value is price sensitive.

Increase in head when SOI>+5		Decrease in head when SOI<-5				
		0	-100	-200	-300	-400
Forecast	0	$0	$3,970	$962	-$2,537	-$6,692
	50	$444	$8,297	$6,994	$2,782	$449
	100	$856	$7,744	$6,543	$4,834	$2,010
	150	$642	$6,572	$6,711	$6,490	$1,178
	200	$635	$4,829	$6,060	$5,142	$1,196
Perfect	0	$0	$779	-$1,354	-$3,074	-$3,120
	50	$410	$6,091	$6,361	$4,847	$3,206
	100	-$2,811	$8,008	$9,224	$5,526	$3,688
	150	-$19,360	$6,040	$7,951	$7,446	$4,401
	200	-$24,077	-$1,193	$4,621	$5,674	$3,424

TABLE 3. Level of adjustment found to be near-optimal for each forecast/perfect knowledge combination (in general Perfect tended to prefer a slightly greater adjustment than Forecast [see Figure 1] so a compromise allowing for the flatness of the optimisation surface was sometimes selected).

Strategy and Forecast type	Increase used when SOI>+5	Decrease used when SOI<-5
Constant and 6 months	+100 head	-100 head
Constant and 12 months	+200 head	-200 head
Reactor and 6 months	1.0 * target utilisation	0.5 * target utilisation
Reactor and 12 months	1.3 * target utilisation	0.45 * target utilisation

4.3 VALUE AND IMPACTS OF FORECASTS RELATIVE TO BASELINE STRATEGIES

The key patterns found for each combination of baseline strategy and forecast are shown in Figure 4 and Table 4. Both 6 and 12 month forecasts provide some modest benefits in terms of long-term mean annual cash flow over a *Constant* strategy, although perfect knowledge does not provide much additional benefit over the respective forecast.

The 12 month forecast is a little more useful than a 6 month one for the prices and decision-making simulated here. A 12 month forecast adds a great deal of value to the *Reactor* strategy, with a substantial gain still to be realised from increased reliability as indicated by perfect knowledge. For the prices simulated, a 6 month forecast for *Reactor* is actually deleterious with the present level of skill. A second decision point seems to have no intrinsic benefits in the simulated price regime, suggesting that trading in stock twice a year provides no margin; the additional trading introduced with forecasts is again deleterious for these prices, and perfect knowledge is only marginally beneficial. Further exploration of buying and selling tactics might change this finding, as would increases in the trading margin between buying and selling prices (see below).

Accounting for tax averaging (Table 4) made at most a small change in the optimum stocking rate target (usually none, but it caused a small drop in the optimum target for the *Reactor* strategy with 6 month Perfect knowledge). However, net profits were generally noticeably less than annual cash flow, and strategies with more variable cash flow – the trading strategies – generally benefited a little less than the more constant strategies; the difference is only of the order of $1000.

The pattern of riskiness is shown more clearly by cumulative distributions of annual cash flow (Figure 5) for the optimal stocking rates. Here it is apparent that the benefits of forecasting are seen mainly through increased earnings in high cash flow years; however, it must be recalled that these may result both from successfully capitalising on a run of good years *and* from selling large numbers of stock in a bad year. It is noticeable that the forecasts seem to do worse than the baseline strategies in lower earning years, thus identifying a target for future action.

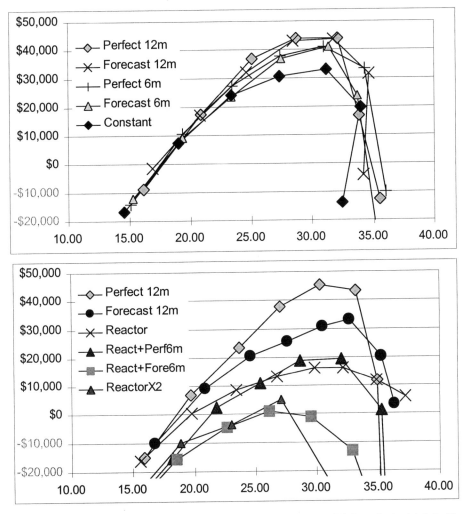

Figure 4. Mean annual cash flow ($) plotted against realised utilisation rate (%) for each simulated stocking target for all strategy combinations - (a) strategies based on *Constant* and (b) strategies based on *Reactor* (the more negative responses here are deliberately cut off to obtain comparable scales).

TABLE 4. Key output statistics for the optimal stocking rate of each management strategy (ie. values pertaining to the top of each curve in Figure 4) for the baseline price scenarios. (* as defined in Table 1.)

Name*	104-year mean at optimal stocking rate				
	Summer Animal Equivalents	Utilisation rate(%)	Soil loss index (kg/ha)	Cash flow ($)	Post-tax profit ($)
Constant	3358	31.2	1097	33,000	1,315
Const+Fore6m	3370	31.4	1059	40,744	6,400
Constant+Perf6m	3319	31.0	1044	41,008	5,556
Constant+Fore12m	3426	31.8	1071	43,640	7,334
Constant+Perf12m	3094	28.8	892	43,880	7,077
Reactor	3455	32.1	1021	16,133	-10,899
Reactor+Fore12m	3513	32.6	1052	33,205	729
Reactor+Perf12m	3267	30.3	888	45,272	7,366
React+React	2925	27.1	868	5,038	-15,384
React+Fore6m	2764	26.1	794	975	-18,538
React+Perf6m	3431	32.0	964	19,247	-8,553

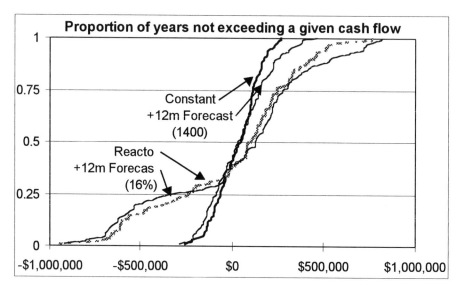

Figure 5. Cumulative distributions of annual cash flow ($) for some strategies at their financially optimal stocking rate (Constant with a target of 1400 cows, plus a 12 month forecast on this, and Reactor with a target utilisation of 16%, also with the comparable 12 month forecast). Note the lack of first order stochastic dominance in either pair.

The utilisation level at which the optimal cash flow is achieved, and the value of the soil loss index at that level (Table 4), give some idea of the likely relative environmental impact of different strategies (noting that small differences can arise just from the choice of particular stocking targets to simulate). When cash flow is compared to soil loss (Figure 6), it is apparent that the same level of cash flow can be achieved at very much lower soil loss risks with forecasting, but that the optimal levels of cash flow are achieved at generally comparable levels with and without forecasting.

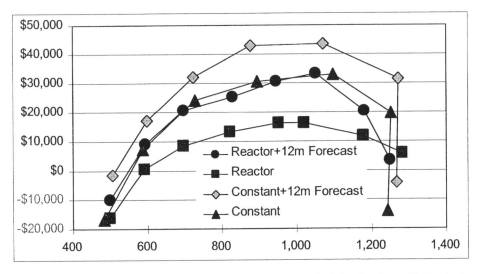

Figure 6. Mean annual cash flow ($) plotted against soil loss index (kg/ha) rather than utilisation (as in Figure 4) for key strategies.

4.4 EFFECT OF PRICES

Not surprisingly, price changes can have an enormous impact on the relative benefits of strategies (Table 5). As the differential between buying and selling increases (ie. selling prices become relatively more attractive), trading strategies come to dominate. The baseline Reactor becomes more attractive than Constant when the differential increases by about 15%. The forecasting strategies, which involve additional trading, also become increasingly attractive with an increasing differential. Here we have assumed consistent prices over the 100-year simulation; the real effect of price fluctuations is to introduce an additional difficult set of decisions into the operation, since different strategies will be optimal for different price regimes. In particular, there are likely to be systematic patterns of price differential changes associated with dry and wet periods, and these may be exacerbated if all producers respond to forecasting information in the same way. This issue should be studied using stochastic modelling approaches (eg. Milham *et al.,* 1993). Here it is sufficient to note that the prices used are typical at the time of writing, but can easily and rapidly move by 20%, thus altering any absolute results dramatically.

TABLE 5. Effects on mean annual cash flow ($) at optimal stocking rate of changing absolute and relative purchase and sale prices. The last three rows give differences between specific management strategies. (* as defined in Table 1.)

Name*	Increased differential between buying and selling					Decreased differential	
	Buy+0%			Buy-5%	Buy-10%	Buy+5%	Buy+10%
	Sell +0%	Sell+5%	Sell+10%	Sell+5%	Sell+10%	Sell-5%	Sell-10%
Constant	$33,000	$46,879	$60,759	$48,146	$63,291	$17,854	$2,911[b]
Const+Fore6m	$40,744	$55,839	$70,934	$57,975	$75,205	$23,513	$6,283
Const+Perf6m	$41,008	$55,957	$70,905	$58,046	$75,084	$23,971	$6,933
Const+Fore12m	$43,640	$59,763	$75,885	$62,749	$81,857	$24,531	$6,818[b]
Const+Perf12m	$43,880	$59,885[a]	$76,071[a]	$62,873[a]	$82,047[a]	$25,697	$7,514
Reactor	$16,133	$36,684	$57,234	$44,697	$73,261	-$10,069[b]	-$34,325[b]
React+Fore12m	$33,205	$55,435	$77,665	$64,304	$95,403	$2,105	-$25,138[b]
React+Perf12m	$45,272	$65,113	$86,596[a]	$72,800[a]	$102,258[a]	$18,864	-$7,545
React+React	$5,038	$23,434	$41,831	$30,495	$55,953	-$20,420	-$45,877
React+Fore6m	$975	$21,294[a]	$43,641[a]	$31,934[a]	$64,920[a]	-$26,563	-$50,394[b]
React+Perf6m	$19,247	$42,601	$65,954	$53,005	$86,764	-$10,560[b]	-$39,870[b]
Constant – Reactor	$16,867	$10,195	$3,525	$3,449	-$9,970	$27,923	$37,236
Const+Fore12m – Constant	$10,640	$12,884	$15,126	$14,603	$18,566	$6,677	$3,907
React+Fore12m – Reactor	$17,072	$18,751	$20,431	$19,607	$22,142	$12,174	$9,187

[a,b] in these cases, the optimal stocking rate increased (a) or decreased (b)

Conclusions

The key finding of this study is that production benefits achieved through forecasting (McKeon et al., 2000; Ash et al., 2000) do not simply translate into economic benefits at the whole enterprise level, consistent with the conclusions of Bowman et al. (1995). Simulation approaches such as that described here enable us to explore what factors are most influential in this translation.

The first point is that identifying the scale of response to make to a forecast is as important as identifying the forecast itself. In general, the greater the reliability of the forecast in terms of both accuracy and lead time, the greater the response that is optimal. Over-reacting to a forecast can actually result in a net loss of profits in the long term, and the differences induced in outcomes by inappropriate responses can exceed the benefits arising from using forecasting. Fortunately, in some cases the optimisation surface is fairly flat so a broad range of responses is near optimal. A much greater focus is needed on this issue in future work since advice could otherwise be very misleading.

The profitability of a reactive strategy relative to a constant one is very market dependent, as are forecast responses, since all imply trading stock. In general trading strategies become more profitable as prices rise, and particularly if sale prices rise more than purchase prices. However, cumulative distribution functions for some cases (Figure 5) suggest that there may be considerable scope to look for trading tactics that avoid the severe downside risks of the worst years.

In terms of the likely acceptance of forecasting by graziers, short-term paybacks naturally outweigh long-term ones, so that forecasts which do not provide benefits reasonably often and substantially will inevitably be regarded with suspicion. The simulations reported here presume a long-term view on benefits. In practice, producers who experience a success early in their use of forecasts (or who fail to benefit by not acting) are likely to be over-supportive of the concept, while those who experience a failure will be unduly negative. The payback from using forecasts is not as rapid as that of taking advantage of better prices. It is likely that the benefits of forecasting are also strongly related to factors such as the degree of climatic variability experienced (cf. Ash *et al.*, 2000), and, though not yet tested, the resilience of the pasture type.

The results from the simulations are generally intuitively sensible. They have serious implications for giving advice about forecasts in particular regions or for particular enterprises, and especially under particular (varying) price regimes. The results highlight the need for extension about the economic responses to forecasting so that graziers are conscious of them when incorporating forecasts into their decisions, and recognise forecasts as simply one more element to include in their own decision-making. Self-reliance is ultimately about managing climatic variability and risk *per se*, in which seasonal forecasting is but one small element. Most critically, a necessary precursor to self-reliance is having a management system in place that is sensitive to pasture conditions and hence using the appropriate stocking rate strategies over time. Without such a system no forecast will help.

Despite all this, a seasonal forecast even with current skill does have some value for production and resource protection when used to trigger appropriate responses; future developments promise more. The risk of soil loss can be reduced; and, although not simulated here, decisions with small downside risks such as fire management can be significantly improved. As Bowman *et al.* (1995) have observed, there are also potential benefits in terms of animal welfare by reducing animal mortalities in poor years.

The key thesis of this paper is that it is important to assess not only whether a response is worthwhile, but also what level of response is most appropriate. To do this, it is crucial to analyse the entire linked system, including management decision-making;

having a forecast index is only the first step in this process. A high degree of communication is therefore needed between those working on creating forecasting indices, those assessing their potential value, and the managers who may eventually use them.

Acknowledgments

We thank Graeme Hammer for the effort put in to organise the forum at which these issues could be aired. Joe Breen assisted with simulations and analysis. We are grateful for prompt and insightful comments from a number of reviewers, in particular David White and Mark Howden.

References

Ash, A.J., O'Reagain, P.J., McKeon, G.M. and Stafford Smith, M. (2000) Managing climate variability in grazing enterprises: a case study of Dalrymple Shire, in G.L. Hammer, N. Nicholls, and C. Mitchell (eds.), *Applications of Seasonal Climate Forecasting in Agricultural and Natural Ecosystems – The Australian Experience.* Kluwer Academic, The Netherlands. (this volume)

Bowman, P.J., McKeon, G.M. and White, D.H. (1995) The impact of long range climate forecasting on the performance of sheep flocks in Victoria. *Australian Journal of Agricultural Research* **46**, 687-702.

Buxton, R. (1997) Management of cattle enterprises in the Charters Towers region: impressions from a visit to the region in Nov-Dec 1996. CSIRO Wildlife and Ecology, Alice Springs.

Buxton, R. and Stafford Smith, M. (1996) Managing drought in Australia's rangelands: four weddings and a funeral. *Rangeland Journal* **18**, 292-308.

Clewett, J.F., Clarkson, N.M., Owens, D.T. and Abrecht, D.G. (1994) AUSTRALIAN RAINMAN: rainfall information for better management. Computer software package and manual, Queensland Department of Primary Industries, Brisbane.

Day, K.A., McKeon, G.M. and Carter, J.O. (1997) Evaluating the risks of pasture and land degradation in native pasture in Queensland (DAQ124A). Final Project Report to Rural Industries Research and Development Corporation. Queensland Department of Natural Resources, Brisbane.

Drosdowsky, W. and Allan, R. (2000) The potential for improved statistical seasonal climate forecasts, in G.L. Hammer, N. Nicholls, and C. Mitchell (eds.), *Applications of Seasonal Climate Forecasting in Agricultural and Natural Ecosystems – The Australian Experience.* Kluwer Academic, The Netherlands. (this volume)

Gillard, P. and Monypenny, R. (1988) A decision support approach for the beef cattle industry of tropical Australia. *Agricultural Systems* **26**, 179-190.

Glantz, M.H. (1996) *Currents of change: El Niño's impact on climate and society.* Cambridge University Press, Cambridge, UK. pp194 .

Hinton, A.W. (1993) Economics of Beef Production in the Dalrymple Shire. Project Report Q093005, Queensland Department of Primary Industries, Brisbane.

Hunt, B.G. and Hirst, A.C. (2000) Global climatic models and their potential for seasonal climatic forecasting, in G.L. Hammer, N. Nicholls, and C. Mitchell (eds.), *Applications of Seasonal Climate Forecasting in Agricultural and Natural Ecosystems – The Australian Experience.* Kluwer Academic, The Netherlands. (this volume)

Kane, R.P. (1997) On the relationship of ENSO with rainfall over different parts of Australia. *Australian Meteorological Magazine* **46**, 39-49

McKeon, G.M., Ash, A.J., Hall, W.B. and Stafford Smith, D.M. (2000) Simulation of grazing strategies for beef production in north-east Queensland, in G.L. Hammer, N. Nicholls, and C. Mitchell (eds.), *Applications of Seasonal Climate Forecasting in Agricultural and Natural Ecosystems – The Australian Experience.* Kluwer Academic, The Netherlands. (this volume)

McKeon, G.M., Day, K.A., Howden, S.M., Mott, J.J., Orr, D.M., Scattini, W.J. and Weston, E.J. (1990) Northern Australian savannas: management for pastoral production. *Journal of Biogeography* **17**, 355-372.

Milham, N., Hardaker, J.B. and Powell, R. (1993) RISKFARM: A PC-based stochastic whole-farm budgeting system. Centre for Agricultural and Resource Economics, University of New England, Armidale, Australia.

Milham, N., Stafford Smith, M., Douglas, R., Tapp, N., Breen, J., Buxton, R. and McKeon, G. (1995) Farming and the environment: an exercise in eco-economic modelling at the farm level in the NSW rangelands. *Proceedings of the International Congress on Modelling and Simulation* **4**, 221–228. University of Newcastle, November, 1995.

Nicholls, N., Frederiksen, C. and Kleeman, R. (2000) Operational experience with climate model predictions, in G.L. Hammer, N. Nicholls, and C. Mitchell (eds.), *Applications of Seasonal Climate Forecasting in Agricultural and Natural Ecosystems – The Australian Experience*. Kluwer Academic, The Netherlands. (this volume)

Pepper, P., Mayer, D., McKeon, G.M. and Moore, A.D. (1996) Validation of reproduction and mortality rates for northern Australian beef and sheep breeders. DroughtPlan Working Paper No.7. Queensland Department of Primary Industries, Brisbane.

Scanlan, J.C. and McIvor, J.G. (1993) Pasture composition influences soil erosion in *Eucalyptus* woodlands of northern Queensland. *Proceedings XVII International Grasslands Congress*, New Zealand, pp.65-66.

Stafford Smith, D.M. and Foran, B.D. (1992) An approach to assessing the economic risk of different drought management tactics on a South Australian pastoral sheep station. *Agricultural Systems* **39**, 83-105.

Stafford Smith, M. and McKeon, G.M. (1998) Assessing the historical frequency of drought events on grazing properties in Australian rangelands. *Agricultural Systems* **57**, 271-299.

Stone, R.C., Smith, I. and McIntosh, P. (2000) Statistical methods for deriving seasonal climate forecasts from GCMs, in G.L. Hammer, N. Nicholls, and C. Mitchell (eds.), *Applications of Seasonal Climate Forecasting in Agricultural and Natural Ecosystems – The Australian Experience*. Kluwer Academic, The Netherlands. (this volume)

Zeide, B. (1991) Quality as a characteristic of ecological models. *Ecological Modelling* **55**, 161-174.

BUILDING KNOWLEDGE AND SKILLS TO USE SEASONAL CLIMATE FORECASTS IN PROPERTY MANAGEMENT PLANNING

JEFF CLEWETT[1], NEIL CLIFFE[1], LYNDA DROSDOWSKY[2], DAVID GEORGE[1], DAMIEN O'SULLIVAN[1], COL PAULL[1], IAN PARTRIDGE[1] AND ROD SAAL[1]

[1] *Queensland Centre for Climate Applications*
Queensland Departments Primary Industries and Natural Resources
PO Box 102
Toowoomba, Qld 4350, Australia

[2] *Bureau of Meteorology*
PO Box 1289K
Melbourne, Vic 3000, Australia

Abstract

Few in Australian agriculture had heard of the El Niño/Southern Oscillation (ENSO) phenomenon and its impact upon Australian climate or agriculture before 1990. Today, few Australian farmers, agri-business people or scientists are not aware of its impact. This change was partly stimulated by revisions to the National Drought Policy in 1990, which introduced a major change in the role of government services. Emphasis was shifted from providing subsidies to providing information so that farmers could be more self-sufficient in managing the climatic risks they faced. Training initiatives for farmers and land managers were initiated through the National Property Management Planning Program. This paper outlines the property management planning process, details how information on seasonal climate forecasting was made available so that the necessary knowledge and skills to use it effectively in farm management could be developed, and discusses what has been learnt in the process and the major issues for the immediate future.

The Property Management Planning (PMP) process embraces all aspects of property management and is a strategic planning approach to property management. PMP has an emphasis on participative problem solving and includes: reviewing and developing personal goals, natural resource assessment and physical planning, sustainable agricultural production, ecological considerations, business and financial management, marketing, drought preparedness and risk management. The PMP process provides the context for developing seasonal climate knowledge and skills within a strategic planning and farm decision-making framework. Climate and seasonal forecasting information has been made available in a diverse range of forms ranging from print and

G.L. Hammer et al. (eds.), The Australian Experience, 291–307.

electronic media to structured workshops. The communication and training processes instigated were very successful at building knowledge and understanding of seasonal climate forecasting. However, this is a necessary but not sufficient condition for its effective application in farm management. Effective application of forecasts involves their use to improve decisions. Feedback from surveys indicates that many primary producers now understand the process of ENSO and believe that seasonal climate outlook information can be used profitably in their business. Many are now considering this information in their decision-making processes. The challenge for the future is to develop training tools/processes that enable analysis of decision options in conjunction with seasonal forecast information and other factors, so that management skill in using forecasts effectively can be enhanced. The lessons we have learnt throughout this process and the insights gained into adult learning are discussed and appear to conform with experiences of others.

1. Introduction

Climate and in particular climatic variability is central to determining the productivity and sustainability of farming systems in Australia and is thus central to both the short and long-term profitability of farming enterprises (Maunder, 1970, 1986). However, few in Australian agriculture had heard of the El Niño/Southern Oscillation (ENSO) phenomenon and its impact upon Australian climate or agriculture before 1990. Today, few Australian farmers, agri-business people or scientists are not aware of its impact.

The period since 1990 has been marked by development of many ENSO related resource materials and education processes. These cover a diversity of approaches ranging from comprehensive courses and workshops, to electronic delivery systems, and direct information about indicators such as the Southern Oscillation Index (SOI) and Sea Surface Temperatures (SSTs) in the media. Several organisations in Australia, but especially the Queensland Department of Primary Industries (QDPI), the Queensland Department of Natural Resources (QDNR), New South Wales Agriculture (NSW Agriculture), and the Bureau of Meteorology have been working collaboratively with primary producers to build knowledge and skills in relation to use of seasonal climate forecasts in farm decision making.

Apart from the impetus provided by improved scientific understanding of ENSO (Cane, 2000) and increasing awareness of its potential use in improving agricultural systems (Hammer et al., 1991), these initiatives were stimulated by a review of drought policy in Australia, as documented in the National Drought Policy (Anon, 1990). The National Drought Policy of 1990 introduced a major change in the role of government services in that emphasis was shifted from providing subsidies to providing information so that farmers could be more self-sufficient in managing the climatic risks they faced (Anon 1990).

A client base of farmers and rural managers, hungering not only for information, but for better understanding of the climate variability they had to live with while competing in global markets, rapidly emerged (Childs *et al.,* 1991; Clewett *et al.,* 1995, 1996). Training initiatives for farmers and land managers were initiated through the National Property Management Planning Program. This emphasised holistic and strategic planning for property management with flexible delivery according to client needs. It also raised the profile of climate variability and how to best manage with it. Many primary producers were keen to develop a better understanding of climatic systems, the climate variability being dealt with, and how to use seasonal climate forecast information in making decisions in the context of property management planning (Anon, 1995; Ridge and Wylie, 1996).

Hence, there was a client base looking for climate information and how to use it to make better decisions, there were organisations broadly supplying the up to date seasonal climate forecasts, and there were now organisations providing educational based forums to deliver and create a learning environment nationally (Anon 1995). It was all these developments occurring in parallel that set the scene for building knowledge and skills to use seasonal climate forecasts in property management planning. This paper outlines the property management planning process, details how information on seasonal climate forecasting was made available so that the necessary knowledge and skills to use it effectively in farm management could be developed, and discusses what has been learnt in the process and the major issues for the immediate future.

2. Property Management Planning Process

The Property Management Planning (PMP) process (detailed in Anon, 1995, pp 3-15), embraces all aspects of property management and is a strategic planning approach to property management. It focuses on landholders reviewing their current management practices and developing long-term sustainable production strategies for better property management. PMP has an emphasis on participative problem solving and includes: reviewing and developing personal goals, natural resource assessment and physical planning, sustainable agricultural production, ecological considerations, business and financial management, marketing, drought preparedness and risk management (Anon, 1995; George, 1993). The PMP process provides the context for developing seasonal climate knowledge and skills within a strategic planning and farm decision-making framework.

In determining the processes for developing knowledge and skills, it is of paramount importance to strategically work out what knowledge and skills are to be developed and then plan how it is best able to be delivered. The PMP process uses the broad framework of:

- identify needs
- develop appropriate educational / technical material in response to the needs
- deliver material according to needs in terms of workshops, information sessions, and relevant materials clients can access eg. Internet, poll fax, etc.

A participative problem solving approach (as expanded in Spencer (1989) facilitates cooperative learning and may allow for the different learning styles of participants (Honey and Mumford, 1986). The information is not intended to help just in short term decision making (even though it does this), but rather to be an integral part of each stakeholders strategic planning process. The PMP process uses information to assist producers to participate in problem solving activities to work out:

- **where do I want to go?** (goal setting)
- **where am I now?** (this is an inventory of personal, financial and physical resources of which weather and climate information is an integral part)
- **how will I get there?** (setting strategies and actions, and comparing alternatives, which includes assessing impacts of climate variability on enterprise outcomes)
- **how is it going?** (monitoring and evaluating progress, readjusting goals and strategies by what works and what hasn't worked etc.).
- **... and returning to 'where am I now...?** ... continuing the process with improved knowledge and skills.

The PMP process provides the general framework in which we have worked to build knowledge and skills in relation to use of seasonal climate forecasts in farm management. Seasonal climate forecasting is one part of this comprehensive planning process. However, in this paper, we will focus on issues dealing with information on climate and seasonal forecasting and how this has been applied to improve decisions. Our approach will be based on the relevance to building appropriate knowledge and skills among farmers.

3. Climate and Seasonal Climate Forecasting Information

Understanding is a vital part of adult learning. Thus, in developing knowledge about seasonal climate forecasts that are based on the El Niño Southern Oscillation (ENSO) phenomenon, some understanding of the processes and mechanisms of the Walker circulation is important. A knowledge of the statistical relationship between indices such as the Southern Oscillation Index (SOI) and rainfall and how to use that information is what will ultimately be useful, but an understanding of how ENSO interacts with the world's ocean/atmosphere circulation is important in acceptance of such relationships. On the one hand this is relatively simple because the Walker circulation consists of a set of physical processes. However, it is also difficult because it requires an understanding of abstract issues such as probability, when effects on rainfall likelihood are considered.

Farmers and scientists learn differently and respond to information in different ways. For example, in communicating knowledge of risk we have found that cumulative probability distributions provide an effective mechanism for scientists to communicate with each other. However, in communicating with the farming community we have found that other diagrams and ways of expressing risk are more effective. The simplest statement for communicating risk has been the percent chance that seasonal rainfall will

be above or below the median rainfall (or above or below the average) (Stone *et al.*, 1996a). This very simple statement of risk is now widely used in the agricultural media. The next simple level for communicating risk is to consider the chance (probability) that seasonal rainfall will be: above average (highest 30%), about average, or below average (lowest 30%). The Bureau of Meteorology used this tercile method in their seasonal forecasts for many years. They have recently converted to percent chance above or below the median.

The fullest understanding of climate risk often occurs where people have been able to view all of the historical rainfall (eg. 100 years of data) as a time series histogram. Time series can be particularly useful because they give an analogue representation (rather than digital) of people's chronological memory patterns. Understanding is maximised where different colours are assigned to different year types (such as using red bars for El Niño year types and blue bars in the histogram for La Nina year types). Such a diagram clearly shows the relative frequency and spatial separation of El Niño and La Nina events. The time series analogue also shows how often above and below median (or above and below average) rainfall occurs without recourse to frequency distribution data or abstract mathematical/statistical representations. Simplicity is often the key to comprehension. Comprehension empowers people with ownership of the forecast and thus confidence in moving towards using it in their decision making. George *et al.* (1998) expand on this concept.

TABLE 1. Seasonal climate forecast information services, analytical tools and educational courses (References to examples are given in the text)

	Method	Example
1	Pamphlets	Climate Variability & El Niño
2	Books	Will It Rain ?
3	Video and audio	Farming a Sunburnt Country
4	Subscription	Seasonal Climate Outlook (National Climate Centre)
5	Mass Media	TV Services, radio and newspaper
6	Phone Services	SOI Hotline
7	Fax Services	Weather by Fax, FarmWeather, SOI Hotline, Farmfax
8	PC Software	AUSTRALIAN RAINMAN , Wheatman, Barleyplan, Feedman, GrazeOn
9	Internet	The Long Paddock (www.dnr.qld.gov.au/longpdk), Met Net (www.bom.gov.au), SILO (www.bom.gov.au/silo/), NOAA
10	Drought Alerts	Aussie GRASS products
11	Workshops and Education Courses	Climate and Your Farm Home Study Course, PMP workshops, Managing for Climate workshops

Organisations in Australia involved with disseminating climatic information have recognised that people gain information, knowledge and skills in many ways and have thus sought to build a rich diversity of information pathways. These pathways range from simple awareness raising and development of aspiration to comprehensive development of knowledge and skills through books, analytical tools and educational courses (see Table 1).

The Bureau of Meteorology provides an extensive range of weather/climate information and services to agriculture. Some of the services are provided nationally, while others are provided by the individual regions covering a State or Territory. National Services include general meteorological and climatic information, satellite pictures, weather charts, ocean forecasts (including sea surface temperatures) and climatic outlooks. QDPI and QDNR provide a large range of climate information products, agricultural decision support products and related information, to a wide range of clients using various communication methods - printed materials, poll-fax, telephone recordings, PC software, video, Internet sites, e-mail and the mass media.

3.1   MASS MEDIA INFORMATION

Seasonal climate forecast information from the Bureau of Meteorology is regularly used by the national and rural press to keep people informed of developing issues. In addition, QCCA also releases a summary of the seasonal climate forecast for the public and media at the start of each month. While media reports and interviews with scientific staff on ENSO are now a regular feature of the rural press it is not so long ago that information on ENSO was very new. The article in the National Geographic magazine in 1984 (Canby, 1984) on the world wide effects of the 1982/83 El Niño may have been the first full length feature article published on ENSO for the general public. The article by Conroy and Clewett *Riding the Southern Oscillation See Saw* published in the Queensland Country Life in 1989 was possibly the first substantive article on ENSO to be published in the Queensland rural press. In Queensland, QDPI initiated weekly television broadcasts of the 30-day average SOI and seasonal climate outlook in 1993. Climate information is now presented regularly on a national agricultural interest weekly television program. There are frequent and regular rural radio broadcasts particularly in association with scientific guest speakers and climate information is reported regularly in major rural newspapers.

3.2   PRINTED MATERIALS

Prior to 1990 there was very little information available to the general public on seasonal climate forecasting and descriptions of the ENSO phenomenon. Since 1990 several publications have dealt with the subject so that a useful amount of printed information is now available. Some key examples are -

- Climate Variability & El Niño - a four page coloured pamphlet published by the Bureau of Meteorology describing ENSO and its impact on climate in Australia (Anon 1994).

- Will it Rain? Effects of the Southern Oscillation and El Niño on Australia – a booklet produced by QDPI (Partridge 1994). At the time of first printing (1991), this was the first book produced on the topic of ENSO for the general public in Australia. Importantly, this book sought to develop understanding of ENSO processes and recognised the potential for seasonal forecasting in agricultural management. The second edition was printed in full colour and contains chapters describing the EL Niño / Southern Oscillation (ENSO), the impacts of ENSO on the Australian climate, and results of bio-economic studies showing the impact of ENSO on runoff, grain and pasture production, and cattle performance. Some 4000 copies of "Will It Rain?" have been distributed.

- Seasonal Climate Outlook - a booklet produced monthly by the National Climate Centre of the Bureau of Meteorology (see Stone and de Hoedt, 2000). The outlook provides the Bureau's seasonal climate forecast and a comprehensive and up to date description of recent Australian climate.

3.3   TELEPHONE AND POLL FAX SERVICES

A comprehensive range of information is available by telephone and poll fax –

- Weather by Fax - a poll fax service for climate and meteorological data provided by the Bureau of Meteorology. Products include information on El Niño, seasonal climate outlook, SOI and SST data, and the latest weather charts, satellite pictures, forecasts, warnings, weather and rainfall reports and educational material

- Farmweather - a specialist part of the Weather by Fax service, this is a group of fax products which have been designed to meet the needs of rural communities. In many cases the material has been tailored to the specific enterprise activities of a particular region. Farmweather provides short-term weather information needed for making better informed decisions regarding such operations as spraying, planting, harvesting and short-term irrigation.

- SOI Phone Hotline – initiated by QDPI in January 1992, this service consists of a two-minute recorded message. The 30-day average SOI (updated daily) and associated information is given.

- SOI Fax Hotline – initiated by QDPI in April 1993, this service enables clients with a facsimile machine to collect selected pages of seasonal climate outlook information.

- FARMFAX - operated by QDPI, this system currently provides eleven pages of climate information, together with a wide range of other technical information required to make sound weather-related decisions.

Over 57,000 enquiries were received on the QDPI phone and fax hotlines in the 5 years to November 1997. In July 1995, a survey was conducted to obtain feedback from users of these hotlines. The main results from the survey were -

- location of users: Queensland 45%, NSW 42%, other states 11%, overseas 2%.
- business operated: primary producers 70%, agribusiness 16%, QDPI staff 8%, and the remainder from other government staff, education and media
- most useful products: SOI fax message (92% of respondents said fairly useful to very useful), fax of sea-surface temperature map (78%), fax of SOI data page (66%), and telephone hotline recorded message (36%)
- respondents reported the information to be used in making a wide range of decisions in primary production including production and marketing strategies, crop selection, crop planting decisions, crop fertiliser rates, herbicide choice, stock sales, stock purchases, feed budgeting, planning sown pasture establishment, estimating crop yields/returns, frost prediction, and insect/pest management. In addition, quantitative data was obtained by some clients for use in software packages such as Australian Rainman, Wheatman and Barley Plan.

3.4   PC SOFTWARE

Computer use in the agricultural sector is increasing. An Australian farmer group (Kondinin) that surveys its members regularly for data on a number of issues, found that 72% of its members surveyed in 1998 use computers (jumping from 40% in 1990) (Anon, 1998). Kondinin have a membership of over 20,000 farmers, which is a significant proportion of the total number of Australian farmers (145,000 – Australian Bureau of Statistics - 31/3/97). Other research completed in 1993 (Anon, 1995) reported that 30% of the top 20% of producers use a computer, while 18% of the middle 60% of producers use a computer.

The PC computer package AUSTRALIAN RAINMAN (Clewett *et al.* 1992, 1994) contains a comprehensive range of daily, monthly and seasonal rainfall analyses that have been tailored to the needs of decision makers. Climatic risks and opportunities, including changes in risk that are associated with the El Niño/Southern Oscillation are highlighted with displays as tables, graphs and maps. The package enables users (decision makers) to not only make seasonal forecasts for their own specific location and season of choice, but to also have ownership of the forecast and an appreciation of its strengths and weaknesses. The standard package contains a manual, the book 'Will It Rain? Effects of the Southern Oscillation and El Niño on Australia' (Partridge 1994), and an extensive set of climatic data from throughout Australia including.

AUSTRALIAN RAINMAN is used by many people including: farmers and pastoralists, business people, researchers, students and teachers . Over 1000 packages have been sold with many packages used by groups (eg. Landcare groups, rural businesses, schools) so that the estimated number of users exceeds 10,000. Feedback from users shows that the package is relevant to needs and:

- addresses issues that are of great topical interest (rainfall, El Niño) and of major importance to sustainable production and the chances of business success,

- builds knowledge and skills for managing climatic risks
- empowers users to do their own forecasts of seasonal conditions
- presents comprehensive information that is easy to understand and pertinent to decisions
- is useful as a learning tool in workshops such as the property management planning workshop *Managing for Climate* in Qld, and *Agriculture – Coping in a Variable Climate* in NSW.

3.5   INTERNET SERVICES

Kondinin's survey of farmers using computers (Anon, 1998), also found that 21% of its members surveyed use the Internet (a 400% increase from 1996). This demonstrates that there is a huge increase in computer and internet demand. Climate and weather information was one of the major uses for the internet. As phone lines in rural areas are upgraded, it is expected that this demand will continue to grow. Relevant internet sites include -

- Met Net – the internet site of the Bureau of Meteorology (http://www.bom.gov.au) contains all of the information provided by their Weather by Fax service plus additional educational and climatic information.

- The Long Paddock - internet site operated by QDPI and QDNR that provides a rich mixture of climatic information that is useful for farm and natural resource management. Products include a weekly summary of seasonal climate outlook, SOI data, and a large range of maps covering rainfall, rainfall probabilities, sea-surface temperature, drought situation and satellite imagery. The site received over 18,000 'hits' during the first 12 months of operation (commencing 19 December 1995).

- Other Sites - several international internet sites provide comprehensive information on El Niño events, seasonal climate forecasts, and meteorological and oceanographic data. Examples are the web sites operated by the International Institute for Climate Predictions (http://iri.ldeo.columbia.edu), the National Oceanographic Atmospheric Administration (http://www.noaa.gov), and the World Meteorological Organisation (http:// www.wmo.ch).

3.6   EDUCATIONAL COURSES, SEMINARS AND WORKSHOPS

A number of structured educational activities have been designed to build knowledge and skills in relation to seasonal climate forecasts. The main ones are -

- PMP workshops - The Futureprofit PMP program (Anon 1997) is implemented as a series of eight integrated workshops: (1) setting direction, (2) identifying natural resources, (3) property layout, (4) human resources, (5) financial analysis, (6) enterprise analysis, (7) testing management options, and (8) future planning. Climate and seasonal forecast information are very relevant to the inventory of natural resources in workshop number 2. However, it is the application of climate and seasonal forecast information that is of greatest importance and, hence, relevance to the financial and enterprise analyses in workshops 5 and 6, and the testing of options in workshop 7. Several analytical packages for cropping and grazing are now available to help in assessing the benefits to be gained from using seasonal climate forecast information (Hammer *et al.*, 1996; Meinke and Hochman, 2000; Stafford-Smith *et al.*, 2000)

- Rainman workshops - The Rainman workshops started in 1988 with cattle and grain producers in the Central Highlands region of Queensland in response to interest in ENSO generated by papers presented at a local symposium (Clewett *et al.*, 1988; Coughlan, 1988). At the initial Rainman workshops the impacts of ENSO on rainfall probabilities were calculated using a spreadsheet. Each of these workshops generated a large amount of interest in the community but the method of calculating changes in the chance of rain with changing ENSO conditions was cumbersome. This created the need to develop the RAINMAN computer package (Clewett *et al.*, 1992, 1994). Well over 150 Rainman workshops have now been conducted in all states of Australia in the 10 years 1988 to 1998. The emphasis of these workshops has been to empower people with (a) a knowledge of climate variability on a seasonal and decadal basis, (b) knowledge of ENSO processes and mechanisms underlying seasonal climate forecasting, (c) the impacts of ENSO on rainfall at their location and how this varies throughout the year, and (d) how to develop skills to apply climatic risk and ENSO information to their farm management issues.

- 'Managing for Climate' Workshops - These educational workshops (O'Sullivan and Paull, 1995) were designed for use with farmers and pastoralists in the property management planning process. The one-day workshop was developed in 1994 by QDPI in collaboration with the Bureau of Meteorology. The workshop aims are to improve understanding of weather maps/influences, ENSO processes, and seasonal climate forecasting methods. It also aims to improve skills in making weather and climate related decisions through the use of practical exercises. A *Workshop Manual* and a *Facilitators Manual* were produced. QDPI staff from each region have been trained with the intention of being able to deliver, market and facilitate workshops. A total of 35 workshops (560 participants) were conducted during 1994 and 1995. These workshops have helped to produce a strong increase in the demand for decision-support information in relation to how to make use of the climate information highlighted. Participants in a 'Managing for Climate' workshop were surveyed one month after the workshop. A total of 78% reported that they had continued to use information in the workshop manual when making management decisions, while 42% specifically mentioned stocking rate decisions.

- 'Agriculture/Coping in a Variable Climate' Workshops - These workshops were developed by the NSW Department of Agriculture within the NSW Farming for the Future PMP program. The workshop has similar aims and format to the Queensland Managing for Climate workshop. A total of 31 workshops/information sessions (involving 581 participants) were held in the period to November 1997. Main stakeholders involved were farmers, professional colleagues, agribusiness, educationists, and meteorologists (observers and forecasters). When feedback was sought from farmers in response to the question "what do you intend to do in your farming as a result of this session?" responses ranged from general increased awareness of weather and climate to specific use of that awareness in making important decisions. The range of responses indicated that adults learn differently and take on board areas of interest or relevance to them. Any workshop activity with producers needs to be flexible to allow for the different learning styles of its participants, to recognise and build on the range of understandings already held, and to enable the greatest learning opportunity for the participants. This concept is reinforced by numerous other cases (Ridge and Wylie, 1996; Zemke and Zemke, 1981; Davey, 1987; Killen, 1996; Pollard, 1992; George et al., 1998; Thorp, 1999; Underwood and Salmon, 1980)

- Climate and Your Farm - This is a stand alone home study package (Brouwer and George, 1995) offered by the CB Alexander Agricultural College (Tocal, NSW), which is part of New South Wales Agriculture Department. The course is available for external students as part of the nationally accredited Landcare Certificate, or as a text. Short residentials enable students to come together and discuss issues and exercises raised in the text and from what they've found in their own experiences. Two of the four books in the course have a focus on use of climate information in farm planning and management and the other two books focus on weather issues.

4. Application of Information to Improve Decisions

Building knowledge and understanding of seasonal climate forecasting is a necessary but not sufficient condition for its effective application in farm management. Effective application of forecasts involves their use to improve decisions (Hammer, 2000). This requires development of management skills based on an understanding of how the forecast might interact with potential decision options. Weather and climate related decisions in agriculture occur on both short and long time-scales. Very short time-scale decisions (such as the tactical timing of hay making, shearing, planting) involve consideration of the weather over one day to a week. The main information required for such decisions includes 4 to 5 day weather forecasts, synoptic charts, upper-air charts and cloud maps. In contrast, long-term decisions in the property plan such as enterprise selection, investment in real estate or major capital works (land development, machinery) relate to time-scales of 5 to 15 years and are of a structural nature. Seasonal climate forecasts are not so relevant to these very short and long time-scale decisions although the frequency of planting opportunities in a season, and the prospects for immediate payback on investment is of importance and thus of some relevance.

Feedback from the workshops described above has revealed that seasonal climate outlook information is considered by producers in making a wide range of decisions in primary production including production and marketing strategies, crop selection, crop planting decisions, crop fertiliser rates, herbicide choice, stock sales, stock purchases, feed budgeting, planning sown pasture establishment, estimating crop yields/returns, frost prediction, and insect/pest management. Seasonal climate outlook information can also be valuable in providing warnings of extreme climatic events. For example, there are some useful relationships between climatic patterns used in seasonal climate outlook forecasting and frosts, heatwaves, cyclones and floods (eg. Stone *et al.*, 1996b).

Feedback on concepts and rules of thumb used by producers include -
- rainfall statistics show past rainfall patterns, and bring realistic expectations into decision-making
- probabilities help in predicting likely rainfall, allow financial outcomes of different options to be calculated, and increase confidence when selecting management options
- identifying historical records that are similar to the current situation helps to determine the outcomes of different management options and to clarify the need to 'hedge bets'
- seasonal climate forecast indicators enable the calculation of rainfall probabilities from historical years in the current forecast category rather than from all years on record
- the relationship between seasonal climate forecast indicators and future rainfall varies for different locations, and for different times of the year.
- seasonal forecasts are based on probability and hence the most likely climate scenario may not occur. Thus, to be confident of increased profitability, seasonal climate forecast indicators need to be used over a number of years.

It is clear from this feedback that many primary producers now understand the process of ENSO and believe that seasonal climate outlook information can be used profitably in their business. However, the number of producers in this category is probably still in the minority and the management skill to take the belief into improved decisions remains poorly developed. How can this be enhanced? The following factors/actions will help to realise the potential value of seasonal climate outlook information.

- There is a need to continue interactions with producers to identify critical climatic parameters and thresholds for a broad range of production problems for different industries, and to develop a range of appropriate new information products. Feedback from producers will be required on the products needed, presentation format and the most convenient delivery methods.

- There is a need to foster processes that enable early integration of seasonal climate forecast information into whole-property management decisions, and to continue fostering use of seasonal climate information within the PMP process.

- There is a continuing and increasing need to develop tools that enable analysis of decision options in conjunction with seasonal forecast information and other factors (eg. prices, soil loss). Such tools build on system modelling capability in agricultural simulation software such as such as APSIM (McCown *et al.*, 1996) and GRASP (McKeon *et al.*, 1990; Day *et al.*, 1997). Such tools might be software products that have been derived from the simulators to facilitate scenario comparison, such as WHEATMAN (Woodruff, 1992) and WHOPPER_CROPPER (Hammer *et al.*, 1996), or spreadsheet tools to enable the financial outcomes of each option to be compared using processes such as decision-tree analyses.

- Additional case studies and practical decision-making exercises for use in training are needed to asses the value, in economic and environmental terms, of seasonal climate forecasts to property managers in different industries. Within the PMP workshop series the emphasis on seasonal forecast information is probably more frequently dealt with in describing the property resources (workshop number 2). There is a need to build capacity for dealing with seasonal forecast information in the later stage workshops such as workshop 7 (testing management options) in the Futureprofit (Anon, 1997) PMP workshops. Studies on crop and pasture systems reported elsewhere in this volume (eg. Carberry *et al.*, 2000; Stafford-Smith *et al.*, 2000) will provide a basis for developing robust case studies.

- There is a continuing need to monitor the ways in which seasonal climate forecast information is being used by primary producers and agri-businesses in property management. There is also a need to evaluate how communication and extension programs concerned with seasonal forecasting are contributing to the knowledge, attitudes, skills and aspirations of the rural community. By improving on our efforts and results, it will ensure sustainable agriculture becomes more achievable.

- There is a need to continue to improve the design and effectiveness of any learning program (de Bono, 1992), dealing with climate variability, seasonal climate forecasts, and their applications to decision-making. This will involve evaluating how effective any program is against established learning outcomes. There has not been enough effort in this to date.

5. What Have We Learned?

Over the past decade, we have pursued a comprehensive program of communication and education about climate variability, ENSO, and seasonal climate forecasting. The scientific language of climate is now firmly entrenched in the farming community and we are progressing on the subsequent demand generated in relation to effective use of seasonal forecasts in decision-making. Throughout this process, a number of lessons have emerged. The major ones are -

- Those involved in agriculture have a genuine concern about climate and weather information and it ranks highly significantly as an influencing factor (Maunder, 1970, 1986; Childs *et al.*, 1991; Hammer and Nicholls, 1996).

- Farmers, agencies, agribusiness and communities need to work together to overcome problems associated with climate variability and make the best sense possible of seasonal climate outlooks. The PMP process is a good way to start this

(Ridge and Wylie, 1996; Daniels and Chamala, 1989; Grimson, 1999; Clewett *et al.*, 1995)

- The National Drought Policy Review (Anon, 1990) highlights that government policy is more likely to succeed when greater emphasis is placed on adult learning and information rather than drought subsidies. This has some supporting evidence as documented in Grimson (1999).

- There needs to be a greater effort on increasing the understanding of the uncertainty in seasonal climate forecasts in the community (Childs *et al.*, 1991; Clewett *et al.*, 1995; Hammer *et al.*, 1996). This will involve emphasising whole farm property management planning and working through the ramifications of a seasonal forecast and the benefits of how planning helps to make the best decisions before and, as the season unfolds.

- Researchers and extension staff need to examine and evaluate the relevance and adoption of their work (Pollard, 1992). Alternatives leading to successful results need to be sought out.

- Misjudging climate variability and balancing property management and enterprise is complex. Information and resources need to allow for this to enable thinking and lifelong learning skills for the immediate and long term benefit (Clewett *et al.*, 1995; Hammer *et al.*, 1996).

- Emphasis needs to be placed on providing opportunities that facilitate the process of learning about the variable climate and seasonal climate forecasts amongst primary producers. This needs to then be followed by how to package information (or even just provide information) to fit the process (Ridge and Wylie, 1996).

We have also gained insight into how people have learned. Our observations and those of others suggest that people have learnt best where -
- an immediate problem must be solved, a task done, or a decision made that needs a new skill (Davey, 1987; Zemke and Zemke, 1981; Pollard, 1992).
- all stakeholders are open to learning (Zemke and Zemke, 1981; Thorp, 1999).
- the learning process recognises and builds on the understandings producers already have (Ridge and Wylie, 1996; Killen, 1996; Huffman, 1974).
- there is an emphasis on motivating others to come to a deeper understanding (as opposed to learning for action) of climate for the purpose of improving management (Ridge and Wylie, 1996).
- information delivery/learning process is flexible to allow for differences among individuals, their backgrounds, interests and aspirations as well as family and business circumstances (Zemke and Zemke, 1981; George *et al.*, 1998).
- adults use a variety of methods in any learning project (thinking, reading, listening, asking, etc). Any learning processes need to be aware of this and deliver according to the participants needs (Davey, 1987; Killen, 1996).

- resources most used in self-directed learning are other people (mostly people they know well) and even 'expert opinion'. It is critically important that these not be underestimated (Underwood and Salmon, 1980). Farmers that have 'survived' the variable climate bring to any learning experience a wealth of knowledge, not only of their climate but how their environment and enterprises responded to that particular situation. This needs to be acknowledged.
- learning situations provide opportunities for reflection and to identify what they do know, and, what they 'don't know they didn't know' (Thorp 1999).

Conclusion

Use of seasonal forecasts looking at climate variability and El-Niño Southern Oscillation (ENSO) information is new to Australian agriculture with few people having used this information prior to 1990. Our paper reviews the value of resource materials and adult education processes that have been developed to build knowledge and skills in rural communities concerning use of seasonal climate forecasts in agricultural decisions and property management planning. This is supported by feedback from primary producers, which shows good awareness of seasonal climate forecast information and how it is beginning to be used in agricultural decisions, with positive results. There is a need to consolidate what has already been done and to build on this into the future.

To build the knowledge and skills of rural communities to use seasonal climate forecasts in farm management requires a continuing commitment to developing in parallel a rich diversity of climate application products and information sources, as well as excellent and continuing learning opportunities. This diversity is needed to meet the needs of many different people and the many different levels of farm decision making. Special emphasis is now required on management skills and relevant tools for effective evaluation of decision options that might be considered in responding to a seasonal forecast.

References

Anon (1990) Australia Drought Relief Policy Review Task Force 1990. Final Report, Drought Review Task Force, May 1990. Australian Government Publishing Service, Canberra.

Anon (1994) Climate Variability and El Niño. Bureau of Meteorology, Dept. of the Environment Sport and Territories, Melbourne, Australia.

Anon (1995) Managing for the Future. Report of the Land Management Task Force, Rural Division, Commonwealth Department of Primary Industries and Energy, Canberra.

Anon (1997) Futureprofit – Integrated Workshop Series. Workshop Manual, Queensland Department of Primary Industries, Brisbane.

Anon (1998) Superhighway draws closer to the Farm Gate. Kondinin Research Report, Farming Ahead with the Kondinin Group. No.77, p. 28. Kondinin, WA.

Brouwer, D. and George, D. (1995) Climate and Your Farm. New South Wales Agriculture Home Study Program, New South Wales Agriculture

Canby, T. (1984) El Niño's Ill Wind. National Geographic 165(2), 144-183.

Cane, M.A. (2000) Understanding and predicting the world's climate system, in G.L. Hammer, N. Nicholls, and C. Mitchell (eds.), Applications of Seasonal Climate Forecasting in Agricultural and Natural Ecosystems – The Australian Experience. Kluwer Academic, The Netherlands. (this volume)

Carberry, P.S., Hammer, G.L., Meinke, H. and Bange, M. (2000) The potential value of seasonal climate forecasting in managing cropping systems, in G.L. Hammer, N. Nicholls, and C. Mitchell (eds.), *Applications of Seasonal Climate Forecasting in Agricultural and Natural Ecosystems – The Australian Experience.* Kluwer Academic, The Netherlands. (this volume)

Childs, I.R., Hastings, P. A. and Auliciems, A. (1991) The acceptance of long range weather forecasts: a question of perception? *Australian Met. Magazine* **39**,105 – 112

Clewett, J.F., Clarkson, N.M., Owens, D.T. and Abrecht, D.G. (1994). AUSTRALIAN RAINMAN: Rainfall Information for Better Management, (A computer software package). Department of Primary Industries, Brisbane.

Clewett, J.F., Clarkson, N.M. and Owens, D.T. and Abrecht, D. G. (1996) Australian Rainman - progress and prospects. Proceedings of Managing with Climate Variability Conference. Occasional Paper CV03/96. Land and Water Resources Research and Development Corporation, Canberra. pp 35-40.

Clewett, J.F., Clarkson, N.M., Owens, D.T. and Partridge, I.J. (1992). RAINMAN: Using El Niño and Australia's rainfall history for better management today, in S. Field (ed.), Harnessing Information for a Smarter Agriculture, Proceedings Tasmanian Zone of the Australian Institute of Agricultural Science National Conference. pp. 16-20.

Clewett, J. F., Kininmonth, W. R. and White, B J. (1995) Sustainable agriculture: A framework for improving management of climatic risks and opportunities, in, Sustaining the Agricultural Resource Base. 12th meeting of the Prime Ministers Science and Engineering Council, Office of the Chief Scientist, Dept of Prime Minister and Cabinet, Canberra.

Clewett, J.F., Young P.D. and Willcocks, J.R. (1988). Effect of climate change on agriculture in Central Queensland I. Rainfall variability analysis, in, The Changing Climate and Central Queensland Agriculture. Proceedings of the fifth symposium of the Central Queensland Sub-Branch of the Australian Institute of Agricultural Science. pp 43-52.

Coughlan, M.J. (1988) Seasonal climate outlooks, in, The Changing Climate and Central Queensland Agriculture. Proceedings of the fifth symposium of the Central Queensland Sub-Branch of the Australian Institute of Agricultural Science. pp 17-26.

Daniels, J. and Chamala, S. (1989) Practical men or Dreamers? A study of how farmers learn. *Australian Journal of Adult Education* **29**, 25 – 31

Davey, P. C. (1987) The role of traditional and alternative extension methods in facilitating information transfer and client change, in M. Littmann (ed.), Rural Extension in an Era of Change. Australasian Agricultural Extension Conference Proceedings. Brisbane. pp. 661 – 668

Day, K.A., McKeon, G.M. and Carter, J.O. (1997) Evaluating the risks of pasture and land degradation in native pasture in Queensland (DAQ124A). Final Project Report to Rural Industries Research and Development Corporation. Queensland Department of Natural Resources, Brisbane.

de Bono, E. (1992) *Teach your child how to think.* Penguin Books. London.

George, D. (1993) A Glossary of Property Management Planning Terms. NSW Agriculture.

George, D., Buckley, D. and Carberry, P. (1998) A Summary of evaluations from participant surveys following workshops and information sessions (on the topic of "Agriculture – Coping in a Variable Climate") Paper presented at the 12th Australia New Zealand Climate Forum. University of Western Australia, Perth.

Grimson, M. (1999) Taking Control: Farmers talk about Property Management Planning. National Farmers Federation. Canberra

Hammer, G.L. (2000) A general systems approach to applying seasonal climate forecasts, in G.L. Hammer, N. Nicholls, and C. Mitchell (eds.), *Applications of Seasonal Climate Forecasting in Agricultural and Natural Ecosystems – The Australian Experience.* Kluwer Academic, The Netherlands. (this volume)

Hammer, G.L. and Nicholls, N. (1996) Managing for climate variability - The role of seasonal climate forecasting in improving agricultural systems. Proc. Second Australian Conference on Agricultural Meteorology. Bureau of Meteorology, Commonwealth of Australia, Melbourne. pp.19-27.

Hammer, G.L., Chapman, S.C. and Muchow, R.C. (1996) Modelling Sorghum in Australia: The State of the Science and its role in the pursuit of improved practices, in M. A. Foale, R. G. Henzell and J. F. Kneipp (eds.), Proceedings Third Australian Sorghum Conference. Australian Institute of Agricultural Science, Melbourne. Occasional Publication No. 93. pp 43 – 61.

Hammer, G.L., McKeon, G.M., Clewett, J.F. and Woodruff, D.R. (1991) Usefulness of seasonal climate forecasts in crop and pasture management. Proc. First Australian Conference on Agricultural Meteorology. Bureau of Meteorology, Commonwealth of Australia, Melbourne. pp.15-23.

Honey, P. and Mumford, A. (1986) *The manual of learning styles.* 2nd edition, Maidenhead, England.

Huffman, W.E. (1974) Decision making: the role of education. *American Journal of Agricultural Economics* **56**, 85 – 97.

Killen, R. (1996) *Effective teaching strategies. Lessons from research and practice.* Social Science Press, NSW, Australia.

Maunder, W.J. (1970) *The Value of the Weather.* Methuen. London

Maunder, W.J. (1986) *The Uncertainty Business* Methuen. London

McCown, R.L., Hammer, G.L., Hargreaves, J.N.G., Holzworth, D.P. and Freebairn, D.M. (1996) APSIM: A novel software system for model development, model testing, and simulation in agricultural research. *Agric. Sys.* **50**, 255-271.

McKeon, G.M., Day, K.A., Howden, S.M., Mott, J.J., Orr, D.M., Scattini, W.J. and Weston, E.J. (1990) Northern Australian savannas: management for pastoral production. *Journal of Biogeography* **17**, 355-72.

Meinke, H. and Hochman, Z. (2000) Using seasonal climate forecasts to manage dryland crops in northern Australia, in G.L. Hammer, N. Nicholls, and C. Mitchell (eds.), *Applications of Seasonal Climate Forecasting in Agricultural and Natural Ecosystems – The Australian Experience.* Kluwer Academic, The Netherlands. (this volume)

National Climate Centre. Seasonal Climate Outlook. Incorporating Seasonal Rainfall Probabilities and El Niño Outlook. Bureau of Meteorology, Melbourne. (monthly publication)

O'Sullivan, D.B. and Paull, C. (1995). Managing for Climate. A Property Management Planning Module. Queensland Department of Primary Industies, Brisbane.

Partridge, I. J. (ed.) (1994). Will It Rain ? Effects of the Southern Oscillation and El Niño on Australia. Second Edition, Department of Primary Industries Queensland, Brisbane, Information Series QI94015, 56 pp.

Pollard, V. (1992) The Education and training needs of farmers, in, Management for Sustainable Farming, Proceedings of the 16th National Conference of the Australian Farm Management Society. Gatton, Qld. pp. 232 – 241

Ridge, P. and Wylie, P. (1996) Farmers training needs and learning for improved management of climatic risk. Proceedings of Managing with Climate Variability Conference. Occasional Paper CV03/96. Land and Water Resources Research and Development Corporation, Canberra.

Spencer, L. J. (1989) *Winning through participation.* Kendall/Hunt Publishing Co. Iowa. USA.

Stafford Smith, M., Buxton, R., McKeon, G.M. and Ash, A. (2000) Seasonal climate forecasting and the management of rangelands: do production benefits translate into enterprise profits?, in G.L. Hammer, N. Nicholls, and C. Mitchell (eds.), *Applications of Seasonal Climate Forecasting in Agricultural and Natural Ecosystems – The Australian Experience.* Kluwer Academic, The Netherlands. (this volume)

Stone, R.C. and de Hoedt, G. (2000) The development and delivery of current seasonal climate forecasting capabilities in Australia, in G.L. Hammer, N. Nicholls, and C. Mitchell (eds.), *Applications of Seasonal Climate Forecasting in Agricultural and Natural Ecosystems – The Australian Experience.* Kluwer Academic, The Netherlands. (this volume).

Stone, R.C., Hammer, G.L. and Marcussen, T. (1996a) Prediction of global rainfall probabilities using phases of the Southern Oscillation Index. *Nature* **384**, 252-256.

Stone, R.C., Nicholls, N. and Hammer, G.L. (1996b) Frost in north-east Australia: trends and influences of phases of the Southern Oscillation. *J. Climate* **9**, 1896-1909.

Thorp, D. (1999) Our best demands more effort from teachers. The Weekend Australian, July 24 – 25, 1999

Underwood, C. A. and Salmon, P. W. (1980) Nature and extent of self directed learning in agriculture. School of Agriculture and Forestry, University of Melbourne, Melbourne.

Woodruff, D.R. (1992). 'WHEATMAN' a decision support system for wheat management in subtropical Australia. *Aust. J. Agric. Res.* **43**,1483-1499.

Zemke, R. and Zemke S. (1981) 30 things we know for sure about adult learning. *Training* **18**, 45 – 52

OPPORTUNITIES TO IMPROVE THE USE OF SEASONAL CLIMATE FORECASTS

NEVILLE NICHOLLS

Bureau of Meteorology Research Centre
PO Box 1289K
Melbourne, Vic 3001, Australia

Abstract

The numerous impediments obstructing the optimal use of seasonal climate forecasts are reviewed. These include

- Scientific problems (eg. limited skill)
- Inappropriate content (eg. not forecasting what users need)
- External constraints (eg. inability of users to change decisions)
- Complexity of target system (eg. the impacts of a predictable climate anomaly may be unpredictable)
- Communication problems (eg. confusion due to multiple forecasts)
- User resistance or misuse (eg. user conservatism)
- Cognitive biases (eg. probability illusions)

Strategies to overcome these problems and thereby improve the use of seasonal climate forecasts are discussed.

1. Introduction

More than two decades ago, Glantz (1977) in a study of the value of drought forecasts for the African Sahel, noted that a variety of social, economic, environmental, and political constraints would limit the value of even a perfect drought forecast, at that time. He concluded that a drought forecast might not even be desirable, let alone useful, until some adjustments to existing social, political and economic practices had been made. However, the value of such a forecast would be greatly enhanced if its implementation was coupled with the removal of some of these constraints.

A few years later Lamb (1981) questioned whether we even knew what sort of climate variables we should be predicting. He pointed out that the achievement of the ultimate goal of improved climate prediction – the minimisation of the adverse socioeconomic consequences of climatic variability – had two prerequisites. First, the activities and

309

G.L. Hammer et al. (eds.), The Australian Experience, 309–327.
© 2000 *Kluwer Academic Publishers. Printed in the Netherlands.*

regions most impacted by climatic variations required identification. Second, the regional economies with the flexibility to adjust to capitalise on the availability of skilful climate forecasts needed to be determined. Only after these prerequisites had been met could we focus on the development of climate prediction systems likely to result in real benefits.

Since these two papers were published there have been substantial advances in our understanding of parts of the climate system, especially the El Niño - Southern Oscillation which provides some climate predictability in many parts of the world (Trenberth, 1997). There has also been research aimed at determining the value of climate forecasts, in a variety of settings (eg. Hulme et al., 1992; Adams et al., 1995; Hammer et al., 1996; Mjelde et al., 1997; Pulwarty and Redmond, 1997). It is clear however, that the climate predictability provided by the El Niño - Southern Oscillation is still not being used comprehensively or optimally (eg. Nicholls and Kestin, 1998).

There have been a few studies aimed at identifying why potential users were not incorporating climate predictions in the decision-making processes. Changnon et al. (1995) surveyed decision-makers in power utilities, to discern the needs and uses of climate forecasts. Only three of the 56 decision-makers surveyed used forecasts. Major hindrances to the use of forecasts included:

- Forecasts difficult to interpret
- Lack of models to integrate information
- Uncertainty over accuracy
- Additional information necessary
- Proof of value necessary
- Lack of access to expertise
- Difficult to assess forecasts

A similar list of difficulties was assembled by Pulwarty and Redmond (1997), to account for the lack of use of climate forecasts involved in salmon management issues in the Columbia River basin, despite the significant influence of precipitation and its subsequent hydrologic impacts on the region. The reasons advanced by decision-makers here included:

- Forecasts not "accurate" enough
- Fluctuation of successive forecasts
- What is the nature of the forecast?
- External constraints (eg. legal requirements) forbid a flexible response to forecast information
- Procedures for acquiring knowledge and implementing decisions by incorporating forecasts have not been defined
- Availability of locally specific information may be more important
- "Value" may not have been demonstrated (by a credible organisation or competitor)

- Required information may not have been provided
- Competing or conflicting forecast information
- Lack of information regarding how well the forecasts are "tracking" the actual climate
- History of previous forecasts not available

Hulme *et al.* (1992), in a study of the potential use of climate forecasts in Africa, suggested that forecasts may be useful at the national and international level (for instance in alerting food agencies to possible supply problems), although they concluded (repeating the comments of Glantz, 1977) that improvements in institutional efficiency and interaction are needed before the potential benefits of the forecasts could be realised. They observed that rural communities (eg. individual farmers) were the group least likely to obtain direct benefits from forecasts. This was because:

- The forecast information is not precise enough to influence local decisions such as where to plant or where to move animals.
- All producers are restricted in their flexibility to respond to forecast information; the poorer and more vulnerable the producer, the greater the restriction.
- Decisions are based on a range of factors of which climate is only one. A climate forecast in isolation from other information is unlikely to improve on existing knowledge systems.

Research from various sources has been used here to identify the impediments to the use of climate prediction. These sources include surveys of potential users (such as those quoted above), as well as psychological and sociological studies. Anecdotal and illustrative evidence gathered from Australian efforts to encourage incorporation of the use of climate forecasts in decision making is also used.

Impediments to the use of climate forecasts can be sorted into seven groupings:
- Scientific problems (eg. limited skill)
- Inappropriate content (eg. not forecasting what users need)
- External constraints (eg. inability of users to change decisions)
- Complexity of target system (eg. the impacts of a predictable climate anomaly may be unpredictable)
- Communication problems (eg. confusion due to multiple forecasts)
- User resistance or mis-use (eg. user conservatism)
- Cognitive bias (eg. probability illusions)

Each of these groups will be discussed below. The focus will be on impediments that have received little attention in the climate literature. The field of cognitive bias, for instance, has many lessons for climate prediction but is little known by climate scientists. Similarly, little attention has been given to the complexity of the systems for which climate forecasts are intended. As will be seen, major unexpected changes can arise from short-term climate anomalies. The introduction of forecasts may complicate a complex system even further. Although the focus of this paper is on the impediments to the use of predictions of seasonal to interannual climate variations, similar problems arise in the use of climate change predictions, and some of the examples will be drawn from climate change studies.

2. Impediments Due to Scientific Problems

The limited skill obtainable, or even expected, in climate predictions is well known. This is an impediment often cited as a reason for the lack of use of climate predictions. In some regions, however, considerable skill has been achievable for a long time, using simple, well-verified forecast systems. One example is the use of the El Niño - Southern Oscillation to predict tropical cyclone activity in the Australian region (eg. Nicholls, 1992). Other factors, discussed later, may restrict the utility of such forecasts, even when the skill obtainable is quite high.

The relatively low levels of forecast skill in most areas, however, has led to searches for more or better predictors. These searches can lead to artificial skill, skill that appears in the "training" data set, but disappears when the forecast system is used on new, independent data. This problem of artificial skill or selection bias has been known to statisticians and some climate scientists for many years (eg. Grant, 1954), but is ignored by many developers of forecast systems. Elsner and Schmertmann (1994) and Chatfield (1995) provide useful reviews of this problem and of strategies to overcome it. Cross-validation is one such strategy.

Some current seasonal forecast systems use cross-validation to estimate their expected skill; others do not. The problem of artificial skill is very great, because of the large number of potential predictors and the small number of realisations available to develop a statistical forecast system. So, any skill comparison between cross-validated systems and those systems not cross-validated will be biased. This complicates the comparison, by users, of the expected skill of various statistical forecast systems.

Although this problem of artificial skill or selection bias is mainly a problem with statistical forecast systems, it can also occur with dynamical models. The Cane-Zebiak model for predicting the El Niño is the most thoroughly assessed model available. The model did not predict the 1997 El Niño, although other models did. Does this mean that we will now select, for future years, only the models that did well in 1997? This process can also lead to selection bias.

A second scientific problem, which has arisen in recent years, is the need to combine forecasts from various models or sources. Workshops in Africa, southeast Asia, and South America, during 1997 combined forecasts from various sources into a single, probability forecast. The forecasts were combined subjectively. There is an extensive literature on the combination of forecasts from various sources, using objective approaches (eg. Thompson, 1977; Winkler and Makridakis, 1983), which is largely ignored by the climate community. Maines (1996) showed that subjective combination of forecasts often leads to conservative forecasts, and that individuals often do not take into account dependence between forecasts or use this dependence incorrectly, when combining forecasts. This suggests that subjective combination of climate forecasts may not result in the optimal combination. Australian Bureau of Meteorology has used seasonal climate outlooks produced by combining forecasts from various prediction schemes in a statistically-optimal fashion (Casey, 1995). The increasing availability of climate forecasts from statistical and model approaches necessitates the adoption of objective combination techniques. Combining model forecasts through ensemble techniques may be the most appropriate approach. However combining statistical forecasts with model ensembles is not easy.

The need to combine forecasts arises from the increasing availability of climate forecasts from different organisations and models. The different models or statistical approaches ensure that these multiple forecasts will differ in their predictions. It is difficult for users to assimilate such multiple sources of information. Redelmeier and Shafir (1995) reported on a study designed to reveal how doctors respond as the number of possible options for treatment is increased. The introduction of additional options (cf. different forecasts) raised the probability that the doctors would choose to either adopt the first option, or decide to do nothing. So the additional information, in the climate case extra forecasts, can distort decisions. DeBondt and Thaler (1986) found that when new information arrives, investors revise their beliefs (in this case about stock market performance) by overweighting the new information and underweighting the earlier information. This causes share prices to systematically overshoot. A similar problem (overweighting of newly-arrived information such as a new forecast) may affect the use of climate forecasts.

Another problem arises from the artificial separation of time-scales in climate predictions. Climate predictions, at present, focus on either interannual variability or climate change time-scales, rather than combining the time-scales and physical processes affecting them into a single model. Thus, models for seasonal climate prediction largely ignore the consequences of increasing concentrations of greenhouse gases. Yet there is evidence that decadal scale climate variability is affecting the relationships used for seasonal prediction (eg. Nicholls *et al.*, 1996) or the El Niño - Southern Oscillation (Trenberth and Hoar, 1996). Some means for avoiding this artificial separation of time-scales is necessary, if we are to include all the factors affecting seasonal climate. As well, other factors such as volcanic activity will also need to be included.

3. Impediments Due to Inappropriate Content

The example of Australian region tropical cyclone activity noted earlier is an example of content being inappropriate to many users. The simple number of tropical cyclones affecting a very large region is a very rudimentary threat index which would not be of use to many organisations (one group that does use this forecast is ocean-going sailors who can move to a different sailing area, based on seasonal cyclone forecasts). An indication of the likelihood of coastal crossings, especially for a limited length of coast, may be more useful. Pielke and Pielke (1997) observe that the use and value of Atlantic seasonal hurricane forecasts which have been available for many years, has not been assessed. The lesson from the tropical cyclone case is that some climate variables we can predict may be of little use to anyone.

Rainfall is, presumably, a different matter. For instance, rainfall (and other climate) variations have a substantial impact on the Australian wheat crop (by far Australia's most important crop). So, Australia satisfies the first pre-requisite identified by Lamb (1981): the country is strongly impacted by climate variations. We can also forecast seasonal anomalies of rainfall in Australia, for at least part of the country, for part of the year, using the El Niño - Southern Oscillation. A consequence of this is that we can also forecast crop yields (Nicholls, 1985, 1986; Rimmington and Nicholls, 1993).

Murphy (1994) noted, however, that:

> Weather forecasts possess no intrinsic value in an economic sense. They acquire value by influencing the behaviour of individuals and organisations ... whose activities are sensitive to weather conditions. Thus weather sensitivity on the part of users is a necessary – but not a sufficient – condition for forecasts to be of positive value.

We need to determine whether a user can take advantage of a climate forecast. This is the second pre-requisite identified by Lamb (1981): users need to have the flexibility to adjust their decisions based on climate forecasts, if the forecasts are to be valuable. This is the notion re-iterated by Hammer (2000) in the introduction to this volume and is the main intent of many of the other contributions.

Further, Murphy (1993) noted that forecasts must reflect our uncertainty in order to satisfy the basic maxim of forecasting, that a forecast should always correspond to a forecaster's best judgement. This means that forecasts must be expressed in probabilistic terms, since the atmosphere is not completely deterministic. In addition, the degree of uncertainty expressed in the forecast must correspond with that embodied in the preparation of the forecast. Since our forecast models do not include all relevant factors affecting seasonal climate it seems likely that any probabilistic forecasts derived from these models (dynamical or statistical) will underestimate the inherent uncertainty. To the extent that subjective judgement is used in estimating these probabilities, there is extensive evidence that forecasters can produce biased estimates of uncertainty, because of motivational factors (Fischhoff, 1994). For instance, there is a tendency to exaggerate one's confidence, if attention goes to those who appear most knowledgable. In climate forecasting, issuing a forecast with wide uncertainties can lead to users seeking the forecasters who are willing to provide the narrowest uncertainties (even if the narrowing is unjustified scientifically).

Fischhoff (1994) has noted that forecasts can be ambiguous in content. Such ambiguity can arise if those issuing forecasts and the likely users of these forecasts have different linguistic norms or thresholds of concerns about the forecast event. Murphy *et al.* (1980), for instance, note that confusion regarding daily rainfall forecasts concerned the event forecast and not the probability. Thus participants in a study were divided fairly evenly over whether 70% referred to the area that would experience rain, the time that it would rain, or the chance of at least some rain somewhere. In climate forecasting care needs to be taken to remove the possibility of ambiguities leading to misuse or even a tendency for potential users to ignore the forecasts completely.

4. Impediments Due to External Constraints

One aspect of the question of the utility of a forecast depends on its availability and lead-time. Even if a user's activities are weather sensitive, the forecast will not be of value unless it is available prior to the time when the user needs to make relevant decisions. For Australian wheat, the lead-time is insufficient for some potential uses. The "predictability barrier" occurring in the El Niño - Southern Oscillation around April (Webster and Yang, 1992) generally precludes skilful climate forecasts until about June (at the moment anyway). This is after the wheat crop has been planted in many parts of Australia, and so perhaps the decision most important to a farmer, which cultivar to plant, cannot be influenced by our current, limited El Niño - Southern Oscillation based predictability. Other (later) decisions, such as the amount of nitrogen fertiliser to use, may be able to take advantage of the predictability (eg. Hammer et al., 1996). Other crops may also be timed in such a way as to take more advantage of the predictability. Sorghum, for instance, is planted from October onwards. This is sufficiently past the "predictability barrier" as to allow good forecasts of average Australian yield before the crop is planted (Nicholls, 1986). So it is possible for farmers to use information from the El Niño - Southern Oscillation to decide how much sorghum to plant or how to manage it (Hammer and Muchow, 1991).

Another example of external constraints to the use of climate forecasts is provided by the water resource industry. Climate scientists in Australia have attempted to convince water managers that seasonal predictions based on the El Niño - Southern Oscillation could be useful in reducing risk or operating costs in water management. These efforts have largely gone unrewarded. Part of the reason for this is that Australian water resource management is often focussed on irrigation, and many managers adopt an extremely conservative policy to ensure their dams do not run out of water. The usual operating criterion is to assume that the input to the dams in the relevant period will be the lowest historical inflow. Use of the El Niño - Southern Oscillation based forecasts could reduce the risk of flooding (eg. by increasing reductions in dam water levels in the early stages of a La Niña) but this would come at the price of an increased risk of running out of water (eg. if the La Niña event has little effect in that particular year). This increased risk may be unacceptable to the managers, so they may not use the forecasts.

It is clear that for many systems affected by the El Niño - Southern Oscillation and therefore likely to be amenable to changes in decision-making based on seasonal climate forecast, external constraints still prohibit the effective use of forecasts. This was the problem identified by Glantz (1977). The problem remains, even in developed countries.

5. Impediments Associated with Communication

Felts and Smith (1997) noted that many decision-makers receive climate information through secondary sources, such as the media or through professional or trade journals, rather than from original sources. Nicholls and Kestin (1998) discuss the communication problems associated with the Australian Bureau of Meteorology's Seasonal Climate Outlooks during the 1997/98 El Niño. The Bureau of Meteorology has been preparing these each month for the past decade (Stone and de Hoedt, 2000). The Outlook consists of a comprehensive booklet (including tabulations of probabilistic predictions of rainfall in terciles: dry, near normal, wet) and a summary-media release. It is prepared at the start of the three-month period, after a meeting involving meteorologists, oceanographers, and some representatives of the agricultural sector (the main user sector for these forecasts). From May 1997 the Bureau had been including indications of a likely El Niño event, and hence of an increased probability of low rainfall over eastern Australia (the part of the country usually affected by the El Niño) in the media releases.

The Outlook issued in early August is representative of the Outlooks through the period May-November. Its headline was "El Niño persists: Dry weather likely to continue over southeastern Australia". The summary went on to say that "…there is a strong likelihood of significantly drier than normal conditions persisting and expanding across much of eastern and southern Australia". The tables included in the August Outlook indicated that rainfall in the dry tercile was, typically, two to three times more likely than the wet tercile. In the event, although there were areas where the August-October period was dry, there were also considerable areas with rainfall much above the average (and well into the "wet" tercile). Moreover, rainfall was good through much of the region in September, a critical time for crops.

Towards the end of the year, criticism of the Bureau's Outlooks was reported in the media. The thrust of this criticism was that the Bureau should have been clearer in describing the limits to the accuracy of its predictions. It was suggested that the Bureau had been exaggerating its confidence in the forecasts. This criticism came as a surprise to those involved in the preparation of the forecasts and media releases. The use of words such as "likely", it was thought, indicated that the forecasts should not be interpreted as a categorical prediction of drought. However, the media criticism, and discussions with forecast users, indicated that there was a wide gap between what the Bureau was attempting to say (ie. an increased likelihood of drier than normal) and the message received by users (ie. definitely dry conditions, perhaps the worst drought in living memory). We have anecdotal evidence of farmers making decisions under the assumption that the Bureau had been forecasting the "drought of the century" – even though the Bureau at no stage made such a prediction.

How did this communication gap arise? Some of the problem is attributable to the different emphases placed by forecasters and users on certain critical words. It appears that users and forecasters interpret "likely" in different ways (Fischhoff, 1994). Those involved in preparing the forecasts and media releases intended this to indicate that dry conditions were more probable than wet conditions, but that there was still a finite chance that wet conditions would occur. Many users, it appears, interpreted "likely" as "almost certainly dry, and even if it wasn't dry then it would certainly not be wet". This interpretation appears to have been reinforced by the plethora of media reports on the El Niño, many detailing possible, or actual, grave consequences in other parts of the world. Many of these reports were sourced from overseas or other organisations, but anecdotal evidence indicates that many users attributed the reports to the Bureau, and accepted their reinforcing message. Inappropriate headlines sometimes distorted messages, even if the text of an article was accurate.

One problem with media releases for seasonal predictions is that they are, by necessity, short. Much of the detail included in the Outlooks (eg. the tables of probabilities) cannot be included in the media release. So words such as "likely" are used to replace the calculated probabilities. Fischhoff (1994) suggests that providing the media and potential users with the explicit probabilities would be less likely to lead to misinterpretation.

Similar communication gaps have been reported with regard to weather forecasts. Fischhoff (1994) examined the reasons why, despite an excellent forecast of a massive winter storm that hit the eastern seaboard of the United States in March 1993, the storm still resulted in a high death toll. The study was undertaken for the Weather Channel, a 24-hour-a-day cable television station devoted to weather forecasting. The Weather Channel staff believed that its experts had seldom had such clear indications of a major weather event, nor expressed themselves with such confidence. Nevertheless, "hundreds of people still died, from exposure, heart attacks, falling tree limbs, road accidents, and the like" (Fischhoff, 1994). Fischhoff concluded that:

> "A forecast is just a set of probabilities attached to a set of future events. In order to understand a forecast, all one needs to do is to interpret those two bits of information. Unfortunately, there are pitfalls to communicating each element, so that the user of a forecast understands what its producer means. One source of potential problems is ambiguity regarding the event being predicted and what is exactly being said about it. Another is the difficulty of determining the relevance of the problem that the forecaster has solved for the problem that the user is facing. Problems can also arise out of epistemological and sociological issues of trust and context".

Fischhoff points out that the best way to avoid such pitfalls is for forecasters and users to recognise the potential pitfalls and to form a partnership to try to improve communication, and to share the "frustration with a reality that poses complexity, ambiguity, and uncertainty". He notes that "Unless forecasters say how confident they are in forecasts, recipients are left to guess". Even if forecasters, in their original predictions, include confidence intervals, by expressing the forecasts in terms of probabilities, dissemination through secondary sources will often lead to stripping of such confidence intervals. Modern technology, such as web sites, provides an opportunity to allow the restoration of such intervals. Thus the media might provide an overall impression of the current and expected state of the climate, but provide address information where detailed forecasts and confidence intervals can be accessed, through the internet.

6. Impediments Due to System Complexity

The social, ecological, and economic impacts of climate anomalies such as droughts can be mediated or aggravated by interactions between the climate system, human actions, and the ecosystem. These interactions can even result in climate impacts unforeseeable and unpredictable from a simple consideration of the climate system alone. Such unpredictability of climate impacts makes the valuation and use of climate predictions, on various time-scales, problematic. Two case studies from 1877 will be used to illustrate the complex nature of these interactions, and the problems these interactions raise for climate prediction use.

The very strong 1877 El Niño - Southern Oscillation episode had severe impacts in many parts of the world. Drought-related famine in India caused over eight million deaths. In many districts a quarter of the population died. Many of these deaths occurred after the end of the El Niño – related drought (Whitcombe, 1993). The proximate cause of many of these deaths was malaria. Malaria incidence increased because of several factors. The drought had probably reduced the numbers of predators of mosquitoes, so that when mosquito breeding increased after the drought-breaking rains there was nothing to restrict the population. The drought would have weakened the human population, making them more vulnerable to malaria. Perhaps as important, however, large numbers of people had assembled in relief centres and towns where food was available, encouraged by the Government. The concentration of people increased the pressure on water supplies, further weakening the population and increasing their vulnerability to malaria. As well, the concentration of population supplied the breeding mosquitoes with an adequate food supply for breeding. Government actions to alleviate the effects of the drought and famine, therefore, ultimately led to increased loss of life after the end of the drought. Such a consequence could only have been avoided if the interactions between crowded labour camps, mosquitoes, and malaria were understood. Even now the use of relief camps to alleviate famines seems an appropriate action. But even today, heavy drought-breaking rains could lead to similar upsurge in deaths from malaria. It is feasible that if El Niño - Southern Oscillation based forecasts of Indian rainfall were available in 1877, these might actually have resulted in more deaths – the forecast might have been used to congregate more people into the areas where malaria eventually caused many deaths.

The 1877 El Niño had a less dramatic impact in Australia than in India, although it did lead to drought. Interactions between European land management techniques and the El Niño - Southern Oscillation led to rapid and undesirable long-term changes in vegetation

(Nicholls, 1991). The best known of these changes is probably the area now known as the Pilliga Scrub. Much of this area of 400,000 ha was open grassy country with only about eight large trees per hectare when Europeans arrived in the 1830s. Frequent burning by Aboriginals, and grazing by indigenous marsupials, restricted the opportunities for trees and shrubs to establish. Fire germinated the seed of the trees and shrubs but rat kangaroos ate many of the resulting seedlings before they could establish. The introduction of sheep reduced the numbers of rat kangaroos, by destroying their cover and their food. A severe drought during the 1877 El Niño further reduced the numbers of indigenous marsupials. The following year, a major La Niña event, was very wet. The few large trees seeded well and when stock owners burnt to destroy grasses with seeds that got into their sheep's wool, seedlings came up thickly, unhindered by the grasses which would usually compete with them for space. This time there were no rat kangaroos to eat the seedlings either and the trees grew unchecked. Over the next decade there were several further periods of establishment, again synchronised with El Niño – La Niña oscillations. Thus the area was unintentionally transformed from grazing land into the dense Pilliga Scrub supporting sustained timber harvesting. This illustrates the difficulty of anticipating the consequences of climate anomalies, especially when human behaviour and ecosystem reactions need to be considered. Without an understanding of the likely interactions between climate anomalies, human actions, and ecosystem reactions, it is difficult to predict the ultimate impacts of the climate anomalies. This means that it is difficult to use forecasts of these climate anomalies in an optimal fashion. As was the case in India in 1877, what appears to be appropriate action may aggravate the adverse impacts. Or, as was the case in Australia, a completely unforeseen impact (long-term vegetation change) may arise from the human-climate-ecosystem interactions.

7. Impediments Due to User Resistance/Misuse

At the start of the 1997 El Niño there was considerable evidence that some Australian farmers in areas known to be regularly affected by droughts during El Niño events were not aware of this, or did not believe the forecast of an El Niño. We have press articles from the middle of 1997 with farmers stating that the El Niño did not affect their area, but then stating that the worst year they had faced was 1982 (the last strong El Niño!). Even in areas where the impact of El Niño was common knowledge, there seemed to be a strong tendency for farmers to prefer a forecast from anyone who was *not* predicting an El Niño or was willing to predict that it would not affect the region. At the end of the 1997/98 El Niño, in June 1998, the opposite effect occurred (in the same areas!). This time farmers were disputing the scientific advice that the El Niño was weakening. So, there seemed a widespread tendency to either ignore or downplay forecasts early in the event, and then to downplay the possibility that the event may be ending.

Once the 1997 El Niño was underway, however, it became clear that some farmers and graziers made substantial changes in management as a result of the forecasts. Despite the strength of the event, the probabilities associated with even the "wet" tercile were still typically predicted to be about 20% (compared with typically 50% for the "dry" tercile). This shift in probabilities from the climatologically expected 33.3% do not appear to be large enough to justify some of the actions undertaken by farmers (eg. selling large portions of their herds; not planting a crop).

Users may "under react' to a forecast. Considerable effort has gone into studies of how individuals process hazards-related information and deal with probabilities. Existing research suggests that many tend to downplay the likelihood of disasters (Felts and Smith, 1997). Thus, faced with a forecast in mid-1997 that Australia faced an El Niño related drought, it appears that many farmers ignored this advice. There is certainly anecdotal information supporting this view. Some of the reasons a potential user may ignore a forecast are discussed later, but these reasons include confusion about the areas in Australia affected by the El Niño, and confusion and lack of knowledge about the expected accuracy of any forecast based on El Niño - Southern Oscillation. Another reason often expressed by farmers is that the expected accuracy of the forecasts is not sufficiently high. When asked how good a forecast of dry/wet conditions would have to be to be useful the typical answer is 80%. Yet forecasts with lower accuracy, expressed and used rationally, should be valuable. An analogy to the farmer request for 80% before they used the forecasts occurs naturally in gambling. If you are told a coin is biased, would you use the information about the bias if the bias only meant heads would come up 60% of the time, or would you ignore the bias unless it was sufficiently biased to result in heads 80% of the time? The often-expressed demand that forecasts be "80%" accurate before their adoption is equivalent to a gambler ignoring a 70% bias in gambling on a coin toss.

At a policy level, one might assume that potential users of climate forecasts might be more knowledgable about the basis and accuracy of climate prediction, and its potential value, compared with the average individual user such as a farmer. However, we have encountered many examples, within Australia, where policy-level organisations have summarily dismissed the potential use of climate predictions, either on anecdotal evidence or after cursory examination of the evidence. Such a tendency to "rush to judgement" is apparently not restricted to seasonal climate prediction. For example, Felts and Smith (1997) found that many policymakers have already judged, despite the unclear nature of the evidence, that climate changes will not cause significant problems in the near future. They also cite examples from other fields where policymakers have been unable to avoid hasty judgements based on incomplete analyses. Within the climate prediction field it seems a tendency exists for at least some policy level users to dismiss the potential of predictions because of some uncertainty about the accuracy of the forecasts, because of confusion arising from possibly conflicting forecasts from different sources, or because their cursory analyses found no potential value.

8. Impediments Due to Cognitive Biases

Most of the impediments discussed so far represent lack of knowledge of either the forecast system or the impact system, or of forecast delivery problems. Another group of impediments to the correct use of forecasts exist: cognitive biases. These problems arise because the capacity of the human mind for solving complex problems is very limited compared with the size of problems whose solution is required for objectively rational behaviour in the real world. So, people use simple rules of thumb or heuristics to simplify decision-making. In general, heuristics are helpful, but they do lead to biases in many situations, especially in uncertain situations where probabilities are encountered. It is well established that many people are unable to reason probabilistically. Piattelli-Palmarini (1994) states that "Any probabilistic intuition by anyone not specifically tutored in probability calculus has a greater than 50 percent chance of being wrong". There is also ample evidence that people do not usually make decisions under uncertainty that accord with a rational expectations model. As well, decision making under uncertainty and risk depends on the way a problem is posed. These problems all have clear ramifications for those issuing weather or climate forecasts, and those using the forecasts or attempting to assess the value of the forecasts. Yet almost no work has been done to establish the effect of these "cognitive biases" on climate and weather forecasts, or on how to ensure that these biases do not subvert the value of the forecasts. The following are some of the well-known sources of cognitive bias. Many of these have been studied in the economic and medical area. Much of the work on these problems stems from work by psychologists Amos Tversky and Daniel Kahneman in the early 1970s. Piattelli-Palmarining (1994) and Bazerman (1994) provide introductions to cognitive illusions and decision making, and most of the examples here are drawn from them. Most of the examples are from laboratory experiments. Stewart (1997) and others, have urged caution in generalising from these studies to decision-making in a real world context. I have tried, where possible, to illustrate how these problems may have affected the use and preparation of climate forecasts during the 1997/98 El Niño, to demonstrate their potential impact in real-world decision making with climate forecasts.

8.1 FRAMING EFFECT

The way a problem is posed can affect the decision. As an example, imagine that Australia faces an outbreak of an unusual disease that is expected to kill 600 people. Two alternative programs to combat the disease have been proposed. Which program would you favour?
- If A is adopted, 200 people will be saved
- If B is adopted there is a 1/3 probability that 600 people will be saved and a 2/3 probability that nobody will be saved

Tests indicate that a majority (72%) would select program A.

What if we change the wording slightly? Now, which of these two programs would you favour?
- If C is adopted 400 people will die
- If D is adopted there is a 1/3 probability that nobody will die, and 2/3 probability that 600 people will die

In this case, a clear majority (78%) of respondents prefer program D. Even though Programs A and C are essentially identical, as are Programs B and D, a slight change in the "framing" of the questions leads to a substantial shift in decision making.

The above question may seem rather abstract. There is, however, considerable evidence that framing is a very severe impediment to rational decision making and that even professionals experienced in decision making are still affected by framing. For instance, if doctors were told there is a mortality rate of 7% within 5 years for a certain operation, they hesitated to recommend it to their patients. If they were told it had a survival rate after 5 years of 93%, they were more inclined to recommend it to their patients. In climate prediction, this framing effect suggests that forecasts expressed in terms of the likelihood of drought may lead to different decisions to forecasts expressed as the non-likelihood of wet conditions. We need to recognise that simple changes in the phrasing (framing) of forecasts could lead to substantial changes in their use.

8.2 AVAILABILITY

Which of the following causes more deaths in the USA each year?
- Stomach cancer
- Motor vehicle accidents

Most respondents select motor vehicle accidents, but stomach cancer causes twice as many deaths. The "availability" of media stories about motor vehicle deaths biases our perception of the frequency of events. Similarly, the high profile of the 1982/83 and other intense El Niño events in media stories, and often mentioned in forecasts, may bias a farmer's (or other user's) expectations of an El Niño to the 1982/83. This appeared to happen in Australia during the 1997/98 event: many stories about the El Niño and the forecast situation for 1997 mentioned the disastrous effects of the 1982/83 El Niño, without indicating that other El Niño events have had less deleterious consequences. Tactics need to be developed to ensure that users do not "anchor" on a particular event (eg. the 1982/83 event) and fail to recognise that not all El Niño events are as severe. We are particularly prone to overestimating unlikely events with vivid impacts. Slovic and Fischhoff (1977) discuss the effect of this "availability" heuristic on perceived risks of nuclear power. They point out that any discussion of the potential hazards, regardless of likelihood, will increase the memorability of these hazards, and increase their perceived risks. A similar situation probably affects public perception of the risks of serious impacts from an enhanced greenhouse effect.

8.3 ASYMMETRY BETWEEN LOSSES AND GAINS

First we are offered a bonus of $300. Then choose between:
- Receiving $100 for sure; or
- Toss a coin. If we win the toss we get $200; if we lose we receive nothing.
There is a strong preference for the first (sure) gain.

This time we first are offered a bonus of $500. Then choose between:
- Losing $100 for sure; or
- Toss a coin. If we lose we pay $200; if we win we don't pay anything.

This time, the majority preference is for the gamble. There is a strong asymmetry between decision making in the face of losses and in the face of gains – we tend to be conservative when offered gains, and adventurers when we face loss. Such an asymmetry will clearly affect decision making by those affected by climate variations. Yet studies of the value of climate forecasts usually do not take such asymmetry into account. Nor is the fact that people react differently to risky situations if they face a possible loss to when they face a possible gain, taken into account in the framing of forecasts.

8.4 CONJUNCTION EFFECT

We know that people have difficulties with adopting rational reactions to uncertain situations, partly because of the difficulties they have with probabilities (eg. Murphy *et al.*, 1980). But the difficulties are rather more complex than are recognised by climate forecasters. Imagine that 1998 was a La Niña. A farmer in eastern Australia, where drought often accompanies an El Niño, is asked to order the following from most likely to least likely:

- 1999 will be a drought year on my farm
- 1999 will be an El Niño year
- 1999 will be an El Niño year and a drought year on my farm

There is much evidence (from similar questions asked in other situations) that the farmer (or other decision maker) will consider option three, the conjunction of the two events, more likely than one of the other two possibilities, even though elementary probability theory indicates that it cannot be more likely than the two events of which it is composed. The likelihood that these two propositions (El Niño and drought) are simultaneously true is always and necessarily inferior to the probability of either of the two single propositions alone. This, and other examples, illustrates the difficulty users have encapsulating probability concepts.

8.5 OVERCONFIDENCE

People tend to be overconfident in their estimates (eg. of the world population in 1998), or of their answers to "Trivial Pursuit" type questions. Fischoff *et al.,* (1977) found that people who assigned odds of 1,000:1 of being correct in their estimates were correct less than 90% of the time. For odds of 1,000,000:1 their answers were correct less than 96% of the time. Sniezek and Henry (1989) found that groups are just as susceptible to unreasonably high levels of confidence in their judgements. Some 98% of subjects believed that their group judgements were in the top half of all group judgements with respect to accuracy. In the climate prediction situation this suggests that a group working to derive a consensus forecast from several forecast systems may well over-estimate the confidence in the group forecast. This would apply in cases when a consensus seasonal forecast is being prepared, but also when "experts" are asked to estimate the probability of a serious impact from an enhanced greenhouse effect. Similarly, it is likely that seasonal climate forecasters, unless they are constrained by objective systems, will quote unreasonably high confidence limits on their forecasts.

Even when forecast preparation is done in a group, group overconfidence will probably lead to an overconfident forecast.

8.6 CERTAINTY EFFECT

Imagine you are exposed to a fatal virus. The possibility that you will get the disease is 1/1000 but there is no known cure once you have it. We could give you an injection now that will stop the development of the disease. What would you pay for it? Respondents to this question in a test produced an average payment of $800. The same group were then asked a simple question, but now the probability of catching illness is 4/1000. The injection works 25% of the time. What would you pay in this situation? The group average payment this time was $250. In both cases the expected reduction in risk is 1/1000, but the certainty offered in the first situation resulted in a higher bid value. In general, people value a reduction in likelihood (of a drought for instance) more when the drought is certain, than when it is probable. This effect would affect the actual value of a climate forecast. It would also tend to bias people more towards forecast systems that completely remove the likelihood of an unlikely event, relative to a system designed to shift the likelihood of a less-rare event.

8.7 UNCERTAINTY EFFECT

Imagine you have just handed in a difficult exam. You will know the day after tomorrow if you have passed or failed. You are offered a real bargain holiday in Hawaii, but have to decide by tomorrow and pay a non-refundable deposit. You can put off the decision by a day (in which case you will know for sure whether you have passed or failed) for an additional non-refundable $15 (the $15 is not deductible from the price of the holiday).

What would you decide to do if:
- You knew you had passed;
- Knew you had failed
- Did not know whether you have passed or failed

Just over 50% of respondents decide to go in either of the first two cases. But less than a third decide to go in the case when they do not know the result of the exam. That is, even though they are willing to go irrespective of whether they have passed or failed, many people will not decide in the face of uncertainty, even if it has no bearing on their final decision. Since climate forecasts are always going to be uncertain, this suggests that many people will never use them. Alternatively, they may, because of the uncertainty effect, delay action based on a forecast until they start to see the effects (of a drought for instance). This happened in South America in mid-1997, with decision makers unwilling to react to forecasts of an impending El Niño until they had clearer evidence that it was underway (see *Nature*, vol., 388, page 108).

Nicholls (1999) and White (2000) discuss other examples of the problems cognitive biases may provide in seasonal climate prediction, and Nicholls (1999) proposes strategies for minimising their impact.

9. What Can We Do To Overcome These Impediments?

The above description of problems in encouraging the use of climate predictions probably seems depressing, to those involved in the development of improved prediction models. The problem of ensuring that a climate forecast is useful is considerably more complicated and multi-faceted than is generally believed by climate scientists. If, however, the lessons presented by sociological and other research, and the lessons in areas such as Australia where attempts have been made to utilise climate predictions, are understood, then the adoption of climate prediction can be facilitated. There are a number of critical steps (sometimes called a systems approach) which need to be undertaken to ensure useful adoption of climate prediction. Some of these were identified in the papers by Glantz and Lamb about two decades ago – others have been added from Australian experience and from evidence in a wide range of studies. The systems approach advocated by Hammer (2000) is a step in the right direction to ensure that the forecast system is considered holistically. However, there are other facets of the use and misuse of climate forecasts, noted above, which are yet to be included in research or operational forecast programs. Thus the cognitive problems people have in dealing with probabilistic concepts are usually ignored, and the communication problems mentioned earlier do not appear to have been widely recognised, except in a very general sense. The breadth of the problems associated with climate prediction, outside of the "simple" scientific problem of the need for more skill, must be brought into the research framework, solved in the context of climate prediction, and then applied, if we are to get the true value from our scientific advances in prediction. The first step must be the recognition that these impediments exist, and need to be overcome. One approach to overcoming such problems is the linking of the impact system with the forecast system. Many of the papers in Section III of this book discuss the ways forecasts and outcomes can be linked. These show how we can change the forecast process to take cognisance of the complexity of the forecast-impact-user system. Much more remains to be done, if we are to optimise the use of climate forecasts.

Acknowledgments

Much of the above reflects experiences, in the past two decades, in helping the Australian community accept and use climate predictions based largely on the El Niño - Southern Oscillation. This effort has involved many individuals, especially Graeme Hammer, Greg McKeon, Roger Stone, and Jeff Clewett (Queensland Department of Primary Industry), Mary Voice, Grant Beard, Mike Coughlan, and Bill Kininmonth (National Climate Centre, Bureau of Meteorology), David White (ex Federal Department of Primary Industry and Energy), Jim Simpson (Lamont-Doherty Earth

Observatory), and Barry White (Land & Water Resources R & D Corporation). My BMRC colleagues and my family patiently sat through many tests on cognitive biases.

References

Adams, R.M., Bryant, K.J., McCarl, B.A., Legler, D.M., O'Brien, J., Solow, A. and Weiher, R. (1995) Value of improved long-range weather information. *Contemporary Economic Policy* **13**, 10-19.

Bazerman, M.H. (1994) *Judgement in managerial decision making.* 3rd ed., Wiley, New York. 226 pp.

Casey, T. (1995) Optimal linear combination of seasonal forecasts. *Aust. Meteor. Mag.* **44**, 219-224.

Changnon, S.A., Changnon, J.M. and Changnon, D. (1995) Uses and applications of climate forecasts for power utilities. *Bull. Amer. Meteorol. Soc.* **76**, 711-720.

Chatfield, C. (1995) Model uncertainty, data mining and statistical inference. *J. R. Statis. Soc. A* **158**, 419-466 (including discussion).

DeBondt, W. and Tahler, R.H. (1986) Does the stock market overreact? *J. Finance* **40**, 793-807.

Elsner, J.B. and Schmertmann, C.P. (1994) Assessing forecast skill through cross-validation. *Weather and Forecasting* **9**, 619-624.

Felts, A.A. and Smith, D.J. (1997) Communicating climate research to policy makers, in H.F. Diaz and R.S. Pulwarty (eds.), *Hurricanes. Climate and socioeconomic impacts.* Springer-Verlag, Berlin, pp.234-249.

Fischhoff, B. (1994) What forecasts (seem to) mean. *Int. J. Forecasting* **10**, 387-403.

Fischhoff, B., Slovic, P. and Lichtenstein (1977) Knowing with certainty: The appropriateness of extreme confidence. *J. Experimental Psychology: Human perception and performance* **3**, 552-564.

Glantz, M. (1977) The value of a long-range weather forecast for the West African Sahel. *Bull. Amer. Meteorol. Soc.* **58**, 150-158.

Grant, A.M. (1954) The application of correlation and regression to forecasting. Meteorological Study No. 7, Bureau of meteorology, Melbourne, pp.14.

Hammer, G.L. (2000) A general approach to applying seasonal climate forecasts, in G.L. Hammer, N. Nicholls, and C. Mitchell (eds.), *Applications of Seasonal Climate Forecasting in Agricultural and Natural Ecosystems – The Australian Experience.* Kluwer Academic, The Netherlands. (this volume).

Hammer, G.L. and Muchow, R.C. (1991). Quantifying climatic risk to sorghum in Australia's semiarid tropics and subtropics: Model development and simulation, in R.C. Muchow and J.A. Bellamy (eds.), *Climatic Risk in Crop Production: Models and Management for the Semiarid Tropics and Subtropics.* CAB International. Wallingford, UK. pp. 205-232.

Hammer, G.L., Holzworth, D.P. and Stone, R. (1996) The value of skill in seasonal climate forecasting to wheat crop management in a region of high climatic variability. *Australian J. Agricultural Research* **47**, 717-737.

Hulme, M., Biot, Y., Borton, J., Buchanan-Smith, M., Davies, S., Folland, C., Nicholds, N., Seddon, D. and Ward, N. (1992) Seasonal rainfall forecasting for Africa. Part II – Application and impact assessment. *Inter. J. Environmental Studies* **40**, 103-121.

Lamb, P.J. (1981) Do we know what we should be trying to forecast – climatically? *Bull. Amer. Meteorol. Soc.* **62**, 1000-1001.

Maines, L.A. (1996) An experimental examination of subjective forecast combination. *Int. J. Forecasting*, **12** 223-233.

Mjelde, J.W., Thompson, T.N., Nixon, C.J. and Lamb, P.J. (1997) Utilising a farm-level decision model to help prioritise future climate prediction needs. *Meteorol. Appl.* **4**, 161-170.

Murphy, A.H. (1993) What is a good forecast? An essay on the nature of goodness in weather forecasting. *Weath. & Forecasting* **8**, 281-293.

Murphy, A.H. (1994) Assessing the economic value of weather forecasts: an overview of methods, results and issues. *Met. Appls.* **1**, 69-73.

Murphy, A.H., Lichtenstein, S., Fischhoff, B. and Winkler, R.L. (1980) Misinterpretations of precipitation probability forecasts. *Bull. Amer. Meteorol. Soc.* **61**, 695-701.

Nicholls, N. (1985) Impact of the Southern Oscillation on Australian crops. *J. Climatology* **5**, 553-560.

Nicholls, N. (1986) Use of the Southern Oscillation to predict Australian sorghum yield. *Agric. & Forest Meteorol.* **38**, 9-15.

Nicholls, N. (1991) The El Niño - Southern Oscillation and Australian vegetation, in A. Henderson-Sellers and A. J. Pitman (eds.), *Vegetation and climate interactions in semi-arid regions*. Kluwer, Dordrecht, pp.238.

Nicholls, N. (1992) Recent performance of a method for forecasting Australian seasonal tropical cyclone activity. *Aust. Meteorol. Mag.* **40**, 105-110.

Nicholls, N. (1999) Cognitive illusions, heuristics, and climate prediction. *Bull. Amer. Meteorol. Soc.* (in press).

Nicholls, N. and Kestin, T. (1998) Communicating Climate. *Climatic Change*, in press.

Nicholls, N., Lavery, B., Frederiksen, C., Drosdowsky, W. and Torok, S. (1996) Recent apparent changes in relationships between the El Niño - Southern Oscillation and Australian rainfall and temperature. *Geophys. Res. Letts.* **23**, 3357-3360.

Piattelli-Palmarini, M. (1994) *Inevitable illusions*. Wiley, New York. pp.242.

Pielke, R.A. and Pielke, R.A. (1997) Vulnerability to hurricanes along the U.S. Atlantic and Gulf coasts: Considerations of the use of long-term forecasts, in H.F. Diaz and R.S. Pulwarty (eds.), *Hurricanes. Climate and socioeconomic impacts*. Springer-Verlag, Berlin. pp.147-184.

Pulwarty, R.S. and Redmond, K.T. (1997) Climate and salmon restoration in the Columbia River basin: The role and usability of seasonal forecasts. *Bull. Amer. Meteorol. Soc.* **78**, 381-397.

Redelmeier, D.A., and Shafir, E. (1995) Medical decision making in situations that offer multiple alternatives. *J. American Medical Assoc.* **273**, 302-305.

Rimmington, G.M. and Nicholls, N. (1993) Forecasting wheat yields in Australia with the Southern oscillation Index. *Australian J. Agricultural Res.* **44**, 625-632.

Slovic, P. and Fischoff, B. (1977) On the psychology of experimental surprises. *J. Experimental Psychology: Human perception and performance* **3**, 544-551.

Sniezek, J.A. and Henry, R.A. (1989) Accuracy and confidence in group judgement. *Organisational behaviour and human decision processes* **43**, 1-28.

Stewart, T.R. (1997) Forecast value: Descriptive decision studies, in R.W. Katz and A.H. Murphy (eds.), *Economic Value Of Weather And Climate Forecasts*. Cambridge University Press. pp.222.

Stone, R.C. and de Hoedt, G. (2000) The development and delivery of current seasonal climate forecasting capabilities in Australia, in G.L. Hammer, N. Nicholls, and C. Mitchell (eds.), *Applications of Seasonal Climate Forecasting in Agricultural and Natural Ecosystems – The Australian Experience*. Kluwer Academic, The Netherlands. (this volume).

Thompson, P.D. (1977) How to improve accuracy by combining independent forecasts. *Mon. Weath. Rev.* **105**, 228-9.

Trenberth, K.E. (1997) Short-term climate variations: Recent accomplishments and issues for future progress. *Bull. Amer. Meteor. Soc.* **78**, 1081-1096.

Trenberth, K.E. and Hoar, T.J. (1996) The 1990-1995 El Niño - Southern Oscillation event: Longest on record. *Geophys. Res. Letts.* **23**, 57-60.

Webster, P.J. and Yang, S. (1992) Monsoon and ENSO: selectively interactive systems. *Quart. J. Roy. Meteorol. Soc.* **118**, 877-926.

Whitcombe, E. (1993) Famine mortality. Economic and Political Weekly, June 5, pp.1169-1179.

White, B. (2000) The importance of climate variability and seasonal forecasting to the Australian economy, in G.L. Hammer, N. Nicholls, and C. Mitchell (eds.), *Applications of Seasonal Climate Forecasting in Agricultural and Natural Ecosystems – The Australian Experience*. Kluwer Academic, The Netherlands. (this volume)

Winkler, R.L. and Makridakis, S. (1983). The combination of forecasts. *J. Roy. Stat. Soc., A* **146**, 150-157.

AUSSIE GRASS: AUSTRALIAN GRASSLAND AND RANGELAND ASSESSMENT BY SPATIAL SIMULATION

J.O. CARTER, W.B. HALL, K.D. BROOK, G.M. MCKEON, K.A. DAY
AND C.J. PAULL

Queensland Centre for Climate Applications
Queensland Department of Natural Resources
80 Meiers Rd
Indooroopilly, QLD 4068, Australia

Abstract

Defining drought, categorising current droughts, and assessing grassland and rangeland sustainability in a quantitative and scientific manner are important national issues for Australian State and Commonwealth governments, landholders and agribusiness. A challenge for ecologists and modellers of Australia's grasslands and rangelands is to integrate biological models, geographic information systems, satellite imagery, economics, climatology and visual high-performance computing into readily available products that can provide monitoring and prediction advice in near real-time.

The QDNR systems approach to the management of native grasslands recognises that drought occurs at a regional scale, and that impacts on livestock and natural resources can be forecast using simple models of soil water, plant growth and animal performance. Our vision for a comprehensive Australian Grassland and Rangeland Assessment System (Aussie GRASS) is one that consists of the best combination of rainfall analyses, seasonal climate forecasts, satellite and terrestrial monitoring, and simulation models of relevant biological processes. This will provide a rational basis for large-scale management decisions by graziers, extension workers, land resource managers, bureaucrats and politicians. Aussie GRASS products are currently used within the Queensland government for drought declaration assessments and applications for Drought Exceptional Circumstances.

The Aussie GRASS national spatial modelling framework allows agricultural simulation models to be run at a continental scale on a 0.05 degree (~5 km) grid. The simulation model currently in use by the Aussie GRASS project is the GRASP pasture model developed for tropical native pastures in Queensland by QDPI and QDNR. In the latest Aussie GRASS project, other regional models are being examined for their applicability to areas such as the southern winter perennial grass zone, chenopod shrublands or the high rainfall temperate zone.

G.L. Hammer et al. (eds.), The Australian Experience, 329–349.

The Queensland version of the Aussie GRASS model is currently used to produce data for a monthly report - *A Summary of Seasonal Conditions in Queensland*. Model outputs are used in conjunction with recorded and forecast rainfall, satellite imagery, Southern Oscillation Index and current drought declarations to build a comprehensive picture of the current and future seasonal conditions impacting on primary producers. Other numerous outputs from the model can be produced and tailored as required.

1. Introduction

The quantitative assessment of the condition of Australia's grasslands and rangelands continues to be a difficult problem because of the complexity of the biophysical system, which includes the interaction of high spatial and temporal variability in resources and climate. Pastoralism in Australia occurs in zones of high year-to-year rainfall variation. Heavy utilisation by animals during drought years causes the loss of desirable perennial plants (McKeon *et al.*, 1990) and reduces protection for the soil surface (eg. Gardener *et al.*, 1990). It is only when data on animal numbers and pasture growth are combined that important indices of sustainable resource use and condition (pasture utilisation and plant cover) can be derived (McKeon *et al.*, 1990; Johnston *et al.*, 1996a, b).

For Queensland, a near-real time capability for calculating pasture utilisation and cover has been developed (Brook, 1996). The work currently being undertaken for the Australian Grassland and Rangeland Assessment by Spatial Simulation (Aussie GRASS) project is aimed at extending the methodologies and modelling capabilities from the Queensland experience to the collaborating States and Territory. The Aussie GRASS research team is made up scientists from Queensland Departments of Natural Resources and Primary Industries, Northern Territory Department of Primary Industries and Fisheries, West Australia Agriculture, Primary Industries and Resources South Australia, Department of Environment, Heritage and Aboriginal Affairs South Australia, and the New South Wales Departments of Land and Water Conservation and Agriculture. CSIRO is also involved in the project on a consultancy basis.

The Aussie GRASS project aims to provide answers to rangeland and grassland issues such as –
- what is the current condition of our grazing systems?
- are our grazing lands in a condition consistent with sustainable use?
- what is the likely future trend in their condition?

Before these issues can be tackled it is necessary to understand the many driving forces in the grazing industries including rainfall and other climatic variables, prices received/paid, technology (supplements, feedlots), land use change (tree clearing, pasture improvements), animal genetics, fire regime and vegetation change. Some of these drivers are completely within the control of graziers (eg. land use change, genetics), others partially (eg. fire regime), whilst there are many drivers over which graziers have little or no control (eg. prices received/paid). Climate and its inherent temporal variability is an important driver that falls into the latter category.

An example of how Aussie GRASS provides a framework for investigating rangeland issues is Dalrymple, an important beef-producing shire in northern Queensland. In the latter half of the 1970s Dalrymple suffered marked resource degradation due to overgrazing. A recent land resource study (L.G. Rogers, pers. comm.) surveyed 2559 sites across the shire between January 1990 and December 1994 and reported 43% of sites were degraded in terms of pasture condition (ie. grass basal area $\leq 1.5\%$) and 26% of sites were marginal (grass basal area between 1.5 and 2.0%). Soil erosion was found at 69% of sites with sheet erosion the most dominant type (34% of sites). Figure 1 shows the mean Dalrymple stocking rate calculated from Australian Bureau of Statistics (ABS) data and an estimated 'safe' stocking rate derived for the period 1945-63 using Charters Towers rainfall. The two stocking rates were similar until the late 1970s when the safe stocking rate was reduced relative to that calculated from the ABS data. Graziers appear to have responded inadequately to the prevailing seasonal conditions and as a result widespread resource degradation was observed in the shire. Factors contributing to this period of relative overstocking have included (Ash et al., 1997) –

- predominance of drought and tick tolerant *Bos indicus* cattle
- widespread use of urea and molasses supplementation, allowing poor quality feed to be utilised so that cattle could be retained on property during dry periods
- low cattle prices in 1974 due to export problems with the US market resulting in more cattle being retained on property (prices did not begin to rise until late 1978 (Figure 2)), and
- eradication campaigns for tuberculosis and brucellosis had led to improved property infrastructure allowing for better management of herds.

There have been many other land and pasture degradation episodes in the history of Australia's rangelands (Table 1). Some are likely to be the result of the interaction of high grazing pressure and episodic rainfall deficits (eg. events 2, 4, 7, 8, 9, 10 and 12 in Table 1) as evidenced by loss of surface cover. Others involve the increase of undesirable species, such as unpalatable grasses and woody weeds, associated with high rainfall periods or lack of fires (eg. events 3, 5 and 11). In some cases the change in vegetation does not result in loss of animal production (eg. event 1) or is readily reversible (Bisset, 1962). A feature common to all these episodes is the fact that surveys and government enquiries are held after them. The goal of the Aussie GRASS project is to forecast the probability of these episodes occurring. We hypothesise that seasonal forecasting will have a major role to play in anticipating the impact of climatic extremes. The work conducted within the Aussie GRASS project will help to prevent degradation episodes from occurring by -

- providing information on pasture growth that will allow estimates of long-term safe stocking rates to be calculated, and
- producing 'degradation alerts' that identify areas where the resource base is at risk due to combinations of relatively low rainfall, low pasture growth, and relatively high utilisation rates.

This paper sets out the Aussie GRASS approach to integrating technologies in a spatial modelling framework to generate relevant information products to government and industry on drought and resource monitoring and prediction in near real-time.

332

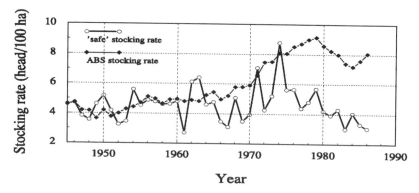

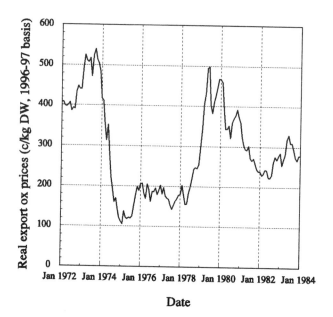

Figure 1. Comparison of estimated 'safe' and actual stocking rates in the Dalrymple shire for 1945-86. Safe stocking rate was derived as a function of simulated pasture growth (using the GRASP model), which was calibrated according to actual shire stocking rate for 1945-63 (a period of pasture stability). The actual shire stocking rate was calculated as the ratio of total beef cattle to the shire area (taken from McKeon *et al.,* 1990). This approach allows the change in animal numbers over time to be expressed relative to the 1950/60s period, before the adoption of new technologies and overcomes to some extent possible errors in ABS data (Mortiss, 1995).

Figure 2. Brisbane export ox (301-320 kg dressed weight) prices (c/kg DW) corrected for inflation to a 1996-97 basis using annual (July-June) CPI values (source - AMLC Annual Statistical Reviews). From 1981, prices were reported for cattle with a dressed weight of 300-350 kg.

TABLE 1. A history of selected degradation episodes in Australia's grazing lands.

	Period	Location	Impacts	References
1	1880s	S.E. Queensland	Loss of pasture species eg. kangaroo grass (*Themeda triandra*). Increase in black speargrass (*Heteropogon contortus*). Loss of sheep production.	Deutsher (1959) Shaw (1957) Bisset (1962) McKeon & Howden (1991)
2	1960s	S.W. Queensland	Loss of pasture and surface cover. Soil loss. Woody weed invasion. Woodland thickening.	Miles (1993) Mills (1984) Burrows (1995) Burrows et al.(1990)
3	1970s to 1980s	S.E. Queensland	Invasion by grass weeds eg. *Aristida* and couch. Loss of black speargrass *(Heteropogon contortus)*.	Orr *et al.* (1991) Orr and Paton (1993) Orr *et al.* (1994) Paton and Rickert (1989) Campbell (1995)
4	1980s to 1990s	N.E. Queensland	Loss of surface cover. Soil loss. Loss of desirable grass species. Invasion of exotic weeds.	Scanlan *et al.* (1996a,b) Gardener *et al.* (1990) Mortiss (1995)
5	1980s to 1990s	N.W. Queensland	Woody weed invasion.	Burrows (1995) Burrows *et al.* (1990) Carter (1994)
6	1980s to 1990s	Central Queensland	Woodland thickening.	Burrows (1995) Burrows *et al.* (1990,1997)
7	1890s	Western New South Wales	Loss of some pasture species. Loss of cover. Soil loss. Woody weeds invasion.	Condon (1968,1986a,b) Beadle (1948) Noble (1997)
8	1930s	N.E. South Australia	Loss of shrubs. Loss of surface cover. Soil loss.	Ratcliffe (1937)
9	1930s	Gascoyne, West Australia	Loss of shrubs. Loss of surface cover. Soil loss.	Wilcox and McKinnon (1972)
10	1930s to 1940s	Western New South Wales	Soil loss.	Beadle (1948)
11	1950s	Western New South Wales	Woody weed invasion.	Condon (1986a, b) Anon (1969) Noble (1997)
12	1960s	Central Australia	Soil loss.	Condon *et al.* (1969a,b,c,d)

2. Why Models?

An important question is why spatially simulate pasture growth when remotely sensed vegetation indices, such as NDVI, and rainfall data are readily available?

Although NOAA satellite-derived imagery has shown to be somewhat useful in the broad-scale spatial assessment of green cover, especially the spatial response of vegetation to rainfall events (Smith, 1994; Dudgeon *et al.*, 1990; Filet *et al.*, 1990), it has inherent limitations in providing a total solution for drought and rangeland monitoring. These limitations include poor relationships with dry pasture biomass, tree cover confounding the signal, and that future projections of the current situation are not inherent. In the tropical regions of Australia prolonged cloud cover at certain times of the year may also limit the use of remotely sensed data. Past experience has shown that satellite data may not be available for approximately 14-28 days of each wet season due to the effect of clouds. NOAA satellite data also suffers from data consistency problems as a result of –
- different satellites being used over time
- each satellite sensor having its own calibration curve and 'decay' over time, and
- variation in dust and water vapour content of the atmosphere, requiring correction.

The short time series of available NOAA data (from 1981) fails to adequately provide boundaries to the 'envelope' of possible data and thus limits the usefulness of comparing and ranking imagery. However, ongoing research into the use of NDVI and other remotely sensed indicators to assess rangeland condition suggests a potential still exists.

Work by Day *et al.* (1997) has shown that for a range of pasture communities in Queensland, observed rainfall explained less variation (r^2=0.40) in seasonal pasture growth than the GRASP pasture growth model (r^2=0.70, Table 2). On a spatial scale the benefits of modelling are highlighted by the recent 1991-95 Queensland drought. Rainfall analyses did not map the drought-declared south-western areas of the State as droughted and conversely, coastal areas of the state were classed as droughted by rainfall analyses when there was no community push for their declaration. Thus we are better able to understand the interaction of critical processes such as climate variability and stocking rate, monitor rangeland condition and predict future trends through the use of pasture simulation models such as GRASP.

TABLE 2. Correlation of rainfall, simulated evapotranspiration and simulated pasture growth with seasonal pasture growth for a range of land systems across Queensland (from Day *et al.*, 1997).

Variable	Correlation (r^2)
Rainfall	0.4
Evapotranspiration	0.6
Simulated pasture growth[1]	0.92
Simulated pasture growth[2]	0.72

[1] Fit to data from exclosures used to calibrate pasture model, GRASP.
[2] Fit to data from exclosures not used in model calibration.

3. Aussie GRASS Modelling Framework

The Aussie GRASS modelling framework is shown in Figure 3. The major inputs are daily rainfall and other climatic information, soil type and associated parameters, pasture community and associated parameters, tree basal area and stocking rate information. Spatial coverage is achieved by dividing Australia into pixels on a 0.05 degree basis (~5 km grid). The spatial model is zero order in that there is no communication between the pixels. This in effect means that each pixel is a point and the model runs independently for each of the 250 000 points that cover Australia. Such an approach reduces the model complexity and model run-times.

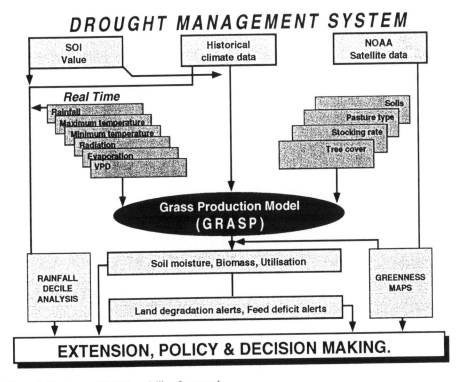

Figure 3. The Aussie GRASS modelling framework.

The Aussie GRASS model is run operationally for all Australia on a monthly basis following receipt of monthly rainfall totals and climatic data (temperature, radiation, vapour pressure, evaporation) from the Bureau of Meteorology. Rainfall totals from the observer network are interpolated across the continent using the kriging technique (Tabios and Salas, 1985) and the rainfall normalisation process of Hutchinson (1991). For each pixel the interpolated monthly total is then distributed throughout the month on a daily basis using the daily rainfall observations of the nearest recording station. As yet there is no technique that adequately interpolates daily rainfall. Other climate variables are also interpolated between recording stations. The full methodology is detailed in Carter *et al.* (1996a).

Probability of Exceeding Median Growth
QLD - February to April 1998

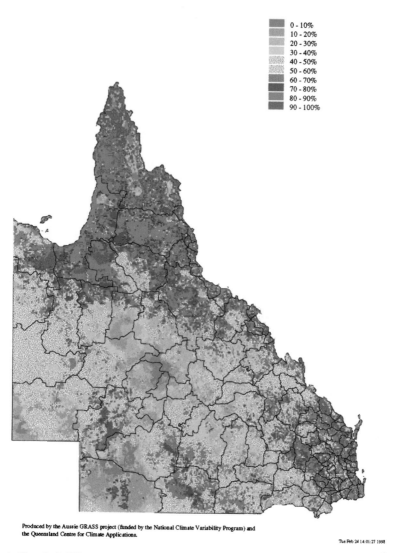

Figure 4 (C). Probability of exceeding median pasture growth for the period February-April 1998. Calculations were performed at the beginning of February 1998 using analogue climate years from the historical climate record selected according to the Southern Oscillation Index phase system of Stone and Auliciems (1992). (This figure is reproduced in the colour section at the end of the book as plate C)

As well as running in near-real time, analogue climate years are selected using the SOI phase system (Stone and Auliciems, 1992; Stone *et al.,* 1996). The model is run forward for the next three months using the ensemble of analogue years to produce a distribution of state variables and numerous accumulators including pasture growth (Figure 4). Analogue years are a simple and easily understood methodology by which seasonal climate forecasting information is able to be incorporated in daily time-step agricultural simulation models. New seasonal forecasting methodologies such as the Bureau of Meteorology's Sea Surface Temperature system (Drosdowsky and Allan, 2000) will also be utilised within Aussie GRASS for those regions where they can be shown to have greater skill than the SOI phase system. It is also essential to recognise that forecasting systems vary in their skill level throughout the year. Currently the SOI phase system and therefore the model outputs simulated using analogue years do not provide information on the skill level of forecasts. This is an issue that needs to be addressed so that users of products are able to use the information more appropriately in their decision making processes.

Products from the operational model run are available within 5-7 days of the end of each month.

4. Non-climatic Data Requirements

Data collection, assimilation and management are probably the most time consuming and difficult aspects of a project such as Aussie GRASS. Rarely are data available at the necessary resolution, in the same format for all of Australia, or over the necessary timeframe. Soil and pasture parameters such as plant available water capacity and transpiration use efficiency are lacking for most soil classes and pasture communities outside of Queensland. Development of the national input data sets is detailed in Carter *et al.* (1996a).

This lack of available data has lead to some innovative research. For example, tree density is a major source of spatial variation in pasture growth because of the competitive effect of trees for water and nutrients. Within the GRASP model this competitive effect is simulated using relationships between tree basal area and tree transpiration. Using the principle that the green cover of grasses varied more throughout the year than it did for trees, it was found that tree basal area for each pixel could be estimated from an analysis of a time series of NOAA NDVI images and field measurements of tree basal area over a range of vegetation types (Danaher *et al*, 1992).

Data standardisation and disaggregation also becomes an issue when analysing stock numbers. Stocking rate calculations require that the various classes of stock for each Statistical Local Area (SLA) are converted to a common unit (animal unit = 400 kg) and then distributed to pixels throughout the SLAs based on heuristics. These heuristics take into account information such as predators (dingos), pasture communities and national parks (eg. sheep are excluded from areas such as those in which dingos are a common problem and/or black spear grass (*Heteropogon contortus*) pastures occur). Areas outside of Queensland are yet to develop the heuristics necessary to distribute stock within SLAs. However, there is the potential that some states (eg. WA) will have access to property level stock data for leasehold lands, which will better enable them to estimate stocking rates and/or distribute ABS stock data across a SLA.

To allow objective assessment of current seasonal conditions relative to the past, Aussie GRASS model outputs are often presented in a ranked or percentile format. For example, March monthly pasture growth for this year is ranked against all other March monthly pasture growths from previous years back to 1957 (the earliest year in which all climatic information is available in a digital format). The percentile format therefore requires that all inputs such as stocking rate are available back to 1957. Currently stock data is available in a digital format for all SLAs from 1983 although for Queensland data have been digitised back to 1952.

Ongoing maintenance of established databases in near-real time is also a time consuming task. For example, currently the only source of national stock information is the annual ABS Agricultural Census. Unfortunately data from the census only become available approximately 12 months after the census has been completed. The recent decision by ABS to downgrade the annual census to a survey for four out of every five years will exacerbate the problem. Data will be available at far less resolution than are currently available and ABS predictions of relative standard errors for some commodities indicate the data are likely to be less accurate. Options under consideration to overcome limitations to current and future stock data include –

- linking with individual producers to provide both stock and rainfall data via the internet
- development of models of regional herd and flock population dynamics incorporating reproduction, mortality and management decisions, which would use both biological (eg. pasture growth) and economic inputs (eg. stock and wool prices), and
- using information from other sources (eg. saleyards, abattoirs, and stock agents) to supplement available ABS data.

Accurate information on density of macropods and feral animals is extremely limited on both a spatial and temporal scale. The recent release of the calicivirus, which causes Rabbit Haemorrhagic Disease (Studdert, 1994), has had variable effects on rabbit populations throughout Australia and highlights the paucity of accurate near-real time spatially complete data on non-domesticated herbivores.

5. Model Description

The GRASP tropical native pasture production model is currently the sole model used within the Aussie GRASP framework. However, for the purpose of initial simulation of the whole continent parameters have been estimated for other pasture communities. A full description of the model is provided in the GRASP manual (Day *et al.*, 1997). The conceptual framework is based upon a daily soil water budget using four soil layers and the processes of run-off, drainage, soil evaporation, and grass and tree transpiration from inputs of rainfall and other climate variables (McKeon *et al.*, 2000). Soil parameters include infiltration parameters, available soil water range for each layer and maximum rate of bare soil evaporation. A plant growth index is calculated from a soil water index, plant growth response to average temperature, solar radiation, and nutrient availability. At low green cover, plant growth is calculated as a function of this growth index, plant density and potential regrowth rate. As green cover increases, plant growth is calculated from a combination of radiation interception, transpiration and vapour pressure deficit.

Because of the high spatial variation in climate and resources (vegetation, soils, topography), regional ecosystems differ in terms of which physical and biological processes are most important. Over the last 30 years different simulation models of soil water balance and plant growth have been developed by various groups allowing simulation of specific regional grazing systems. These models have typically been derived from agronomic experiments, resource monitoring and grazing trials and thus represent the best available regional knowledge of agricultural research organised in a systematic and predictive way. A major component of the current Aussie GRASS work is to assess several regional models using a process which –

- identifies suitable vegetation types for each model
- quantifies each model's parameter and input requirements
- examines the ease with which models or model components can be integrated into the existing computing environment, and
- compares performance of selected models using existing data sets.

The models being examined are IMAGES (Hacker *et al.*, 1991), ARIDGROW (Hobbs *et al.*, 1994), SEESAW (S. Marsden, CSIRO, Canberra), GRASSGRO (Moore *et al.*, 1997) and DYNAMOF (White *et al.*, 1983; Bowman *et al.*, 1982). Involvement of regional researchers in Aussie GRASS provides local knowledge and expertise and thus allows for close 'reality' checking of inputs and outputs. This local expertise should be an essential component of all simulation-based research of complex systems and also helps to promote the uptake and use of results.

6. Model Calibration

To date, model calibration has been conducted only for Queensland. Calibration of the Aussie GRASS model for other areas of Australia is a major component of on-going work.

The main plant and soil parameters for the GRASP model can be derived by calibrating with data collected using a standard field methodology (Day *et al.*, 1997). In Queensland GRASP has been calibrated on a point basis for a broad range of pasture communities (>40), soil types and climatic conditions for >75 grazing exclosures over a number of years resulting in 175 site by year combinations. GRASP has also been evaluated over a range of grazing pressures on ten major grazing trials in Queensland (Day *et al.*, 1997). However there are still many soil classes and pasture communities in Queensland for which the appropriate data has not been collected to allow for calibration.

Soil parameters in Aussie GRASS were estimated for each soil class based on information derived from extensive soil surveys together with sampled profile data (Day et al., 1997). Pasture parameters for GRASP were derived by developing general parameter sets for each pasture community based on data from these same grazing exclosures.

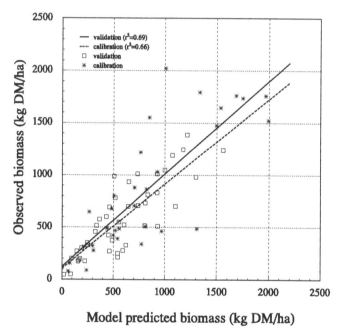

Figure 5. Calibration and validation results of the GRASP model predictions for biomass in eleven cells on six occasions using the field data of Wood et al. (1996).

A further source of data with which to calibrate pasture parameters resulted from an extensive field survey of Queensland undertaken by two field officers from January 1994 to August 1995 (Wood et al., 1996). Over 220 000 visual estimates of standing pasture yield were recorded during 122 000 km of travel in 256 days of field work. Harvests to calibrate visual estimates for each operator were carried out during each day. Pasture parameters for each pasture community were derived by optimising the model with a subset of this extensive field survey data. The remainder of the data not used in the calibration procedure were used for validation. For eleven cells (areas of ~30 km radius with high density of sampling) the spatial model explained 66% of the observed variation in the data used in its calibration (Figure 5). Close agreement was also found between these pasture community growth parameters (eg. transpiration efficiency) derived from the extensive field survey data and those derived from the grazing exclosure data set (Day *et al.*, 1997).

7. Model Validation

Validation of the Aussie GRASS model has been conducted only in Queensland and included the use of two spatial data sets. The first was the subset of the field survey data collected by Wood *et al.* (1996) that was not used in the calibration exercise. Thus these data provide an independent validation of the model output. For the eleven cells with a high density of sampling the spatial model explained 69% of the observed variation in these data (Figure 5).

A second independent data set is the NOAA composite NDVI images for 1988 to 1994, a total of 89 images. Information from these images was correlated with a synthetic NDVI signal calculated using the model on both a spatial and temporal basis (Carter *et al.*, 1996b). For the spatial analysis, the NOAA NDVI signal for each pixel was compared to the synthetic NDVI signal for all images (Figure 6). Areas where there were negative correlations were the Darling Downs and the Channel Country. The Darling Downs is an important winter cropping area and as a result the winter NOAA NDVI signal is still green whereas the GRASP model is simulating these areas as pastures that would hay off or frost during this period and thus have a markedly different NDVI signal. A crop mask is available to eliminate these areas from the model but they are included on the basis that not necessarily all land is devoted to cropping and the output represents likely pasture growth response. The Channel Country in south west Queensland is unique in that it often benefits from widespread flooding as a result of heavy rains in the upper catchments of the rivers that meander through the region. This flooding, or run-on, is not simulated by the model and thus the NOAA and synthetic NDVI signals were not correlated.

Based on the results of the model validation using the NOAA imagery, research has begun into the use of this data to further calibrate pasture parameters. This process has marked advantages in terms of providing complete coverage of all pixels within Australia, relatively low cost and the growing time series of imagery available to researchers. In a preliminary analysis for northern Mitchell grass the parameters for this community were optimised using a genetic algorithm. The possible range of each parameter was tightly constrained to biologically sensible values slightly outside the range derived from direct field measurement at a number of sites (Day *et al.*, 1997). To reduce run-time only a subset of all pixels in the northern Mitchell grass zone were used to calculate the correlation between the satellite and model derived data. The optimisation of parameters resulted in a large improvement in model predictions (Figure 7). This technique may also be useful for identification and correction of inadequate model functionality. For example, there is no set of parameters that adequately simulates the maximum NDVI recorded in 1991 without reducing model performance at other times. A spatial analysis of correlation between satellite and model data for the northern Mitchell grass zone also identified localised areas where long term correlations were lower than the average for the entire region. This suggests that these areas within the same pasture community have different characteristics and thus correspond to climate differently.

COMBINING SATELLITE DATA AND MODEL OUTPUTS
" SPATIAL VIEW "

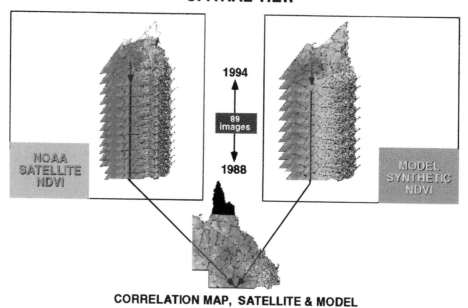

CORRELATION MAP, SATELLITE & MODEL

Figure 6 (D). Spatial view of Aussie GRASS validation using NOAA NDVI imagery (1988-1994). Model calculated NDVI values for each pixel have been compared against the time series of 89 NOAA images. The black area on Cape York represents an area of no coverage. (This figure is reproduced in the colour section at the end of the book as plate D)

8. Outputs and Products

The information available from the GRASP model includes soil water, pasture growth, biomass (leaf, stem, green, dead, total), cover (green, dead, total), synthetic NDVI, and pasture utilisation. Use of the SOI phase system and analogue years enables the model outputs and products to be produced in both near-real time and predictive modes. The resultant products and delivery systems currently available for Queensland are listed in Table 3. The *Summary of Seasonal Conditions in Queensland* report combines a range of information including rainfall, pasture growth, remote sensing and SOI into a four page colour leaflet which is mailed to over 450 individuals and organisations each month. Other numerous outputs from the model can and have been tailored into products as required.

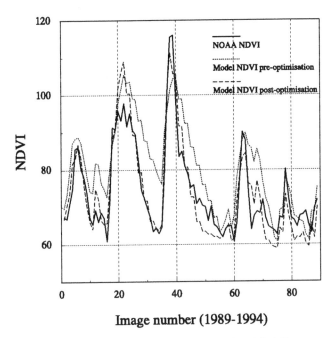

Image number (1989-1994)

Figure 7. Predicted and observed NDVI signals for the northern Mitchell grass community (1989-1994) prior to and following calibration using a genetic algorithm. The calibrated NDVI signal accounted for 84% of the observed variation.

TABLE 3. Aussie GRASS products and delivery systems currently available for Queensland.

Product	Delivery System
Rainmaps - monthly totals - percentiles on 1, 3, 6, 12 and 24 month basis	Direct mail out Long Paddock web site

Remote sensing - NOAA NDVI imagery
          - relative greenness maps
          - Sea Surface Temperature maps
	Direct mail out
	Long Paddock web site

Pasture growth percentile maps - percentiles on 1, 3, 6, 12 and 24 month basis	Direct mail out Long Paddock web site (proposed)
Summary of Seasonal Conditions in Queensland. This report combines information on rainfall, simulated pasture growth , NDVI, SST, SOI, probability of exceeding median rainfall and median pasture growth for next 3 months, drought declared shires and exceptional circumstances shires.	Direct mail out Long Paddock web site (proposed)
SOI and seasonal climate outlook information.	Poll fax service Recorded phone message Long Paddock web site

9. Users

Aussie GRASS products have in the past been focused primarily at policy makers and officers in both State and Federal governments. This has occurred because, as a result of the seasonal conditions in eastern Australia during the first half of the 1990s, there was a strong demand by these groups for the type of objective information and products that Aussie GRASS delivers. Further, the broad-scale approach of Aussie GRASS integrated easily with their approaches and decision-making processes. However, current users of Aussie GRASS products are diverse and include stock inspectors and local drought committees, extension officers, agribusiness, Rural Fire Boards, producers, industry organisations and other researchers.

A specialist extension team currently operates across regional Queensland exploring ways in which graziers and local drought committees might better use Aussie GRASS information and derive new products. This work will soon advance to a national scale by the involvement of multi-state collaborating organisations. However, the lack of high resolution digital data (especially stock numbers) limits the applicability of Aussie GRASS products to individual properties.

In Queensland, Aussie GRASS products have been delivered operationally since November 1991 to Queensland government officers including briefings to the Queensland government cabinet. Currently all Queensland drought declaration assessments and applications for Drought Exceptional Circumstances incorporate Aussie GRASS outputs. More recently, Aussie GRASS products were provided to government officers from New South Wales, South Australia and Victoria for inclusion in their applications for Drought Exceptional Circumstances.

10. Further Research and Implementation

Several key issues limiting the applicability of grazing system simulation studies to the real world and requiring further research have been identified. These include –
- the effects of overgrazing on soil loss and thus water infiltration rates, available water range and nutrient availability
- changes in species composition including woody weed invasion and woodland thickening in response to different stocking and fire regimes
- integration of the effects of (a) and (b) on animal production and grazing economics
- identification of burnt areas using satellite data so as to reset biomass in the spatial model, and
- bare ground and thus degradation assessment using satellite data.

The implementation of crop models in the spatial simulation framework was examined by Hammer et al. (1996). They found that for yield forecasting in wheat at a regional scale, simpler approaches tailored to that scale, such as those detailed by Stephens et al. (2000), were more accurate, robust and easier to implement than connecting crop simulation models to the spatial grid. Integrated implementation of this crop forecast with that for pasture systems is planned as part of the summary of seasonal conditions.

The use of General Circulation Models with nested Regional Climate Models from which daily climate data is downscaled for use in Aussie GRASS is also being investigated (Young *et al.*, 1997). It is hoped that as GCMs are further developed they may increase the lead-time available for seasonal climate forecasting from the current 3-6 months using the SOI phase system to 6-18 months (Hunt and Hirst, 2000).

As stated previously, NOAA NDVI imagery is currently being investigated for its potential as a relatively cheap spatially complete data source against which to calibrate pasture growth parameters.

Past research, which is yet to be implemented operationally, has included work focused on the extension of information available from the model to include animal production and the economics associated with this production. For example, Hall (1996) examined data from five western Queensland sheep grazing trials covering a fifty year period (~1940-90). Multiple regressions were developed using variables from GRASP which explained a high proportion (eg. r^2=0.752 P<0.01) of the variation in annual fleece production across both Mitchell and mulga grassland systems. These simple models were combined with ABS stocking rate data and output from Aussie GRASS to calculate spatial wool production for the sheep grazing areas of Queensland. Spatial wool production data was combined with information on fibre diameter, micron specific indicator prices and industry based variable costs to estimate gross margins available from wool production in Queensland (Figure 8).

11. Summary

Aussie GRASS is a project, which in its first phase has delivered a near-real time system for monitoring the condition and trend of Queensland's rangelands and grasslands. The Aussie GRASS modelling framework successfully integrates the GRASP pasture production model, geographic information systems, satellite imagery, seasonal climate forecasting and visual high-performance computing. The Queensland system has been extensively calibrated and validated (with more planned) and currently operates in near-real time on a monthly basis producing a range of products that are available through a number of delivery systems to a varied user group.

The current phase of the Aussie GRASS project is concerned with refining the existing Australia wide modelling framework. This will be done through the identification and testing of suitable regional pasture and browse production models, the upgrading of data inputs both spatially and temporally, and comprehensive calibration and validation of the Aussie GRASS model using both field and remotely sensed data.

Wool Gross Margins - 1995
(cents / ha, Base 1992-93)

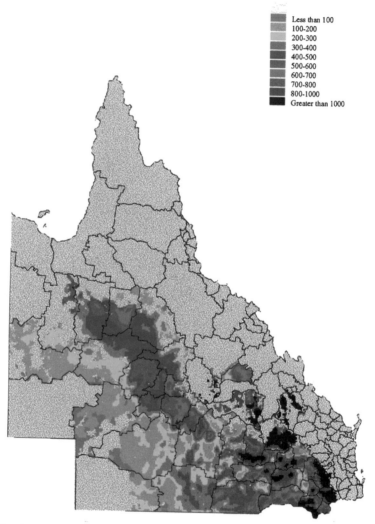

Figure 8 (E). Gross margin for Queensland wool production for 1995 calculated using the Aussie GRASS spatial analysis system. (This figure is reproduced in the colour section at the end of the book as plate E)

Acknowledgments

The authors gratefully acknowledge the support of RIRDC, NCVP, LWRRDC, MRC, IWS, BoM and Environment Australia. We are also grateful to our colleagues for the access to data used in the development of the GRASP model.

References

Anon. (1969) Report of the inter-departmental committee on scrub and timber regrowth in the Cobar-Byrock district and other areas of the Western Division of N.S.W. Government Printer, Sydney.

Ash, A.J., McIvor, J.G., Mott, J.J and Andrew, M.H. (1997) Building grass castles: Integrating ecology and management of Australia's tropical tallgrass rangelands. *Rangeland Journal* **19**, 123-44.

Beadle, N.C.W. (1948) The vegetation and pastures of Western New South Wales with special reference to soil erosion. Department of Conservation of New South Wales, Sydney.

Bisset, W.J. (1962) The black speargrass *(Heteropogon contortus)* problem of the sheep country in central-western Queensland. *Queensland Journal of Agricultural Science* **19**, 189-207.

Bowman, P.J., White, D.H., Cayley, J.W.D. and Bird, P.R. (1982) Predicting rates of pasture growth, senescence and decomposition. *Proc. Australian Society of Animal Production* **14**, 36-37.

Brook, K.D. (1996) Development of a national drought alert strategic information system. Volume 1, Research summary. Final report on QPI20 to Land and Water Resources Research and Development Corporation. Qld. Dept. Natural Resources. pp.11.

Burrows, W.H. (1995) Greenhouse revisited - land-use change from a Queensland perspective. *Climate Change Newsletter* **7**, 6-7.

Burrows, W.H., Anderson, E.R., Back, P.V. and Hoffmann, M.B. (1997) Regrowth and woody plant thickening/invasion impacts on the land use change and forestry inventory. IPCC Workshop on Biomass Burning, Land-Use Change and Forestry, Rockhampton, Australia.

Burrows, W.H., Carter, J.O. Scanlan, J.C. and Anderson, E.R. (1990) Management of savannas for livestock production in northeast Australia: contrasts across the treegrass continuum. *J. Biogeog.* **17**, 503-12.

Campbell, S.D. (1995) Plant mechanisms that influence the balance between *Heteropogon contortus* and *Aristida ramosa* in spring burnt pastures. Unpub. Ph. D. Thesis. University of Queensland, Brisbane.

Carter, J.O. (1994) *Acacia nilotica*: a tree legume out of control, in R.C. Gutteridge and H.M. Shelton (eds.), *Forage Tree Legumes in Tropical Agriculture*. CAB International, U.K. pp.338-351.

Carter, J.O., Flood, N.F., Danaher, et al. (1996a). Development of a national drought alert strategic information system. Volume 3, Development of data rasters for model inputs. Final Report on QPI 20 to Land and Water Resources Research and Development Corporation. Qld. Dept. Natural Resources. pp.76.

Carter, J.O., Flood, N.F., McKeon, G.M., Peacock, A. and Beswick, A. (1996b) Development of a national drought alert strategic information system. Volume 4, Model framework, parameter derivation, model calibration, model validation, model outputs, and web technology. Final Report on QPI 20 to Land and Water Resources Research and Development Corporation. Qld. Dept. Natural Resources. pp. 42.

Condon, R.W. (1968) Estimation of grazing capacity on arid grazing lands, in G.A. Stewart (ed.), Land Evaluation: Papers of a CSIRO symposium organised in cooperation with UNESCO. CSIRO Division of Land Research, Canberra. pp. 112-124.

Condon, R.W. (1986a) Recovery from catastrophic erosion in western New South Wales, in 'Rangelands: A Resource Under Siege'. Proceedings of the Second International Rangeland Congress, Adelaide. p.39.

Condon, R.W. (1986b) Scrub invasion on semi-arid grazing lands - causes and effects, in 'Rangelands: A Resource Under Siege'. Proceedings of the Second International Rangeland Congress, Adelaide. p.40.

Condon, R.W., Newman, J.C., and Cunningham, G.M. (1969a) Soil erosion and pasture degeneration in Central Australia. Part I - Soil erosion and degeneration of pastures and topfeeds. *Journal of the Soil Conservation Service of New South Wales* **25**, 47-92.

Condon, R.W., Newman, J.C. and Cunningham, G.M. (1969b) Soil erosion and pasture degeneration in Central Australia. Part II - Prevention and control of soil erosion and pasture degeneration. *Journal of the Soil Conservation Service of New South Wales* **25**, 161-82.

Condon, R.W., Newman, J.C. and Cunningham, G.M. (1969c) Soil erosion and pasture degeneration in Central Australia. Part III -The assessment of grazing capacity. *Journal of the Soil Conservation Service of New South Wales* **25**, 225-50.

Condon, R.W., Newman, J.C. and Cunningham, G.M. (1969d) Soil erosion and pasture degeneration in Central Australia. Part IV. *Journal of the Soil Conservation Service of New South Wales* **25**, 295-321.

Danaher, T.J., Carter, J.O., Brook, K.D., Peacock, A. and Dudgeon, G.S. (1992) Broad scale vegetation mapping using NOAA-AVHRR imagery. Proceedings of the Sixth Australasian Remote Sensing Conference, Wellington, New Zealand. Volume 3, pp.128-37.

Day, K.A., McKeon, G.M. and Carter, J.O. (1997) Evaluating the risks of pasture and land degradation in native pasture in Queensland. Final Project Report for Rural Industries and Research Development Corporation project DAQ124A. Qld. Dept. Natural Resources. pp.119.

Deutscher, N.C. (1959) The development of the cattle industry in Queensland 1840-1890. Unpub. B.A. (Hons.) Thesis, University of Queensland, Brisbane.

Drosdowsky, W. and Allan, R. (2000) The potential for improved statistical seasonal climate forecasts, in G.L. Hammer, N. Nicholls, and C. Mitchell (eds.), *Applications of Seasonal Climate Forecasting in Agricultural and Natural Ecosystems – The Australian Experience*. Kluwer Academic, The Netherlands. (this volume).

Dudgeon, G.S., Nilsson, C.S. and Fry, W.B. (1990) Rangeland vegetation monitoring using NOAA-AVHRR data: 1. Improving the spatial and radiometric accuracy of NDVI data. Proceedings of the Fifth Australasian Remote Sensing Conference, Perth, Western Australia. pp.208-17.

Filet, P., Dudgeon, G., Scanlan, J., Elmes, N., Bushell, J., Quirk, M., Wilson, R. and Kelly, A. (1990) Rangeland vegetation monitoring using NOAA-AVHRR data: 2. Ground truthing NDVI data. Proceedings of the Fifth Australasian Remote Sensing Conference, Perth, Western Australia. pp.218-27.

Gardener, C.J., McIvor, J.G. and Williams, J.C. (1990) Dry tropical rangelands: solving one problem and creating another. *Proc. Ecol. Soc. Aust.* **16**, 279-80.

Hacker, R.B., Wang, K., Richmond, G.S. and Lindner, R.K. (1991) IMAGES: An integrated model of an arid grazing ecological system. *Agricultural Systems* **37**, 119-63.

Hall, W.B. (1996) Near-real time financial assessment of the Queensland wool industry on a regional basis. Unpub. Ph.D. Thesis, University of Queensland, Brisbane. pp.465.

Hammer, G., Stephens, D. and Butler, D. (1996) Development of a national drought alert strategic information system. Volume 6, Wheat modelling sub-project: Development of predictive models of wheat production. Final Report on QPI 20 to Land and Water Resources Research and Development Corporation. Qld. Dept. Natural Resources. pp.41.

Hobbs, T.J., Sparrow, A.D. and Landsberg, J.J. (1994) A model of soil moisture balance and herbage growth in the rangelands of central Australia. *Journal of Arid Environments* **28**, 281-98.

Hunt, B.G. and Hirst, A.C. (2000) Global climatic models and their potential for seasonal climate forecasting, in G.L. Hammer, N. Nicholls, and C. Mitchell (eds.), *Applications of Seasonal Climate Forecasting in Agricultural and Natural Ecosystems – The Australian Experience*. Kluwer Academic, The Netherlands. (this volume).

Hutchinson, M.F. (1991) The application of thin plate smoothing splines to continent-wide data assimilation, in J.D. Jasper (ed.), *Data Assimilation Systems*. Bureau of Meteorology, Melbourne. pp.104-13.

Johnston, P.W., McKeon, G.M. and Day, K.A. (1996a) Objective "safe" grazing capacities for south-west Queensland Australia: development of a model for individual properties. *Rangeland Journal* **18**, 244-58.

Johnston, P.W., Tannock, P.R., and Beale, I.F. (1996b) Objective "safe" grazing capacities for south-west Queensland Australia: a model application and evaluation. *Rangeland Journal* **18**, 259-69.

McKeon, G. M., Day, K.A., Howden, S.M., Mott, J.J., Orr, D.M., Scattini, W.J. and Weston, E.J. (1990) Northern Australian savannas: management for pastoral production. *J. Biogeog.* **17**, 355-72.

McKeon, G.M. and Howden, S.M. (1991) Adapting northern Australian grazing systems to climate change. *Climate Change Newsletter* **3**, 5-8.

McKeon, G.M., Ash, A.J., Hall, W.B. and Stafford Smith, D.M. (2000) Simulation of grazing strategies for beef production in north-east Queensland, in G.L. Hammer, N. Nicholls, and C. Mitchell (eds.), *Applications of Seasonal Climate Forecasting in Agricultural and Natural Ecosystems – The Australian Experience*. Kluwer Academic, The Netherlands. (this volume).

Miles, R.L. (1993) The rates, processes and effects of soil erosion on semi-arid woodland. Unpub. Ph.D. Thesis. Griffith University, Brisbane.

Mills, J.R. (1984) Landuse and its effects on soil erosion and water quality, in I.F. Beale, J.R. Mills and A.J. Pressland (eds.), *Use of Conservation of Arid Rangeland in Queensland*. Queensland Department of Primary Industries. pp.9-14.

Moore, A.D., Donnelly, J.R. and Freer, M. (1997) GRAZPLAN: Decision support systems for Australian grazing enterprises - III. Pasture growth and soil moisture submodels and the Grassgro DSS. *Agricultural Systems* **55**, 535-82.

Mortiss, P.D. (1995) The environmental issues of the Upper Burdekin catchment. Project report Q095017, Land and Water Resources Research and Development Corporation. Queensland Department of Primary Industries, Brisbane.

Noble, J.C. (1997) The delicate and noxious scrub: CSIRO studies on native tree and shrub proliferation in the semi-arid woodlands of Eastern Australia. CSIRO, Canberra, Australia.

Orr, D.M. and Paton, C.J. (1993) Fire and grazing interact to manipulate pasture composition in black speargrass (*Heteropogon contortus*) pastures. Proc. XVII International Grassland Congress. pp. 1910-11.

Orr, D.M., McKeon, G.M. and Day, K.A. (1991) Burning and exclosure can rehabilitate degraded black speargrass (*Heteropogon contortus*) pastures. *Tropical Grasslands* **25**, 233-36.

Orr, D.M., Paton, C.J. and McIntyre, S. (1994) State and transition models for rangelands. 10. A state and transition model for the southern black speargrass zone of Queensland. *Tropical Grasslands* **28**, 266-69.

Paton, C.J. and Rickert, K.G. (1989) Burning, then resting, reduces wiregrass (*Aristida* spp.) in black speargrass pastures. *Tropical Grasslands* **23**, 211-18.

Ratcliffe, F.N. (1937) Further observations on soil erosion and sand drift, with special reference to south-western Queensland. Council for Scientific and Industrial Research Pamphlet no. 64, Melbourne.

Scanlan, J.C., Pressland, A.J. and Myles, D.J. (1996a) Runoff and soil movement on mid-slopes in north-east Queensland grazed woodlands. *Rangeland Journal* **18**, 33-46.

Scanlan, J.C., Pressland, A.J. and Myles, D.J. (1996b) Grazing modifies woody and herbaceous components of north Queensland woodlands. *Rangeland Journal* **18**, 47-57.

Shaw, W.H. (1957) Bunch spear grass dominance in burnt pastures in south eastern Queensland. *Australian Journal of Agricultural Research* **8**, 325-34.

Smith, R.C.G. (1994) Australian vegetation watch. Final Report on DOL-1A to the Rural Industries Research and Development Corporation. CSIRO, Canberra, Australia.

Stephens, D.J., Butler, D.G. and Hammer, G.L. (2000) Using seasonal climate forecasts in forecasting the Australian wheat crop, in G.L. Hammer, N. Nicholls, and C. Mitchell (eds.), *Applications of Seasonal Climate Forecasting in Agricultural and Natural Ecosystems – The Australian Experience*. Kluwer Academic, The Netherlands. (this volume).

Stone, R.C. and Auliciems, A. (1992) SOI phase relationships with rainfall in eastern Australia. *International Journal of Climatology* **12**, 625-36.

Stone, R.C., Hammer, G.L. and Marcussen, T. (1996a) Prediction of global rainfall probabilities using phases of the Southern Oscillation Index. *Nature* **384**, 252-255.

Studdert, M.J. (1994) Rabbit Haemorrhagic Disease: a calicivirus with differences. *Australian Veterinary Journal* **71**, 264-66.

Tabios, G.Q. and Salas, J.D. (1985) A comparative analysis of techniques for spatial interpolation of precipitation. *Water Resource Bulletin* **21**, 365-80.

White, D.H., Bowman, P.J., Morley, F.H.W., McMannus, W.R. and Filan, S.J. (1983) A simulation model of a breeding ewe flock. *Agricultural Systems* **10**, 149-89.

Wilcox, D.G. and McKinnon, E.A. (1972) A report on the condition of the Gascoyne catchment. Department of Agriculture and Department of Lands & Surveys, Western Australia.

Wood, H., Hassett, R., Carter, J. and Danaher, T. (1996) Development of a national drought alert strategic information system. Volume 2, Field validation of pasture biomass and tree cover. Final Report on QPI 20 to Land and Water Resources Research and Development Corporation. Qld. Dept. Natural Resources. pp. 50.

Young, R., Lau, L., Hutchinson, M., Hunt, B., Graham, N., Goddard, L., Duncalfe, F., Burrage, K. and McKeon, G. (1997) Integrating advanced GCM seasonal climate forecasts, fine resolution spatial systems, and dynamic web visualisation for sustainable environmental management. Proc. Mission Earth '97: Modeling and Simulation of the Earth System. The Society for Computer Simulation, Phoenix, Arizona, U.S.A. pp. 43-52.

USING SEASONAL CLIMATE FORECASTS IN FORECASTING THE AUSTRALIAN WHEAT CROP

DAVID STEPHENS[1], DAVID BUTLER[2] AND GRAEME HAMMER[3]

[1] Spatial Resource Information Group
Agriculture Western Australia
South Perth, WA 6151, Australia

[2] Queensland Department of Primary Industries
PO Box 102
Toowoomba, Qld 4350, Australia

Abstract

Recent research has pointed to the potential to forecast the Australian wheat crop using seasonal climate forecasts. Over the last decade seasonal climate outlooks have been issued for the wheat-growing season and regional crop yield modelling has progressed to an operational stage. To quantify the potential benefits to crop production forecasting of combining seasonal climate forecasts and regional crop yield models, long-term wheat production across the Australian wheatbelt was simulated using an agroclimatic stress index model (STIN) and grouped according to the phase of the SOI in April-May (around planting). Throughout the wheatbelt there was a clear tendency for lower production in SOI phase 1 (consistently negative SOI) or phase 3 (rapidly falling SOI) years and for higher production in phase 2 or 4 years (consistently positive or rapidly rising SOI). There was greater spread in the simulated yields associated with some phases and greater certainty of obtaining production anomalies with others. This differentiation of productivity early in the crop season suggested that seasonal climate forecasts based on SOI phases would likely provide useful skill when used in real-time yield forecasting.

To test this capability in a forecasting mode, an operational yield forecasting model (Weighted Rainfall Index - WRI) was tested with hindcasts using 1986-95 rainfall and compared with forecasts issued by the Australian Bureau of Agricultural and Resource Economics (ABARE). These hindcasts showed that use of the WRI model generally improved on ABARE forecasts, with mean prediction errors reduced 25-30% when average rainfall was assumed throughout the growing season. The addition of rainfall forecasts based on SOI phases, however, gave little improvement in yield forecasts. This may have been due to the unrepresentativeness of the short hindcast test period and differences between the WRI and STIN models. A single analogue forecast system was also tested, but this highlighted the need for a number of analogues to stabilise mid-season

G.L. Hammer et al. (eds.), The Australian Experience, 351–366.

forecasts and mean prediction errors. Further studies to examine these aspects are needed to better test the utility of current seasonal forecasting capability, before definitive conclusions are possible.

1. Introduction

The Australian wheat crop is an important, but variable, contributor to world grain markets. Recently, total production has ranged from 8.9 million tonnes in 1994, to 23.7 million tonnes in 1996 (ABARE, 1997). Efficient management of this grain is only possible if marketing and transport agencies have accurate early-season production forecasts to assist them in all aspects of decision making (Walker, 1989). In other grain exporting countries, crop weather models have been used operationally in crop assessments (Motha and Heddinghaus, 1986; Walker, 1989), but early season forecasts were impeded by the uncertainty of future weather (Walker, 1989).

As a result of its semi-arid environment, Australian wheat yield is largely dependent on rainfall amount and distribution (Nix, 1975; Stephens and Lyons, 1998). Rainfall in turn is strongly related to the broad-scale atmosphere-ocean phenomenon ENSO (El Nino/Southern Oscillation) (McBride and Nicholls, 1983; Stone and Auliciems, 1992). Fortuitously, ENSO events switch into extreme phases around planting time (April-June), with wet or dry rainfall anomalies tending to persist through to harvest (December) (Nicholls, 1990; Ropelewski and Halpert, 1987; 1989). Consequently, Australian wheat yields are significantly correlated to the magnitude (Nicholls, 1985) and trend (Rimmington and Nicholls, 1993) in the Southern Oscillation Index (SOI), more so than any other major grain crop in the world (Garnett and Khandekar, 1992).

Prior to 1996, production forecasts issued by the Australian Bureau of Agricultural and Resource Economics (ABARE) did not utilise this information. Forecasts in May and June were based on yield in the previous year, a subjective estimate of how pre-growing season rains had affected soil moisture, and an assumption that subsequent rainfall would be near to average (AACM, 1991). Later in the season forecasts were based on a survey of agencies involved in the grain industry. This conservative approach tended to underestimate yields, except in major drought years such as 1982 and 1994 when the SOI was very low in the winter growing season (June to Sept) (Figure 1). However, not all years with low SOI in winter had low yield eg. 1987 and 1993.

Seasonal outlooks have been issued in Australia for a number of years. The Bureau of Meteorology issued categorical outlooks from 1989 to 1991 for regions that had significant relationships between the SOI and rainfall, but subsequently switched to a weighted probability forecast of above or below average rainfall (Smith, 1994). The seasonal outlooks provided by the Bureau are described further by Stone and de Hoedt (2000). Lamond Weather Services (formerly Austweather Pty Ltd) have issued categorical forecasts based on a single selected analogue year at the beginning of each year since 1984 for southern Australia. These outlooks have up to a 12-month lead-time and have exhibited significant skill for Western Australia (Smith, 1994). A third series of outlooks have combined information about the magnitude and trend in the SOI. Stone and Auliciems (1992) defined five SOI 'phases' using SOI values over two adjacent months (mean values for 110 years in brackets):

Phase 1 - consistently negative SOI values (-12.2);
Phase 2 - consistently positive SOI values (9.5);
Phase 3 - rapidly falling SOI (month one 2.7, month two -9.9);
Phase 4 - rapidly rising SOI (month one -4.4, month two 6.6); and
Phase 5 - SOI values close to zero with little change (mean -1.7).

This forecasting scheme identifies shifts in seasonal rainfall probabilities based on groups of analogue years associated with each phase (Stone *et al.*, 1996). The analogues can be used with crop models to simulate likely outcome distributions and consequences on risky decisions at both field and regional scales (eg. Hammer and Muchow, 1991; Hammer *et al.*, 1996).

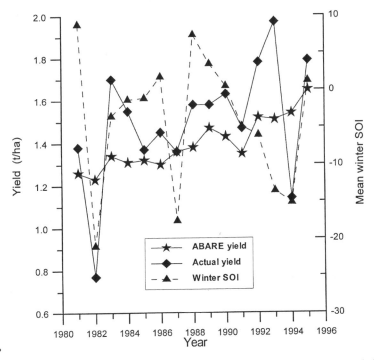

Figure 1. Australian average wheat yield forecast by ABARE in May/June, the actual yield recorded by ABS, and the mean winter (June, July, August) SOI from 1981 to 1995.

The potential of agroclimatic yield models to explain most of the variation in wheat yield across Australia has been demonstrated by Stephens (1995). The stress index model (STIN; Stephens, 1995) uses a weekly soil water balance to determine the degree of water stress experienced by the crop. This index is used in a simple regression model to predict wheat yield for each wheat-producing shire in Australia. The index is similar in concept to that proposed by Nix and Fitzpatrick (1969). It utilises biophysical knowledge of the crop, allows consideration of soil type effects, and derives the stress index by contrasting soil water supply with crop demand. The weighted rainfall index model (WRI; Stephens *et al.*, 1994) employs weightings on monthly rainfall to predict State and national yields. Stephens *et al.* (1989) showed that using the STIN model produced improved yield forecasts at the shire level in Western Australia for the period 1984 to 1986 if categorical forecasts from Lamond Weather Services were utilised. Stephens and Lyons (1993) used the WRI model to derive yield predictions in Queensland utilising the Bureau of Meteorology's 3-month seasonal outlooks. Hindcasts showed improved forecast accuracy in the drought year of 1991, however, the weighted probability forecasting scheme introduced by the Bureau was more difficult to incorporate into the commodity forecasting models and was more conservative and of less value in extreme conditions. Stephens (1996) showed that if the WRI model was used with forecast median rainfall, derived from all analogue years associated with the SOI phase of the previous month, a clear indication of very low yields was made between June to September in 1994.

Since September 1995, ABARE have incorporated the WRI model and the median rainfall derived from the SOI phase forecasting system to assist in forecasting of State and national yields (ABARE, 1995-1997). With only three years of operational model use it is unclear about potential advantages associated with (i) using a crop model in place of traditional surveys, and (ii) using rainfall derived from SOI phase analogue years instead of average rainfall over all years.

The aim of this paper is to examine the potential of seasonal forecasts to improve wheat yield predictions in an operational environment in the Australian context. Official Government yield predictions are compared to those based on crop models, which have the powerful capability of being able to combine current rainfall to date within a season, future average rainfall or future rainfall based on seasonality indicators. In the first instance, to determine the impact of a seasonal forecast, the STIN model was run on rainfall data from across the Australian wheatbelt for a long time series (1915 – 1993) and predicted production distributions associated with each SOI phase at around wheat planting were derived and compared. We then used hindcasts over the last 10 years to compare yield forecasts derived from the WRI model with conventional ABARE forecasts and the actual final yield recorded by the Australian Bureau of Statistics (ABS). Comparisons are made at both State and national levels. The WRI model was run for the 10-year period using either long-term average rainfall or rainfall derived from the forecasting systems examined.

2. Methods

2.1 HISTORICAL YIELD AND PRODUCTION ANALYSIS

To determine long term relationships between crop production and the phase of the SOI around planting (May), long-term runs of the STIN agroclimatic model were made using historical data from across the wheatbelt. The STIN agroclimatic model combines the soil water balance routines from the CERES-style crop models (Ritchie and Otter, 1985) with the FAO (Food and Agriculture Organisation) crop monitoring method (Frere and Popov, 1979) and is detailed in Stephens *et al.* (1989). Soil moisture at sowing and growing season rainfall are used to determine an accumulated crop moisture-stress index, which is related to yield. A district level (shire) sowing date is calculated on the basis of sowing rules determined from a national sowing date survey of farmers (Stephens, 1995). Data needed for this model is daily rainfall, average climate data (maximum temperature, minimum temperature, and solar radiation) and representative soil profile parameters.

The long term historical climate data from across the wheatbelt was collated by examining length of records for relevant stations. Stations falling within within the cropping boundary defined by Hamblin and Kyneur (1993) and with daily rainfall from 1 January 1915 to 31 December 1993 were identified. Stations were selected if their area of influence included shires with continuous wheat production from 1975 to 1993. Historical daily rainfall data was sourced from the Rainman project (Clewett *et al.*, 1994). Average climate data were generated for each station from the ESOCLIM package (Hutchinson, 1991). Representative soil parameters were assigned for each shire on the basis of whether the predominant soil type was a light, medium or heavy texture (Northcote *et al.*, 1975; Forrest *et al.*, 1985).

The STIN model was then run at each station on every year of rainfall data from 1915 to 1993. Within each shire an aggregate stress index was calculated from a weighted summation of values from individual stations. Weighting was based on the area of influence of the individual station in relation to the cropping area of the total shire. Relative shire production for each year was then calculated using the following three steps:

(1)   The Stress Index (SI) in STIN was calibrated with observed Australian Bureau of Statistics (ABS) yields for the years 1975 to 1993 using multiple regression. A predictive equation is formed where predicted yield Y is calculated:

$$Y = a + b*(SI) + c*(year) \tag{1}$$

where a, b and c are population regression coefficients estimated by the method of least squares. The last term accounts for the technological increase in yields with time, with c being the slope of yield increase in kg/ha/year.

(2)   Predicted (de-trended) shire wheat yields were calculated for each year from 1915 to 1993 for each shire using Equation (1) with the SI value calculated for each year, with the year term fixed at 1990 (ie. current technology).

(3)   Predicted production for each shire was calculated by multiplying the predicted yield with the average area of wheat planted in that shire. Planted area data for the period 1975-90 was used to derive the average area planted.

Production was then aggregated to the State level and years were grouped into one of five groups depending on the phase of the SOI in May.

2.2 HINDCAST TESTING OF OPERATIONAL PRODUCTION FORECASTS

The Weighted Rainfall Index (WRI) model (Stephens *et al.*, 1994) was used to perform the hindcast testing of a number of production forecasting systems. The WRI model requires only monthly rainfall from 45 meteorological districts covering the wheatbelt, but explains 89 to 93% of variation in wheat yield at a state and national level (1980-1995). This index is calculated nationally as follows:

$$WRI = \sum_{m=-3}^{11} \frac{\sum_{d=1}^{45} R_{md} * W_{md} * W_d}{\sum_{d=1}^{45} W_d} \tag{2}$$

where:

m	=	month (-3 (previous October) to 11 (November))
d	=	district (1 to 45)
R_{md}	=	rainfall during month m in district d
W_{md}	=	monthly weighting factor for month m in district d
W_d	=	percentage of the national crop area in district d

Thus, rainfall (R_{md}) is added and weighted from October in the previous year through to November of the crop year. The district weighting factors (W_d) were determined from a database containing the most recent years of shire area statistics. Walker (1989) found that changes in area weightings made little difference to prediction accuracy at the large scale when many stations were contributing. Hence, to simplify data manipulations, the individual W_d values were kept constant over years in model calibration and testing. The monthly weighting factors W_{md} were assigned on the basis of physical and physiological reasons (Stephens *et al.*, 1994).

For use in hindcasts the WRI model was calibrated using ABS yield data in a similar manner to equation (1). However, the WRI model was re-calibrated each year on the 13 years of data preceding the year of prediction by two years (eg. the 1994 season is predicted with a model calibrated on data from 1980-92). This delay occurs because final yield data from ABS are not available until two years after the crop is harvested. Due to the lack of droughts in Western Australia after 1980, the equations for 1995 were based on the 14 years 1980-93, while for the first three years (1986-88) equations were derived with data prior to 1975 removed due to the widespread rust epidemic of 1973 and 1974 (Watson and Butler, 1984). Equations thus formed explained 76 to 95% of the yield variance for the 10-year analysis period 1986 to 1995. To derive total production estimates from estimates of yield per unit area, actual ABS planted area data were used. Hence, potential errors associated with planted area estimates were not considered in this analysis.

Predictions based on runs of the WRI model were updated progressively through the crop season by replacing future rainfall (ie. long-term average or forecast) with actual rainfall once it had occurred. Predictions from the beginning of the year can be made using projections based on long term average rainfall or on the single historical analogue issued as the forecast by Lamond Weather Services. For rainfall derived from

SOI phase analogue years, predictions are recommended only from the beginning of June when phase relationships show significant forecasting skill (Stone, pers. comm., 1996). Yield predictions from these three forecasting approaches were compared with official ABARE yield forecasts published in routinely released Crop Reports (ABARE, 1986-95). Model predictions and ABARE forecasts are directly comparable, as district rainfall is available in real-time when ABARE forecasts are finalised.

3. Results and Discussion

3.1 HISTORICAL YIELD AND PRODUCTION ANALYSIS

When the simulated shire wheat yields derived from the STIN model for each year from 1915 to 1993 were converted into production and aggregated at a state and national level, a large variation in production was found. The impact of ENSO on Australian wheat and its likely utility in crop forecasting are clearly illustrated when these annual production estimates were grouped according to the phase of the SOI in May (Figure 2), which is prior to the main growing season of the crop. The resulting production distributions are presented as box plots (Figure 2), which show the median (50 percentile), 25% and 75% production, with the range in error bars extending to the 10 and 90% values. Individual extreme outliers are also shown in some states. While these data represent predicted production, annual variation in area planted is not included, as an average area estimated from historical data was used in all years. Hence, the variability shown reflects only the variability in yield per unit area. The additional effect of annual variability in crop area in each season type is not known, but should be larger in north-eastern Australia compared to Western Australia (O'Donnell et al., 1998).

Clear differences occur in the median production for different phases and these were consistent throughout the country. A tendency for lower yields in phases one and three (SOI consistently negative; rapidly falling) and higher yields in phases two and four (SOI consistently positive, rapidly rising) were noticeable. When the SOI was close to zero the median yields were close to average nationally, but appeared above average in New South Wales (NSW) and Queensland (Qld), and more below average in South Australia (SA) and Victoria (Vic.). There was, however, considerable spread in production in each phase and this varied for different states. In Qld (Figure 2) there was a low median and a small range in production around the median in Phase 1, so there was a high probability of obtaining low production in this case. In other states, the shift in median was not as great and/or the range in production was larger. Hence, while there was an increased chance of low yields in phase 1 years, there remained about 20-30% of those years with above long-term median production. In phase 2 years, there was a high median in all states and a restricted range in NSW. Phase 3 years responded similarly to phase 1, although there was a restricted range in WA. Phase 4 years responded similarly to phase 2 years. In phase 5 years, all states showed a wide range in production, indicating the likely diverse type of season associated with that phase.

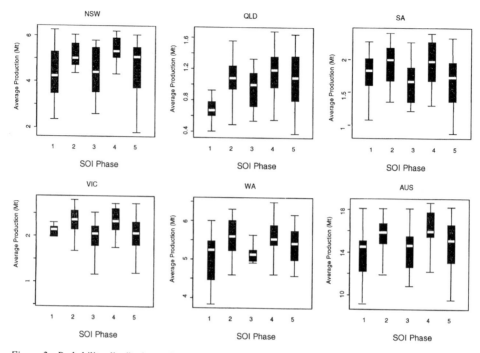

Figure 2. Probability distributions of simulated State and national wheat production for 1915-93 grouped according to the SOI phase in April-May. The boxplots show lowest, 25%, Median (50%), 75%, and highest values.

3.2 HINDCAST TESTING OF OPERATIONAL PRODUCTION FORECASTS

The differentiation of future production anomalies as early as May, suggests that the SOI phase approach would likely provide useful skill for operational production forecasting early in the crop season. To test this the WRI model was updated progressively through the crop season using each of the analog years associated with the SOI phase in the previous month. The model was run in this updating mode for the ten-year period from 1986 to 1995. The expected distribution in yield was derived from this approach, so that yield likelihoods could be assigned. Sample hindcasts for 1993 and 1994 are shown in Figure 3. When the WRI was calculated using average rainfall as the estimate for the remainder of the season, only one estimate of yield was derived. However, when the WRI was calculated from analogue years based on SOI phases, a range of yield estimates was derived. The median of these estimates and the highest and lowest values are shown in Figure 3. As expected, the range narrowed as the season progressed. The estimate given by ABARE is also shown. The ABARE estimate issued in June is not updated until September, and in both 1993 and 1994, the other procedures had shifted considerably closer to the outcome reported by ABS at the end of the year, during the June to September period. That is, both forecasting procedures responded to the season quicker than the conventional approach.

(a)

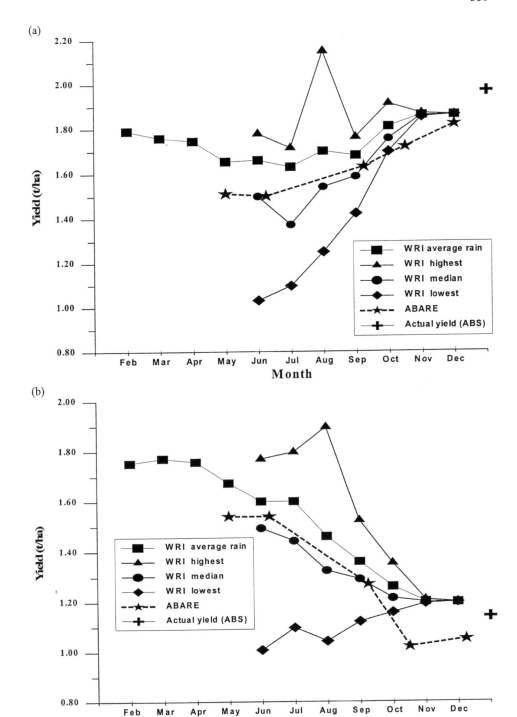

(b)

Figure 3. Yield forecasts updated each month of the year using the Weighted Rainfall Index (WRI) model with average rainfall for future months, or rainfall derived from an SOI phase forecast for future months. This is compared with traditional ABARE forecasts for (a) 1993 and (b) 1994.

In both 1993 and 1994, the SOI was consistently negative (Phase 1) from May to October, but contrasting seasons eventuated. This highlights the probabilistic nature of seasonal forecasting. While in such seasons, a below average season is more likely, there remains a significant chance that this will not occur. In both years, the yield forecast based on average rain exceeded the median of the yield forecasts derived from the SOI phase analogue years as those analogues contained a higher proportion of low rainfall years. In 1994, progressive yield predictions fell dramatically from April to November as low rainfall, in fact one of the worst droughts on record, did eventuate. The median yield based on the SOI phase analogues was lower for each month between May-October as the forecast projections were for higher chance of drier than normal conditions. As this eventuated in 1994, this yield forecasting system appeared superior as it predicted the yield decline sooner than the other systems. The range of outcomes associated with the analogue years (Figure 3) showed, however, that considerable uncertainty around the yield forecast remained until September, when the range narrowed considerably, ultimately moving towards the lower end of the expected range defined at the outset of the season. A major feature of the SOI phase system is the specification of the distribution of likely outcomes, which quantifies the uncertainty in the prediction, giving a more realistic assessment based on information available (and its uncertainty) at any point in time through the season. Using the WRI model with average rainfall gives no information about the uncertainty in the prediction. This could be remedied by using all years singly and then examining the distribution of predictions to quantify confidence limits. The notion of a 'right' or 'wrong' prediction is inappropriate in such a variable climate. The more realistic prediction of a shift in the likely distribution of production is more useful to risk management.

In 1993, despite the negative SOI phase, above average rainfall occurred from July through to November. While this scenario had a low probability, it was not unprecedented. In this year, as the wetter season unfolded, the yield forecast based on the SOI phase analogues moved towards the upper range of the forecast made early in the season. It was, however, slower in responding than the forecast based on average rainfall, which started at a higher level (Figure 3).

For the ten years 1986-95, the average absolute difference between predicted and actual yields for each relevant month of the year is plotted for each prediction scheme and state in Figure 4. The median value of the set of annual yields predicted was used as the predicted yield for forecasts based on the SOI phase analogue years. The Lamond system identifies the most likely analogue for each year, so there is only a single estimate of yield each month of any year. In general, as the season progressed, absolute error decreased and stabilised at near its final value (0.1 – 0.2 t/ha depending on State) by the end of September. This reinforces the known importance of August-September rainfall to the Australian wheat crop (Stephens and Lyons, 1998). The estimates based on the WRI model updated each month using average future rainfall gave the lowest mean error on most occasions at the national scale, but there was some variation in performance from State to State (Figure 4). On average, at the national scale, these estimates had errors 25-30% less than did those of the ABARE forecasts. Neither of the systems using seasonal forecasts improved on these estimates using average rainfall, although estimates based on the SOI phase system approximated the average rainfall system in many instances. Both forecasting systems tended to have greater prediction error early in the season.

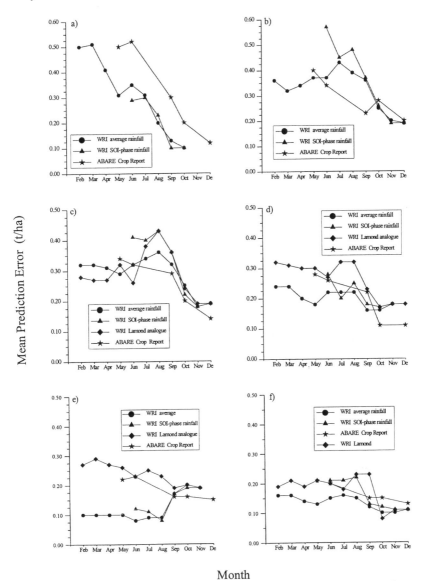

Month

Figure 4. Mean absolute yield prediction error for the period 1986-95 for ABARE forecasts (★), forecasts using the WRI model with average rainfall (●), forecast rainfall using SOI phases (▲), or forecast rainfall using Lamond analogues (◆), for a) Queensland, b) New South Wales, c) Victoria, d) South Australia, e) Western Australia, and f) Australia.

The considerable improvement in moving from forecasts based on past average yield (ABARE) to those based on past average rainfall (WRI model estimates) was related to the inherent capacity of the WRI model to respond sooner to unusual pre-season and seasonal rainfall conditions and to incorporate the effects of technology trends on yield. In Queensland for example (Figure 4a), where there were many below average years in the test period (seven years with yield below 1.12 t/ha), estimates derived from the WRI model responded more rapidly due to inclusion of rainfall effects in the preceding summer that cause low soil moisture reserves in drought years. At a national level, however, there was only one severe drought in the 1986-95 interval (1994), with the first six years having little yield variation (1.36-1.63 t/ha) (Figure 1). In average years, there is likely to be little difference between ABARE estimates and those derived from average rainfall. This may explain the generally similar outcomes of the two systems in NSW, Vic, and SA (Figure 4b,c,d).

In Western Australia (Figure 4e), the WRI model gave better estimates than the ABARE forecasts early in the season even though many average years were recorded. Significant technological yield increases in this State since the early 1980s (Hamblin and Kyneur, 1993) probably explain this. The calibration procedure used for WRI is based on equation 1 and utilises the preceding 13 years of yield data. This inherently captures such technological advances, whereas it appears that ABARE forecasts failed to take this into account. The difference in prediction error, however, diminishes later in the season. This was caused by the increase in error associated with the estimates from the WRI model. Most of this additional error was generated in two years (1986, 1992) that had large crop losses from either excessive rain, and/or disease, late in the season. Operational use of the WRI model would need to adjust estimates based solely on rainfall to account for these factors.

The lack of improvement in forecast error associated with use of seasonal forecasts may relate to areal influence of the forecast, the length and representativeness of the test period, and the use of the less robust WRI model. Although earlier studies (Pittock, 1975; McBride and Nicholls, 1983) showed high correlations between rainfall and the SOI only for NE Australia, more recent studies using pattern analysis procedures (Stone and Auliciems, 1992; Stone et al., 1996) have indicated that significant effects on rainfall likelihood across the entire Australian wheatbelt are associated with particular SOI phases. Hence, the areal influence is unlikely to be relevant. A relatively short testing period like 10 years, however, allows a large error in one or two years to dominate the average error, as noted above for Western Australia. Much of the additional 10-year average error associated with estimates derived from the SOI phase system early in the season (Figure 4) could be attributed to the unusual seasons in 1992 and 1993 (Figures 1 and 3). The SOI was in phase 1 throughout the season in 1993, but varied through the season in 1992. However, a good season in a year like 1993 is not impossible, just less likely. The other two seasons of similar type in this period (1987 and 1994) recorded well below average yield (Figure 1). A longer test period may be required to lessen the influence of years generating high error, such as 1992 and 1993.

The use of a single analogue system has the potential to give the most accurate yield predictions if the analogue and prediction year are similar in extreme years, but also be most inaccurate if the opposite anomaly occurs. Accordingly, the Lamond analogue system was most affected by a large deviation in yield in one or two years, which counteracted other more accurate forecasts. Also, yield prediction errors based on one analogue are magnified mid-season if the rainfall distribution is markedly different between the analogue and prediction year. This can happen in two ways. Firstly, if the predicted sequence of above and then below average rainfall is actually reversed in order within the observed growing season, and secondly, if the observed early and late season anomaly are both more extreme than predicted when the opposite anomaly is forecasted at each end of the growing season. In these cases, growing season rainfall may be accurately predicted, but mid-season yield predictions differ markedly from the final yield. To overcome these problems a set of analogues derived from this yield forecasting system should be investigated.

To put the results for this ten year period in context, a key indicator of Australian yields, the mean SOI in winter (Rimmington and Nicholls, 1993; Garnett and Khandekar, 1992), was correlated to average Australian and Queensland wheat yields for 10-year intervals from 1936-45 to 1986-95 (Figure 5). While linear correlation is an overly simplistic approach for yield prediction, the results of the association are useful to highlight the considerable decadal variability that is possible. Figure 5 shows that the study period chosen had the lowest correlation between SOI and national wheat yield over the last six decades, whereas the opposite occurred when the analysis was restricted to Queensland. This may be reflected in the slight improvement in prediction error for Queensland in June when the SOI phase forecast rainfall is used. Decadal variation in the SOI is well known (Trenberth and Shea, 1987) and current research is exploring possible causes (Latif et al., 1997). The results found in the long-term simulation with the STIN model (Figure 2) suggest clear shifts in yield distribution are associated with the April-May SOI phase when applied over a longer period. The range of outcomes in each phase remains broad, however, and this may limit the utility of this finding in operational forecasting.

Although the results of this study suggest little or no improvement was gained by employing seasonal climate forecasts in forecasting the Australian wheat crop, it is premature to reach a general conclusion. The STIN agroclimatic index model is more biophysically robust than the WRI model and the hindcast study using it showed significant shifts in yield distributions associated with differing phases. A similar study using a longer test period and the STIN model is needed before any general conclusions can be reached. The STIN model, however, is more difficult to run operationally than WRI due to the need for reliable real-time individual station rainfall data.

In operational production forecasting, there has been a tendency to seek a categorical estimate, when, clearly, the uncertain season ahead (or what remains of it) has a large impact on the outcome. The best that can be achieved realistically with the current state of knowledge is an estimate of the likely distribution of production. The expected value of such a distribution would be the best estimate for those seeking to avoid

364

considerations of risk, but the full distribution contains significantly more information for those interested in risk management. For marketing and transport agencies that need an actual yield estimate the middle value from a range of selected analogues would be the most appropriate to use and compare with a forecast based on long term average rainfall. The full range of predictions based on each year of historical weather data should be compared to the distribution of forecasts using a set of analogues from a seasonal forecast. This would provide the ultimate test of utility of the forecasting system in production forecasting and is the target of future work.

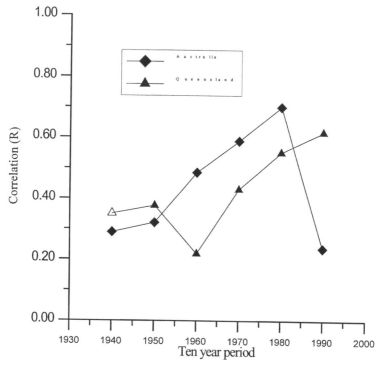

Figure 5. Correlation of either annual average Australian or Queensland wheat yield with mean winter SOI (June, July, August) for consecutive ten year periods from 1936 to 1995.

Conclusions

The hindcast testing of production forecasting methods showed that using the WRI model to update predictions throughout the season using rainfall to date and average rainfall for the remainder of the season was generally superior to the traditional ABARE forecasts. Mean error of the forecast over the 1986-95 period was reduced by 25-30% throughout the season. The additional use of a seasonal rainfall forecast in deriving production estimates via the WRI model produced no clear advantage. However, the simulation analysis of historical production using the STIN model showed clear differences in the median production for different phases of the SOI in April-May and these were consistent throughout the country. There was a tendency for lower yields in phases one and three (SOI consistently negative; rapidly falling) and higher yields in phases two and four (SOI consistently positive, rapidly rising). The inability of this effect to improve forecasts using the WRI model may have been due to the short duration and unrepresentativeness of the test period, differences between the WRI and STIN models, and the focus on mean outcomes rather than distributions of outcomes. Further studies to examine these aspects are needed to better test the utility of current seasonal forecasting capability.

Acknowledgments

We thank Land & Water Resources Research & Development Corporation for partial funding of this work, Dr Roger Stone (APSRU) for providing SOI phases for all years of the historical record, and Mr Mal Lamond (Lamond Weather Services) for providing details of historical analogues.

References

AACM (1991) Review of crop forecasting systems. Australian Agricultural Consulting and Management Company Pty. Ltd., Adelaide. pp.35.

ABARE (1995-1997) Crop Report. No. 35-103. Australian Bureau of Agricultural and Resource Economics, Canberra.

ABARE (1997. Australian Commodity Statistics. Australian Bureau of Agricultural and Resource Economics, Canberra, pp.346.

Clewett, J.F., Clarkson, N.M., Owens, D.T. and Abrecht, D.G. (1994) Australian Rainman: Rainfall information for better management. Queensland Department of Primary Industries, Brisbane.

Forrest, J.A., Beatty, J., Hignett, C.T., Pickering, J. and Williams, R.G.P. (1985) A survey of the physical properties of wheatland soils in eastern Australia. CSIRO Division of Soils, Report No. 78, Adelaide. pp.49.

Frere, M. and Popov, G.F. (1979) Agrometeorological crop monitoring and forecasting. FAO Plant Production and Protection Paper No. 17. FAO, Rome. pp.64.

Garnett, E.R. and Khandekar, M.L. (1992) The impact of large-scale atmospheric circulations and anomalies on Indian monsoon droughts and floods and on world grain yields - a statistical analysis. *Agric. For. Meteorol.* **61**, 113-28.

Hamblin, A.P. and Kyneur, G. (1993) Trends in wheat yields and soil fertility in Australia. Bureau of Resource Sciences, Canberra. Australian Government Publishing Service. pp.141.

Hammer, G.L. and Muchow, R.C. (1991) Quantifying climatic risk to sorghum in Australia's semiarid tropics and subtropics: Model development and simulation, in R.C. Muchow and J.A. Bellamy (eds.), *Climatic Risk in Crop Production: Models and Management for the Semiarid Tropics and Subtropics.* CAB International, Wallingford, UK. pp.205-32.

Hammer, G.L., Holzworth, D.P. and Stone, R. (1996) The value of skill in seasonal forecasting to wheat crop management in a region with high climatic variability. *Aust. J. Agric. Res.* **47**, 717-37.

Hutchinson, M.F. (1991) The application of thin plate smoothing splines to continent-wide data assimilation, in J.D. Jasper (ed.), Data Assimilation Systems. BMRC Research Report No. 27. Bureau of Meteorology, Melbourne. pp.104-13.

Latif, M., Kleeman, R. and Eckert, C. (1997) Greenhouse warming, decadal variability, or El Nino? An attempt to understand the anomalous 1990s. *J. Climate* **10**, 2221-39.

McBride, J.L. and Nicholls, N. (1983) Seasonal relationships between Australian rainfall and the Southern Oscillation. *Mon. Weath. Rev.* **111**, 1898-2004.

Motha, R.P. and Heddinghaus, T.R. (1986) The Joint Agricultural Weather Facility's operational assessment program. *Bull. Amer. Meteor. Soc.* **67**, 1114-22.

366

Nicholls, N. (1985) Impact of the southern oscillation on Australian crops. *J. Climatol.* **5**, 553-60.

Nicholls, N. (1990) Predicting the El Nino-Southern Oscillation. *Search* **21**, 165-7.

Nix, H.A. (1975) The Australian climate and its effects on grain yield and quality, in A. Lazenby and E.M. Matheson (eds.), *Australian Field Crops Vol. 1 Wheat and other Temperate Cereals*. Angus and Robertson. Sydney. pp.183-226.

Nix, H.A. and Fitpatrick, E.A. (1969) An index of crop water stress related to wheat and grain sorghum yields *Agric Meteor.* **6**, 321-7.

Northcote, K.H., Hubble, G.D., Isbell, R.F.,Thompson, C.H. and Bettenay, E. (1975) *A description of Australian soils*. Wilke and Company, Clayton, Vic. pp.170.

O'Donnell, V., Curran, B., Martin, P., Podbury, T. and Knopke, P. (1998) Farm performance and income variability in Australian Agriculture. Proceedings of the National Agricultural and Resources Outlook Conference, Canberra, February 1998, Vol. 2, Agriculture. ABARE, Canberra. pp.3-21.

Pittock, A.B. (1975) Climatic change and the patterns of variation in Australian rainfall. *Search* **6**, 498-504.

Rimmington, G.M. and Nicholls, N. (1993) Forecasting wheat yields in Australia with the southern oscillation index. *Aust. J. Agric. Res.* **44**, 625-32.

Ritchie, J.T. and Otter, S. (1985) Description and performance of CERES-Wheat: A user oriented wheat yield model, in W.D. Willis (ed.), ARS Wheat Yield Project. United States Department of Agriculture, Agricultural Research Service. ARS-38. pp.159-75.

Ropelewski, C.F. and Halpert, M.S. (1987) Global and regional scale precipitation patterns associated with the El Nino-Southern Oscillation. *Mon. Wea. Rev.* **115**, 1606-26.

Ropelewski, C.F. and Halpert, M.S. (1989) Precipitation patterns associated with the high index phase of the Southern Oscillation. *J. Climate* **2**, 268-84.

Smith, I. (1994) Assessments of categorical rainfall predictions. *Aust. Meteor. Mag.* **43**, 143-51.

Stephens, D.J. (1995) Crop yield forecasting over large areas in Australia. Unpubl. Ph.D. Thesis. Murdoch University, Perth, Western Australia. pp.305.

Stephens, D.J. (1996) Using seasonal climate forecasts for national crop outlook. Proceedings of Managing with Climate Variability Conference, Canberra, November, 1995. LWRRDC Occasional Paper CV03/96. pp.111-5.

Stephens, D.J. and Lyons, T.J. (1993) Operational forecasting of the Australian wheat yield using district rainfall and seasonal outlooks. 4th International Conference on Southern Hemisphere Meteorology and Oceanography, Hobart, 29 March - 2 April, 1993. American Meteorological Society. pp.153-4.

Stephens, D.J. and Lyons, T.J. (1998) Rainfall-yield relationships across the Australian wheatbelt. *Aust. J. Agric. Res.* **49**, 211-3.

Stephens, D.J., Lyons, T.J. and Lamond, M.H. (1989) A simple model to forecast wheat yield in Western Australia. *J. Roy. Soc. West. Aust.* **71**, 77-81.

Stephens, D.J., Walker, G.K. and Lyons, T.J. (1994) Forecasting Australian wheat yields with a weighted rainfall index. *Agric. For. Meteor.* **71**, 247-63.

Stone, R.C. and Auliciems, A. (1992) SOI phase relationships with rainfall in eastern Australia. *Int. J. Climatol.* **12**, 625-36.

Stone, R.C. and de Hoedt, G. (2000) The development and delivery of current seasonal climate forecasting capabilities in Australia, in G.L. Hammer, N. Nicholls, and C. Mitchell (eds.), *Applications of Seasonal Climate Forecasting in Agricultural and Natural Ecosystems – The Australian Experience*. Kluwer Academic, The Netherlands. (this volume).

Stone, R.C., Hammer, G.L. and Marcussen, T. (1996) Prediction of global rainfall probabilities using phases of the Southern Oscillation Index. *Nature* **384**, 252-5.

Trenberth, K.E. and Shea, D.J. (1987) On the evolution of the southern oscillation. *Mon. Wea. Rev.* **115**, 3078-96.

Walker, G.K. (1989) Model for operational forecasting of Western Canadian wheat yield. *Agric. For. Meteor.* **44**, 339-51.

Watson, I.A. and Butler, F.C. (1984) Wheat rust control in Australia - National conferences and other initiatives and developments. University of Sydney, Sydney. pp.80.

CAN SEASONAL CLIMATE FORECASTS PREDICT MOVEMENTS IN GRAIN PRICES?

SCOTT CHAPMAN[1], ROBERT IMRAY[2] AND GRAEME HAMMER[3]

[1]CSIRO Tropical Agriculture
306 Carmody Rd
St. Lucia, Qld 4067, Australia

[2]FarMarCo Enterprises (Aust.) Pty Ltd
Herries St
Toowoomba, Qld 4350, Australia

[3]Agricultural Production Systems Research Unit (APSRU)
Queensland Department of Primary Industries
PO Box 102
Toowoomba, Qld 4350, Australia

Abstract

The worldwide supply and demand for internationally traded commodities are important determinants of their market price. Australian producers are increasingly attempting to control price risk as part of their enterprise management. As the El Niño/Southern Oscillation (ENSO) phenomena has been shown to affect weather, (particularly rainfall) and crop production in different countries, one might expect measures of ENSO to have some relationship with supply and demand for agricultural commodities, and therefore price. Information about such relationships might be used to assist decisions about controlling price risk, via price hedging, for example. In this paper we make a preliminary analysis of relationships between an ENSO indicator (the phases of the southern oscillation index (SOI)) and long-term datasets of wheat 'prices received' and wheat 'December futures prices'. Within each year, prices were expressed as a percentage of the price in December.

On an annual basis in the 'prices received' dataset, the average pattern is that price relative to December begins high, falls during June to September (as the US crop comes in) and rises again toward December. The actual pattern varies from year to year, with no apparent consistency over sequences of years. However, in the 13 years (of 88) when the SOI was phase 1 (consistently negative) in June/July, wheat prices always increased 6 months later. This also applied for 11 of the 13 years when the SOI phase was 1 in April/May. This is a critical time for Australian producers to consider price hedging of their 'soon-to-be-planted' wheat crops.

G.L. Hammer et al. (eds.), The Australian Experience, 367–380.

Similar patterns were found in the Chicago Board of Trade (CBOT) December futures contracts for 1949 to 1996. In different years, the January price of the contract was only 50% of, or up to 50% more than, the final (spot) price. The average pattern in the CBOT prices for all years was broadly similar to that in the prices received data, except that price movements began earlier (March compared with May). In years when the SOI was in phase 1 in April/May, the relative price changes in the futures contract also tended to increase until December.

One explanation for the price effects is related to the fact that in years when the SOI phase is 1 in April/May there is a greater probability that many areas of Australia, China and India are likely to receive lower than median rainfall during June to September. The opposite is the case for the USA. If rainfall is lower in Australia, the export crop will be decreased, while the high rainfall at this time in the USA can have a negative effect on grain harvesting and quality - both factors that may raise the actual and futures prices.

In years when the SOI phase is 1 in April/May (e.g. 1995) the futures contracts follow the reverse pattern to that in an ideal year for Australian hedgers (1996). In years like 1995, producers could modify their hedging strategy. For example, they might reduce the amount of crop hedged in anticipation of both a smaller crop from their own farm, and the increased likelihood of a sustained price rise through the season. Alternative price derivative products can take more advantage of these effects to improve the coverage of price risk. In the meantime, other indicators of ENSO could be analysed against relevant price data to accommodate the effects of price support and subsidies, or they could be used to examine likely patterns of commodity supply and demand as inputs to trade models.

1. Introduction

In non-subsistence agriculture, where producers operate in a cash economy, farm profit is determined by both commodity price and production. Climate research as related to agriculture, has concentrated on production and resource impacts of climate variability (eg. Meinke and Hochman, 2000; Ash et al., 2000). However, price variability is globally recognised as a driver of seasonal returns and long-term profitability of modern farms. Evidence for this at the farm level is the increasing use of price hedging strategies to protect returns on current and future crops, and at national and international levels, the continuing use of commodity price subsidies to ensure farms remain viable for production and/or protection of resources.

Variability in climate, principally rainfall, impacts greatly on agricultural production, particularly that of dryland broad-acre commodities. The timing of both positive and negative influences of ENSO on rainfall varies cross the globe, e.g. Stone et al., (1996). For example, in El Niño years, effects on rainfall have been shown to carry through to influence the size of crop production in terms of both area and yield per unit area

(Nicholls, 1986; Garnett and Khondeker, 1992; Cane *et al.*, 1994; Stephens *et al.*, 2000).

If one takes a global view, the supply and demand of a commodity plays a major role in price discovery by the market (Cramer and Heid, 1983). The general level of supply of a crop is affected by price expectations, but at the time of planting, the producer cannot know the final price to be received. After a crop is planted, it is largely the weather that will determine the final crop surplus and the eventual price received. If, over time, all dryland crops experienced the same weather, supply would be relatively stable as would the final price. While global demand for an international commodity may be more predictable than supply, it is the variation (real and expected) in the spatial and temporal distribution of factors affecting world demand and supply that is the key determinant of change in seasonal prices.

At the local level, market participants attempt to anticipate the size and timing of global and national supply and demand and make economic decisions based on the analysis of the likely effects on price and price changes. Many of the top grain producers, especially in northeastern Australia are now using futures contracts or derivatives to stabilise their returns. At the time of writing , we could find no articles that examine direct relationships between ENSO indicators and actual or futures prices of agricultural commodities. This information may assist the decision process of those who use futures markets to manage price risk (hedgers) and those who accept the risk (speculators). Australian producers also stand to benefit, being exporters of dryland-grown, ENSO-influenced commodities.

2. Managing Price Risk

An estimate of the expected market price at harvest needs to be made when considering mixtures of enterprises and decisions made on-farm during the season. One method of managing the potential variability in this expected price is hedging in a commodity futures market. Kaufmann (1986) defined hedging as "establishing a position in a futures contract approximately equal, but opposite to an already existing or anticipated net cash position to protect profit margins against an adverse change in price". Ederington (1979) reasons that "the classic rationale for futures markets is that they ... allow those who deal in a commodity to transfer the risk of price changes in that commodity to speculators who are more willing to bear such risks".

Hedging is done by both producers and users of a commodity to stabilise a selling/buying price at an acceptable level that minimises their exposure to price fluctuations (eg. Imray, 1996). In the case of futures markets, hedgers enter into a future contract prior to delivery in order to "lock in' a price. However, they need to take an offsetting action at the time of physical sale/purchase to create a cashflow, either positive or negative, called 'the hedge'. Price variability is minimised by the fact that the cost of the hedge fluctuates opposite to the corresponding change in the underlying cash value of the commodity over the period of the hedge. The main benefit of using futures is that they substitute for the cash market. For producers (see example later), this usually means that they have protected themselves against a fall in the delivery price since the time of taking the futures contract, but will have foregone additional profit if the price at the time of delivery has risen. The key function is that producers or users can lock into a price with which they are 'comfortable' and which minimises their exposure to adverse market movements. More comprehensive reviews of the futures market literature and the theory of hedging and futures markets are available (Gray and Rutledge, 1971; Kamara, 1982; Kolb, 1997). Note particularly that as hedging is often undertaken in overseas markets, there are many other influences, such as currency exchange rates, that are not discussed here. The 'basis risk' (that due to using different currencies as a basis for a contract) is an additional concern for Australian hedgers, but it is not considered here.

Given that global grains surpluses and the resulting price movements are difficult to predict, we aim to investigate whether relationships exist between ENSO indicators and the market and futures prices of one of the most traded commodities, wheat. Our intention is not to evaluate an entire hedging strategy, but rather to determine relationships between ENSO-based seasonal forecasts and, firstly, wheat prices and, secondly, wheat futures prices. The implications for managing producer price risk and the implications for Australian production are discussed.

3. Relationships Between ENSO Indicators and Actual Prices

3.1 ENSO INDICATOR DATA

The monthly SOI phase classifications from 1900 to 1996 (Stone and Auliciems, 1992) were applied to classify the data into different years to examine possibilities for forecasting percentage changes in prices. These phases classify two consecutive months (e.g. June/July) into one of five categories and have been widely used as predictors of the likelihood of different levels of rainfall over an ensuing three month period (e.g. Stone *et al.*, 1996). The five phases are: SOI consistently negative (1); SOI consistently positive (2); SOI rapidly falling (3); SOI rapidly rising (4) and SOI near zero (5). Prices were then analysed in terms of how they changed given the SOI phase at any time within a year.

3.2 PRICE DATA AND ANALYSIS

The wheat price data used was monthly averages of the "Prices Received by Farmers Historical Index (1908 to 1996)" obtained from the USDA Economics and Statistics Service (http://usda.mannlib.cornell.edu/usda/usda.html). Two types of analyses were performed on the data. In a continuous analysis of the "Prices Received" data, we calculated a percent change in price, given a six month lag. In other words, relative to the price at any time of the year, we calculated how much the price would increase or decrease in the next 6 months. Years of data were grouped into one of five groups according to the SOI phase for each time of year, ie. for each time of year and SOI phase, there was a distribution of price changes (from different years). For each time of year, a Kolmogorov-Smirnov goodness of fit test[1] was used to compare the significance of differences in distributions of price change among SOI phase groups. In a second analysis, we assumed that trading ended in December (the major delivery month for many Australian producers), and determined for any month whether the current price was likely to increase or decrease in the time until December.

Note that in all of the analyses presented, the data were not de-trended for the long-term rise in prices (due to inflation). The assumption was that this long-term effect would be of no consequence as the phases were well distributed over the length of the records, and would be greatly lessened by the conversion of monthly prices to a ratio of the December price. A more detailed analysis may benefit from de-trending, although there are substantial assumptions required (including how to determine the purchasing power of dollars received by farmers) in de-trending of such long records.

3.3 SOI PHASES AND WHEAT PRICES RECEIVED WITH 6 MONTH LAG

After examining each calendar month of data, for the 88 years from 1908 to 1995, we found that there was some forecasting skill at certain times of year when the SOI phase was consistently negative (phase 1). In all 13 years when the SOI phase was 1 in June/July, the wheat prices increased 6 months later (Figure 1a). In all but 2 of these years, the same effect was noticed when the SOI phase was 1 at any time until Oct/Nov (data not shown). The effects were still present when data from years prior to 1960 were excluded; that is when world commodity trade was relatively small. Figure 1b shows density distributions of the Figure 1a data, for each SOI phase in June/July. The distribution (and the mean) of the effect in years when the June/July SOI phase was 1, is significantly different ($p < 0.1$) from each of the other distributions. A supply-related explanation may exist for the positive relationship between phase 1 SOI in the southern hemisphere winter/spring and the 6 month lag US wheat price. The phase 1 SOI during this time is associated with a low likelihood of exceeding median spring rainfall in much of the Australian wheatbelt (particularly the eastern region) and in much of wheat producing areas of southern and northern Asia (Stone *et al.*, 1996). The likelihood of the USA and Canada receiving greater than normal rainfall is also high, potentially causing problems for harvesting and decreasing grain quality. While the effects on importing countries and in reducing Australian supply would not be evident until about September or October, it appears to be indicated by the SOI phase as early as June/July.

3.4 WHEAT PRICES RECEIVED, RELATIVE TO DECEMBER PRICE

Using the same dataset as above, we found that between 1908 and 1995, the number of years in which wheat prices either increased or decreased between January and the following December was the same (44 years). Figure 2 shows the price, relative to December averaged across all years (1st line). This follows a "textbook definition"

---

[1] Using S-Plus 4.5 software, Mathsoft Inc.

(Cramer and Heid, 1983) of price trends for a crop that is being harvested mid-year in the USA: the price begins high as the only supplies are from import or stocks, and then the price decreases under "harvest pressure" in April/May before slowly coming back in the last 3 to 4 months of the year. The month of May is then a critical decision point for Australian producers to be considering price risk, and so we concentrate on the phase value in April/May for the remainder of these analyses.

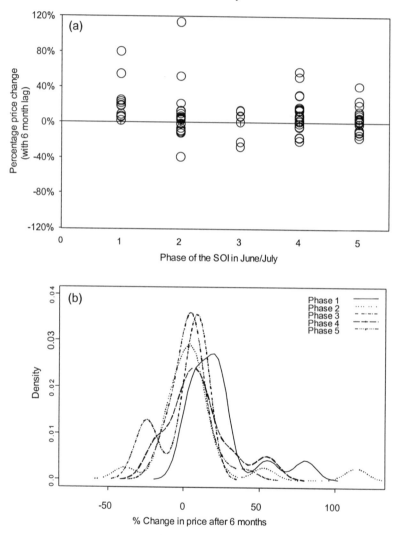

Figure 1. Percentage change in the US 'wheat price received' for each year from 1908 to 1995, grouped by the SOI (Southern Oscillation Index) phase existing in June/July.

(a) The percentage change determined as the difference between the current price and the price in 6 months time (ie. Dec/Jan), as a fraction of the current price. The SOI phases are consistently negative (1); consistently positive (2); falling (3); rising (4) or near zero (5) (Stone and Auliciems, 1992).

(b) Density plot of the distributions of price changes for each SOI phase in (a). The phase 1 distribution is significantly different to phase 4 at p<0.1, and to the other phases at p<0.05

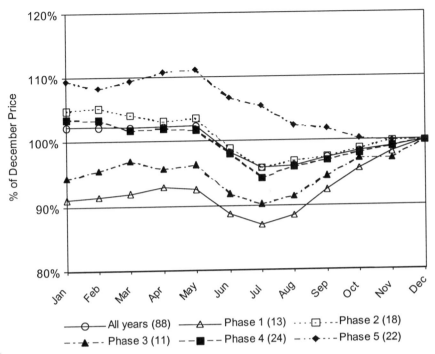

Figure 2. US wheat price received in January as a percentage of the price in the following December averaged across all years from 1908 to 1995 or averaged across years grouped by the SOI phase in April/May. Phases are as in Figure 1, and numbers in parentheses represent the number of years in each phase.

The other curves in Figure 2 are the averages across years as grouped by the phase of the SOI during April/May, with the number of years (out of 88) indicated in the legend. The shapes of the price curves of all phases, except phase 5 (SOI near zero) are similar to that of the textbook "all years" average. When the April/May SOI phase is 5 (25% of years), the average price over years began high, but continued to decrease through all months to December. When the SOI phase is near zero, expectations of receiving median rainfall are about 50% virtually worldwide (Stone *et al.*, 1996). One would expect that local supply would be relatively stable in these years and that imports would be in reasonable supply.

In the 48% of years in which the April/May SOI was phase 2 (positive) or 4 (rising) the price changes tracked the "all years" case. In these types of years, rainfall prospects for the USA are less than optimal although producers in Australia experience a better than 50% chance of receiving median rainfall. In contrast to the effects of phases 2, 4 or 5, in years in which the April/May SOI phase was 1 (negative) or 3 (falling) the price in January to May was actually *below* the final December price, although the prices still experienced a dip in the months of June to August. The occurrence of phases 1 and 3 appears to predict seasons in which there will be a shortage of wheat and hence price is likely to rise continuously during the latter part of the year.

As all of the data in Figure 2 are averages, they do not represent what happens in any single year. In all cases, there were exceptions to the average price patterns (individual year data not presented). The most repeatable price pattern occurred in years in which the April/May SOI was in phase 1. In 11 out of 13 years (excepting 1912 and 1940), the prices rose between January/May and December.

4. Relationships Between ENSO Indicators and Futures Prices

4.1 A TYPICAL HEDGING STRATEGY

One possible hedging strategy of Australian wheat farmers is to use the December contract for Chicago Board of Trade (CBOT) wheat futures[2]. In the simplest application of this strategy, a farmer aims to sell contracts to 'lock-in' at a price acceptable to the farmer and trades out (buys back) these contracts at about the time the physical sale of the commodity takes place. This stabilises their returns at an acceptable price early in the season, whether the price then moves up or down. As farmers rarely actually deliver (grain marketers do) to Chicago, the returns they receive from their wheat crop are a combination of the price received for the crop upon sale (usually similar to the final 'spot' price of the contract) and the cost of hedging, which will reflect any price difference (profit or loss) between the time that a futures contract is taken and when the farmer trades out or buys back the contract. At harvest, the futures contract will be similar to the market price, so the combination of the market price and the hedge gives the farmer the original 'acceptable' price. It is important to realise that the objective of hedging for producers is to stabilise returns, although in years when the price falls during the season, they stand to receive a more favourable return than producers who do not hedge.

Figure 3 shows the price of the December CBOT wheat futures contract in 1995 and 1996. The 1996 season (SOI phase 4 during most of winter and spring) was 'ideal' for hedging as the farmer could sell contracts at higher prices during April to June, prior to planting and prior to the final release of US production figures. These contracts would then be traded out at, or close to, harvest. For 1996, such a strategy would have realised a price differential of about 2.50 (USD per bushel), or about a 50% improvement over the prices received in December at the end of the contract. In contrast to 1996, the price in 1995 continued to increase all year. While a hedge to lock-in a price would still have been appropriate at a price acceptable to a producer, such a hedge would have foregone a price improvement of about 40%. Our analysis examines relationships between SOI indicators and futures prices, when considering the opportunity for hedging near planting time (in Australia).

---

[2] While the Sydney Futures Exchange now also trades wheat futures contracts, we concentrated on the CBOT because of its longer data record.

4.2 PRICE DATA AND ANALYSIS

The wheat futures data were the closing prices of December CBOT Futures contract from 1949 to October 1997. These data were processed[3] to generate a contract price for each trading day. The data were used raw, or were averaged over trading days for each month.

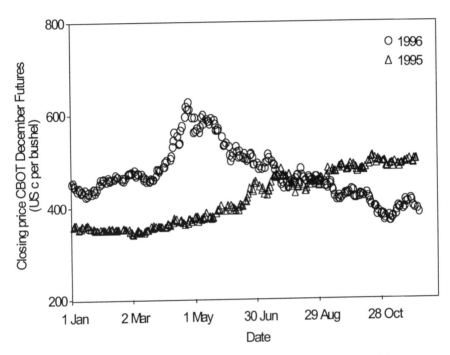

Figure 3. Closing price of the December CBOT wheat futures contract during 1995 and 1996.

As trading in this contract ends in December, we aimed to determine whether the current price of the contracts was likely to increase or decrease in the time between any month and December. To remove the effects of long-term trends, monthly prices and contracts in each year were expressed as a percentage of the price on the 1st December trading day (the approximate spot price) for that year.

4.3 ANNUAL CBOT DECEMBER WHEAT FUTURES CONTRACT, RELATIVE TO FINAL CONTRACT

The average patterns in the CBOT prices were broadly similar to those in the prices received data (Figure 2), but with some price movements beginning earlier (March

---

[3] Commodity Systems Inc. Unfair Advantage® software

compared with May) and a smoothing of the price patterns evident (Figure 4). In different years, the January price of the CBOT December wheat futures contract was up to 50% more than or 50% less than the final spot price (Figure 5). In 5 of the 10 years when the April/May SOI phase was 5 (near zero), the price decreased between May and December while in the other 5 years it changed little throughout the year (Figure 6).

In years when the April/May SOI was phase 1, the relative price changes in the futures contract (Figure 7) followed a similar pattern to that identified in the same years in the prices received data (data not presented). In these years, May prices were below the final price in five years out of seven, but usually dipped lower in July/August, before rising through September to December. As for the prices received data, one general explanation is that price peak normally comes about prior to the US crop being harvested, with the price falling after the US crop is known (this effect can be seen when data are averaged over all years – data not presented). The price does not begin to rise again until Sep/Oct when the market 'realises' that other crops (including Australia) are likely to be lower than normal. The SOI phase 1 in May/June and onwards, has foreseen this effect two months ahead of the first actual forecasts (in August), but in past years, the futures market does not appear to have anticipated the rising market price.

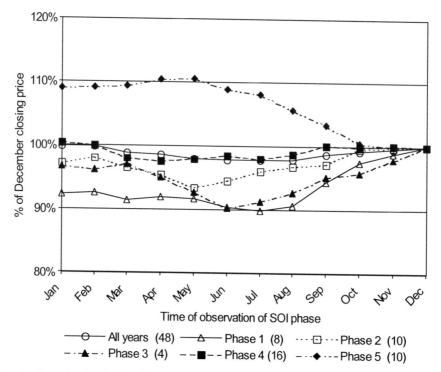

Figure 4. December CBOT wheat futures contract price as a percentage of the price in December averaged across all years from 1949 to 1996 or averaged across years grouped by the SOI phase in April/May. Phases are as in Figure 1, and numbers in parentheses represent the number of years in each phase.

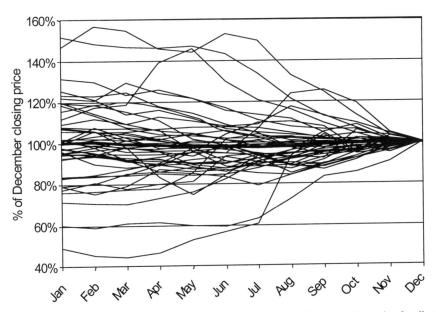

Figure 5. December CBOT wheat futures contract price as a percentage of the price in December for all years from 1949 to 1996.

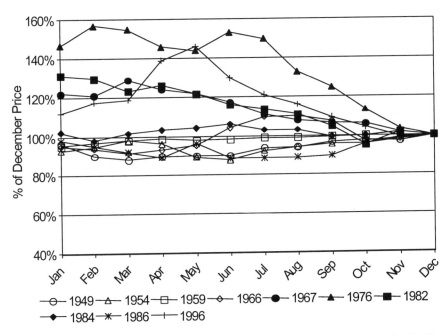

Figure 6. December CBOT wheat futures contract price as a percentage of the price in December for all years from 1949 to 1996 when SOI phase in April/May was 5 (near zero).

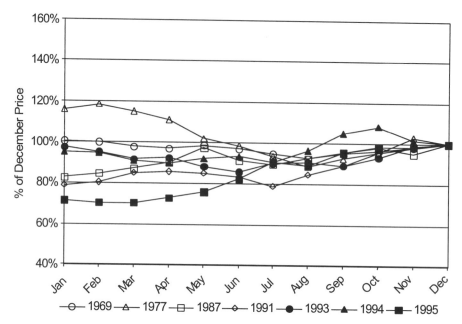

Figure 7. December CBOT wheat futures contract price as a percentage of the price in December for all years from 1949 to 1996 when SOI phase in April/May was 1 (consistently negative).

5. Implications for Managing Price Risk

In years when the SOI phase is 1 in April/May (Figure 3, 1995), the futures contracts follow the reverse pattern to that generally preferred in an ideal year for Australian hedgers (eg. 1996). One reaction in years like 1995 would be to either hedge a lower percentage of production or not to hedge at all in anticipation of higher values at harvest time. This would depend greatly on the futures value at the time. Given that in SOI phase 1 years, production per unit area is more likely to be lower than in 'average' years, then the amount of the crop hedged should be more conservative as the risk of crop failure increases. Hence, producers can use ENSO indicators to modify decisions in anticipation of both reduced production and increased price. Importantly, they can avoid being 'over-hedged'.

There appears to be some consistent skill in forecasting changes in the CBOT futures (December) contract. However, this skill only exists in years in which the SOI phase is near zero or negative during April/May. Skill might also be obtained through the use of other pricing/hedging tools that do not depend on a simple hedge/non-hedge approach. These are not considered here, but could be investigated. Note that while we refer to consistent skill, there appears to be partial skill for some of the other SOI phases that might still be leveraged by market price tools.

The near zero years are ones in which rainfall, and production is likely to be about average in most countries, with the contract price falling throughout the year in about half of the years (Figure 6). In years when the SOI phase is negative, rainfall is likely to be less than average in the areas of wheat production for some US competitors (especially Australia) and some of the major wheat importers (Asia). Other phases and other times of year do not seem to offer consistent skill, as in different years, prices changed in either direction (data not shown). The SOI phase 1 effect on the futures contract has begun to fade by Aug/Sep, but offers opportunities to make decisions in the three months prior to this.

The effects of SOI phase 1 warrant further investigation as it and phase 3 (SOI falling) have been shown to be the most discriminating in terms of changes in the probability of receiving different levels of rainfall around the world (Chapman and Stone, 1997). For example, a consistently negative SOI (phase 1) during April/May or May/June is associated with a low probability of exceeding median rainfall in the ensuing three months in Australia (wheat exporter), China, India, SE Asia and Central America (often importers). At the same time, this phase is associated with a high probability of exceeding median rainfall by harvest time in Europe, USA, Canada and Argentina (all exporters).

Futures contracts are a useful tool in managing price risk and are well suited to analysis against variables like SOI phases. Although the contracts are renewed each year and so appear to be independent, our analysis may be improved by removing long-term price trends. Historical price records of contracts offered at different times of the year could also be investigated, for relationships to SOI phases or other indicators of ENSO. Of course, it may be too late now to take advantages of some of these analyses, as the global markets become more greatly aware of the impact of ENSO during 1997.

In this paper, we have undertaken only a preliminary analysis of the direct relationships between SOI phases and price, having considered that the phases have been shown to relate to potential production in some countries and commodities (Cane et al., 1994; Stephens et al., 2000) and that the balance of supply and demand determines price in a free market. Obviously, commodity markets, including those for wheat are complicated by price-support mechanisms and input subsidies. It may be that SOI phase or other indicators could be better utilised for markets by being used in conjunction with forecasting models to predict supply and demand of commodities, and allowing the price to be investigated by economic models.

References

Ash, A.J., O'Reagain, P.J., McKeon, G.M. and Stafford Smith, M. (2000) Managing climate variability in grazing enterprises: A case study of Dalrymple shire, north-eastern Australia, in G.L. Hammer, N. Nicholls, and C. Mitchell (eds.), *Applications of Seasonal Climate Forecasting in Agricultural and Natural Ecosystems – The Australian Experience*. Kluwer Academic, The Netherlands. (this volume).

Cane, M. A., Eshel, G. and Buckland, R. W. (1994) Forecasting Zimbabwean maize yield using eastern equatorial Pacific sea surface temperature. *Nature* **370**, 204-205.

Chapman, S.C. and Stone, R.C. (1997) Patterns in spatial rainfall forecasts as affected by the El Niño/Southern Oscillation, in G.O. Edmeades, M. Banziger, H.R. Mickelson and C.B. Peña-Valdivia (eds.), Proceedings of Symposium on Developing Drought- and Low N-tolerant Maize, 25-29 March 1996, CIMMYT, El Batan, Mexico.

Cramer, G. L. and Heid, W. G. Jr. (1983) *Grain Marketing Economics*. New York: John Wiley & Sons

Ederington, L. H. (1979) The hedging process of new futures markets. *Journal of Finance* **34**,157-170.

Garnett, E.R. and Khondeker, M.L. (1992) The impact of large scale atmospheric circulations and anomalies on Indian monsoon droughts and floods and on world grain yields – a statistical analysis. *Agricultural and Forest Meteorology* **61**, 113-28.

Gray, R. W. and Rutledge, D. J. S. (1971) The economics of commodity futures markets: A survey. *Review of Marketing and Agricultural Economics* **34**, 57-108.

Imray, R. J. (1996) An empirical analysis of minimum variance hedge ratios for raw sugar in the Far East: An Australian perspective. University of Southern Queensland, Toowoomba, Australia.

Kaufmann, P. J. (1986) *The Concise Handbook of Futures Markets*. New York: John Wiley & Sons

Kamara, A. (1982) Issues in futures markets: A survey. *The Journal of Futures Markets* **2**, 261-294.

Kolb, R. W. (1997) *Understanding Futures Markets*. 5th ed., Blackwell, New York.

Meinke, H. and Hochman, Z. (2000) Using seasonal climate forecasts to manage dryland crops in northern Australia – Experiences from the 1997/98 seasons, in G.L. Hammer, N. Nicholls, and C. Mitchell (eds.), *Applications of Seasonal Climate Forecasting in Agricultural and Natural Ecosystems – The Australian Experience*. Kluwer Academic, The Netherlands. (this volume).

Nicholls, N. (1986) Use of the Southern Oscillation to predict Australian sorghum yield. *Agricultural and Forest Meteorology* **38**, 9-15.

Stephens, D.J., Butler, D.G. and Hammer, G.L. (2000) Using seasonal climate forecasts in forecasting the Australian wheat crop, in G.L. Hammer, N. Nicholls, and C. Mitchell (eds.), *Applications of Seasonal Climate Forecasting in Agricultural and Natural Ecosystems – The Australian Experience*. Kluwer Academic, The Netherlands. (this volume).

Stone, R. C. and Auliciems (1992) SOI phase relationships with rainfall in eastern Australia. *International Journal of Climatology* **12**, 625-636.

Stone, R.C., Hammer, G.L. and Marcussen, T. (1996) Prediction of global rainfall probabilities using phases of the Southern Oscillation Index. Nature **284**, 252-255.

CLIMATE VARIABILITY, SEASONAL FORECASTING AND INVERTEBRATE PESTS – THE NEED FOR A SYNOPTIC VIEW

ROB SUTHERST

CSIRO Entomology
Long Pocket Laboratories, PMB No 3
Indooroopilly, Qld 4068, Australia

Abstract

There has been very limited exploitation in integrated pest management of the recently improved seasonal forecasting capability. Thus, this review of seasonal forecasting applications and opportunities is necessarily reliant on anecdotal observations. The nature of a particular population's response to variation in climate will depend on whereabouts it is located relative to the species' climate envelope. Hence, it is essential when dealing with biological organisms to take a synoptic view of their response to variation in climate.

Information that gives several months warning of an event can be used to improve strategic decisions such as the choice of crop variety, but it is less obvious how it can be used to improve decision making in entomology. The challenge in entomology is to identify situations where a climate-based seasonal forecast can add value to the decision-making process in a cost-effective way by demonstrating its advantage over insect sampling methods.

Some effects of climatic variation and the opportunities that are presented by the new seasonal forecasting capability are illustrated by examples, primarily involving the cattle tick, *Boophilus microplus*, in Australia but also other species. The potential changes in numbers and the most appropriate response strategies vary depending on the location of the pests relative to their geographical range.

1. Introduction

Despite the long history of interest by entomologists in the role of climate in the population dynamics of insects, there has been very little activity related to the exploitation of the recent, seasonal forecasting capability (Drake, 1994). Thus, this review of seasonal forecasting applications and opportunities is necessarily reliant on anecdotal data. The distribution, phenology and, to a lesser extent, relative abundance of insect populations are determined to large degrees by climate. The boundaries of the geographical distribution are determined largely by the limiting effects of climate on the species' population dynamics,

G.L. Hammer et al. (eds.), The Australian Experience, 381–397.

in the absence of limitations due to lack of hosts. The limiting effects become progressively more severe and of longer duration as the distance increases from the climatically benign core area of its range (Sutherst and Maywald, 1985; Sutherst et al., 1995; Sutherst, 1998). Meanwhile, the conditions favourable for growth decline in suitability and duration from the core to the edge of the distribution. Thus the nature of the species' response to variation in climate will depend on whereabouts the population of interest is located relative to the species' climate envelope. Populations in the core of the species' geographical distribution will be exposed to longer periods of more suitable conditions and shorter periods of adverse conditions than populations near the edges of the distribution. For example, populations whose growth and survival are limited by sub-optimal, low temperatures in temperate habitats will respond quite differently under a hot-dry scenario to others limited by high temperatures or dryness in tropical habitats. Hence, it is essential when dealing with biological organisms to take a synoptic view of their response to variation in climate. This implies the need for a constant awareness of the species' geographical distribution.

The year can be divided into two types of season, one that is suitable for population growth and the other during which the species has to survive adverse conditions (Southwood, 1977; Sutherst and Maywald, 1985). The overall abundance of a species is determined by the combined effect of both of these seasons. Seasonal variability of climate can therefore affect reproductive rates of one life-cycle stage in one season and survival rates of an immature stage in another, each of which contributes to the overall population density of the species at a given location. The CLIMEX model (Sutherst and Maywald, 1985; Sutherst et al., 1995) combines descriptions of both of these processes into an Ecoclimatic Index (EI) that summarises the response of a species to the average climate at a given location or specific climate in a given year. It can be used to investigate the effect of different seasonal conditions on insect abundance and geographical distributions.

2. The ENSO and its Climatic Implications for Insects

The potential role of climate variability in pest forecasting needs to be examined in the light of new predictive capabilities on phenomena such as the El Niño Southern Oscillation (ENSO) (Cane, 2000). Climate forecasters are demonstrating increasing levels of skill at forecasting the climate in the medium term, based on the Southern Oscillation Index (SOI), which measures the differential between the atmospheric pressure over Darwin and Tahiti (Stone and Auliciems, 1992). These authors defined five phases in the 2-month patterns of the SOI that related to rainfall patterns in eastern Australia (Stone et al., 1996):

> Phase 1 – Consistently negative SOI
> Phase 2 – Consistently positive SOI
> Phase 3 – Rapidly falling SOI
> Phase 4 – Rapidly rising SOI
> Phase 5 – Near zero SOI

Coupled with the use of sea surface temperatures in the Pacific and Indian oceans, information on the ENSO signal in the form of the SOI is providing opportunities to reduce the risks in agriculture in Australia (Hammer and Nicholls, 1996). Information that gives several months warning of a season type can be used to improve strategic decisions, such the choice of crop variety or whether to plant a crop at all, and can lead to better economic outcomes despite relatively low correlations between the indicators and actual rainfall (Hammer et al., 1996).

It is less obvious how medium term forecasting can be used to improve decision making in crop protection where many decisions are more tactical in nature. The challenge for pest managers is to identify opportunities to improve the management of pests, diseases and weeds in agricultural or environmental fields and parasites in medical and veterinary fields by harnessing this new capability.

Changes in the direction of the SOI have implications for both temperatures and rainfall, both of which are key drivers of pest population dynamics. As shown by McBride and Nicholls (1983), there is a positive correlation between the SOI and rainfall in northern and eastern Australia. Strongly negative SOI values are associated with below average rainfall and higher maximum temperatures, and *vice versa*. Annual changes in climate, as experienced during ENSO cycles, result in the displacement of the climate envelopes that determine the area that is climatically suitable for a species, with resulting impacts on local populations.

The effect of shifts in a species' climate envelope during periods of changed climate caused by ENSO-related phenomena is illustrated in Figure 1. Species 1 has a climate envelope that indicates a preference for a cooler, wetter climate than that required by Species 2. This corresponds in Figure 1 with a geographical distribution that lies to the southeast of Species 2. Given a change in climate to a hotter, drier El Niño season, the climate at the locations of each population will change, triggering quite different population responses at each location as the climate envelope shifts towards the lower right of Figure 1. Consider the 3 stationary populations of Species 1 at geographical locations A, B, C under the influence of an El Niño-type climate shift. Population A, in the core of the climate envelope, will remain in the comfort zone of Species 1 and exhibit a moderate response. Meanwhile population B, which was previously in a location that was marginal due to excessive heat and dryness, will fall out of the species' climate envelope as the latter moves to the lower right hand corner of Figure 1. Concurrently, the climate at the location of population C, which previously lay on the cool-wet extremity of the climate envelope, will become more suitable and the population is likely to increase greatly.

Similarly, in a cooler, wetter La Niña season, the climate envelope for Species 1 will shift to the top left corner of Figure 1. Population A will again remain in the comfort zone for the species, while population B will be exposed to a climate that is more suitable for it as population C is exposed to marginal, cold and wet conditions near the edge of the species' climate envelope. Meanwhile a population of Species 2 at site B will be advantaged by an El Niño season but will be displaced by a La Niña season, which is the converse to the experience of Species 1. These responses emphasise the differences due to both species

384

and locations that are so vital in determining the outcomes of large seasonal or longer shifts in climate caused by ENSO or by the enhanced greenhouse effect.

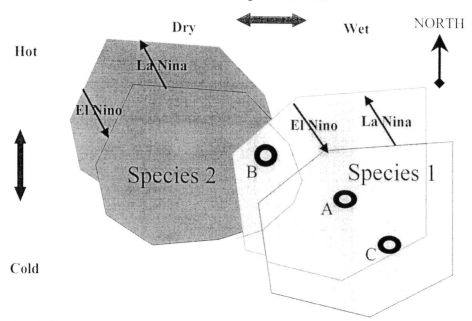

Figure 1. The effect of a shift in the climate space of two species with different climatic requirements during El Niño and La Niña events. A, B, C are fixed geographical locations with populations of Species 1, with species 2 also at Site B.

In years with a negative SOI the pest risk will increase in areas where a species is limited by inadequate thermal accumulation, while it will decrease in areas where it is limited by the availability of moisture or by excessive temperatures. By contrast, the pest will increase in years with a positive SOI in those parts of its range where moisture or high temperatures are limiting, while populations in areas made marginal by low temperatures will be reduced. Thus, it is possible that the SOI can be used to forecast high or low risk years for different pest species at specific locations where the limiting effect of temperature or moisture is mitigated by the change in climate. Hence, pest managers are faced with a range of responses that vary depending on both species and location. This provides a compelling reason for Australia, with its vast landmass, wide range of climates, numerous species of noxious species, and scarce resources, to adopt a national approach to understanding and managing its major pests, diseases and weeds.

There are a number of options available to entomologists for forecasting insect abundance at some future date. Traditionally, such forecasts have involved the identification of breeding populations of migratory insects in areas with low agricultural productivity prior to their invasion of valuable cropping areas (Hamilton *et al.*, 1994; Gregg *et al.*, 1995; McDonald, 1995; Walden, 1995; Maelzer *et al.*, 1996; Maelzer *et al.*, in press). In view of the dearth of other applications related to climate variability, some of the climate variability scenarios related to pest management are illustrated with examples from the cattle tick, *Boophilus microplus*, in Australia. Analogies with other pest systems are identified under each scenario and opportunities for improving the quality of decision making are identified.

2.1 CLIMATE SCENARIO 1: LA NIÑA, SOI POSITIVE, STONE AND AULICIEMS SOI PHASE 2

2.1.1 Summer Cool/Wet
A number of examples of likely impacts of a La Niña event on ticks and some other pests in Australia in summer, and potential responses are shown in Table 1. Seasonal variation in climate in tropical Australia is more likely to cause changes in tick numbers in the areas where temperatures and moisture are sub-optimal, ie. in inland areas. On the coast, temperatures are more moderate and rainfall is higher, so they would have to change more than inland to cause major fluctuations in tick numbers. While tick numbers on European breed cattle respond to changes in the tropical environment (Wharton *et al.*, 1969), variation in numbers on zebu-based breeds (Wharton *et al.*, 1969; Sutherst *et al.*, 1979) are dampened by strong negative feedback mechanisms (Bourne *et al.*, 1988) related to host resistance (Sutherst *et al.*, 1973) and animal grazing behaviour (Sutherst *et al.*, 1986). Thus La Niña forecasts would have less value for zebu-based breeds on the coast than for European breeds inland.

TABLE 1. Biological and management responses for climate scenario 1 - Summer Cool/Wet

Biological Response	Management Response
Location - Tropical Australia	
Cattle tick reproduction increases	Intensify surveillance & treatment on European breeds. Vaccinate zebu breeds for tick-borne diseases.
Buffalo Fly (*Haematobia irritans exigua*) numbers increase and spread	Treat moving cattle to prevent spread
Vector-borne diseases of humans and livestock increase in frequency	Prepare supplies of prophylactics
Location - Throughout Australia	
Arbovirus vectors increase	Increase surveillance for human arbovirus infections Vaccinate cattle against ephemeral fever.
Sheep blowfly (*Lucilia cuprina*) strike increases	Plan for pre-emptive treatment of sheep

During the mid-1970s there was a period when the SOI was strongly positive for three successive years with associated high summer and winter rainfall. During this period the buffalo fly, *Haematobia irritans exigua*, barrier that had existed at Bororen for decades was dismantled. Soon afterwards the fly spread far into south-western Queensland and gained a permanent foothold in the coastal area of southern Queensland and northern New South Wales, where it has now become permanently established (Williams *et al.*, 1985). Awareness of the abnormal SOI and its implications might have influenced the decision to remove the barrier and stop treating moving cattle to prevent the spread of buffalo fly. Agricultural industries need to be more receptive to measures aimed at preventing the spread of pests and weeds in particular.

The incidence of sheep blowfly strike is related to the weather, particularly persistent rainfall in the previous month, which wets the fleece to the skin (Wardhaugh and Morton, 1990). Fly numbers are also weather-dependent - but with a lag time of one generation (Vogt et al., 1985). Hence medium term La Niña forecasts would provide agricultural advisors and sheep graziers with some quantitative estimate of likely trends - especially if they depart markedly from the norm. If the La Niña forecasts were sufficiently reliable, the graziers could take pre-emptive action and jet their sheep earlier than normal to prevent an increase in fly numbers.

Human vector-borne diseases are more frequent in years with heavy summer rainfall that is often associated with La Niña events in Australia (Nicholls, 1986, 1993). Arboviruses in Australia are transmitted by a range of mosquito and midge species that increase in numbers in wet years when open water or high soil moisture increases survival and reproduction rates (Standfast and Maywald, 1992; Sutherst, 1994; Lindsay and McKenzie, 1997). Forecasts of light or heavy rainfall years could guide decisions on when and where to place most surveillance and vaccination efforts. However, caution is needed because the vectors and disease organisms are affected differently by temperature and moisture in different parts of their ranges as discussed below. In addition, irrigation is associated with transmission of vector-borne diseases (WHO, 1990), so knowledge of how much irrigation will be needed will assist managers target local high-risk areas. Forewarning of low rainfall can enable planning to divert parasite control effort into other areas.

2.1.2 Autumn-Winter Cool/Wet
Some examples of likely impacts of a La Niña event on ticks and some other pests in Australia in autumn-winter, and potential responses are shown in Table 2. An understanding of the ecology of pests in relation to climate is an essential foundation for reliable pest forecasting systems. The cattle tick provides an example of the value of such an understanding in providing a forecast several months in advance. The free-living stages of ticks develop on the soil surface in pastures. They are more successful in warm and wet weather. In winter, temperatures are limiting in southern areas and tick survival depends on receiving sufficient radiant heat to enable them to maintain a minimum metabolic rate. In wet winters, cloud cover combined with greater ground cover from pasture growth reduces the success of ticks breeding (Sutherst, unpublished data). At the same time, the rainfall improves the nutritional status of the cattle that are then able to maintain a higher level of host resistance (Sutherst et al., 1983). The result is that fewer ticks mature on cattle in the spring generation following high winter rainfall, with a consequent lower demand for acaricide. This has significant implications for management of chemical inventories as illustrated by the incident below.

The costs of storage of agricultural chemicals have a significant impact on profitability of agrochemical companies. If advance warning was available, supplies could be tailored more efficiently to meet differences in demand from year to year. In one incident an acaricide company shipped large quantities of product to Queensland on the expectation (based on advice received from another source) that winter rains would increase tick numbers in the following spring, only to find that sales were particularly low because the tick numbers were lower than usual. The resultant holding costs of stored inventories could have been averted if the company had had an understanding of the relationship between climate and tick population dynamics. Given that winter rains give more than adequate warning of the likely low tick numbers in spring, a seasonal climate forecast would add little value to that information. The value is in the knowledge of the relationship between tick survival, host immunity and winter rainfall. There is a need to disseminate such information to the cattle industry.

Unfortunately, agro-chemical companies claim that they need a forecast with an 85% accuracy to be available 6-12 months in advance before it will be useful for inventory management in the field crop industry dealing with *Helicoverpa* infestations (Davis *et al.*, 1997). Such a forecast would need to be based on the SOI or similar indicator but current accuracy and reliability is far below the required level.

TABLE 2. Biological and management responses for climate scenario 1 - Autumn-Winter Cool/Wet

Biological Response	Management Response
Location – central & southern Queensland, New South Wales	
Cattle tick reproduction decreases	Reduce surveillance
Host resistance to ticks increases	Minimise treatments
Helicoverpa breeding increases	Increase surveillance in spring
Location – southern Australia	
Exotic snails increase	Precision targeting of treatments
Early planting helps aphids increase	Pre-emptive treatments

In South Australia, the exotic white and conical snails (*Theba pisana, Cernuella* spp.) contaminate grain in spring and early summer, mostly as newly recruited juvenile snails (Baker and Hawke, 1988). Their numbers appear to be strongly dependent upon autumn rainfall, which dictates the success of the adult breeding season. High autumn rainfall indicates potential problems six months later and hence can be used to identify "optimal" times to use costly and environmentally harmful molluscicides. This example, like that of the cattle tick above, highlights the lag times between signals and responses that are common in integrated pest management (IPM). They provide inherent time lags that enable biological knowledge to compete with climate-based seasonal forecasts as early warning systems. Traditional entomological approaches that rely on inferences about populations of pests, based on observed prior seasonal climate and its relationships with the population dynamics of pests, can be further improved by using simulation models (Sutherst, 1993).

Early autumn rains cause early germination of subclover that can lead to outbreaks of spotted clover aphid, a subspecies of *Therioaphis trifolii* (Monell) (Hemiptera: Aphididae), in south-western Western Australia, resulting in substantial yield loss

(Mike Grimm, pers. comm.). The same is true in parts of New South Wales in early-irrigated pastures (Andrew Storrie, pers. comm.; Wendy Milne, pers. comm.). Early autumn rains can also increase the likelihood of problems with aphids in crops. They enable farmers to plant crops such as wheat in April and May and *Rhopalosiphum padi* (the oat aphid) then has several weeks in which to invade the crop and infect it with Barley Yellow Dwarf Virus (McKirdy and Jones, 1997). A seasonal forecast of early autumn rainfall would assist industry to prepare for such problems.

Winter rainfall in central Australia leads to growth of annual herbs, which are hosts of *Helicoverpa* spp.. The moths emerging from winter breeding grounds migrate eastwards in spring and are largely responsible for the initiation of damaging infestations on cotton and other crops. Maelzer *et al.* (1996, in press) showed that high winter rainfall in central Australia and low rainfall in the east in September resulted in large populations of moths in December. However, the short-term nature of the forecast meant that it was too late to influence inventory management of insecticides, but did have the potential to reduce the level of anxiety and surveillance effort of growers.

2.2 CLIMATE SCENARIO 2: EL NIÑO, SOI NEGATIVE, STONE AND AULICIEMS SOI PHASE 1.

2.2.1 Summer - Hot/Dry

A few examples of likely impacts of an El Niño event on ticks and some other pests in Australia in summer, and potential responses are shown in Table 3. In El Niño years there is a higher than average probability of reduced rainfall and higher temperatures in northern Australia. Such conditions would reduce the reproductive success of cattle ticks in the resultant hot and dry season. The effects on subsequent tick populations on cattle vary depending on where the affected area is relative to the tick's geographical range.

In the parts of the tick's distribution – such as along a Hughenden-Mt Isa transect - that are marginal due to inadequate average rainfall and excessive temperatures, the lower rainfall in El Niño years will have the effect of reducing the geographical range of the tick. These occasions provide opportunities to target local eradication efforts in those marginal habitats, provided that the nutritional stress of the cattle is not so excessive that it prevents handling them for treatment.

On the other hand, El Niño conditions in areas between the coast and the marginal inland areas, referred to above, reduce tick reproduction and survival in pastures but do not kill the whole population. This leaves a residue of tick larvae on the pastures in the spring. If the dry season is extended by the El Niño conditions, cattle lose more weight than normal (McCown *et al.,* 1981) and their resistance to ticks declines severely (Sutherst *et al.,* 1983; Sutherst, unpubl.). The result is that 4-5 times more tick larvae than normal may mature to adults on cattle and lead to excessive infestations that stress cattle at a time when they are already in poor condition. As the recovery of resistance can take up to six months (Sutherst unpubl.), the animals can be expected to experience a full tick season of high infestation following the collapse of resistance at the end of the dry season in spring.

This phenomenon is the primary cause of high tick numbers on zebu-based breeds of cattle in northern Australia and indicates an opportunity to anticipate the problems and implement pre-emptive treatments where possible.

TABLE 3. Biological and management responses for climate scenario 2 - Summer Hot/Dry

Biological Response	Management Response
Northern Australia	
Reproduction of cattle ticks declines	Focus surveillance on cattle in poor condition
Cattle resistance to ticks declines	Pre-emptive treatments in endemic area. Intensify eradication on boundaries
Throughout Australia	
Biocontrol agents decrease	Delay release of new agents until +ve SOI
Southern Queensland and northern New South Wales	
Cattle ticks increase	Increase surveillance on European breeds Vaccinate for tick-borne diseases
Tropical highlands & some lowlands (e.g. Venezuela)	
Malaria transmission increases causing epidemics	Increase surveillance Prepare supplies of prophylactics and treatments

Given the great ability of cattle owners to judge the condition of their cattle, communication of the information on the relationship between nutrition and host resistance may be adequate to enable them to take the necessary steps to prevent excessive tick burdens on stressed cattle. Such relationships may have application to helminth parasites in livestock where nutrition is also important.

This relationship between the tick resistance and field tick population size is so strong in central Queensland that it is possible to use it to predict annual differences in the size of the first generation of ticks from a single measurement of host resistance (Sutherst *et al.,* 1979). Unfortunately, it has not yet been possible to identify a reliable climatic predictor of host resistance in the area as other factors such as breeding may have a strong effect. Any advance indicator based on climatic data could extend the warning period to give more opportunity for cattle owners to carry out pre-emptive control measures. It would also remove the need to test the resistance status of cattle for use as a predictor, an approach that has potential but has not yet been used because it is logistically difficult.

The cattle tick requires hot-wet conditions to maximise its reproductive rates in pastures. In the south the tick is limited by low temperatures that slow development rates in summer and so limit the number of generations in high altitude areas to 3 (Sutherst, 1983), compared with 4-5 further north. In addition, winter temperatures in the south are too low to maintain metabolism and the tick's over-wintering success is greatly reduced. The sensitivity of the tick populations to changes in temperature is shown in Figure 2 (after

Sutherst, 1983). In the summer of 1972/73 the average temperature was 1.6°C higher than in the previous year. The result was a major increase in tick numbers, with an outbreak of tick-borne disease, *Babesia bovis,* in the herd of cattle. *Babesia* is preventable by vaccination but many farmers in marginal areas that experience intermittent outbreaks do not regularly vaccinate their cattle. A seasonal forecast would be invaluable in warning the cattle industry in the cool areas that they should vaccinate cattle for tick-borne diseases in years of high temperatures. The developing seasonal forecasting capability promises to be a suitable tool with which to make a seasonal forecast, as in this case (Figure 2), the negative relationship between the SOI and temperature would have provided a reliable indicator of the need to vaccinate. In the final four months of 1971 the SOI was +10.5 while in the same period in 1972 it was −10.5, so these values are consistent with the higher temperatures in the second year.

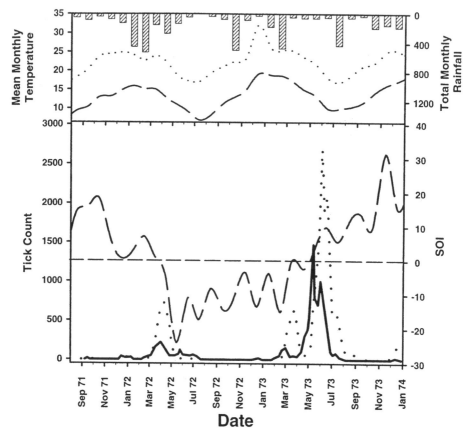

Figure 2. Observed (solid line lower panel) and simulated (dotted line lower panel) numbers of cattle ticks on European breed cattle at Mt Tamborine in southern Queensland in years with contrasting SOI (heavy dashed line in lower panel), weekly maximum (dotted line top panel) and minimum (dashed line top panel) temperatures and monthly rainfall (histogram) values.

A radio and press (Anon, 1998) alert of possible high tick numbers and associated risks of outbreaks of tick fevers caused by *Babesia* spp., which themselves multiply faster at higher temperatures in unfed ticks (Dalgleish and Stewart, 1976) was given in November 1997 following the record negative SOI readings. The statement read as follows:

> *"At a meeting on seasonal forecasting and agriculture last month, Dr Bob Sutherst of the CSIRO Division of Entomology warned cattle owners in the southern coastal areas of Queensland and high altitude districts, such as Atherton and Eungella, that the current El Niño event had a high probability of causing above-average temperatures as well as lower-than-average rainfall this summer. Dr Sutherst said that one consequence of this would be that cattle ticks in wet coastal and high altitude areas will undergo more development than usual to produce an extra partial-generation of ticks. This will pose a greater threat than normal in the late-summer and autumn, and result in an abnormally high risk of tick fever. Cattle owners in these areas are advised to vaccinate stock against tick fever this year, and to treat their cattle for ticks early in the season.*
>
> *However, Dr Sutherst said, the high temperatures and low rainfall would assist tick eradication efforts in the western parts of the tick area where the tick is normally limited by dryness and heat. In northern areas he said an extended dry season could result in nutritional stress to beef cattle, with a resulting upsurge of ticks in the spring each year. As it takes cattle some months to recover their resistance after such events, there will be benefits in controlling ticks on stressed cattle early in the season."*

The weather actually experienced between November 1997 and April 1998 is shown in Table 4. It shows above-average temperatures throughout Queensland, but not in northern New South Wales where much higher than average rainfall was also experienced. Rainfall in Queensland varied from below average in the south, to above average in the north. By early March 1998, there appeared to be an exceptional number of outbreaks of tick fever in southern Queensland (A. De Vos, pers. comm.). However, exceptionally high temperatures made cattle-owners reluctant to handle their cattle. A further advisory mailing was posted by the Queensland Tick Fever Research Centre in March, timed to remind graziers to vaccinate their cattle as soon as possible. While the eradication areas around Taroom were subjected to hot conditions with a resulting moisture deficit as forecast, the far north of Queensland was inundated by a cyclone. Thus, there is a need to emphasise the uncertainty associated with the El Niño forecasts and the local variation in weather that is likely.

In New South Wales the cattle tick is contained by a vigorous search-and-destroy program aimed at eradication of local infestations with strict quarantine surveillance on the border with tick infested country in Queensland. The results from Mt Tamborine (above) have direct application to NSW in interpreting the annual variation in risk of detectable infestations. In hot and wet years surviving ticks will multiply faster than in cool years and will become evident in the summer and autumn when inspections of cattle are timed to detect them. High rainfall may also increase tick numbers in warmer, low-lying areas. An analysis of historical infestation rates in relation to climate would assist interpretation of annual variation in tick numbers and be of potential value in identifying high-risk years for increased surveillance effort.

TABLE 4. Mean daily maximum temperatures (Tmax) and mean monthly rainfall for Lismore in northern New South Wales and Amberley, Taroom and Charters Towers in southern, central and northern Queensland respectively, from November 1 1997 to April 30 1998, compared with long-term averages.

Location	Tmax (^{0}C)	Rain (mm)
Lismore Average	28.4	146.2
Lismore 1997/8	27.5	91.3
Amberley average	29.6	98.7
Amberley 1997/8	32.4	73.5
Taroom Average	31.9	76.3
Taroom 1997/8	34.3	78.2
Charters Towers Average	32.6	90.2
Charters Towers 1997/8	34.1	103.7

Not only pests are affected by adverse seasonal conditions associated with El Niño in northern Australia. Several species of African dung beetles were imported into northern Australia in the 1970s and 1980s to dispose of cattle dung and reduce the breeding of the damaging buffalo fly (Waterhouse, 1974). The final release had to be expedited by the cessation of beef industry funding for the program in 1982. Unfortunately, it coincided with the onset of one of the most severe El Niño events this century and most of the species that were released at that time failed to survive the drought and have not been recovered since (A. Macqueen, pers. comm.). If cognisance had been taken of the SOI readings in Queensland in 1982-83, steps could have been taken to delay the release of the newly imported species of dung beetles.

G. Baker *et al.* (pers. comm.) are in the process of developing mass-rearing and inoculation techniques for the deep-burrowing earthworm, *Aporrectodea longa*, which appears suited to areas of mainland Australia with > 600 mm annual rainfall. *A. longa* is currently restricted to Tasmania, but several trials have indicated that it will establish on the mainland if introduced. The experience with the futile release of the dung beetles into a forecast drought, has alerted the earthworm researchers to ensure that the same fate does not befall the earthworms.

Bouma *et al.* (1997) reviewed the global impacts of natural disasters on human health and related them to the ENSO cycle. They found that disasters were more likely in the first and second year of an El Niño event. Bouma and Dye (1997) observed a lagged correlation between the incidence of malaria in Venezuela and El Niño events. The incidence increased in the year after El Niño years, which are dry in Venezuela. The authors suggested that high mortality of mosquito predators in dry weather could have led to the increase in frequency of this mosquito-borne disease in years of low rainfall. An alternative explanation, based on nutritional stress of the human population – analogous to the situation with ticks on cattle - was not considered to be applicable in Venezuela where food shortages do not occur. Thus it seems that second order interactions between different trophic levels may need to be taken into account when applying seasonal forecasts to some pests or diseases.

Temperatures in high altitude areas of the tropics are limiting for both vector populations and for disease development in the vector (Lindsay and Martens, 1998). In such places the transmission of diseases such as malaria, dengue fever and arboviruses is accelerated by high temperatures. Indeed, the reported correlation between the incidence of malaria in South America and El Niño events, referred to above, may be explainable in part by the accelerated development of the parasites in the vectors at higher temperatures. The regional variation in correlation between El Niño events and rainfall means that the phenomenon is not always a useful predictor of epidemics as seems to be the case in East Africa (Lindsay and Martens, 1998).

3. Discussion

The anecdotal nature of the results above indicates the dearth of quantitative research on the relationships between seasonal climate, pest incidence and host susceptibility to parasites. Management of pests to reduce the impact of foreign chemicals in the environment demands that these relationships be defined. Unfortunately, there is a lack of data on the biological processes of even the most costly insect pests in Australia, which hinders the development of mechanistic models. In addition, long time-series of data, such as those on *Helicoverpa* spp. (Maelzer *et al.*, 1996) are rarely available for use in identifying relationships between pest numbers and climate as a basis for climate-based forecasting tools. The pest community does not have the advantage of crop managers in having access to time-series of data equivalent to crop production statistics. There is an opportunity to exploit the vast numbers of pest reports to be found in the archives of State government offices. This will require an appreciation by policy makers of the value of historical data sets and the use of qualitative analytical tools to process the information.

Some effects of climatic variation and the opportunities that are presented by the new seasonal forecasting capability are illustrated by examples for the cattle tick, and other species, in Australia. The potential changes in numbers and the most appropriate response strategies vary depending on the location of the ticks relative to their geographical range. For example, in years with low rainfall, high temperatures and negative SOIs, vaccination against tick-borne diseases is most appropriate in the south at the same time as increased eradication is appropriate in the northwest and minimal response is needed in the central region. These observations emphasise the importance of considering where the problem exists as well as when. They also illustrate that for any species it is safer to live in the core of the distribution because living near the margin can be more suitable in favourable years, with the potential for causing epidemics, but more risky in unfavourable years with the threat of extinction.

Information that gives several months warning of an event can be used to improve strategic decisions such as the choice of crop variety, but it is less obvious how it can be used to improve decision making in entomology. Seasonal forecasts, that rely on insect traps to detect the first arrival of migratory pests or the first emergence of a new generation of insects, are well established in entomology. The challenge for seasonal forecasting is to demonstrate the value of such a climate-based forecast in adding value to the pest management decision-making process in a cost-effective way.

The advent of modular modelling tools such as DYMEX (Maywald et al., 1997) is alleviating the problem caused by a shortage of skilled personnel. The adoption of a national, collaborative approach to IPM could help to overcome the shortage of suitable data for establishing meaningful relationships between pests and climate. A national approach has also been greatly facilitated by the development of modern software tools, such as the CLIMEX climate-matching model and the DYMEX model building toolkit for developing models of pest populations. The latter has had the great benefit of enabling models to be built 'on the fly' during Modelling Workshops involving all interested parties in Australia (Maywald et al., 1997; Sutherst and Maywald, 1998). The Workshop outcomes include long-term collaborative linkages between modellers, field entomologists, teachers and policy makers, with shared responsibilities for further testing and data collection, and joint access to the resulting state-of-the-art models. The models then provide a common medium for collaboration and guidance on research direction, with benefits flowing on to end users through better problem solving, field application and communication.

Hosts respond to climate as much as insects, so there is a need to define the relationships between hosts and pests under different climatic scenarios. This demands that closer links be established between pest and production disciplines. Successful definition of the relationship between prior climate and tick resistance in cattle could lead to a better understanding of the nature of host immunology while also leading to the development of a useful seasonal forecasting tool for cattle tick control.

The pest management community will also need to come to grips with probabilistic forecasts in order to convey meaningful information on which managers can make their decisions. The overlapping distributions of rainfall with ENSO phases and the variable signal around Australia mean that it is a large challenge to the climate science community to provide information in a form that avoids risks of major crop losses while also reducing costs.

The use of the SOI needs to be made with caution as the correlation between the index and rainfall has varied greatly over the past century (Nicholls et al., 1997). Also, there is a problem with the vast amount of autocorrelation in both observational data and SOI series that tends to generate spurious correlations. In addition, the frequency of extreme seasons is likely to change under the influence of the enhanced greenhouse effect so it will be necessary to use the statistics of extremes (Gaines and Denny, 1993) to estimate changed probabilities and costs of disease outbreaks.

The SOI signal has to be related not only to a meteorological variable but also to the specific location and how that relates to the geographical distribution of the species. This is illustrated by the different responses of ticks to the El Niño in different parts of its range. The consequence of this is that IPM workers need to adopt a national or at least, regional perspective of their problems. They also need to be able to adapt the message to the local level around Australia. These opportunities for improved intervention in IPM suggest that precision targeting of control measures in time and space have considerable potential to reduce costs and environmental damage. This in turn calls for high resolution meteorological databases and simulation models that can deliver locally applicable information of high quality to industry.

Effective, locally applicable information delivery mechanisms are needed to deliver customised advice, derived from nationally applicable models, to clients in different parts of Australia. The need for pest-related ecological research to be based on a synoptic framework is self-evident, but is hampered by a lack of co-ordination and co-operation between researchers in different parts of Australia. Better links are needed between researchers and end users and this calls for the establishment of a national, collaborative approach to the management of pests, diseases and weeds in Australia (Sutherst and Maywald, 1998).

Acknowledgements

Thanks to Drs Geoff Baker, Wendy Milne and Keith Wardhaugh for information on responses of snails, aphids and sheep blow flies to climate. Drs Tania Yonow and James Ridsdill Smith gave helpful feedback on the manuscript.

References

Anon. (1998) El Niño and cattle tick control. Queensland Dairy Farmer. December 1997/January 1998 pp.14.

Baker, G.H. and Hawke, B.G. (1988) Life history and population dynamics of Theba pisana (Mollusca: Helicidae) in a cereal–pasture rotation. J. Appl. Ecol. **26**, 16-29.

Bouma, M.J. and Dye, C. (1997) Cycles of malaria associated with El Niño in Venezuela. The Journal of the American Medical Association **278**, 177-180.

Bouma, M.J., Kovats, R.S., Goubet, S.A., Cox, J.St.H. and Haines, A. (1997) Global assessment of El Niño's disaster burden. The Lancet **350**, 1435-9.

Bourne, A.S., Sutherst, R.W., Sutherland, I.D., Maywald, G.F. and Stegeman. D.A. (1988) Ecology of the cattle tick (Boophilus microplus) in subtropical Australia. III. Modelling populations on different breeds of cattle. Australian Journal of Agricultural Research **39**, 309-318.

Cane, M.A. (2000) Understanding and predicting the world's climate system, in G.L. Hammer, N. Nicholls, and C. Mitchell (eds.), Applications of Seasonal Climate Forecasting in Agricultural and Natural Ecosystems – The Australian Experience. Kluwer Academic, The Netherlands. (this volume).

Dalgleish, R.J. and Stewart, N.P. (1976) Stimulation of the development of infective *Babesia bovis* (*B. argentina*) in unfed *Boophilus microplus* larvae. *Australian Veterinary Journal* **52**, 54.

Davis, E., Adamson, D. and Rochester, W. (1997) The implementation of a *Helicoverpa* forecasting and information service. Workshop report. CRC for Tropical Pest Management, University of Queensland, Brisbane. pp. 44.

Drake, V. A. (1994) The influence of weather and climate on agriculturally important insects: An Australian perspective. *Australian Journal of Agricultural Research* **45**, 487-509.

Gaines, S.D. and Denny, M.W. (1993) The largest, smallest, highest, lowest, longest, and shortest: extremes in ecology. *Ecology* **74**,1677-1692.

Gregg, P.C., Fitt, G.P., Zalucki, M.P. and Murray, D.A.H. (1995) Insect migration in an arid continent. II. *Helicoverpa* species in eastern Australia, in Drake, V.A. and Gatehouse, A.G. (eds.), *Insect Migration. Tracking Resources Through Space and Time.* Cambridge University Press. pp.151-172.

Hamilton, J.G., Rochester, W.A. and Gregg, P.C. (1994) Predicting long-distance migration of insect pests in eastern Australia. Proceedings 21st Conference on Agricultural and Forest Meteorology and 11th Conference on Biometeorology and Aerobiology. Boston: American Meteorological Society. pp.431-434.

Hammer, G.L. and Nicholls, N.N. (1996) Managing for climate variability - The role of seasonal climate forecasting in improving agricultural systems. The Second Australian Conference on Agricultural Meteorology. Conference Proceedings: The Impact of Weather and Climate on Agriculture. The University of Queensland, Brisbane. 1-4 Oct 1996. pp.19-27.

Hammer, G.L., Holzworth, D.P. and Stone, R.C. (1996) The value of skill in seasonal climate forecasting to wheat crop management in a region with high climatic variability. *Aust. J. Agric. Res.* **47**, 717-737.

Lindsay, M. and McKenzie, J. (1997) Vector-borne viral diseases and climate change in the Australasian region: Major concerns and the public health response, in Curson, P., Guest, C. and Jackson, E. (eds.), *Climate Change and Human Health in the Asia-Pacific Region.* AMA & Greenpeace Int. pp.109.

Lindsay, S.W. and Martens, W.J.M. (1998) Malaria in the African highlands: past, present and future. Bulletin of the World Health Organization **76**, pp.33-45.

Maelzer, D., Zalucki, M.P. and Laughlin, R. (1996) Analysis and interpretation of long term light trap data for *Helicoverpa punctigera* (Lepidoptera; Noctuidae) in Australia: population changes and forecasting pest pressure. *Bulletin of Entomological Research* **86**, 547-557.

Maelzer, D.A., Zalucki, M.P. and Laughlin. R. (in press) Analysis of long term light trap data for *Helicoverpa armigera* (Lepidoptera; Noctuidae) in Australia: the effect of climate and crop host plants. *Bulletin of Entomological Research* (in press)

Maywald, G.F., Sutherst, R.W. and Zalucki, M.P. (1997) Generic modelling for integrated pest management, in McDonald, A.D. and McAleer, M. (eds.), *MODSIM '97* Hobart, volume 3, pp.1115-1116.

McBride, J..L. and Nicholls, N. (1983) Seasonal relationships between Australian rainfall and the southern oscillation. *Monthly Weather Review* **111**,1998-2004.

McCown, R.L., Gillard, P., Winks, L. and Williams, W.T. (1981) The climatic potential for beef cattle production in tropical Australia: Part II – liveweight change in relation to agro-climatic variables. *Agricultural Systems* **7**, 1-10.

McDonald, G. (1995) Insect migration in an arid continent. I. The common armyworm *Mythimna convecta* in eastern Australia, in Drake, V.A. and Gatehouse, A.G. (eds.), *Insect Migration. Tracking Resources Through Space and Time.* Cambridge University Press. pp.131-150.

McKirdy, S.J. and Jones, R.A.C. (1997) Effect of sowing time on barley yellow dwarf virus infection in wheat: virus incidence and grain yield losses. *Australian Journal of Agricultural Research* **48**, 199-206.

Nicholls, N. (1986) A method for predicting Murray Valley encephalitis in southeast Australia using the Southern Oscillation. *Australian Journal of Experimental Biology and Medical Science* **64**, 587-594.

Nicholls, N. (1993) El Niño-Southern Oscillation and vector-borne disease. *The Lancet* **342**, 1284-1285.

Nicholls, N., Drosdowsky, W. and Lavery, B. (1997) Australian rainfall variability and change. *Weather* **52**, 66-72.

Southwood, T.R.E. (1977) Habitat, the templet for ecological strategies? *J. Anim. Ecol.* **46**, 337-365.

Standfast, H.A. and Maywald, G.F. (1992) Modelling for global changes affecting arboviruses, in Walton, T.E. and Osburn, B.I (eds.), *Bluetongue, African Horse Sickness and Related Orbiviruses.* Proc. 2nd Int. Symp., CRC Press, Boca Raton. pp16-20.

Stone, R. and Auliciems, A. (1992) SOI phase relationships with rainfall in eastern Australia. *International Journal of Climatology* **12**, 625-636.

Stone, R.C., Hammer G.L., and Marcussen, T. (1996a) Prediction of global rainfall probabilities using phases of the Southern Oscillation Index. *Nature* **384**, 252-55.

Sutherst, R.W. (1983) Variation in the numbers of the cattle tick, Boophilus microplus (Canestrini), in a moist habitat made marginal by low temperatures. *Journal of the Australian Entomological Society* **22**, 1-5.

Sutherst, R.W. (1993) Role of modelling in sustainable pest management, in Corey, S.A., Dall, D.J. and Milne, W.M. (eds.), *Pest Control and Sustainable Agriculture*. CSIRO, Melbourne. pp.64-69.

Sutherst, R.W. (1994) Climate change and arbovirus transmission, in Uren, M.F. and Kay, B.H. (eds.), 6th Symposium Arbovirus Research in Australia. pp.297-299.

Sutherst, R.W. (1998) Implications of global change and climate variability for vector-borne diseases: generic approaches to impact assessments. *International Journal for Parasitology* **28**, 935-945.

Sutherst R.W. and Maywald G.F. (1985) A computerised system for matching climates in ecology. *Agriculture Ecosystems and Environment* **13**, 281-99.

Sutherst, R.W. and Maywald, G.F. (1998) DYMEX modelling workshops: a national, collaborative approach to pest risk analysis and IPM in Australia, in Myron Zalucki and Richard Drew (eds.), *Pest Management - Future Challenges*. Proc. 6th Australasian Applied Entomological Research Conference, University of Queensland Press. pp.57-62.

Sutherst, R.W., Floyd, R.B., Bourne, A.S. and Dallwitz, M.J. (1986) Cattle grazing behaviour regulates tick populations. *Experientia* **42**, 194-196.

Sutherst, R.W., Kerr, J.D., Maywald, G.F. and Stegeman, D.A. (1983) The effect of season and nutrition on the resistance of cattle to the tick Boophilus microplus. *Australian Journal of Agricultural Research* **34**, 329-339.

Sutherst, R.W., Maywald, G.F. and Skarratt, D.B. (1995) Predicting insect distributions in a changed climate. in Harrington, R. and Stork, N.E. (eds). *Insects in a Changing Environment*. Academic Press, London. pp.59-91.

Sutherst, R.W., Utech, K.B.W., Dallwitz, M.J. and Kerr, J.D. (1973) Intra-specific competition of Boophilus microplus (Canestrini) on cattle. *Journal of Applied Ecology* **10**, 855-62.

Sutherst, R.W., Wharton, R.H., Cook, I.M. Sutherland, I.D., and Bourne, A.S. (1979) Long-term population studies on the cattle tick (Boophilus microplus) on untreated cattle selected for different levels of tick resistance. *Australian Journal of Agricultural Research* **30**, 353-68.

Vogt WG, Woodburn TL, and Foster G.G. (1985) Ecological analysis of field trials conducted to assess the potential of sex-linked translocations strains for genetic control of the Australian sheep blowfly, *Lucilia cuprina* (Wiedemann). *Aust J. Biol. Sci.* **38**, 259-73.

Walden, K.J. (1995) Insect migration in an arid continent. III The Australian plague locust *Chortoicetes terminifera* and the native budworm *Helicoverpa punctigera* in Western Australia, in Drake, V.A. and Gatehouse, A.G. (eds.), *Insect Migration. Tracking Resources Through Space and Time*. Cambridge University Press. pp 173-190

Wardhaugh. K.G., and Morton R. (1990) The incidence of flystrike in sheep in relation to weather conditions, sheep husbandry and the abundance of the Australian sheep blowfly, *Lucilia cuprina* (Wiedemann) (Diptera: Calliphoridae). *Australian Journal of Agricultural Research* **41**, 155-67.

Waterhouse, D.F. (1974) The biological control of dung. *Scientific American* **230**, 101-109.

Wharton, R.H., Harley, K.L.S., Wilkinson, P.R., Utech, K.B.W. and Kelley, B.M. (1969) A comparison of cattle tick control by pasture spelling, planned dipping, and tick-resistant cattle. *Australian Journal of Agricultural Research* **20**, 783-797.

WHO (1990) Potential health effects of climatic change. WHO Report WHO/PEP/90/10, Geneva. pp.58.

Williams, J.D., Sutherst, R.W., Maywald, G.F. and Petherbridge, C.T. (1985) The southward spread of buffalo fly (*Haematobia irritans exigua*) in Eastern Australia and its survival through a severe winter. *Australian Veterinary Journal* **62**, 367-369.

ENSO REGULATION OF INDO-PACIFIC GREEN TURTLE POPULATIONS

COLIN LIMPUS[1] AND NEVILLE NICHOLLS[2]

[1]*Queensland Parks and Wildlife Service*
PO Box 155
Brisbane Albert St, Qld 4002, Australia

[2]*Bureau of Meteorology Research Centre*
PO Box 1289K
Melbourne, Vic 3000, Australia

Abstract

The green turtle, *Chelonia mydas*, is a large, long-lived, herbivorous marine reptile that grazes on the marine macrophytes (seagrass, algae, mangrove fruit). Although green turtles are resident in shallow coastal waters for feeding, they migrate up to 2,600km to breed at traditional nesting beaches. Female green turtles do not breed annually. It takes in excess of a year for a female green turtle to lay down her fat reserves and deposit the yolk stores in her ovaries in preparation for the breeding migration. The annual number of breeding green turtles has been recorded in a monitoring program encompassing 23 years at nesting beaches within two genetically independent stocks in eastern Australia (Heron Island within the southern Great Barrier Reef breeding stock; Raine Island within the northern Great Barrier Reef breeding stock). The number of females recorded annually at the nesting beaches varies widely across years - sometimes varying up to 3 orders of magnitude in successive years. Approximately synchronous fluctuations in annual breeding numbers occur at the eastern Australian nesting beaches.

An examination of the green turtles in their home feeding grounds prior to the commencement of the breeding migration shows that the inter-annual fluctuations in breeding numbers are not due to fluctuations in the number of adult turtles in the feeding areas. Rather, the annual fluctuations in breeding numbers are the result of fluctuations in the proportion of adult females that prepare to breed in a particular year.

There is a significant correlation between the SOI index two years before the breeding season and the number of females recorded at the nesting beach. In the extremes, massed nesting occurs two years following major El Niño events and extremely low nesting numbers occur two years after major anti El Niño events. It is now possible to predict within reasonable confidence limits the size of the annual nesting population at key eastern Australian green turtle rookeries based on the SOI two years before the

G.L. Hammer et al. (eds.), The Australian Experience, 399–408.

commencement of the breeding season. No comparable relationship has been identified between ENSO and any of the other species of marine turtles.

Additional research suggests that there is a nutritional basis to this annual fluctuation in green turtle population parameters. Regional breadth of the phenomenon is demonstrated by the synchrony of function of green turtles that feed almost exclusively on algae on the coral reefs of the outer Great Barrier Reef with those that feed primarily on seagrass in inshore bays. However, there is no suitable database available to determine whether the fluctuations are driven by changes in quantity or quality of the food resource. Changes in water temperature alone do not appear to be sufficiently large to account for the effects.

Without the linkage to ENSO regulation of breeding numbers, it has not been possible to use short term nesting census data to determine green turtle population stability. If the changes in green turtle population dynamics as a result of ENSO events are nutritionally based, then green turtle populations have the potential to provide parameters that can be monitored and compared with the performance of seagrass and algal pastures in response to environmental change. This would improve our understanding of the functioning of marine grazing ecosystems. This could apply to both changes from natural events such as climate fluctuation as well as from anthropogenic induced changes.

1. Introduction

In 1974, Dr. Colin Limpus took over leadership of the turtle research program at Heron Island at the request of the then Director of Fisheries in Queensland. Previous research by Dr. Robert Bustard had indicated that some 200-600 green turtles, *Chelonia mydas*, would nest on Heron Island each summer. To our surprise, we tagged ~1200 nesting females in the 1974-1975 breeding season and the local residents spoke of the greatest number of nesting turtles in their memory. We returned in the following summer equipped to tag thousands of nesting green turtles only to be met by a total of 21 females breeding on Heron Island for the entire summer. Understandably we were puzzled. Some locals suggested that "Limpus has scared the turtles away". But when we checked the nearby islands where large numbers of green turtles had nested in the 1974-1975 breeding season, we found that all islands had few nesting green turtles in the 1975-1976 breeding season. Discussions with indigenous hunters at the time gave no suggestion that there had been any large reduction in the availability of green turtles in their home feeding areas. These two summers set the scene for our ongoing search for an underlying cause of the fluctuations in nesting numbers that we saw each year (Figure 1).

Ten years were to pass before we again saw the very high nesting densities in 1984-1985. With the 1974 and the 1984 seasons both occurring two years after major El Niño events, we turned to the regional climate data in search of answers. A strong correlation was found between the mean atmospheric pressure measurement at Darwin during November to January, as an index of the El Niño Southern Oscillation (ENSO), and the number of green turtles breeding at Heron Island two years later (Limpus and Nicholls, 1988). This paper summarises continuing developments in this investigation.

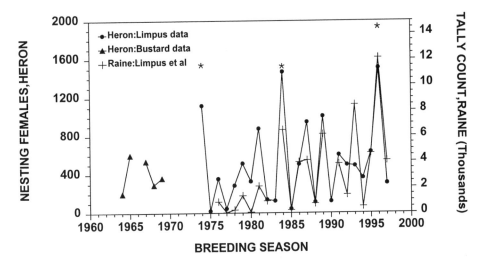

Figure 1. Annual census data from representative rookeries from each of the eastern Australian stocks of green turtle. The Heron Island data is based on a total tagging census of the annual nesting population. The Raine Island data is an index of the size of the annual nesting population using the mean nightly numbers of turtles ashore within the first two weeks of December. Those years at Raine Island when a nightly count of nesting turtles exceeded 10000 females are denoted by *.

2. Green Turtle Biology

The green turtle is a large, long-lived, marine reptile. They are one of the few large herbivorous animals of the oceans. Green turtles graze on the marine macrophytes (seagrass, algae, mangrove fruit) in shallow tropical and temperate coastal waters (Lanyon *et al.*, 1989). When they breed, they migrate from their dispersed feeding areas and travel up to 2,600km to lay eggs on traditional nesting beaches within the tropics (Limpus *et al.*, 1992). On a global scale, the major breeding aggregations form genetically discrete breeding units, with each turtle returning to breed at the area of its birth (Bowen *et al.*, 1992; Norman *et al.*, 1994). Female green turtles do not breed annually. It takes in excess of a year for a female green turtle to lay down her fat reserves and deposit the yolk stores in her ovaries in preparation for the breeding migration. In eastern Australia, green turtles average 5-6 years between breeding seasons (Limpus *et al.*, 1994b). The Queensland Parks and Wildlife Service has monitored the annual number of breeding green turtles for 24 years at nesting beaches within the two genetically independent stocks in eastern Australia. At Heron Island (23°26'S, 151°55'E) within the southern Great Barrier Reef breeding stock, effectively every nesting female green turtle has been tagged each summer since 1974 (Figure 1). At other islands in the southern Great Barrier Reef, the size of the annual nesting population has been estimated by extrapolation from the nightly number of turtles nesting at the peak of the nesting season during the last two weeks of December. The estimated annual nesting population on these islands can be compared with the total tagging census data from Heron Island in the same years (Figure 2). At Raine Island (11°36'S, 144°01'E) within the northern Great Barrier Reef breeding stock, the average number of females ashore for nesting per night during the first two weeks of December has been recorded in all but two years (1983, 1990) since 1976 (Figure 1). This measure provides an index of the size of the annual nesting population at Raine Island. These census data demonstrate several characteristics of green turtle reproduction:

- the number of females recorded annually at the nesting beaches varies widely across years - sometimes varying up to 3 orders of magnitude in successive years;
- there is a high level of synchrony of nesting fluctuations for rookeries within the same breeding unit.
- approximately synchronous fluctuations in annual breeding numbers occur at rookeries within each of the eastern Australian breeding units.

An examination of the adult female green turtles in their home feeding grounds prior to the commencement of the breeding migration shows that the inter-annual fluctuations in breeding numbers at the nesting beaches are not due to fluctuations in the number of adult turtles in the feeding areas. Rather, the annual fluctuations in breeding numbers are the result of fluctuations in the proportion of adult females that prepare to breed in a particular year (Figure 3).

3. Climate Impact on Green Turtle Biology

In testing for the best correlation between breeding numbers and ENSO events, a significant correlation was detected only for the time period approximately two years before the breeding season (Limpus and Nicholls 1988, 1994). Biologically, this fits well with the time it takes for a female to prepare for a breeding season. With census data now available from 23 breeding seasons, there continues to be a significant linear correlation between the SOI two years before the breeding season and the log of the annual green turtle census data (census) from representative rookeries from each of the eastern Australian genetically distinct stocks (Table 1, Figure 4).

Heron Island: $\ln(\text{census}) = -0.086 * \text{MOSOI} + 5.602$ ($F_{1,22} = 22.3$, $p < 0.001$; $r^2 = 0.50$)
Raine Island: $\ln(\text{census}) = -0.118 * \text{MOSOI} + 6.858$ ($F_{1,18} = 13.3$, $p < 0.001$; $r^2 = 0.42$)

where MOSOI is the mean May to October SOI.

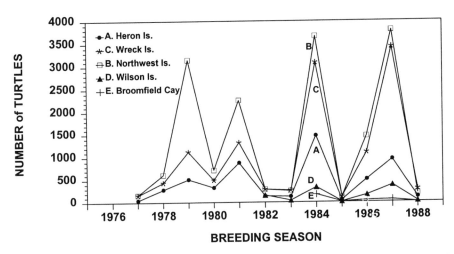

Figure 2. Variation in the size of the annual nesting population of green turtles, *Chelonia mydas*, at rookeries within the southern Great Barrier Reef. The Heron Island data is based on a total tagging census of the annual nesting population. At other islands the size of the annual nesting population has been estimated from the number of turtles ashore for nesting during the peak of the nesting season in the last two weeks of December.

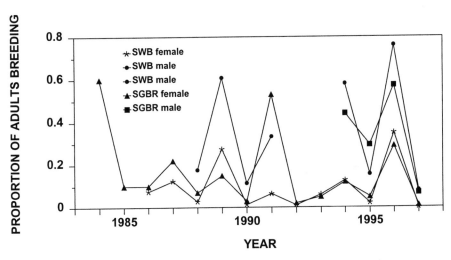

Figure 3. Comparison of the proportion of adult green turtles, *Chelonia mydas*, that annually prepared for breeding from each of two feeding areas in eastern Australia over the 15 year period, 1983-1997. SWB denotes the Shoalwater Bay area of the central Queensland coast. SGBR denotes Heron Island Reef and adjacent Wistari Reef of the southern Great Barrier Reef. Breeding condition was determined by gonad examination (Limpus *et al.,* 1994a) or by recording the turtle at courtship or nesting.

TABLE 1. Annual census data for nesting green turtles, *Chelonia mydas*, at Heron Island representing the southern Great Barrier Reef genetic stock and Raine Island representing the northern Great Barrier Reef genetic stock. MOSOI denotes the mean May to October SOI. The Heron Island census data is derived from a tagging census of the total annual nesting population. The Raine Island census data is the mean nightly number of nesting females counted in one walk of the island for the first two weeks of December. Annual turtle census data are presented in conjunction with the SOI data from 2yr before the respective breeding seasons.

Year	MOSOI	Breeding Season	Heron Island		Raine Island	
			Census	ln(census)	census	ln(census)
1972	-13.6	1974	1121	7.02	-	-
1973	9.5	1975	21	3.04	-	-
1974	8.8	1976	355	5.87	872	6.77
1975	17.3	1977	46	3.83	32	3.47
1976	-5.4	1978	285	5.65	243	5.49
1977	-13.0	1979	513	6.24	1347	7.21
1978	4.0	1980	327	5.79	59	4.08
1979	-0.8	1981	878	6.78	2049	7.63
1980	-2.6	1982	139	4.93	925	6.83
1981	6.1	1983	125	4.83	-	-
1982	-18.8	1984	1471	7.29	6493	8.78
1983	1.6	1985	42	3.74	227	5.42
1984	-1.2	1986	505	6.22	3956	8.28
1985	-1.0	1987	947	6.85	4102	8.32
1986	-0.1	1988	111	4.71	731	6.59
1987	-15.2	1989	1009	6.92	6144	8.72
1988	11.2	1990	120	4.79	-	-
1989	6.4	1991	602	6.40	3828	8.25
1990	1.5	1992	498	6.21	1417	7.26
1991	-10.6	1993	488	6.19	8462	9.04
1992	-5.7	1994	370	5.91	500	6.21
1993	-11.7	1995	632	6.45	4647	8.44
1994	-15.0	1996	1509	7.32	12100	9.40
1995	-0.6	1997	289	5.67	4067	8.31

In the extremes, massed nesting occurs two years following major El Niño events (eg. 1972 and 1982) and crashes in nesting numbers occur two years after major anti El Niño events (eg. 1973 and 1983). It is now possible to predict within broad limits the size of the annual nesting population at key eastern Australian green turtle rookeries based on the SOI two years before the commencement of the breeding season. No similar relationship has been identified between ENSO and any of the other species of marine turtles, all of which are carnivorous. Other parameters that vary in synchrony with SOI in the southwest Pacific, such as sea surface temperature and tropical cyclone activity, will also correlate well with annual green turtle breeding numbers (Limpus and Nicholls, 1994).

An examination of adult and immature green turtles of both sexes in their home feeding grounds demonstrates that the ENSO climate events are impacting more than the adult females:

- The proportion of adult males that are in spermatogenesis in preparation for their annual breeding migrations to the courtship areas fluctuates approximately synchronously with fluctuations in the proportion of adult females in vitellogenesis in the same feeding grounds (Figure 3).
- In the same years that a large proportion of the adults is preparing for breeding, immature turtles show increases in growth rates (Limpus and Chaloupka, 1997).

Climate is likely to be impacting a suite of parameters that encompasses fat deposition, vitellogenesis, spermatogenesis and growth within a total population of green turtles. This strongly suggests that there is a nutritional basis to these annual fluctuations in green turtle population parameters. However, there is no suitable database available to determine whether the fluctuations are driven by changes in quantity or quality of the food resource or changes in the physiology of the turtles. Regional breadth of the phenomenon is demonstrated by the synchrony of function (Figure 3) of green turtle populations that feed almost exclusively on algae on the coral reefs of the outer Great Barrier Reef (reefs near Heron Island) with those that feed primarily on seagrass in inshore bays (Shoalwater Bay). Green turtles are poikilothermic animals and their body temperature is approximately that of the surrounding water (Read et al., 1996). Annual fluctuations in water temperature operating through changes in metabolic rates of the turtles do not appear to be sufficiently large to cause the above fluctuations in turtle breeding rates.

The climate interaction with green turtle breeding numbers is not exclusively an Australian phenomenon. Examination of egg production as an index of the size of the nesting population at green turtle rookeries from widely scattered sites in South East Asia (Thailand, Philippines, Malaysia) shows a synchrony of fluctuations in the sizes of the annual nesting populations (Figure 5). These Asian nesting populations (mid year peak of nesting) breed six months out of phase with the eastern Australian populations (November- February peak of nesting). The fluctuations in annual egg production at the Asian beaches largely parallel the fluctuations in nesting numbers at the eastern Australian beaches six months later. Given the large geographical scale of the total ENSO phenomenon, it is not surprising that ENSO impacts green turtle nesting populations at the rookeries throughout the southeastern Asian and western Pacific region.

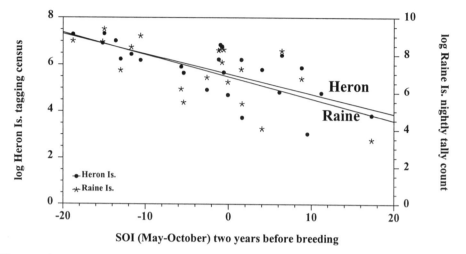

Figure 4. Correlation between SOI data and the green turtle, *Chelonia mydas*, census data at nesting beaches in eastern Australia. See Table 1 for definition of data. These data encompass 23 turtle breeding seasons from 1974-1975 to 1997-1998. Trend lines are presented for each population.

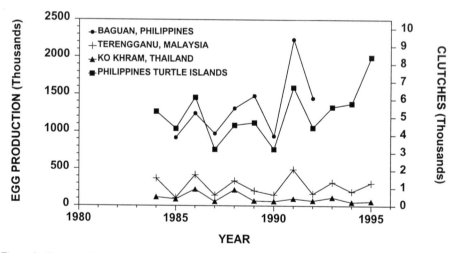

Figure 5. Season influence on egg production at green turtle, *Chelonia mydas*, rookeries in southeast Asia. Annual egg production or the number of clutches laid provides a good index of the size of the nesting population.

4. Discussion

The dynamics of the Australian marine environment in response to ENSO is complex with positive and negative correlations with changes in a range of parameters including sea level, water temperature, salinity and concentrations of oxygen, nitrate and inorganic phosphate (Hsieh and Hamon, 1991). This is further complicated by variable timing of the onset of ENSO events with respect to seasonal cycles (Hsieh and Hamon, 1991). This translates to variable biological responses across marine species. In particular, the impact of ENSO on green turtle populations does not appear to parallel the impacts of ENSO reported for other marine species. With associated shifts in the strength and position of currents, ENSO has been reported to influence population size for a wide range of invertebrates and fish in Western Australia through the dispersal and or recruitment of planktonic larvae (Pearce and Phillips, 1988). ENSO events can be linked to the location of water masses of suitable temperature and nutrient characteristics and hence the location of schools of jack mackerel that determines the success or failure of jack mackerel fisheries off Tasmania (Harris et al., 1992). For the green turtle however, the climate effect is operating on a large, widely dispersed, relatively sedentary herbivore (at least while it is in its feeding mode prior to migration) that grazes on sedentary plant crops. Up to the present, the mechanism of the ENSO linkage to green turtle population function has not been established. Potential linkages could range from damage to pastures during cyclones, or changes in growth of pastures in response to flood runoffs, to changes in nutrient quality of water brought to the pastures by onshore currents or upwellings.

Without the linkage to ENSO regulation of breeding numbers, it had not been possible to use short term nesting census data to determine green turtle population stability. Theoretically, it should now be possible to use an analysis of ENSO climate data and the nesting beach census data, corrected for the proportion of females that prepared for breeding in the home feeding grounds, to detect change or stability in the size of the total adult female population using a breeding area.

Green turtles, dugong and manatee are the only large marine animals that graze on seagrass pastures (Lanyon et al., 1989) and fluctuations in the abundance or quality of the pasture (Preen et al., 1993) can be expected to impact on the performance of the herds. At present there have been no studies that quantify both abundance and nutritional quality of seagrass at a scale that would be informative for understanding the functioning of populations of these threatened species. If the changes in green turtle population dynamics as a result of ENSO events are nutritionally based, then green turtle populations have the potential to provide readily measurable parameters (annual breeding rates, growth rates) that can be monitored and compared with the performance of seagrass and algal pastures in response to environmental change. This could apply to both changes from natural events such as climate fluctuation as well as from anthropogenic induced changes. A multidisciplinary study that examined seagrass pastures and the associated green turtle and dugong populations has the potential for providing valuable insights into the responses of marine grazing communities to environmental changes.

408

References

Bowen, B.W., Meylan, A.B., Ross, J.P., Limpus, C.J., Balazs, G.H. and Avise, J.C. (1992) Global population structure and natural history of the green turtle (*Chelonia mydas*) in terms of matriarchal phylogeny. *Evolution* **46**, 865-81.

Harris, G.P., Griffiths, F.B. and Clementson, L.A. (1992) Climate and the fisheries off Tasmania - interaction of physics, food chains and fish. *South African Journal of Marine Science* **12**,585-97.

Hsieh, W.W. and Hamon, B.V. (1991) The El Niño-Southern Oscillation in south-eastern Australian waters. *Australian Journal of Marine and Freshwater Research* **42**,263-75.

Lanyon, J., Limpus, C.J. and Marsh, H. (1989) Dugongs and turtles grazers in the seagrass system, in A.W.D. Larkum, A.J. McComb and S.A. Shepherd (eds.), *Biology of Seagrasses*. Elsevier, Amsterdam. pp.610-34.

Limpus, C.J. and Chaloupka, M. (1997) Nonparametric regression modelling of green sea turtle growth rates (southern Great Barrier Reef). *Marine Ecology Progress Series* **149**, 23-34.

Limpus, C.J. and Nicholls, N. (1988) The southern oscillation regulates the number of green turtles (*Chelonia mydas*) breeding around northern Australia. *Australian Wildlife Research* **15**, 157-161.

Limpus, C.J. and Nicholls, N. (1994) Progress report on the study of the interaction of the El Niño Southern Oscillation on annual *Chelonia mydas* numbers at the southern Great Barrier Reef rookeries, Proceedings of the Australian Marine Turtle Conservation Workshop. Queensland Department of Environment and Heritage and Australian Nature Conservation Agency, Canberra. pp. 73-8.

Limpus, C.J., Couper, P.J. and Read, M.A. (1994a) The green turtle, *Chelonia mydas*, in Queensland: population structure in a warm temperate feeding area. *Memoirs of the Queensland Museum* **35**, 139-54.

Limpus, C.J., Eggler, P. and Miller, J.D. (1994b) Long interval remigration in eastern Australian Chelonia. National Oceanographic and Atmospheric Administration, National Marine Fisheries Service, Southeast Fisheries Science Center, Technical Memorandum 341. pp.85-8.

Limpus, C.J., Miller, J.D., Parmenter, C.J., Reimer, D., McLachlan, N. and Webb, R. (1992) Migration of green (*Chelonia mydas*) and loggerhead (*Caretta caretta*) turtles to and from eastern Australian rookeries. *Australian Wildlife Research* **19**,347-58.

Norman, J.A., Moritz, C. and Limpus, C.J. (1994) Mitochondrial DNA control region polymorphisms: genetic markers for ecological studies in marine turtles. *Molecular Ecology* **3**,363-73.

Pearce, A.F. and Phillips, B.F. (1988) ENSO events, the Leeuwin Current, and larval recruitment of the western rock lobster. *J. Const. Int. Explor. Mer.* **45**,13-21.

Preen, A., Lee Long, W.J. and Coles, R.G. (1993) Widespread loss of sea grasses in Hervey Bay. Unpublished report to Queensland Department of Environment and Heritage. pp. 1-12.

Read, M.A., Grigg, G.C. and Limpus, C.J. (1996) Body temperature and winter feeding in immature green turtles, *Chelonia mydas*, in Moreton Bay, Southeastern Queensland. *Journal of Herpetology* **30**, 262-5.

STREAMFLOW VARIABILITY, SEASONAL FORECASTING AND WATER RESOURCES SYSTEMS

FRANCIS CHIEW[1], TOM MCMAHON[1], SEN-LIN ZHOU[1] AND TOM PIECHOTA[2]

[1] *Department of Civil and Environmental Engineering*
University of Melbourne
Parkville, Vic 3052, Australia

[2] *Civil and Environmental Engineering Department*
University of California
Los Angeles, California, USA

Abstract

The interannual variability of streamflow in Australia is greater than elsewhere in the world. Reliable forecasts of streamflow would go a long way towards improving the management of water resources systems by enabling them to cope better with the high inter-annual variability. In the first part of this paper, a clear link between runoff in Australian catchments and ENSO is established using lag-correlation analysis. This suggests that runoff in many areas in Australia can be forecast over certain seasons with some success using indicators of ENSO and the serial correlation in runoff. The second part of the paper presents statistical methods for forecasting streamflow. Probabilistic methods for forecasting categorical values of streamflow are well developed, although their use is limited by the short historical data and the varying strength in the streamflow-ENSO relationship over the last century. There have been far fewer studies on forecasting probabilities of exceedance of streamflow amounts, although these forecasts are required to evaluate the benefits and risks of using ENSO-based forecasts for different levels of conservatism in water resources management. The third part of the paper discusses the present and potential uses of seasonal streamflow forecasts for managing urban water and rural irrigation systems.

1. Introduction

River runoff can vary considerably from year to year. Figure 1 shows the time series of annual runoff in two typical Australian rivers. The mean annual runoff in Herbert Creek in the north-east coast of Australia is 200mm, but runoff varies from almost zero in some years to more than 600mm in other years. In the Campaspe River in the Murray River Basin, the coefficient of variation (standard deviation divided by the

G.L. Hammer et al. (eds.), The Australian Experience, 409–428.

410

mean) of annual runoff is about 65%. The management of water resources systems involves design features and operating characteristics that must cope with these levels of variability.

The inter-annual variability of Australian and Southern African streamflow volumes and peak discharges are about twice that of rivers elsewhere in the world. This is illustrated by the plots in Figure 2 showing the L-Cv (linear moment equivalent of the coefficient of variation - see Wallis *et al.* (1974) and Hosking (1990)) of annual runoff volume and peak discharge of Australian and Southern African rivers and rivers in other parts of the world. Haines *et al.* (1988), Finlayson and McMahon (1991), and McMahon *et al.* (1992) provide a detailed comparison of continental flows and peak discharges. As a consequence of the higher variability, floods in large catchments are more severe and yields of water resources systems are likely to be smaller in Australia and Southern Africa than elsewhere. As the reservoir design storage size is roughly proportional to the square of the coefficient of variation (see McMahon and Mein, 1986), reservoir storages in Australia and Southern Africa must be about four times larger than those elsewhere in the world, for a given reliability and degree of river regulation. The reservoir storage problem is further compounded by the lower runoff and higher evapotranspirative demand in Australia.

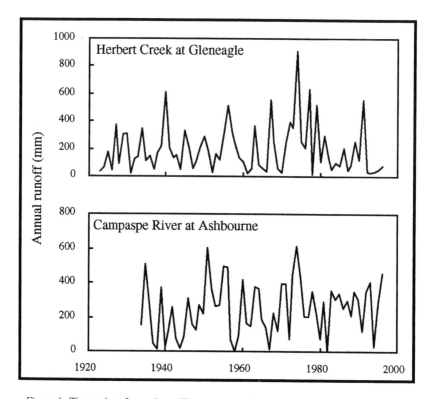

Figure 1. Time series of annual runoff in two Australian catchments

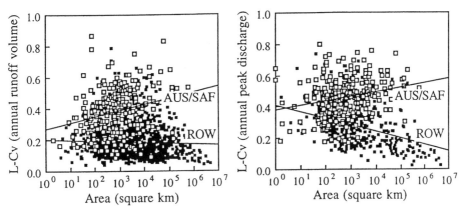

Figure 2. Linear moment equivalent of the coefficient of variation (L-Cv) of annual runoff volume and peak discharge versus catchment area for catchments in Australia and Southern Africa (AUS/SAF; filled squares) and catchments in the rest of the world (ROW; open squares). Data for runoff are from 259 catchments in AUS/SAF and 1023 catchments in ROW and data for peak discharge are from 252 catchments in AUS/SAF and 374 catchments in ROW.

Because of the high inter-annual variability, forecasts of streamflow several months ahead would be invaluable to the management of water resources systems in Australia. For example, reliable forecasts of reservoir inflows can be used to determine irrigation and environmental water allocations, as well as appropriate types of crops to be planted. It is possible that streamflow can be forecast from the serial correlation in streamflow and the association between streamflow and El Nino/Southern Oscillation (ENSO). The streamflow-ENSO association is illustrated by the box plots in Figure 3, which show the means and various runoff percentiles over the three consecutive months with the highest average runoff for the two locations shown in Figure 1, for three categories of Troup Southern Oscillation Index (SOI) values in the preceding three months. The plots indicate that runoff is generally higher when the average SOI is above +5 compared to when it is below -5. However, the considerable overlap in the runoff values in the three SOI categories suggests that the streamflow-ENSO link is relatively weak. As such, streamflow forecasts should be used cautiously, and the benefits and risks of using such forecasts must be properly taken into account.

The association between climate and ENSO has been realised for more than 50 years, but research into the link between ENSO and streamflow has only begun in earnest several years ago. Studies using Australian data include those by Simpson *et al.* (1993a, 1993b), Chiew *et al.* (1994, 1998) and Piechota *et al.* (1998).

The aim of this paper is to discuss seasonal streamflow forecasting and its use in water resources management. There are three parts to the paper. The first part explores the potential for forecasting streamflow by investigating the serial correlation in runoff and the lag correlation between Australian runoff and indicators of ENSO. The second part presents methods for forecasting streamflow for water resources management. The third part discusses the present and potential uses of seasonal streamflow forecasts to improve the management of land and water resources in Australia.

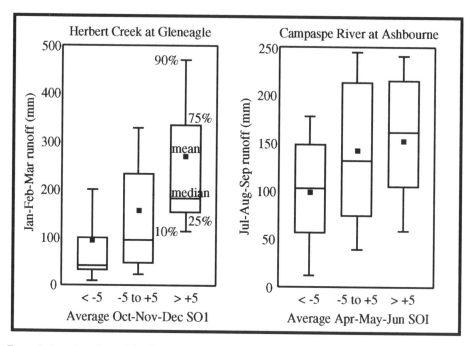

Figure 3. Box plots of runoff distributions from two Australian catchments showing means and percentiles of runoff over the three month periods Jan-Mar and Jul-Sep for three categories of average SOI in the preceding three month period.

2. Relationships Between Runoff and ENSO

2.1 DATA

Continuous monthly streamflow data from 45 unimpaired Australian catchments are used here (Figure 4). All sites have at least 48 years of data from 1949 to 1996. The catchments are located in east, south-east and south-west Australia. Apart from these locations, there are practically no other unimpaired sites in Australia with more than 48 years of recorded streamflow data. The catchment areas range from 25 to 6900 km^2 (mean of 1200 km^2 and median of 500 km^2). Concurrent monthly rainfall data for these catchments are estimated from 0.25 degree latitude by 0.25 degree longitude rainfall data gridded by the Australian Bureau of Meteorology. For the smaller catchments, the rainfall data estimated here will not be the same as the rainfall averaged over the catchments, but the temporal characteristics will be sufficiently similar for the purpose of this study.

The two most widely used indicators of ENSO, the Southern Oscillation Index (SOI) and sea surface temperature (SST) anomaly, are used here. The Troup SOI (Troup, 1965) series compiled by the Bureau of Meteorology is used. The Troup SOI is defined as ten times the standardised value of the Tahiti minus Darwin mean sea level pressure. Allan *et al.* (1996) compared several Troup and other SOI series, and concluded that they are all very highly correlated.

Twelve SST series derived by the Bureau of Meteorology Research Centre are used here. The different series represent the first 12 components of an empirical orthogonal function (EOF) analysis on Pacific and Indian Ocean SSTs (Drosdowsky and Chambers, 1998). Previous studies generally relate rainfall data to either the Wright or NINO3 and NINO4 SST series (see Reynolds, 1988; Wright, 1989; Chiew *et al.*, 1998), which give average SST anomalies over defined areas in the equatorial Pacific Ocean. The relationship between Australian rainfall and SOI is usually better than the relationship with the Wright, NINO3 or NINO4 series, but the analysis of streamflow and rainfall data here and a more detailed analysis of Australian rainfall data by Drosdowsky and Chambers (1998) indicate that the hydroclimate data are more strongly correlated to some of the SST EOFs than to the SOI, particularly for seasons and regions that are not strongly affected by the SOI.

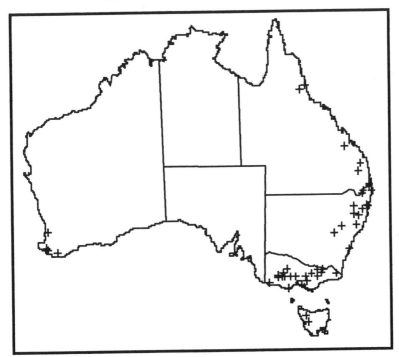

Figure 4. Locations of 45 unimpaired catchments in Australia with long-term streamflow data.

2.2 LAG CORRELATIONS

The lag correlations of the multiple linear regression between rainfall and streamflow versus values of explanatory variables (Troup SOI, 12 SST series, rainfall or streamflow) several months before are calculated to investigate the potential for forecasting rainfall and streamflow. Lag-one and lag-three correlations between the total rainfall and streamflow amounts in the four seasons and the explanatory variables are presented here, with the lag-n correlation defined as the highest correlation that can be obtained using information available n months before. For example, the lag-one correlation between spring (Sep-Oct-Nov) runoff and SOI is defined as the highest of the correlations between spring runoff versus Aug SOI, average Jul-Aug SOI, average Jun-Jul-Aug SOI, ..., and average Mar-Apr-May-Jun-Jul-Aug SOI. It should be noted that the lag-one correlation provides an indication of the potential for forecasting more than one month ahead, as the explanatory variables are related to the hydroclimate variables over the next three months.

Figures 5 to 8 summarise the results by presenting the median value of the lag correlations in catchments within each of the seven arbitrarily chosen regions (see Figure 4) - Queensland coast (seven catchments), New South Wales coast (eleven), New South Wales (Murray-Darling) (three), Victorian coast (nine), Victoria (Murray) (nine), Tasmania (two) and south-west coast of Western Australia (four). The plots present the median value of the serial correlation in rainfall and runoff, the lag correlation between rainfall/runoff versus Troup SOI, the highest of the lag correlations between rainfall/runoff versus the 12 SST series, the highest of the lag-correlations between rainfall/runoff versus two explanatory variables (previous month(s) rainfall/runoff and SOI, previous month(s) rainfall/runoff and SST, or SOI and SST) and the lag correlation between rainfall/runoff versus pervious month(s) rainfall/runoff, SOI and SST. The correlations vary among catchments, but the median values presented here provide a reasonable overview of the strength of the relationship between the variables in different parts of Australia.

It is possible that a non-linear relationship can explain the data better, and further fitting of the data using common non-linear regressions occasionally led to significantly higher correlations. The correlation coefficient is only one of many variables that describe the goodness of fit of the data, and for a more detailed analysis, other statistical terms (for example, those which account for the error terms in the linear regression) should also be considered. Other statistical methods have also been used to study the association between hydroclimate variables and ENSO (examples include the harmonic analysis method of Ropelewski and Halpert (1986) and the canonical correlation method in Nicholls (1987)), but they merely provide different interpretations of the same data. Compared to other statistical methods, the lag correlation coefficients of the linear regression presented here provide a more direct and consistent measure for exploring the potential for forecasting rainfall and streamflow several months ahead. Figure 9 shows typical correlations of 0.4 to 0.7. It should also be noted that, with the 48 years of data used in the analysis, the correlation is only statistically significant at the 5% level if it is greater than 0.28, 0.35 and 0.40 when one, two and three explanatory variables are used, respectively. These are shown as horizontal lines in the plots in Figures 5 to 8.

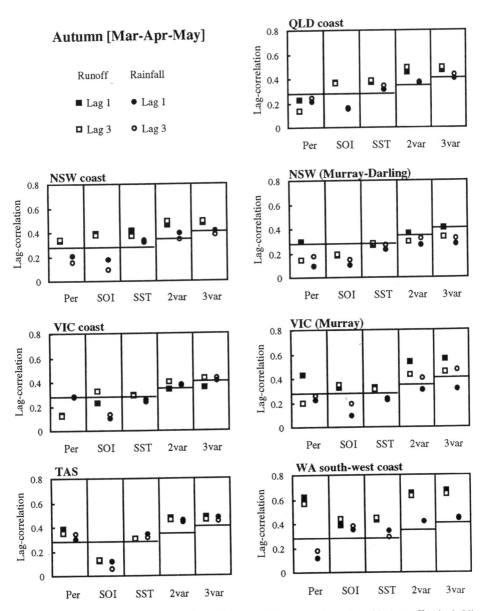

Figure 5. Median values of 1-month and 3-month lag correlations for autumn (Mar-May) runoff and rainfall with each of five explanatory variables (see text) by geographical catchment groupings

416

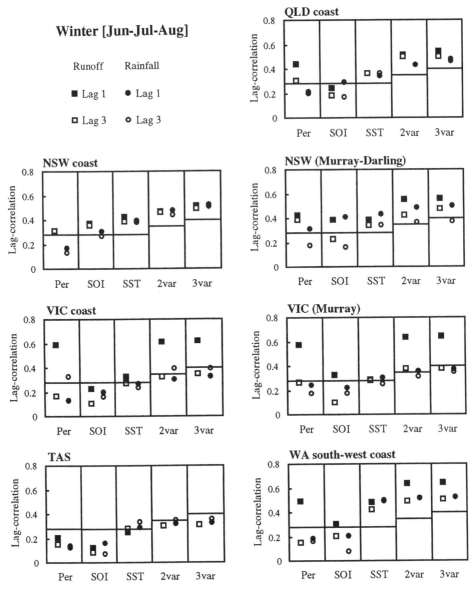

Figure 6. Median values of 1-month and 3-month lag correlations for winter (Jun–Aug) runoff and rainfall with each of five explanatory variables (see text) by geographical catchment groupings

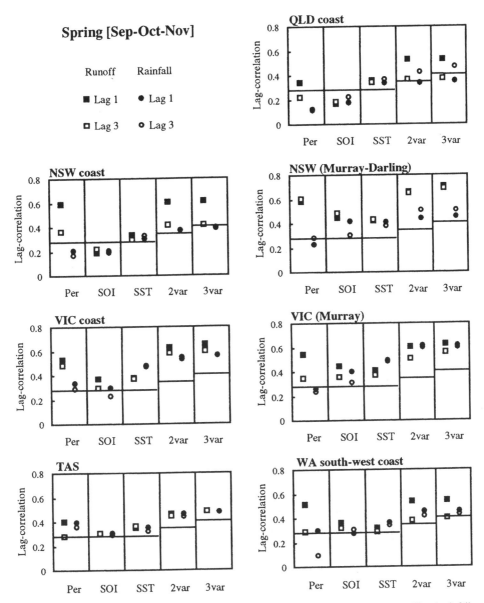

Figure 7. Median values of 1-month and 3-month lag correlations for spring (Sep-Nov) runoff and rainfall with each of five explanatory variables (see text) by geographical catchment groupings

418

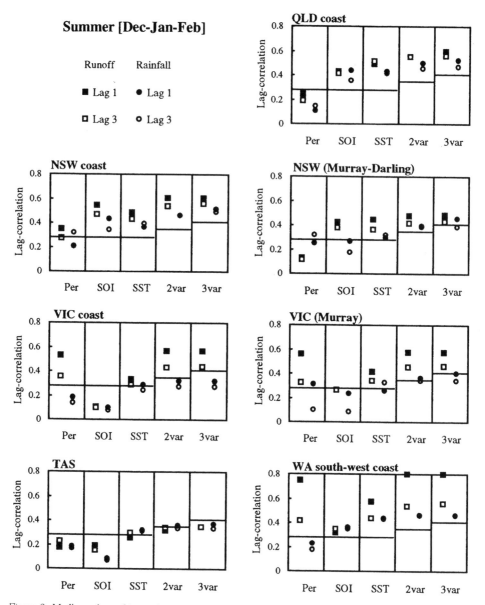

Figure 8. Median values of 1-month and 3-month lag correlations for summer (Dec-Feb) runoff and rainfall with each of five explanatory variables (see text) by geographical catchment groupings.

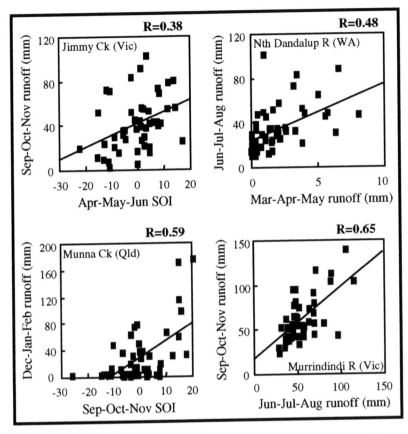

Figure 9. Sample scatter plots showing typical lag correlations for seasonal runoff.

The analysis indicates that the serial correlation (also referred to as persistence) in runoff is much higher than the persistence in rainfall. The persistence in rainfall is rarely significant (statistically significant at the 5% level of confidence), while the lag-one serial correlation in runoff is usually significant throughout the year. The persistence in runoff results from the delayed response in the rainfall-runoff process due to soil and groundwater storage, giving the streamflow data a memory of conditions over several months. This "memory" in the streamflow data disappears relatively quickly, with the difference between the lag-one and lag-three serial correlations being much greater than that between the lag-one and lag-three runoff-SOI or runoff-SST correlation.

The runoff-SOI correlations are generally greater than the rainfall-SOI correlations, but the runoff-SST and rainfall-SST correlations are relatively similar. The runoff-SST correlations are also generally higher than the runoff-SOI correlations. However, the SST series (SST EOFs in the Pacific and Indian Oceans) that gives the best correlation may be different for different seasons and parts of Australia. In the remainder of this section, only the runoff correlations are discussed. Nevertheless, the presentation of the rainfall correlations in Figures 5 to 8 is important, because there are very few unimpaired catchments with long streamflow records, and where these are not available, there may be a need to estimate streamflow from rainfall forecasts.

In **Victoria**, the lag-one runoff serial correlations are usually greater than 0.55 except in autumn. The median values of the lag-one and lag-three runoff-SST correlations in spring and summer and the runoff-SOI correlation in spring are about 0.4. The runoff-ENSO correlation is barely significant in autumn and winter. The lag correlation between spring runoff and two explanatory variables (runoff and SOI, or runoff and SST) is significantly higher than the persistence in runoff. The results suggest that the persistence in runoff can be used to forecast runoff with some success in winter, spring and summer, and that the use of SOI or SST with runoff data can give better forecasts of spring runoff than the use of runoff persistence alone, particularly for longer lead times.

In the **New South Wales coast**, the median value of the lag-one spring runoff serial correlation is 0.6. At other times of the year, the runoff-SOI and runoff-SST correlations are higher than the runoff serial correlation. The lag-one and lag-three runoff-SOI and runoff-SST correlations in several catchments are greater than 0.5 in summer. In the other seasons, although the correlations are significant, they are relatively low. The results suggest that the persistence in runoff can be used to forecast spring runoff with some success, while in summer, runoff may be forecast from the indicators of ENSO.

In the **Queensland coast**, practically all the runoff occurs in summer and autumn. Unlike the south-east coast, the persistence in runoff is not significant in these two seasons. In autumn, the median values of the lag-one and lag-three runoff-SOI and runoff-SST correlations are about 0.4. In summer, the lag-one and lag-three runoff-SST correlations in several catchments are greater than 0.5, and the runoff-SOI correlations are greater than 0.4. The results suggest that summer and autumn runoff can be forecast with some success using the SST or SOI.

In **Tasmania**, the lag-correlations are not significant in summer and winter. The persistence and the runoff-SST correlations are significant in autumn and spring, but are less than 0.4.

In the **south-west coast of Western Australia**, practically all the runoff occurs in winter and spring. The lag-one runoff serial correlation is greater than 0.5 throughout the year, but the lag-three serial correlation is much lower. The runoff-SST correlations are almost always higher than the runoff-SOI correlations, and are significant throughout the year. In the high flow seasons of winter and spring, the median values of the runoff-SST correlations (both lag-one and lag-three) are about 0.45 and 0.35 respectively. The results suggest that the high flows in winter and spring may be forecast using both persistence and SST data.

The analysis here shows a clear link between runoff and ENSO. The results suggest that runoff in many areas in Australia over certain seasons can be forecast with some success using SOI, SST or the persistence in the runoff. Although the explanatory variables are correlated, particularly the SOI and SST, the results indicate that the use of two or three variables can improve the runoff estimates compared to the use of a single variable. However, although the correlations are significant, they are relatively low, and forecasts derived from these data should be used with caution. Nevertheless, it should be noted that only the median values in each region are presented here, and the lag correlations in some catchments can be much higher than the median value.

3. Seasonal Streamflow Forecasting

The association between ENSO and climate is the scientific basis of long-range weather forecasts provided by research institutions and meteorological agencies in several countries. These forecasts are derived using either statistical models based on historical data or physical models of the atmosphere-ocean-land systems (see Barnston *et al.*, 1994; and Latif *et al.*, 1994). The physical coupled models have reasonable skill in forecasting equatorial Pacific SST (see Zebiak and Cane, 1987; Barnett *et al.*, 1993; and Chen *et al.*, 1995), but the routine use of such models for forecasting rainfall over small local areas is still many years away. The physical models also do not represent the surface hydrology adequately to estimate streamflow directly. For these reasons, only the statistical methods are discussed here.

The seasonal forecasts are usually derived from conditional probabilities between categories of hydroclimate and ENSO variables. For example, the historical streamflow and SOI data can be divided into categories (eg. high, medium, low), and the number of occurrences in each category are calculated. If there are 30 occasions with low SOI, and 24 of these occasions led to low runoff, then there is an 80% chance that a low SOI value will lead to low runoff. The forecasts derived from different statistical models can be combined to provide a consensus forecast. An optimal linear combination of the different forecasts is used, where the probability that a particular runoff category will occur is determined by applying weighted factors to the forecasts from the different models (see Fraedrich and Leslie, 1987; and Casey, 1995). The factors are chosen to optimise a skill score that measures some difference between the estimated and observed runoff. The half-Brier (Brier and Allen, 1951) and LEPS (Ward and Folland, 1991; and Potts *et al.*, 1996) are two commonly used skill scores.

The most commonly used explanatory variables are the previous month(s) SOI, SST and runoff, and forecast SST. The ENSO parameters can also be considered in different forms. These include the actual value, the trend, and a linear discriminant analysis (LDA) of the ENSO and runoff data. The LDA model evaluates the shift in the ENSO probability distributions for the different categories of runoff (see Ward and Folland, 1991). For example, the historical runoff data can be separated into several categories, and the corresponding SOI values from the prior season are fitted to a statistical distribution. The posterior probabilities of each runoff category are then conditioned upon the previous season's SOIs, to give the probability of observing the SOI prior to a runoff of a particular category.

In Australia, the SOI "phases" of Stone *et al.* (1996) have been used with some success. In this method, the historical hydroclimate data are divided into five categories, defined by the SOI values one month and two months ago (consistently positive SOI, consistently negative SOI, rapid SOI rise, rapid SOI fall and near zero). For a given category, the historical hydroclimate data in the category give a direct probability distribution for the forecast. Several hindsight studies of agricultural applications in various parts of Australia have shown the benefits of using seasonal rainfall forecasts based on the SOI phases (Meinke and Hochman, 2000). It is possible that seasonal forecasts derived from the different forms of ENSO indicators (like the SOI phase and LDAs) are better than forecasts estimated from the actual values of the indicators (as presented in the lag-correlation analysis in the previous section).

The following paragraphs present results of categorical streamflow forecasts for six catchments in the central east coast of Australia, derived using an optimal linear combination of models based on climatology, persistence LDA, SOI LDA and SST LDA. The Troup SOI and SST over the equatorial Pacific region defined by Wright (1989) were used. The forecasts are for Jan-Feb-Mar runoff, which are the three consecutive months with the largest flows. Sixty-six years of data (1927 to 1992) are used and forecasts are derived for below normal (lowest 30 percentile), normal (30 to 70 percentile) and above normal (upper 30 percentile) runoff categories. The weights in the optimal linear combination are optimised to minimise the half-Brier score. The analysis is carried out separately for each catchment, but results from all six catchments are presented together here in the same plots.

Figure 10 shows the observed streamflow percentiles and the forecast probabilities for below normal and above normal flows. The plots indicate that when the forecast probability for above normal runoff is high, the runoff is usually above normal, and likewise, high forecast probabilities for below normal runoff generally coincide with below normal runoff amounts (note that with climatology alone, there is a 30% expectation that runoff will fall in the above normal category, a 40% expectation of normal runoff and a 30% expectation of below normal runoff).

The plots in Figure 10 show the results when all 66 years of data are used to calibrate the model. Figure 11 presents the results for 20 years of independent testing period (1973 to 1992) using models developed from 46 years of historical data (1927 to 1972). The agreement between high forecast probability of above normal Jan-Feb-Mar runoff with above normal runoff amount and vice versa in Fgure 11, is less obvious than that in Figure 10. In fact, a comparison of the half-Brier scores indicate that the Jan-Feb-Mar runoff forecasts derived from the model are only slightly better than those based on climatology.

There are several possible explanations for the above result. It is possible that with the use of many forecast models based on explanatory variables that are correlated to one another, runoff estimates during the calibration period reflect only an artificial skill in the forecast. It is also possible that the poorer forecasts in the testing period compared to the calibration period may be due to different ENSO-runoff relationships over different periods of this century. However, the main difficulty in assessing the forecasting skill is because of the short historical data. An independent testing with 20 data points may not be sufficient, while the use of a longer testing period will result in a shorter record available for training the model. As a result of these limitations, it is likely that with increasing understanding of the atmosphere-ocean-land systems, physical models will eventually give better forecasts than statistical methods. However, at present, where there is sufficient local data, statistical methods provide better forecasts of rainfall and runoff for lead times of less than four or five months, compared to physical models.

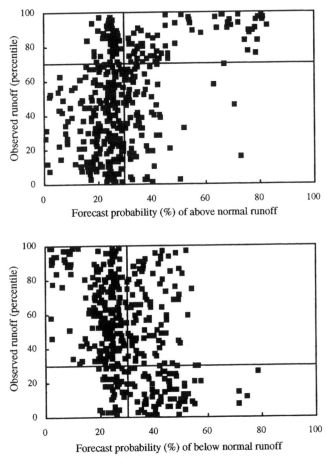

Figure 10. Observed percentile for Jan-Mar runoff versus forecast probability of above normal (top 30%) or below normal (bottom 30%) runoff for six catchments in the central east coast region of Australia (based on data for 66 years of calibration period).

424

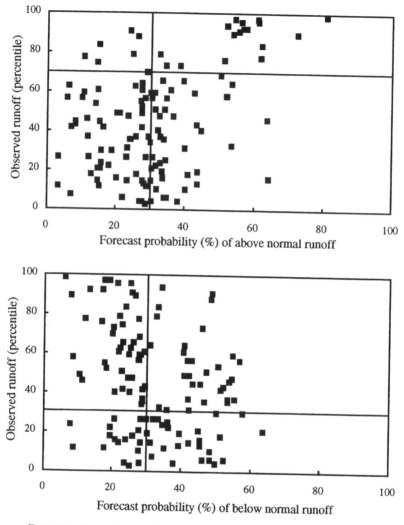

Figure 11. Observed percentile for Jan-Mar runoff versus forecast probability of above normal (top 30%) or below normal (bottom 30%) runoff for six catchments in the central east coast region of Australia (based on independent data for 20 years, with forecast model developed using 46 years of prior data).

Figure 12 summarises the statistical methods that can be used to forecast rainfall and runoff. Two points in Figure 12 are worth highlighting. First, the forecasts can also be presented as a continuous distribution (eg. probability that any particular runoff amount will be exceeded), rather than the discrete categories discussed above. The continuous forecast distribution can be determined using discrete phase categories (Stone *et al.,* 1996) using a LDA model that considers the runoff amounts over a range values of the explanatory variable (see Moss *et al.,* 1994), or by evaluating the error terms in a regression between runoff and the explanatory variables. Second, monthly streamflow is strongly correlated to rainfall, and as such, streamflow forecasts can also be estimated directly from rainfall forecasts via a rainfall-runoff model. This is particularly important because there are few unimpaired catchments in Australia with sufficiently long streamflow records to develop statistical streamflow forecast models directly. The authors are currently developing and assessing methods for forecasting probability of exceedance of runoff amounts (see Figure 13) using data from the catchments in Figure 4.

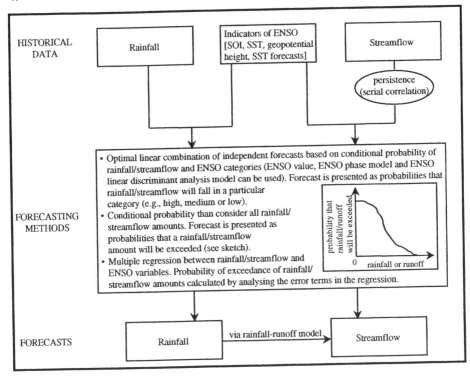

Figure 12. Statistical methods for forecasting rainfall and streamflow.

4. Seasonal Streamflow Forecasts for Land and Water Resources Management

Australian water authorities at present take a very conservative approach to managing water resources systems. Most authorities assume that the water resources available for the coming months consist only of the present stored water, less evaporative and distribution losses, plus the minimum historically observed inflows. The conservative approach is adopted mainly because of concerns about water shortfalls (Long and McMahon, 1996).

Serial correlation in streamflow has long been recognised, and some authorities have made use of it in managing the water resources. For example, the Murray-Darling Basin Commission (MDBC) use the serial correlation between spring and winter inflows into the Hume and Dartmouth reservoirs in their water allocation procedure. However, in 1995, the unusually large winter flows did not lead to large inflows in

426

spring. Despite MDBC's conservative use of the serial correlation, a water shortfall resulted in late spring and summer. Consternation developed because of the irrigation water shortfall, and as a result, most water authorities are now wary of using seasonal forecasts to determine water allocations.

The MDBC's experience is an example of the risk associated with using seasonal forecasts for water resources management. Nevertheless, managing land and water resources involves taking risks to maximise profits during good years and minimise losses during bad years. It is therefore unwise to ignore the additional information provided by the serial correlation in streamflow and the association between streamflow and ENSO when making water resources management decisions. It is likely that seasonal forecasts will improve so that decisions on water allocation for irrigation and environmental flows will be more realistically based. However, the consequences of using seasonal streamflow forecasts should be properly investigated so that authorities can make educated decisions. Here, forecasts of probability of exceedance of runoff amounts (see Figure 13) are more useful than forecast probabilities of runoff occurrences in discrete categories. The authors are currently involved in a collaborative project with several research organisations and rural water agencies that will assess the benefits and risks of using seasonal streamflow forecasts for different levels of conservatism of water resources management (for example, setting irrigation water allocation based on a 90% probability of exceedance of future inflows). The study will develop appropriate models for forecasting probability of exceedance of runoff amounts and investigate the benefits and risks of using the forecasts for reservoir management in several urban water and rural irrigation systems.

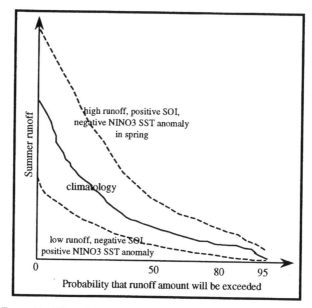

Figure 13. Schematic showing forecasts of probability of exceedance of runoff amounts for typical years with different hydroclimate conditions.

Conclusions

The interannual variability of runoff volumes and peak discharges in Australian rivers is greater than elsewhere in the world. As the design and management of water resources systems involves coping with this variability, forecasts of streamflow several months ahead would be invaluable. The analysis here indicates that the serial correlation in runoff and the association between runoff and ENSO can be used to forecast streamflow with some success. In particular, the persistence in runoff, and to a lesser extent, the SOI and SST, can be used to forecast winter, spring and summer runoff in the Murray and the south-east and south-west coasts of Australia. Spring and summer runoff in the east coast may be forecast from SOI, SST and the persistence in runoff, and the summer and autumn flows in the north-east coast can be forecast using SOI and SST.

It is likely that seasonal streamflow forecasts can benefit the management of water resources and allow decisions on irrigation water allocation and environmental flows to be more realistically based. However, because the runoff serial correlation and runoff-ENSO association are not sufficiently strong to consistently predict runoff accurately, the benefits and risks of using seasonal forecasts must be properly assessed. There are two main research issues to be addressed. The first is the development of appropriate models for forecasting probabilities of exceedance of runoff amounts and the second involves evaluating the benefits and risks of using seasonal forecasts for different levels of conservatism of water resources management.

References

Allan, R.J., Beard, G.S., Close, A., Herczeg, A.L., Jones, P.D. and Simpson, H.J. (1996) Mean sea level pressure indices of the El Nino-Southern Oscialltion: Relevance to discharge in south-eastern Australia. Divisional Report 96/1, CSIRO Division of Water Resources, Canberra. pp.23.

Barnston, A.G., van den Dool, H.M., Zebiak, S.E., Barnett, T.P. and others (1994) Long-lead seasonal forecasts - Where do we stand? *Bulletin of the American Meteorological Society* 75, 2097-2114.

Bartnett, T.P., Latif, M., Graham, N, Flugel, M., Pazan, S. and White, W. (1993) ENSO and ENSO-related predictability. 1. Prediction of equatorial Pacific sea surface temperatures with a hybrid coupled ocean-atmosphere model. *Journal of Climate* 6, 1545-1566.

Brier, G.W. and Allen, R.A. (1951) Compendium of Meteorology. American Meteorological Society. pp.841-848.

Casey, T. (1995) Optimal linear combination of seasonal forecasts. *Australian Meteorological Magazine* **44**, 219-224.

Chen, D., Zebiak, S.E., Busalacchi, A.J. and Cane, M.A. (1995) An improved procedure for El Nino forecasting: implications for predictability. *Science* **269**, 1699-1702.

Chiew, F.H.S., McMahon, T.A., Dracup, J.A. and Piechota, T.A. (1994) El Nino/Southern Oscillation and streamflow patterns in south-east Australia. Civil Engineering Transactions, Institution of Engineers, Australia, CE36(4). pp.285-291.

Chiew, F.H.S., Piechota, T.C., Dracup, J.A. and McMahon, T.A. (1998) El Nino/Southern Oscillation and Australian rainfall, streamflow and drought: links and potential for forecasting, *Journal of Hydrology* **204**, 138-149.

Drosdowsky, W. and Chambers, L. (1998) Near global sea surface temperature anomalies as predictors of Australian seasonal rainfall. Research Report 65, Bureau of Meteorology Research Centre, Melbourne, Australia. pp39.

Finlayson, B.L. and McMahon, T.A. (1991) Runoff variability in Australia: causes and environmental consequences. Proceedings of the International Hydrology and Water Resources Symposium,

428

October 1991, Perth. Institution of Engineers, Australia, National Conference Publication, 91/22 pp.504-511.

Fraedrich, K. and Leslie, L.M. (1987) Combining predictive schemes in short-term forecasting. *Monthly Weather Review* **115**, 1640-1644.

Haines, A.T., Finlayson, B.L. and McMahon, T.A. (1988) A global classification of river regimes. *Applied Geography* **8**, 255-272.

Hosking, J.R.M. (1990) L-moments: analysis and estimation of distributions of order statistics. *Journal of the Royal Statistics Society* B **52**, 105-124.

Latif, M., Bartnett, T.P., Cane, M.A., Flugel, M., and others (1994) A review of ENSO prediction studies. *Climate Dynamics* **9**, 167-179.

Long, A.B. and McMahon, T.A. (1996) Review of research and development opportunities for using seasonal climate forecasts in the Australian water industry. Land and Water Resources Research and Development Corporation, Occasional Paper CV02/96. pp.46.

McMahon, T.A. and Mein, R.G. (1986) River and Reservoir Yield. Water Resources Publication, Colorado, USA. pp.368.

McMahon, T.A., Finlayson, B.L., Haines, A.T. and Srikanthan, R. (1992) *Global Runoff - Continental Comparisons of Annual Flows and Peak Discharges*. Catena Paperback, Cremlingen-Destedt. pp.166.

Meinke, H. and Hochman, Z. (2000) Using seasonal climate forecasts to manage dryland crops in northern Australia – Experiences from the 1997/98 seasons, in G.L. Hammer, N. Nicholls, and C. Mitchell (eds.), *Applications of Seasonal Climate Forecasting in Agricultural and Natural Ecosystems – The Australian Experience*. Kluwer Academic, The Netherlands. (this volume)

Moss, M.E., Pearson, C.P. and McKerchar, A.I. (1994) The Southern Oscillation Index as a predictor of the probability of low streamflows in New Zealand. *Water Resources Research* **30**, 2717-2723.

Nicholls, N. (1987) The use of canonical correlation to study teleconnections. *Monthly Weather Review* **115**, 393-399.

Piechota, T.C., Chiew, F.H.S., Dracup, J.A and McMahon, T.A. (1998) Seasonal streamflow forecasting in eastern Australia and the El Nino-Southern Oscillation. *Water Resources Research* **34**, 3035-3044.

Potts, J.M., Folland, C.K., Jolliffe, I.T. and Sexton, D. (1996) Revised "LEPS" scores for assessing climate model simulations and long-range forecasts. *Journal of Climate* **9**, 35-53.

Reynolds, R.W. (1988) A real-time global sea surface temperature analysis. *Journal of Climate* **1**, 75-86.

Ropelewski, C.F. and Halpert, M.S. (1986) Global and regional scale precipitation associated with the El Nino/Southern Oscillation. *Monthly Weather Review* **115**, 1606-1626.

Simpson, H.J., Cane, M.A. Herczeg, A.L. and Zebiak, S.E. (1993a) Annual river discharge in southeastern Australia related to El Nino-Southern Oscillation forecasts of sea surface temperatures. *Water Resources Research* **29**, 3671-3680.

Simpson, H.J., Cane, M.A., Lin, S.K., Herczeg, A.L. and Zebiak, S.E. (1993b) Forecasting annual discharge of River Murray, Australia from a geophysical model of ENSO. *Journal of Climate* **6**, 386-390.

Stone, R.C., Hammer, G.L. and Marcussen, T. (1996) Prediction of global rainfall probabilities using phases of the Southern Oscillation Index. *Nature* **384**, 252-255.

Troup, A.J. (1965) The Southern Oscillation. *Quarterly Journal of the Royal Meteorological Society* **91**, 490-506.

Wallis, R.H., Matalas, N.C. and Slack, J.R. (1974) Just a moment. *Water Resources Research* **10**, 211-219.

Ward, M.N. and Folland, C.K. (1991) Prediction of seasonal rainfall in the north Nordeste of Brazil using eigenvectors of sea surface temperature. *International Journal of Climatology* **11**, 711-743.

Wright, P.B. (1989) Homogenised long-period Southern Oscillation indices. *International Journal of Climatology* **9**, 33-54.

Zebiak, S.E. and Cane, M.A. (1987) A model El Nino-Southern Oscillation. *Monthly Weather Review* **115**, 2262-2278.

THE EFFECT OF CLIMATE ON THE INCIDENCE OF VECTOR-BORNE VIRAL DISEASES IN AUSTRALIA: THE POTENTIAL VALUE OF SEASONAL FORECASTING

JOHN MACKENZIE[1], MICHAEL LINDSAY[2] AND PETER DANIELS[3]

[1]*Department of Microbiology and Parasitology*
University of Queensland
St. Lucia, Qld 4072, Australia

[2]*Department of Microbiology*
University of Western Australia
Nedlands, WA 6907, Australia

[3]*CSIRO Australian Animal Health Laboratory*
Private Bag 24
Geelong, Vic 3220, Australia

Abstract

Over 80 vector-borne viruses have been described in Australia, although relatively few have been associated with clinical disease in humans or domestic animals. These viruses are classified as arboviruses (an abbreviation of arthropod-borne viruses), and by definition are transmitted between blood-feeding arthropod vectors (mosquitoes, biting midges and ticks) and vertebrate hosts, undergoing cycles of replication in both hosts alternately. The major viral diseases of humans are mosquito-borne and include Ross River and Barmah Forest viruses, both members of the *Alphavirus* genus of the family *Togaviridae*; and Murray Valley encephalitis, Kunjin, Japanese encephalitis, and dengue viruses of the *Flavivirus* genus of the family *Flaviviridae*. The major viral diseases of livestock are transmitted by biting midges (*Culicoides* sp.) and include bovine ephemeral fever virus, a member of the *Ephemerovirus* genus of the family *Rhabdoviridae*; bluetongue viruses of the *Orbivirus* genus of the family *Reoviridae*; and Akabane virus, a member of the Simbu group in the *Bunyavirus* genus of the family *Bunyaviridae*.

These viruses are normally transmitted through an alternating cycle in arthropods (all insects in the present context) and vertebrates. This cycle consists of a complex and dynamic interaction between the virus, its vertebrate host(s), its insect vector(s) and the environment. Female insects acquire an arbovirus while obtaining a bloodmeal from a vertebrate host. Virus is ingested and replicates in the insect's tissues. It is then

G.L. Hammer et al. (eds.), The Australian Experience, 429–452.

transmitted to new vertebrate host(s) via the saliva each time an infected insect takes another bloodmeal. The virus then replicates in the new vertebrate host during which it is released into the blood stream causing a viraemia, from which other insects may acquire the virus if they feed on that host. The viraemia in the vertebrate host only lasts for a few days, after which the vertebrate host becomes immune, usually for life.

The alternating replicative cycles of the virus in the insect vectors and vertebrate hosts are profoundly linked to a complex web of environmental factors associated principally with the effects of climate (long-term weather patterns) and weather (moisture, temperature, humidity, wind velocity, atmospheric pressure) on the biology of the insect vector, especially breeding, development rate, nutritional status, host-seeking behaviour, diapause, as well as many other factors. The environment also plays a crucial role in migration, breeding, immune status, behaviour, and other factors pertaining to the vertebrate hosts, and may determine the likelihood of human or animal exposure to arboviruses by altering host behaviour and demographics. Finally, the environment also has a significant effect on the replication of the virus in the insect vector.

Thus taking these factors into account, it seems highly likely that only minor changes in climate and weather will alter the potential for outbreaks of human or animal diseases. This paper describes the complex interactions between vectors, vertebrate hosts, environmental factors, and viruses; the influence of climate and weather on outbreaks of arboviral disease, providing examples within the Australian context; and explores the potential for seasonal forecasting and disease risk assessments.

1. Introduction

A number of vector-borne viruses or arboviruses (a shortened form of _ar_thropod-_bo_rne viruses) capable of causing human and animal diseases are known to occur in Australia (Mackenzie _et al._, 1994a). These viruses are maintained in nature by a biological transmission cycle between susceptible vertebrate hosts and haematophagous arthropod vectors, comprising mosquitoes, _Culicoides_ sp. (biting midges), and ticks, with the virus able to replicate in both vector and host.

Of the viruses associated with human diseases, all are mosquito-borne, and belong to three virus families. These families and their major members are: flaviviruses, including Murray Valley encephalitis (MVE), Kunjin, Kokobera, Stratford, Alfuy, Edge Hill, Dengue, and most recently, Japanese encephalitis (JE) viruses; alphaviruses, including Ross River (RR), Barmah Forest (BF) and Sindbis viruses; and two bunyaviruses, Gan Gan and Trubanaman viruses (Mackenzie _et al._, 1994b). The most important of these with respect to human health are MVE, Kunjin, dengue, JE and RR viruses. The major livestock viruses of veterinary importance are bluetongue (BT), bovine ephemeral fever (BEF), and Akabane viruses. They are all transmitted by _Culicoides_ sp., but BEF can also be transmitted by mosquitoes (Mackenzie _et al._, 1994a). In addition, RR, Kunjin and JE viruses can cause infections in horses (Pascoe *et al.*, 1978; Badman *et al.*, 1984; Burke and Leake, 1988; Azuolas, 1997), and JE virus causes abortions and foetal wastage in pigs (Burke and Leake, 1988).

The insect-vertebrate host-insect transmission cycle is a highly complex and dynamic interaction between the virus, its vertebrate host(s), vector(s), and the environment (Woodring *et al.*, 1996). Indeed the ecology of all arboviruses tends to be finely balanced between these four parameters, but in particular, it is intimately associated with, and affected by, environmental conditions. Thus possible changes in climate and weather, which strongly influence environmental conditions, could substantially affect the transmission cycles of mosquito-borne viruses. This has led to a growing concern that changing climatic conditions associated with global warming (the 'Greenhouse Effect') may alter the ecology of these viruses, resulting in a greater geographic spread and an increased incidence of disease. The relationship between arboviral disease and climate has been widely recognised, and has been reviewed extensively both in general terms (eg. Reiter, 1988; Sellers, 1980) and with respect to global warming (Liehne, 1988; Lindsay *et al.*, 1993a; Mackenzie *et al.*, 1993b; Reeves *et al.*, 1994; Patz *et al.*, 1996, 1998; Jetten and Focks, 1997; Lindsay and Mackenzie, 1997; Gubler, 1998; Russell, 1998a). This chapter will therefore describe the major arboviruses causing human and animal diseases, and their complex ecology with respect to arthropod vector and vertebrate host, and will then focus on some of the environmental factors associated with climate that are essential components in the ecology and transmission of these viruses, and discuss the potential role of seasonal forecasting.

2. Vectors and Vertebrate Hosts: Two Sides of the Transmission Cycle

The arboviral transmission cycle involves female insects and vertebrate hosts. Female insects generally require blood as a source of nutrients to produce yolk protein for their eggs, and therefore acquire blood by feeding on various species of vertebrate host. Virus is ingested in the bloodmeals, and then replicates in the tissues of the insects, eventually reaching high titres in the salivary glands. This period of viral replication or 'incubation' in the insects is known as the extrinsic incubation period. Once infected by an arbovirus, the insects generally remain infectious for the rest of their lives. Onward transmission of the virus occurs whenever a female mosquito takes further bloodmeals, during which virus is introduced to new vertebrate hosts in the saliva. During the subsequent replication of the virus in the new vertebrate host, the virus is released into the blood causing a viraemia, from which other insects can acquire the virus if they feed on that host. Viraemia in susceptible hosts usually lasts for a few days, after which the host normally develops life-long immunity to the virus.

Over 45 species of mosquito have been implicated in the transmission of arboviruses associated with human disease in Australia through virus isolation (Russell, 1995). However, only a few have been confirmed as vectors by transmission studies. The most important vector species are: *Culex annulirostris*, a fresh water-breeding species, and the major vector of MVE and Kunjin viruses, and an important vector of RR virus; *Ae. vigilax*, a saltmarsh breeding species in eastern and northern Australia, and a major vector of RR and BF viruses; *Ae. camptorhynchus*, a saltmarsh breeding species in southern Australia, and a major transmitter of RR and BF viruses; and *Ae. aegypti*, the only vector of dengue in Australia.

With respect to arboviruses associated with livestock diseases in Australia, the major vectors are *Culicoides* species. *Culicoides brevitarsis* is the major vector of Akabane and the related Aino viruses, and of BT viruses throughout most of their range. However, in the Northern Territory (NT), BT viruses are also transmitted effectively by *C. wadai*, *C. actoni* and *C. fulvus* (Standfast *et al.*, 1985a; Melville *et al.*, 1996a). *C. brevitarsis* is also a major vector of BEF, but this virus is unusual among livestock arboviruses in that it is also spread by mosquitoes, especially *Cx. annulirostris* and to a lesser extent by *Anopheles bancroftii*. The geographic distribution of these different *Culicoides* species, and of bluetongue, Akabane and BEF viruses is described by Muller (1995). It is clear that *C. brevitarsis* is by far the most important vector of livestock viruses (Muller *et al.*, 1982), with a geographic distribution which extends from the north of Western Australia (WA), the northern half of the NT, most of Queensland, to eastern New South Wales (NSW) (Muller, 1995). It breeds in discrete cow dung pats, and moisture and warmth are essential for breeding. Despite its importance in disease transmission, infection rates of *C. brevitarsis* for bluetongue are very low (0.3%) compared to *C. fulvus* (62%), *C. wadei* (11%), and *C. actoni* (2%), but these latter three species are restricted to areas of northern Australia with an annual summer rainfall in excess of 800mm (Standfast *et al.*, 1985a), whereas *C. brevitarsis* extends through areas with rainfall in excess of 700mm (Standfast *et al.*, 1979; Standfast and Muller, 1989). It has been suggested from modeling that *C. wadai* has not completed colonization of all areas where the climate is favourable (Standfast and Maywald, 1992).

The transmission cycle is relatively specific with respect to the vector and host. For some viruses, only certain insects are receptive to infection with the virus and able to transmit it to a new vertebrate host. Thus, in Australia, only one mosquito species, *Ae. aegypti*, is able to transmit dengue viruses; other species are not able to transmit, either being refractory to infection, or the virus is unable to invade the salivary glands or to replicate to a sufficient titre to enable it to be transmitted to a new host. *Ae. aegypti* mosquitoes are peri-domestic mosquitoes, breeding all year round in empty receptacles, old car tyres, plant pot holders, and blocked gutters around human habitation, with a geographic range restricted to northern Queensland. In contrast to the dengue transmission cycle, RR virus is transmitted seasonally by different mosquito species in different situations, and thus the epidemiology of the disease varies with region and circumstance (Russell, 1998a). RR virus has been isolated from over 30 species of mosquitoes belonging to 6 genera (Mackenzie *et al.*, 1994b; Russell, 1995; Russell, 1998b), many of which have been shown to be capable of transmission in the laboratory, or have been implicated in transmission on epidemiological grounds. The ability to transmit is known as 'competence'. A knowledge of the competent vectors of a given virus is essential for understanding the ecology of that virus and for determining the effects of climate and weather on the incidence and spread of disease.

The vertebrate hosts are equally important in the ecology of arboviruses. To be a successful vertebrate host, the virus must be able to replicate in the host to a high titre in order to provide a sufficient level of viraemia to be able to infect mosquitoes. Thus, pigs are excellent amplifying hosts of JE virus because of the high level of viraemia generated during infection with the virus, whereas the titre in cattle is low and, even although many mosquitoes may prefer cattle for their bloodmeals, there is insufficient virus present in the cattle blood to infect the mosquitoes (Burke and Leake, 1988). For dengue viruses, the only vertebrate hosts in Australia are humans (Gubler, 1988).

Cattle (*Bos* spp.) and the closely related buffaloes (*Bubalus bubalis*) are the main vertebrate hosts for the arboviruses of livestock importance, BT, BEF and Akabane viruses. BT virus also infects small ruminants and causes disease mainly in sheep, particularly European breeds, but is maintained in nature mainly by cattle (Daniels *et al.*, 1996). BT does not cause disease in sheep in Australia, largely because the two serotypes currently circulating in eastern Australia are avirulent and because the vectors do not commonly penetrate sheep farming areas. Akabane virus causes disease in cattle, sheep and goats, but like BT virus, is propagated in cattle. The importance of cattle to *Culicoides* transmission is also associated with vector preferences; the extensive hair coat of cattle offers midges a greater surface area of accessible skin from which to feed than is available on sheep whose fleece they cannot penetrate (Muller and Murray, 1977).

These parameters suggest that only certain mosquitoes are competent to act as vectors of any particular arbovirus, and only certain specific animals can act as vertebrate hosts for that arbovirus. However, arbovirus transmission cycles are more complex than this would suggest; variations in vector competence may occur in different geographic populations of mosquito species, and some mosquito species may be able to act as vectors for some virus strains but not others. Vertebrate hosts also vary in their susceptibility to infection and in their attractiveness to different insect species.

3. The Viruses: A Brief Description of the Major Arboviruses Associated with Human and Animal Disease

There have been a number of reviews of Australian arboviruses. Although some of these are general reviews of arboviruses associated with human infections (e.g. Mackenzie *et al.*, 1994a; 1994b; 1996b; 1998; Russell, 1995; 1998a; 1998b) or animal infections (Doyle, 1992; Mackenzie *et al.*, 1994a), others are more specific for particular viruses or combinations of viruses, including MVE and Kunjin viruses (Marshall, 1988), RR virus (Marshall and Miles, 1984; Kay and Aaskov, 1989), dengue viruses (Mackenzie *et al.*, 1996a), Akabane and Aino viruses (St George and Standfast, 1989b), BEF virus (St George and Standfast, 1989a; Uren, 1989) and BT viruses (Gard and Melville, 1992; Daniels *et al.*, 1996).

All flaviviruses exhibit cross-reacting antigens (Monath and Heinz, 1996). MVE, Kunjin, Alfuy, and Japanese encephalitis viruses are all closely related and belong to the Japanese encephalitis serological complex of flaviviruses. Kokobera and Stratford have also been classified in the Japanese encephalitis serological complex, but are more distantly related (Poidinger *et al.*, 1996). Edge Hill virus is a member of the Uganda S serological complex, and dengue viruses form a single serological complex distinct from other serological groupings. The alphaviruses RR, Sindbis and BF viruses are from different serological groups, and differ significantly in antigenic and genetic characteristics. Gan Gan and Trubanaman viruses are members of the family Bunaviridae, and are in the Mapputta serological complex within an unassigned genus (Mackenzie *et al.*, 1994a).

3.1 MURRAY VALLEY ENCEPHALITIS (MVE) AND KUNJIN VIRUSES

MVE virus is the major Australian encephalogenic flavivirus (Marshall, 1988), and the aetiological agent of a serious and sometimes fatal disease known previously as Murray Valley encephalitis but more recently as Australian encephalitis (Marshall, 1988; Mackenzie *et al.*, 1993a). The virus is endemic in parts of northern Australia, especially in the Kimberley region of WA, NT, and possibly northern Queensland (Mackenzie *et al.*, 1994b; Marshall, 1988). The major transmission cycle is between ardeid birds, particularly the Nankeen night heron (*Nycticorax caledonicus*), and *Cx. annulirostris* mosquitoes (Marshall, 1988). However, MVE virus strains have been isolated from a number of other mosquito species, especially in the north of WA (Mackenzie *et al.*, 1994b; Russell, 1995), and serological studies have indicated that a wide range of domestic and wild animals can be infected by the virus (Marshall, 1988), but the roles of these other vector and host species in natural transmission cycles is not known.

Kunjin virus is a minor cause of encephalitis in Australia, Kunjin encephalitis (Mackenzie *et al.*, 1993a), as well as fever with or without myalgia. It shares a similar natural transmission cycle with MVE virus, and like MVE, it has been associated with a number of mosquito (Mackenzie *et al.*, 1994b; Russell, 1995) and vertebrate species. There appears to be some ecological differences between MVE and Kunjin as the latter occurs more widely than MVE in northern Australia, and incursions occur more commonly in southeastern Australia.

There has been a close relationship between heavy rainfall events associated with flooding and MVE virus activity, as indicated by human cases, sentinel chicken seroconversions, and virus isolations (Broom *et al.*, 1993; Broom *et al.*, 1997). It has been suggested that this increased virus activity in the mosquito population may in turn lead to a spill-over of infected mosquitoes into epizootic areas, and as MVE is believed to be able to persist by vertical transmission in desiccation-resistant eggs of certain *Aedes* sp. (Broom *et al.*, 1995), epizootic areas may thus become seeded with virus-infected mosquito eggs (Mackenzie and Broom, 1998). Subsequently, when the next heavy rainfall event and flooding occurs, the eggs hatch and give rise to infected adults, initiating a new episode of virus activity. In support of this contention, MVE virus has been isolated from male *Ae. tremulus* mosquitoes (Broom *et al.*, 1995). Extreme weather events have also been associated with the movement of MVE virus to southeastern Australia to initiate epidemic activity (Forbes, 1978).

3.2 ALFUY, KOKOBERA AND STRATFORD VIRUSES

These viruses have been associated with human infections, but their role in human disease has still to be confirmed (Mackenzie *et al.*, 1994b). They have been isolated from different parts of Australia, and from different mosquito species. Alfuy virus is closely related to MVE and JE viruses and is believed to have a transmission cycle involving wild birds and *Culicine* mosquitoes. Kokobera and Stratford viruses are closely related to each other but more distant to other members of the serological complex. They appear to have transmission cycles involving marsupials rather than birds.

3.3 EDGE HILL VIRUS

Edge Hill virus has only once been associated with human disease (Aaskov *et al.*, 1993). It is believed to have a natural transmission cycle involving marsupials and various mosquitoes, including *Cx. annulirostris, Ae. normanensis, Ae. vigilax,* and *An. amictus* (Mackenzie *et al.*, 1994b; Russell, 1995).

3.4 DENGUE VIRUSES

Dengue is not endemic in Australia, but rather has to be re-introduced by a viraemic traveller arriving in a dengue-receptive area to initiate new epidemic activity (Mackenzie *et al.*, 1996a; Mackenzie *et al.*, 1998; Russell, 1998b). Humans are the only vertebrate hosts and *Ae. aegypti* the only vector mosquito species in Australia (Mackenzie *et al.*, 1996a). Thus the area believed to be receptive to potential dengue virus activity is restricted to the geographic range of *Ae aegypti* mosquitoes, which is currently believed to extend from the Torres Strait in the north to about Mackay in the south and Mt Isa in the west. *Ae. aegypti* is one of the most efficient mosquito vector species, being the major vector of dengue in all tropical regions of the world, and the major urban transmitter of yellow fever virus in Africa and South America. As described above, it breeds around human habitation. A description of dengue fever and dengue haemorrhagic fever, and complications of the pathology of dengue infections can be found in a review by Gubler (1988).

3.5 JAPANENESE ENCEPHALITIS VIRUS

JE virus is closely related antigenically and genetically to MVE, Alfuy and Kunjin viruses. It differs, however, in one crucial aspect of its ecology; the involvement of pigs as major amplifying hosts (Burke and Leake, 1988). Thus JE has a maintenance cycle between ardeid birds and *Culicine* mosquitoes, and an amplifying cycle between mosquitoes and pigs. *Cx. tritaeniorhynchus* is the major vector species in Asia, and *Cx. annulirostris* is believed to be the major vector species in Australasia (Hanna *et al.*, 1996; Ritchie *et al.*, 1997; Mackenie *et al.*, 1997). It is one of the most important and rapidly spreading of the emerging viruses. Its spread in Asia has been facilitated by changing land use to irrigated agriculture, sometimes assisted by deforestation and dam building, and in localities where intense transmission cycles have been reported, transmission has always been facilitated by the presence of pig pens close to areas of rice fields.

JE was first recognised in northern Australia in 1995 when an outbreak occurred on Badu Island in the Torres Strait which resulted in three cases and two deaths (Hanna *et al.*, 1996; Ritchie *et al.*, 1997). The virus causing the outbreak appears to have originated from Western Province of Papua New Guinea (PNG) (Johansen *et al.*, 1997; Mackenzie *et al.*, 1997; Mackenzie *et al.*, 1998; Mackenzie, 1999a,b). A further case of JE was observed in the Torres Strait in 1998, followed shortly thereafter by the first case on mainland Australia (Hanna *et al.*, 1999). Strains of JE virus isolated from *Cx. annulirostris* mosquitoes and pigs in the Torres Strait, from pigs in the northern Peninsula area of mainland Australia, and from mosquitoes in PNG, were virtually identical indicating a common virus pool. Although JE virus has not yet become enzootic in Australia, its potential to become established and to subsequently spread in mosquito-bird and mosquito-feral pig cycles has been clearly recognised (Mackenzie, 1997; Mackenzie *et al.*, 1998). It thus poses a significant threat to the pig industry and to livestock exports.

3.6 ROSS RIVER AND BARMAH FOREST VIRUSES

RR virus is the most common human arboviral disease in Australia (Mackenzie *et al.*, 1998), and is the aetiological agent of an epidemic polyarthritis now called Ross River virus disease. The symptoms of the disease have been described in detail elsewhere (Kay and Aaskov, 1989; Mackenzie and Smith, 1996; Flexman *et al.*, 1998). It is found in all Australian states, including northern Tasmania, as well as PNG, and possibly the Solomon Islands. Up to 7,000 cases have been reported annually. The largest single outbreak occurred in 1979-80 in various Pacific Island nations with over 50,000 cases; the outbreak was probably initiated by the movement of a viraemic tourist from Australia.

BF virus causes a polyarthritic disease, which closely resembles that of RR virus (Mackenzie and Smith, 1996; Flexman *et al.*, 1998), but to distinguish it from RR virus, it is now referred to as Barmah Forest virus disease. It is believed to account for approximately 10% of cases of mosquito-transmitted epidemic polyarthritis. BF virus is probably found in all mainland states of Australia, although the evidence for South Australia is still sketchy, but it has not been reported from Tasmania or PNG. Recent evidence has suggested that it may only recently have reached WA (Lindsay *et al.*, 1995a; Lindsay *et al.*, 1995b).

The major vertebrate hosts of RR virus are believed to be macropods, especially kangaroos, but a number of other domestic and wild animals have been implicated on serological grounds, including horses, rabbits, and fruit bats. Humans can also act as vertebrate hosts to a limited extent in urban transmission cycles. As described above, RR virus has been isolated from over 30 species of mosquitoes, although only a few have so far been shown to be involved in natural transmission cycles. RR virus is transmitted seasonally by different mosquito species in different situations. In coastal areas of northern and northeastern Australia, RR virus exhibits an endemic pattern of disease with virus activity and human cases occurring all year, but elsewhere in Australia, and particularly in southern areas, virus activity is seasonal and epidemic. Epidemics and outbreaks are driven by environmental conditions, especially heavy rainfall events and flooding, and by tidal inundation of saltmarshes (Lindsay et al., 1993a; Lindsay and Mackenzie, 1997). Virus can persist in the environment during adverse conditions by vertical transmission in desiccation-resistant mosquito eggs (Lindsay et al., 1993b; Dhileepan et al., 1996).

The transmission cycles of BF virus are believed to be similar to those of RR virus, involving the same vectors and vertebrate hosts, but epidemic activity does not always occur at the same time for the two viruses, which would suggest that there are other as yet unknown parameters involved in the genesis of outbreak activity. Much less is known, however, about the role of different animals in BF virus transmission, although the incidence of antibodies appears to be significantly less than for RR virus (Aldred et al., 1990; C.A. Johansen, M.D. Lindsay and J.S. Mackenzie, unpublished results).

3.7 OTHER ARBOVIRUSES ASSOCIATED WITH HUMAN DISEASE

Sindbis virus has the widest geographic range of all arboviruses, extending from Africa to northern Europe, Asia and Australia. It is the virus most frequently isolated from mosquitoes in Australia. The natural transmission cycle involves birds and many different mosquito species, but unlike African and European strains of Sindbis virus, it only rarely causes human disease in Australia (Mackenzie et al., 1994b; Russell, 1998b).

Trubanaman and Gan Gan viruses have been associated with human disease; the former on serological grounds alone, and the latter with a polyarthralgic-like disease. However, serological studies would suggest that most infections are sub-clinical. The natural transmission cycles are not well understood; the viruses have been isolated from several mosquito species, although whether they are actively involved in transmission is unknown, and the vertebrate hosts are believed to be wild and domestic animals (reviewed in Mackenzie et al., 1994b; Russell, 1998b).

3.8 ORBIVIRUSES - BLUETONGUE AND EPIZOOTIC HAEMORRHAGIC DISEASE OF DEER VIRUSES

There are several serogroups of orbiviruses that infect domestic animals in Australia (Doyle, 1992). The main one, the bluetongue virus serogroup, has been mentioned. Epizootic haemorrhagic disease of deer (EHD) viruses are closely related serologically to BT viruses and have been associated with diseases in Japan and North America. However they do not cause disease in Australia, and have only minor importance in relation to trade. They are spread by the vector C. brevitarsis, although other Culicoides may be involved (Gibbs and Greiner, 1994). Of the 24 serotypes of BT virus known globally, 8 have been identified in Australia (Gard and Melville, 1992).

The evidence that these and all major arboviruses of ruminants in Australia are introductions to this country has been reviewed (Daniels *et al.*, 1997). Entry to Australia from the islands to the north has been hypothesised (St George, 1986a). Molecular studies of recent isolates of BT virus near Darwin show that genetically they have been more closely related to BT viruses from Southeast Asia than to those isolated previously from Australia. Furthermore, one genetic strain, or topotype, found first in Indonesia and now established in northern Australia, is more closely related to a BT virus from India than to others from Southeast Asia and Australia. Such observations suggest that there may be a continuing movement of BT virus strains from west to east across the South Asia – Australasian region (Daniels *et al.*, 1997).

In Australia *C. brevitarsis*, *C. wadai*, *C. fulvus* and *C. actoni* are considered competent vectors for BT viruses (Standfast *et al.*, 1985b). However most BT virus serotypes have not distributed beyond the Top End of the NT, where apparently isolated populations of *C. fulvus* and *C. actoni* occur. Only BT virus 1 and BT virus 21 are widely distributed, through to the east coast, in regions where *C. brevitarsis* also occurs. Hence with the bluetongue viruses there may be no simple relationship to be plotted between virus spread and climate, and the possibility of restrictions between strain of virus and species of vector will have to be elucidated.

3.9 BOVINE EPHEMERAL FEVER VIRUS

BEF virus is a member of the *Ephemerovirus* genus of the *Rhabdoviridae* and causes 3-day sickness, ephemeral fever, in cattle. Buffaloes are an alternative vertebrate host. The virus is endemic and causes disease throughout tropical and subtropical Australia. BEF in Australia initially showed epidemic waves spreading from the north to the south of the country over a period of one year to two summers (St George, 1986b). This suggests efficient spread by its vectors among a naïve cattle population. Evidence to identify the vectors has been reviewed (Kirkland, 1993). Infections have been identified in *C. brevitarsis* and various mosquitoes. Since the virus has sometimes been distributed beyond the ecological limits of *C. brevitarsis* active spread by a mosquito seems probable, with *Culex annulirostris* the likely candidate.

3.10 AKABANE VIRUS

Akabane virus is one of 7 members of the Simbu serogroup of the *Bunyaviridae* known in Australia. Akabane causes arthrogryposis and hydranencephaly in cattle, sheep and goats. Virus isolation studies from wild caught insects implicate only *C. brevitarsis* as the vector. Disease occurs where naïve ruminants in the susceptible stage of gestation become infected. Hence the disease is seen mainly in southern Australia, around the normal limits of *C. brevitarsis* distribution (Standfast *et al.*, 1986).

4. The Influence of Environmental Factors on Arbovirus Transmission and Spread

The relationship between environmental factors and arboviral disease has been well established. Most of the environmental factors are influenced by climate (long-term weather patterns) and weather (moisture, precipitation, temperature, humidity, sun-light, wind velocity, atmospheric pressure and other factors) on breeding, development rate, nutritional status, survival, host-seeking behaviour, diapause, and many other aspects of the biology of the vector (Sellers, 1980; Edman and Spielman, 1988; Mitchell, 1988; Reiter, 1988; Lindsay and Mackenzie, 1997). Environmental factors also affect food supplies for both adult (plant nectar) and larval mosquitoes (organic matter or other mosquito larvae), and may play a major role in vector dispersal and thus disease spread. The environment also plays a crucial role in influencing migration, breeding, immune-status, behaviour and other factors pertaining to the vertebrate host (Scott, 1988), and may determine the likelihood of human exposure to arboviruses by altering human behaviour and demographics (Gahlinger et al., 1986; Gregg, 1988). Indeed, as indicated above, the ecology of arboviruses is finely balanced, and it seems very likely that only minor changes in climate and weather could alter this balance and thus the potential for outbreaks of human and animal disease. Some of these factors are discussed in more detail below.

4.1 RAINFALL

Rainfall or precipitation is one of the elements that is predicted to change significantly under 'greenhouse' conditions (Climate Impact Group, 1996; Hennessy and Whetton, 1997). All mosquitoes have aquatic larval and pupal stages and therefore require water for breeding. Considerable evidence has accrued to show that heavy rainfall and flooding can lead to increased mosquito breeding and outbreaks of arboviral disease in Australia. Examples are readily available of RR virus outbreaks in various parts of Australia, and for MVE virus in the Kimberley region of WA (reviewed by Lindsay et al., 1993a; Broom et al., 1997; Lindsay and Mackenzie, 1997; Lindsay et al., 1997; Russell, 1998a; Russell, 1998b; Mackenzie et al., 1998).

It is important to note, however, that not all episodes of heavy rain and flooding lead to outbreaks of disease; studies of RR virus activity in the south-west of WA and of MVE at a remote community in the south-east Kimberley (Broom et al., 1993; Lindsay et al., 1993a) have clearly demonstrated that while heavy rainfall and flooding are important, other factors also play a role. These factors include the recruitment of new young, susceptible vertebrate hosts into the transmission cycles, and the seasonality of the rainfall and flooding. Thus heavy rainfall in the south-west of WA in a year following a major outbreak was found to be insufficient to generate an outbreak of RR virus disease because of the lack of sufficient numbers of non-immune, young grey kangaroos (*Macropus fuliginosus*), the major vertebrate host of RR virus. It appeared that above average winter and spring rains were required to ensure an abundant food supply for the female grey kangaroos in order to increase the survival rate of juveniles, and without this recruitment of new young non-immune animals, the transmission cycles were unable to be maintained (Lindsay and Mackenzie, 1997). Indeed, while heavy unseasonal autumn and winter rainfall and flooding in the southwest may assist in the breeding of potential vertebrate hosts, it does not lead to disease outbreaks, largely due to the low temperatures for mosquito breeding and for extrinsic incubation.

These results demonstrate that although heavy rainfall events and flooding are important for generating arboviral disease outbreaks, they are not sufficient in themselves to ensure an outbreak; other factors are also necessary.

As the opposite of rainfall, the concept of drought as an amplifying parameter for mosquito-borne disease is perhaps difficult to understand, yet under some conditions drought may be important. This was clearly shown in studies of the 1933 outbreak of St Louis encephalitis virus, an American flavivirus closely related to JE and MVE viruses (Lumsden, 1958). Drought can lead to a reduction in the flow of water in streams, resulting in the establishment of a series of stagnant pools, often high in organic matter, making perfect breeding sites for a number of mosquito species. Added to this, such reduced water sources are likely to become central to animal and bird drinking requirements, thus increasing the potential for vertebrate host-insect transmission cycles. This may be particularly important in areas that normally have high humidity. Thus, the apparent increase in activity of Japanese encephalitis virus in Western Province, PNG, leading to the first 4 cases of clinical encephalitis and the spill-over of virus into the Torres Strait and northern mainland Australia early in 1998, is believed to be due to the extensive drought in PNG, and the consequent drying-up of rivers. Indeed, significantly higher numbers of mosquitoes were trapped in 1998 in PNG and the Torres Strait than in previous years. Over 10,000 *Culex* mosquitoes were collected in some light traps in PNG, and there was a 500% increase in *Cx. annulirostris* mosquitoes trapped on Badu Island compared to 1995 during the previous outbreak (S.A. Ritchie, C.A. Johansen. R. Paru, and A. van den Hurk, unpublished results).

Rainfall is also an essential component in the life cycle of *Culicoides* species. Moisture is necessary for the breeding sites (cow dung pats), and the abundance and activity of many Culicoides species is proportional to rainfall (Walker and Davies, 1971; Ward, 1994). Indeed Culicoides species are confined to areas with an annual rainfall of 700-800mm (Standfast and Muller, 1989; Ward 1996). The monitoring of arboviral infections of livestock in northern Australia since 1980 has shown that BT virus infections tend to occur mainly after the peak of the monsoonal wet season has passed, although in some years the virus becomes active with the first rains at the end of the dry season (Gard and Melville, 1992; Melville *et al.*, 1996b). Frequency of isolation becomes less as the dry season progresses, even when naïve vertebrate hosts are introduced into the study area.

BEF infections are also governed by rain, occurring in the NT after the first rains at the end of the dry season. In big wet seasons BEF infections are detected well beyond their normal ranges, as far south as Alice Springs, as occurred in 1997 (Anon, 1998). Similarly BEF can occur throughout much of central and western Queensland and central NSW, depending on the wet season in those areas.

4.2 TIDES AND SEA LEVEL

Tidal inundation of saltmarshes is a major source of water for breeding two of the major vectors of RR and BF viruses, *Ae. vigilax* and *Ae. camptorhynchus* mosquitoes. The former is the major saltmarsh vector extending from southern NSW northwards around the north of the Continent, and down the WA coast to south of Perth. The latter is the major saltmarsh vector along the southern coastline from Gippsland in Victoria to just north of Perth. Large populations of these species can emerge as little as 8 days after a series of spring tides have inundated the saltmarsh, depending on temperature. This can be exacerbated by prevailing winds and storm surges. High tides have been implicated as important factors in the genesis of RR and BF virus outbreaks on both sides of the Continent (Lindsay *et al*, 1993; McManus *et al*., 1992; Phillips *et al*., 1992; Doggett *et al*., 1995), and major efforts have been made by a number of local councils to reduce the risk of epidemic activity by improving drainage in coastal saltmarshes near human habitation by the use of runnelling. It should also be noted that saltmarsh mosquito breeding can be exacerbated by rainfall in the absence of, or in addition to, tidal inundation.

4.3 TEMPERATURE

Environmental temperature has a major effect on the length (Lundstrom *et al*., 1990; Cornel *et al*., 1993; Reeves *et al*., 1994) and efficiency (Kramer *et al*., 1983; Turell, 1993; Reisen *et al*., 1993) of extrinsic incubation of arboviruses in their vectors (Hardy, 1988), and on the survival of adult mosquitoes (Reeves *et al*., 1994). Thus most studies have shown that mosquitoes exposed to higher temperatures after ingestion of virus become 'infectious' more rapidly than mosquitoes of the same species exposed to lower temperatures. This largely reflects the increased replication rate of viruses as temperatures increase. Transmission of an arbovirus may therefore be enhanced under warmer conditions because more vector mosquitoes become infectious within their often-short life span. In addition, within the temperature range that a mosquito species can breed, larvae reared at a higher environmental temperature develop much faster than those reared at lower temperatures. Temperature may therefore dictate whether a generation reaches maturity before breeding sites dry up or are flushed away by a new series of high tides or rains. However, studies have also suggested that an elevated temperature of rearing of larval mosquitoes reduces their vector efficiency as adults (Kramer *et al*., 1983; Turell, 1993).

Temperature has a major effect on the breeding and survival of adult mosquitoes and *Culicoides* species, although this effect is also strongly influenced by humidity. Some species of mosquito are temperature limited in their breeding; for instance, *Cx. annulirostris*, the principle vector of MVE in Australia (Marshall, 1988; Mackenzie *et al*., 1994b; Russell, 1995) and of JE in the Torres Strait (Hanna *et al*., 1996; Ritchie *et al*., 1997), and a major vector of RR, BF (Mackenzie *et al*., 1994b; Russell, 1995) and BEF (Mackenzie *et al*., 1994a) viruses, does not breed when the daily temperature falls below 17.5 degrees C (Kay and Aaskov, 1989). The breeding of other important vector species is also temperature-limited, including the two major saltmarsh breeding species, *Ae. vigilax* and *Ae. camptorhynchus* (Lindsay and Mackenzie, 1997). With respect to survival, one study carried out in California showed that the daily adult mortality rate of *Cx. tarsalis* mosquitoes increased by 1% for each 1 degree C increase in temperature (Reeves *et al.*, 1994), although adult mortality is also affected by humidity (see below). For *Culicoides* species, the metabolism of the vector is dependant on temperature, and *Culicoides* species become active and fly between 13 and 35 degrees C (Sellers, 1980; Sellers and Maarouf, 1989; Ward, 1994). The most important Australian vector, *C. brevitarsis*, is a "tropical" insect as is shown by the relatively high temperature at which it commences to fly, and its overall distribution in Australia (Murray, 1995). In its southerly range, temperature is the limiting factor of winter survival. Indeed, temperature is important for the persistence of viruses through overwintering in *Culicoides* species (Sellers and Mellor, 1993). Warmth is also necessary for hatching the eggs and for larval development (Ward, 1994).

Temperature also has a major physical effect on mosquito breeding by reducing vector breeding sites. Thus breeding sites in summer months may be less abundant, especially in arid or semi-arid areas or where humidity is low, because of rapid evaporation of saltmarshes, floodplains, and temporary ground pools.

4.4 HUMIDITY

Humidity is an important environmental parameter with respect to the survival of mosquitoes and *Culicoides*. Relative humidity is defined as the ratio of water vapour content of the air to its total capacity at a given temperature. Because it affects survival and thus longevity, humidity also affects dispersal, mating, feeding behaviour and oviposition of vector species. Under conditions of high humidity, mosquitoes and *Culicoides* tend to survive for longer periods, which allows them to disperse further and to have a greater opportunity to participate in transmission cycles. In southern Australia summer rain allows survival of C. brevitarsis adults, greatly increasing the proportion and absolute numbers of insects that have had two or more blood meals and that therefore may be active in spreading arboviral infections (Murray, 1986). Humidity also affects the rate of evaporation of water at breeding sites.

4.5 WIND

Wind is central to insect dispersal and thus virus movement, providing it is sufficiently warm and moist for flight. There is a substantial literature concerned with the role of wind in the genesis of epidemic activity of arboviruses through the movement of infected arthropods, especially mosquitoes and *Culicoides* species from studies carried out overseas (eg. Garrett-Jones, 1962; Min and Mei, 1996; Ming *et al.*, 1993; Pedgley, 1982; Reynolds *et al.*, 1996; Sellers, 1980, 1989; Sellers and Maarouf, 1989, 1990, 1991, 1993) and in Australia (e.g. Seddon, 1938; Murray, 1970, 1987; Newton and Wheatley, 1970; Murray and Kirkland, 1995; Ward, 1995; Bishop *et al.*, 1996). Flights of 5-100km have been recorded for *Cx. taeniorhynchus* (a vector of Venezuelan equine encephalitis virus), and 175km for *Ae. sollicitans* (a vector of Venezuelan and Eastern equine encephalitis viruses) (Sellers, 1980). *Culicoides* species have also been carried at least 8km, and there is circumstantial evidence that they have been carried 40-700km (Murray, 1970, 1987; Sellers, 1980; Sellers and Maarouf, 1989). These flights were completed at temperatures of 15-35 degrees C at heights of up to 1.5km (Sellers, 1980). Most studies, however, have tended to capture mosquitoes and *Culicoides* species at lower heights, and most commonly below about 500m (eg. Ming *et al.*, 1993; C.G. Johnson, cited by Sellers, 1980). Wind is also used by *Cx. tritaeniorhynchus* mosquitoes as a means of migration each year in China, flying north in the spring and returning south in the autumn (Min and Mei, 1996; Ming *et al.*, 1993).

In Australia, movement is particularly associated with the warm, tropical northerly and north-westerly winds emanating from the Intertropical Convergence Zone, which is located just to the north of Australia in January, its most southerly latitude (Murray, 1970; Sellers, 1980). These winds are the result of the movement of low or high pressure areas, with carriage of insects in eastern Australia most likely to result from low pressure areas. Recent work by Ritchie has suggested that JE virus may have been introduced into the Torres Strait and northern Australia in tropical monsoonal low pressure systems (S.A. Ritchie, unpublished observations). Outbreaks of Akabane disease (Murray, 1987) and bluetongue infections (Murray and Kirkland, 1995) have been linked to long range wind-borne dispersal.

International dispersal of infected vectors by wind is one of the major factors in veterinary arbovirology in Australia. Most of the viruses and vectors are introduced species (St George, 1986a). New strains of BTV continue to arrive in northern Australia (Daniels, *et al.*, 1997). Since these strains are similar to Indonesian BTV (Daniels *et al.*, 1995), deductions may be made regarding their origin. It would be of greater concern if new vector species were to arrive. There are *Culicoides* in Indonesia of unknown , but suspected, vector potential. Further to the west, in India, *C. imicola* is a major vector. In Africa *C. imicola* is the vector of highly virulent BTV strains, and highly virulent BTV also occurs in India. There is concern that eastward movement of this vector species into Southeast Asia and Australia could be accompanied by emergence of BTV strains of high virulence in these areas also. Monitoring of *Culicoides* spp present in both geographical areas is recommended.

5. Usefulness of Seasonal Forecasting for Predicting Outbreaks of Arbovirus Disease

Models for predicting outbreaks of Australian encephalitis (MVE virus) in southern Australia based on the influence of climate on vector breeding, gonotrophic cycle, survival and extrinsic incubation of the virus (Kay, 1980; Kay, 1982; Kay *et al.*, 1987) and meteorological conditions (Forbes, 1978; Nicholls, 1986; Nicholls, 1993; Voice *et al.*, 1990) have been described. The latter were based on a positive value for the Southern Oscillation Index (SOI) (see below). However, no MVE outbreak has occurred since 1974, and the models have not been seriously tested, although climatic and meteorological conditions have, on occasion, been close to those implicated in the models. Seasonal forecasting may have a role nevertheless for predicting MVE activity in northern Australia where monsoonal weather conditions can often be predicted. Thus if a heavy wet season and possible flooding is predicted, it indicates an increased risk of MVE activity and human cases. Indeed MVE activity can be predicted on the basis of increased rainfall and flooding over a wide area of northwestern Australia.

Models or predictions are less accurate for RR and BF viruses, largely because the pattern of meteorological conditions that predispose to epizootics of the virus appears to be too complex and variable between and within regions to enable construction of accurate predictive models. Also less is known of the basic parameters with respect to vector breeding, identification of vectors and vertebrate hosts, and the influence of climate on vector breeding. However, analysis of rainfall, temperature and tidal patterns using weather reports and seasonal climate outlooks, in association with ongoing vector surveillance programs, may provide some means of predicting outbreaks (Lindsay, 1995). Environmental factors influenced by the El Niño-Southern Oscillation (ENSO) meteorological phenomenon (Cane, 1983, 2000; Voice et al., 1993) may also be a useful indicator of potential RR virus activity on a state-wide or regional basis. For example, the Leeuwin Current, an ocean current that flows down the west coast of WA from the warm tropics to cold southern ocean waters, is weaker in years when negative values of the SOI are recorded (El Niño years). This results in a drop in ocean temperatures and a lower mean sea level off the WA coast. The cooler ocean temperatures are also thought to result in a decreased chance of above average rains over parts of WA. During positive SOI (La Niña) years, the opposite occurs, with a strengthening of the Leeuwin Current, increased ocean temperatures, a rise in the mean sea level and an increased risk of above average rainfall (Pearce and Phillips, 1988). An outbreak of RR virus in the southwest of WA in 1988/89 occurred during a La Niña year when the SOI rose above +15 for several consecutive months. The observed rise in mean sea level, increased late spring rains and warmer coastal temperatures were presumably due to strengthening of the Leeuwin Current. Thus, above average levels of RR virus activity could be anticipated when extreme positive values of the SOI are recorded prior to and during established seasons of RR virus activity in this region. It is important to note, however, that monitoring of extreme positive values of the SOI would only have predicted one of three major southwest RR outbreaks since 1987. SOI values during major outbreaks in the region in 1991/92 and 1995/96 were only weakly positive, neutral or negative. Thus, other methods of forecasting unusual/extreme meteorological patterns are crucial for developing adequate predictive models of RR virus activity. Such methods may become available as studies of regional climate forecasting are carried out, for example the Indian Ocean Climate Initiative, which commenced in February 1998.

Long-term studies of RR virus ecology in WA have clearly indicated that virus activity is influenced by environmental conditions occurring at a local level. Indeed, patterns of virus activity differ markedly even within local government authority areas, depending on rainfall and tidal patterns, the level of activity in the preceding season (which affects the level of immunity in the amplifying host population) and many other factors (Lindsay, 1995). Considerable detailed information about the ecology and epidemiology of RR virus activity and its possible vectors is now available for WA, but the level of information is much less for other States, except in specific locations or areas. Seasonal forecasting and monitoring of the SOI, both of which provide an indication of likely weather patterns over larger geographic regions, do not provide the level of detail about local weather patterns to enable accurate prediction of the potential for RR virus activity. For this reason, a combination of monitoring of local weather patterns (as they occur), and prospective surveillance of vector populations, infection rates of vectors and immunological status of amplifying host populations seems to be the only currently available means for predicting the potential for outbreaks of RR virus disease.

Seasonal forecasting for livestock arboviral diseases is also in its infancy, but as further data become available, it offers the potential for contingency planning and broadly-based (geographically) forecasting as suggested for some invertebrate vectors (Sutherst, 2000). The role of environmental conditions, particularly rainfall and temperature on vector breeding and vector abundance, humidity on vector survival, and winds on vector dispersal, are all well established, but it is clear that many other factors need to be incorporated into modelling systems (Standfast and Maywald, 1992; Ward, 1994; Murray, 1995). Indeed, as described above, vector abundance and dispersal are only one side of a biological equation with the availability of non-immune hosts on the other side, but both sides are influenced in different ways by environmental and climatic parameters. Modelling the possible distribution and abundance of *Culicoides* species under different climatic conditions, and particularly the possible colonisation of new areas, are important in trying to predict longer term climatic effects on the spread and incidence of arboviral diseases. A study of the southern limits of the distribution of *C. brevitarsis* and the epidemiology of arbovirus infections in NSW made use of a computer model (GROWEST) derived for pasture grass as a basis for further model development (Murray and Nix, 1987; Murray, 1986; 1987). Calculations from the model gave the Environmental Indices which were correlated with extensive field data on seasonal abundance and insect distributions, and made it possible to define areas suitable for summer multiplication and winter survival of *C. brevitarsis* (and thus the limits to its distribution), from which regions suitable for survival of breeding populations were determined for the spring of each year. The model therefore defines regions of *potential* transmission, depending on climatic factors. However, the actual overall abundance of the insect varies between localities and seasons, being influenced by hosts and climate, and especially local climatic conditions (Murray, 1995). The use of linear regression analysis of monthly maximum, average and minimum temperatures has been described to determine low temperature thresholds at which *C. brevitarsis* would be absent, and thus provide a risk assessment for farmers of subsequent arboviral disease (Bishop *et al.*, 1995). Two computer programs, BIOCLIM and CLIMEX, have also been employed to model the possible spread of *C. wadai* in Australia (Standfast and Maywald, 1992). The latter program appears to be the most useful and indeed has indicated that *C. wadai* has not completed colonisation of all areas where the climate is presently favourable, and given the change of climate predicted for the "Greenhouse Effect", this species may be capable of colonising some of the cattle- and sheep-rearing areas of southern NSW which are currently free of known *Culicoides* vectors of arboviruses (Standfast and Maywald, 1992).

Use of seasonal forecasting for arboviral disease is further complicated by the fact that virus activity in one region may depend upon or be directly or indirectly affected by activity in other regions. For example, the potential for spread of MVE virus to eastern and south-eastern Australia is likely to depend on climatic conditions in northern and central Australia (Mackenzie *et al.*, 1993b). RR virus transmission in the suburbs of Perth follows high levels of virus activity in rural areas further south, presumably relying on viraemic humans to bring the virus back to the metropolitan area where low-level, inefficient transmission cycles in local vector species can occur (Lindsay, 1995). The potential for incursion of JE virus into northern Australia will almost certainly be influenced by climatic conditions and their effect on JE transmission cycles in countries to the north of Australia in which JE virus is enzootic, as well as auspicious weather patterns assisting wind-borne mosquito dispersal. The incursions of exotic vectors and more virulent BT strains will also be closely dependent on climatic conditions in Indonesia. Thus, accurate seasonal forecasting of outbreaks of arboviral disease will also require monitoring of climatic predictions for a much larger geographical region than just an area of interest. Coupled with this, a more complete understanding of the ecology of the viruses, their vectors and hosts, and the interactions of each of these with the environment, in the area of interest and in other regions from which the viruses or their vectors may be disseminated, is also essential.

Acknowledgments

We would like to thank Dr Scott Ritchie for helpful discussions in the preparation of this manuscript.

References

Aaskov, J.G., Phillips, D.A. and Weimers, M.A. (1993) Possible clinical infection with Edge Hill virus. *Transactions of the Royal Society for Tropical Medicine and Hygiene* **87**, 452-453.

Aldred, J., Campbell, J., Davis, G., Lehmann, N. and Wolstenholme, J. (1990) Barmah Forest virus in the Gippsland Lakes region. *Medical Journal of Australia* **153**, 434.

Anon (1998) National Arbovirus Monitoring Program 1997-1998 Report. Australian Animal Health Council Ltd., Canberra. 4pp.

Azuolas, J.K. (1997) Arboviral diseases of horses and possums, in B.H. Kay, M.D. Brown and J.G. Aaskov, (eds.), *Arbovirus Research in Australia*. Queensland Institute of Medical Research, Brisbane. pp. 5-7.

Badman, R.T., Campbell, J. and Aldred, J. (1984) Arbovirus infections in horses – Victoria, 1984. *Communicable Diseases Intelligence* **17**, 5.

Bishop, A.L., Barchia, I.M. and Harris, A.M. (1995) Last occurrence and survival during winter of the arbovirus vector *Culicoides brevitarsis* at the southern limits of its distribution. *Australian Veterinary Journal* **72**, 53-55.

Bishop, A.L., Kirkland, P.D., McKenzie, H.J. and Barchia, I.M. (1996) The dispersal of *Culicoides brevitarsis* in eastern New South Wales and associations with the occurrences of arbovirus infections in cattle. *Australian Veterinary Journal* **73**, 174-178.

Broom, A.K., Lindsay, M.D., Wright, A.E. and Mackenzie, J.S. (1993) Arbovirus activity in a remote community in the south-east Kimberley. *Arbovirus Research in Australia* **6**, 262-266.

Broom, A.K., Lindsay M.D.A, Johansen, C.A., Wright, A.E. and Mackenzie, J.S. (1995) Two possible mechanisms for survival and initiation of Murray Valley encephalitis virus activity in the Kimberley region of Western Australia. *American Journal of Tropical Medicine and Hygiene* **53**, 95-99.

Broom, A., Lindsay, M., Van Heuzen, B., Wright, T., Mackenzie, J. and Smith, D. (1997) Contrasting patterns of flavivirus activity in the Kimberley region of Western Australia, 1992-1996. *Arbovirus Research in Australia* **7**, 25-30.

Burke, D.S. and Leake, C.J. (1988) Japanese encephalitis, in T.P. Monath (ed.), *The Arboviruses: Epidemiology and Ecology, Volume III*, CRC Press Inc., Boca Raton. pp. 63-92.

Cane, M.A. (1983) Oceanographic events during El Nino. *Science* **222**, 1189-1195.

Cane, M.A. (2000) Understanding and predicting the world's climate system, in G.L. Hammer, N. Nicholls, and C. Mitchell (eds.), *Applications of Seasonal Climate Forecasting in Agricultural and Natural Ecosystems – The Australian Experience*. Kluwer Academic, The Netherlands. (this volume).

Climate Impact Group (1996) Climate change scenarios for the Australian region. Climate Impact Group, CSIRO Division of Atmospheric Research, Melbourne.

Cornel, A.J., Jupp, P.G. and Blackburn, N.K. (1993) Effect of environmental temperature on the vector competence of *Culex univittatus* (Diptera: Culicidae) for West Nile virus. *Journal of Medical Entomology* **30**, 449-456.

Daniels, P.W., Melville, L., Pritchard, I., Sendow, I., Sreenivasulu and B. Eaton (1997) A review of bluetongue viral variability along a South Asia - Eastern Australia transect. *Arbovirus Research in Australia* **7**, 66-71.

Daniels, P.W., Sendow, I. and Melville, L. (1996) Epidemiological considerations in the study of bluetongue viruses, in T.D. St George and Peng Kegao (eds.), *Bluetongue Disease in Southeast Asia and the Pacific*, ACIAR, Canberra. pp.110-119.

Daniels, P.W., Sendow, I., Soleha, E., Sukarsih, Hunt, N.T. and Bahri, S. (1995) Australian-Indonesian collaboration in veterinary arbovirology. A review. *Veterinary Microbiology* **46**, 151-174.

Dhileepan, K., Azoulas, J.K. and Gibson, C.A. (1996) Evidence of vertical transmission of Ross River and Sindbis viruses (Togaviridae: Alphavirus) by mosquitoes (Diptera: Culicidae) in southeastern Australia. *Journal of Medical Entomology* **33**, 180-182.

Doggett, S., Russell, R., Cloonan, M., Clancy, J. and Haniotis, J. (1995) Arbovirus and mosquito activity on the south coast of New South Wales, 1994-95. *Communicable Disease Intelligence* **19**, 473-475.

Doyle, K.A. (1992) An overview and perspective on Orbivirus disease prevalence and occurrence of vectors in Australia and Oceania, in T.E. Walton and B.I. Osburn (eds.), *Bluetongue, African Horse Sickness and Related Orbiviruses*, CRC Press Inc., Boca Raton. pp 44-57.

Edman, J.D. and Spieman, A. (1988) Blood-feeding by vectors: physiology, ecology, behavior, and vertebrate defence, in T.P. Monath (ed.), *The Arboviruses: Epidemiology and Ecology, Volume I*, CRC Press Inc., Boca Raton. pp.153-189

Flexman, J.P., Smith, D.W., Mackenzie, J.S., Fraser, R.E., Bass, S., Hueston, L., Lindsay, M.D.A. and Cunningham, A.L. (1998) A comparison of the diseases caused by Ross River virus and Barmah Forest virus. *Medical Journal of Australia* **169**, 159-163.

Forbes, J.A. (1978) Murray Valley encephalitis 1974, also epidemic variance since 1914 and predisposing rainfall patterns. Australasian Medical Publishing, Sydney.

Gahlinger, P.M., Reeves, W.C. and Milby, M.M. (1986) Air conditioning and television as protective factors in arboviral encephalitis risk. *American Journal of Tropical Medicine and Hygiene* **35**, 601-610.

Gard G.P. and Melville, L.F. (1992) Results of a decade's monitoring for orbiviruses in sentinel cattle pastured in an area of regular arbovirus activity in Northern Australia, in T.E. Walton and B.I. Osburn (eds.), *Bluetongue, African Horse Sickness and Related Orbiviruses*, CRC Press Inc., Boca Raton. pp. 85-90.

Garrett-Jones, C. (1962) The possibility of active long distance migration of *Anopheles pharoensis*. *Bulletin of the World Health Organization* **27**, 299-302.

Gibbs, E.P.J. and Greiner, E.C. (1994) The epidemiology of bluetongue. *Comparative Immunology and Microbiology of Infectious Diseases* **17**, 207-220.

Gregg, M.B. (1988) Epidemiological principles applied to arboviral diseases, in T.P. Monath (ed.), *The Arboviruses: Epidemiology and Ecology, Volume I*, CRC Press Inc., Boca Raton. pp. 291-309.

Gubler, D.J. (1988) Dengue, in T.P. Monath (ed.), *The Arboviruses: Epidemiology and Ecology, Volume II*, CRC Press Inc., Boca Raton. pp. 223-260.

Gubler, D.J. (1998) Resurgent vector-borne diseases as a global health problem. *Emerging Infectious Diseases* **4**, 442-450.

448

Hanna, J.N., Ritchie, S.A., Phillips, D.A., Shield, J., Bailey, M.C., Mackenzie, J.S., Poidinger, M., McCall, B.J. and Mills, P.J. (1996) An outbreak of Japanese encephalitis in the Torres Strait, Australia, 1995. *Medical Journal of Australia* **165**, 256-260.

Hanna, J.N., Ritchie, S.A., Phillips, D.A., Lee, J.M., Hills, S.L., Van Den Hurk, A.F., Pyke, A.T., Johansen, C.A., and Mackenzie, J.S. (1999) Japanese encephalitis in North Queensland, 1998. *Medical Journal of Australia*, in press.

Hardy, J.L. (1988) Susceptibility and resistance of vector mosquitoes, in T.P. Monath (ed.), *The Arboviruses: Epidemiology and Ecology, Volume I*, CRC Press Inc., Boca Raton. pp.87-126.

Hennessy, K.J. and Whetton, P. (1997) Development of Australian climate change scenarios, in P. Curson, C. Guest and E. Jackson (eds.), *Climate Change and Human Health in the Asia-Pacific Region*, Australian Medical Association and Greenpeace International, Canberra. pp. 7-18.

Jetten, T.H. and Focks, D.A. (1997) Potential changes in the distribution of dengue transmission under climate warming. *American Journal of Tropical Medicine and Hygiene* **57**, 285-297.

Johansen, C., Ritchie, S., Van den Hurk, A., Bockarie, M., Hanna, J., Phillips, D., Melrose, W., Poidinger, M., Scherret, J., Hall, R. and Mackenzie, J. (1997) The search for Japanese encephalitis virus in the Western Province of Papua New Guinea. *Arbovirus Research in Australia* **7**, 131-136.

Kay, B.H. (1980) Towards prediction and surveillance of Murray Valley encephalitis activity in Australia. *Australian Journal of Experimental Biology and Medical Science* **58**, 67-76.

Kay, B.H. (1982) MVE virus 'tales from the Aboriginal dreamtime'. *Arbovirus Research in Australia* **3**, 50-56.

Kay, B.H. and Aaskov, J.G. (1989) Ross River virus (epidemic polyarthritis), in T.P. Monath (ed.) *The Arboviruses: Epidemiology and Ecology, Volume IV*, CRC Press Inc., Boca Raton. pp.93-112.

Kay, B.H., Saul, A.J. and McCullagh, A. (1987) A mathematical model for the rural amplification of Murray Valley encephalitis virus in southern Australia. *American Journal of Epidemiology* **125**, 690-705.

Kirkland, P.D. (1993) The epidemiology of bovine ephemeral fever in south-eastern Australia: evidence for a mosquito vector, in T.D. St George, M.F. Uren, P.L. Young and D. Hoffmann (eds.), *Bovine Ephemeral Fever and Related Rhabdoviruses*, ACIAR, Canberra. pp. 33-37.

Kramer, L.D., Hardy, J.L. and Presser, S.B. (1983) Effect of temperature of extrinsic incubation on the vector competence of *Culex tarsalis* for western equine encephalomyelitis virus. *American Journal of Tropical Medicine and Hygiene* **32**, 1130-1139.

Liehne, P.F.S. (1988) Climate influences on mosquito-borne diseases in Australia, in G.I. Pearman (ed.), *Greenhouse – Planning for Climate Change*, CSIRO Australia. pp. 624-637.

Lindsay, M.D. (1995) Ecology and epidemiology of Ross River virus in Western Australia. PhD Thesis, The University of Western Australia, Perth.

Lindsay, M. and Mackenzie, J. (1997) Vector-borne viral diseases and climate change in the Australasian region: major concerns and the public health response, in P. Curson, C. Guest, and E. Jackson, (eds.), *Climate Change and Human Health in the Asia-Pacific Region*, Australian Medical Association and Greenpeace International, Canberra. pp.47-62.

Lindsay, M.D., Broom, A.K., Wright, A.E., Johansen, C.A. and Mackenzie, J.S. (1993b) Ross River virus isolations from mosquitoes in arid regions of Western Australia: implication of vertical transmission as a means of persistence of the virus. *American Journal of Tropical Medicine and Hygiene* **49**, 686-696.

Lindsay, M.D.A, Johansen, C.A, Broom, A.K., Smith, D.W. and Mackenzie, J.S. (1995a) Emergence of Barmah Forest virus in Western Australia. *Emerging Infectious Diseases* **1**, 22-26.

Lindsay, M.D., Johansen, C.A., Smith, D.W. and Mackenzie, J.S. (1995b) An outbreak of Barmah Forest virus disease in the south-west of Western Australia. *Medical Journal of Australia* **162**, 291-294.

Lindsay, M.D., Mackenzie, J.S. and Condon, R.J. (1993a) Ross River virus outbreaks in Western Australia: epidemiological aspects and the role of environmental factors, in C.E. Ewan, E.A. Bryant, G.D. Calvert and J.A. Garrick, (eds.), *Health in the Greenhouse*, Australian Government Publishing Service, Canberra. pp.85-100

Lindsay, M., Oliveira, N., Jasinska, E., Johansen, C., Harrington, S., Wright, T., Smith D. and Shellam, G. (1997) Western Australia's largest recorded outbreak of Ross River virus disease. *Arbovirus Research in Australia* **7**, 147-152.

Lumsden, L.L. (1958) St Louis encephalitis in 1933. Observations on epidemiological features. *Public Health Reports* **73**, 340-352.

Lundstrom, J., Turell, M. and Niklasson, B. (1990) Effect of environmental temperature on the vector competence of *Culex pipiens* and *Cx. torrentium* for Ockelbo virus. *American Journal of Tropical Medicine and Hygiene* **43**, 534-542.

Mackenzie, J.S. (1997) Japanese encephalitis: an emerging disease in the Australasian region, and its potential risk to Australia. *Arbovirus Research in Australia* **7**, 166-170.

Mackenzie, J.S. (1999a) Emerging viral diseases: an Australian perspective. *Emerging Infectious Diseases* **5**, 1-8.

Mackenzie, J.S. (1999b) The ecology of Japanese encephalitis virus in the Australasian region. *Clinical Virology (Japan)* **27**, 1-17.

Mackenzie, J.S. and Broom, A.K. (1998) Ord River irrigation area: the effect of dam construction and irrigation on the incidence of Murray Valley encephalitis virus, in B.H. Kay (ed.), *Water Resources – Health, Environment and Development*, E & FN Spon., London. pp.108-122.

Mackenzie J.S. and Smith, D.W. (1996) Mosquito-borne viruses and epidemic polyarthritis. *Medical Journal of Australia* **164**, 90-93.

Mackenzie, J.S., Broom, A.K., Hall, R.A., Johansen, C.A., Lindsay, M.D., Phillips, D.A., Ritchie, S.A., Russell, R.C. and Smith, D.W. (1998) Arboviruses in the Australian region, 1990-1998. *Communicable Diseases Intelligence* **22**, 93-100.

Mackenzie, J.S., Cunningham, A., Hueston, L. and LaBrooy, J. (1996a) Dengue in Australia. *Journal of Medical Microbiology* **45**, 159-161.

Mackenzie, J.S., Lindsay, M.D. and Broom, A.K. (1993b) Climate changes and vector-borne diseases: potential consequences for human health, in C.E. Ewan, E.A. Bryant, G.D. Calvert and J.A. Garrick, (eds.), *Health in the Greenhouse: the Medical and Environmental Health Effects of Global Climate Change*. Australian Government Publishing Service, Canberra. pp.229-234.

Mackenzie, J.S., Lindsay, M.D., Coelen, R.J., Broom, A.K., Hall, R.A., and Smith, D.W. (1994b) Arboviruses causing human disease in the Australasian region. *Archives of Virology* **136**, 447-467.

Mackenzie, J.S., Poidinger, M., Lindsay, M.D., Hall, R.A. and Sammels, L.M. (1996b) Molecular epidemiology and evolution of mosquito-borne flaviviruses and alphaviruses enzootic in Australia. *Virus Genes* **11**, 225-237.

Mackenzie, J.S., Poidinger, M., Phillips, D., Johansen, C., Hall, R.A., Hanna, J., Ritchie, S., Shield, J. and Graham, R. (1997) Emergence of Japanese encephalitis virus in the Australasian region, in J.F. Saluzzu and B. Dodet, (eds.), *Factors in the Emergence of Arbovirus Diseases*, Elsevier, Paris. pp.191-201.

Mackenzie, J.S., Smith, D.W., Broom, A.K. and Bucens, M.R. (1993a) Australian encephalitis in Western Australia, 1978-1991. *Medical Journal of Australia* **158**, 591-595.

Mackenzie, J.S., Smith, D.W., Ellis, T.M., Lindsay, M.D., Broom, A.K., Coelen, R.J. and Hall, R.A. (1994a) Human and animal arboviral disease in Australia, in G.L. Gilbert (ed.), *Recent Advances in Microbiology, Volume 2*, Australian Society for Microbiology Inc., Melbourne. pp.1-91.

Marshall, I.D (1988) Murray Valley and Kunjin encephalitis, in T.P. Monath (ed.), *The Arboviruses: Epidemiology and Ecology, Volume III*, CRC Press Inc., Boca Raton. pp.151-190.

Marshall, I.D. and Miles, J.A.R. (1984) Ross River virus and epidemic polyarthritis. *Current Topics in Vector Research* **2**, 31-56.

McManus, T.J., Russell, R.C., Wells, P.J., Clancy, J.G., Fennell, M. and Cloonan., M.J. (1992) Further studies on the epidemiology and effects of Ross River virus in Tasmania. *Arbovirus Research in Australia* **6**, 68-72.

Melville, L., Hunt, N.T and Daniels, P.W. (1996a) Application of the polymerase chain reaction (PCR) test with insects in studying bluetongue virus activity, in T.D. St George and Peng Kegao (eds.), *Bluetongue Disease in Southeast Asia and the Pacific*, ACIAR, Canberra. pp.141-145.

Melville, L., Weir, R., Harmsen, N., Walsh, S., Hunt, N.T., Pritchard I and Daniels, P.W. (1996b) Recent experiences with the monitoring of Sentinel herds in northern Australia, in T.D. St George and Peng Kegao, (eds.), *Bluetongue Disease in Southeast Asia and the Pacific*, ACIAR, Canberra. pp.100-105.

Min, J.-G. and Mei, X. (1996) Progress in studies on the overwintering of the mosquito *Culex tritaeniorhynchus*. *Southeast Asian Journal of Tropical Medicine and Public Health* **27**, 810-817.

Ming, J-G., Hua, J., Riley, J.R., Reynolds, D.R., Smith, A.D., Wang, R-L., Cheng, J-Y. and Cheng, X-N. (1993) Autumn southward 'return' migration of the mosquito *Culex tritaeniorhynchus* in China. *Medical and Veterinary Entomology* **7**, 323-327.

Mitchell, C.J. (1988) Occurrence, biology and physiology of diapause in overwintering mosquitoes, in T.P. Monath (ed.), *The Arboviruses: Epidemiology and Ecology, Volume I*. CRC Press Inc., Boca Raton. pp.191-217.

Monath, T.P. and Heinz, F.X. (1996) Flaviviruses, in B.N. Fields, D.M. Knipe, and P.M. Howley, (eds.), *Fields Virology*, Lippencott-Raven Publishers, Philadelphia. pp. 961-1034.

Muller, M.J. (1995) Veterinary arbovirus vectors in Australia – a retrospective. *Veterinary Microbiology* **46**, 101-116.

Muller, M.J. and Murray, M.D. (1977) Blood-sucking flies feeding upon sheep in eastern Australia. *Australian Journal of Zoology* **25**, 75-85.

Muller, M.J., Standfast, H.A., St George, T.D. and Cybinski, D.H. (1982) *Culicoides brevitarsis* (Diptera: Ceratopogonidae) as a vector of arboviruses in Australia. *Arbovirus Research in Australia* **3**, 43-49.

Murray, M.D. (1970) The spread of ephemeral fever of cattle during the 1967-68 epizootic in Australia. *Australian Veterinary Journal* **46**, 77-82.

Murray, M.D. (1986) The influence of abundance and dispersal of *Culicoides brevitarsis* on the epidemiology of arboviruses of livestock in New South Wales. *Arbovirus Research in Australia* **4**, 232-234.

Murray, M.D. (1987) Akabane epizootics in New South Wales: evidence for long-distance dispersal of the biting midge, *Culicoides brevitarsis. Australian Veterinary Journal* **64**, 305-308.

Murray, M.D. (1995) Influences of vector biology on transmission of arboviruses and outbreaks of disease: the *Culicoides brevitarsis* model. *Veterinary Microbiology* **46**, 91-99.

Murray, M.D. and Kirkland, P.D. (1995) Bluetongue and Douglas virus activity in New South Wales in 1989: further evidence for long-distance dispersal of the biting midge *Culicoides brevitarsis. Australian Veterinary Journal* **72**, 56-57.

Murray, M.D. and Nix, H.A. (1987) Southern limits of distribution and abundance of the biting midge *Culicoides brevitarsis* Kieffer (Diptera: Ceratopogonidae) in south-eastern Australia: an application of the GROWEST model. *Australian Journal of Zoology* **35**, 575-585.

Newton, L.G. and Wheatley, C.H. (1970) The occurrence and spread of ephemeral fever of cattle in Queensland. *Australian Veterinary Journal* **46**, 561-568.

Nicholls, N. (1986) A method for predicting Murray Valley encephalitis in southeast Australia using the southern oscillation. *Australian Journal of Experimental Biology and Medical Science* **64**, 578-594.

Nicholls, N. (1993) El Niño-southern oscillation and vector-borne disease. Lancet **342**, 1284-1285.

Pascoe, R.R., St George, T.D. and Cybinski, D.H. (1978) The isolation of a Ross River virus from a horse. *Australian Veterinary Journal* **54**, 600.

Patz, J.A., Epstein, P.R. Burke, T.A. and Balbus, J.M. (1996) Global climate change and emerging infectious diseases. *Journal of the American Medical Association* **275**, 217-223.

Patz, J.A., Martens, W.J.M., Focks, D.A. and Jetten, T.H. (1998) Dengue fever epidemic potential as projected by general circulation models of global climate change. *Environmental Health Perspectives* **106**, 147-153.

Pearce, A.F. and Phillips, B.F. (1988) ENSO events, the Leeuwin Current and larval recruitment of the Western rock lobster. *Journal du Conseil International pour l'Exploration de la Mer* **45**, 13-21.

Pedgley, D.E. (1982) *Windblown pests and diseases*. Ellis Horwood Publishers, Chichester.

Phillips, D.A., Sheridan, J., Aaskov, J., Murray, J. and Weimers, M. (1992) Epidemiology of arbovirus infection in Queensland. *Arbovirus Research in Australia* **6**, 245-248.

Poidinger, M., Hall, R.A. and Mackenzie, J.S. (1996) Molecular characterisation of the Japanese encephalitis serocomplex of the flavivirus genus. *Virology* **218**, 417-421.

Reeves, W.C., Hardy, J.L., Reisen, W.K. and Milby, M.M. (1994) Potential effect of global warming on mosquito-borne arboviruses. *Journal of Medical Entomology* **31**, 323-332.

Reisen, W.K., Meyer, R.P., Presser, S.B. and Hardy, J.L. (1993) Effect of temperature on the transmission of western equine encephalomyelitis and St Louis encephalitis viruses by *Culex tarsalis* (Diptera: Culicidae). *Journal of Medical Entomology* **30**, 151-160.

Reiter, P (1988) Weather, vector biology, and arboviral recrudescence, in T.P. Monath, (ed.), *The Arboviruses: Epidemiology and Ecology, Volume I*, CRC Press Inc., Boca Raton. pp.245-255.

Reynolds, D.R., Smith, A.D., Muhhopadhyay, S., Chowdhury, A.K., De, B.K., Nath, P.S., Mondal, S.K., Das, B.H. and Mukhopadhyay, S. (1996) Atmospheric transport of mosquitoes in northeast India. *Medical and Veterinary Entomology* **10**, 185-186.

Ritchie, S.A., Phillips, D., Broom, A., Mackenzie, J.S., Poidinger, M. and Van den Hurk, A. (1997) Isolation of Japanese encephalitis virus from *Culex annulirostris* mosquitoes in Australia. *American Journal of Tropical Medicine and Hygiene* **56**, 80-84.

Russell, R.C. (1995) Arboviruses and their vectors in Australia: an update on the ecology and epidemiology of some mosquito-borne arboviruses. *Reviews of Medical and Veterinary Entomology* **83**, 141-158.

Russell, R.C. (1998a) Mosquito-borne arboviruses in Australia: the current scene and implications of climate change for human health. *International Journal for Parasitology* **28**, 955-969.

Russell, R.C. (1998b) Vectors vs.humans in Australia – who is on top down under? *Journal of Vector Ecology* **23**, 1-46.

St George, T.D. (1986a) Arboviruses infecting livestock in the Australian region. *Arbovirus Research in Australia* **4**, 23-25.

St George, T.D. (1986b) The epidemiology of bovine ephemeral fever in Australia and its economic effect. *Arbovirus Research in Australia* **4**, 281-286.

St George, T.D. and Standfast, H.A. (1989a) Bovine ephemeral fever, in T.P. Monath, (ed.), *The Arboviruses: Epidemiology and Ecology, Volume II*, CRC Press Inc., Boca Raton. pp.71-86.

St George, T.D. and Standfast, H.A. (1989b) Simbu group viruses with teratogenic potential, in T.P. Monath (ed.), *The Arboviruses: Epidemiology and Ecology, Volume IV*, CRC Press Inc, Boca Raton. pp.145-166.

Scott, T.W. (1988) Vertebrate host ecology, in T.P. Monath (ed.), *The Arboviruses: Epidemiology and Ecology, Volume I*, CRC Press Inc., Boca Raton. pp.257-280.

Seddon, H.R. (1938) The spread of ephemeral fever (three-day sickness) in Australia in 1936-37. *Australian Veterinary Journal* **14**, 90-101.

Sellers, R.F. (1980) Weather, host and vector – their interplay in the spread of insect-borne animal virus diseases. *Journal of Hygiene* **85**, 65-102.

Sellers, R.F. (1989) Eastern equine encephalitis in Quebec and Connecticut, 1072: introduction by infected mosquitoes on the wind? *Canadian Journal of Veterinary Research* **53**, 76-79.

Sellers, R.F. and Maarouf, A.R. (1989) Trajectory analysis and bluetongue virus serotype 2 in Florida, 1082. *Canadian Journal of Veterinary Research* **53**, 100-102.

Sellers, R.F. and Maarouf, A.R. (1990) Trajectory analysis of winds and eastern equine encephalitis in USA, 1980-85. Epidemiology and Infection **104**, 329-343.

Sellers, R.F. and Maarouf, A.R. (1991) Possible introduction of epizootic hemorrhagic disease of deer virus (serotype 2) and bluetongue virus (serotype 11) into British Columbia in 1987 and 1988 by infected Culicoides carried on the wind. Canadian *Journal of Veterinary Research* **55**, 367-370.

Sellers, R.F. and Maarouf, A.R. (1993) Weather factors in the prediction of western equine encephalitis epidemics in Manitoba. *Epidemiology and Infection* **111**, 373-390.

Sellers, R.F. and Mellor, P.S. (1993) Temperature and the persistence of viruses in *Culicoides* spp. during adverse conditions. *Revue Scientifique et Technique de l'Office International des Epizooties* **12**, 733-755.

Standfast, H.A. and Muller, M.J. (1989) Bluetongue in Australia – an entomologist's view. *Australian Veterinary Journal* **48**, 77-80.

Standfast, H.A. and Maywald, G.F. (1992) Modelling for global changes affecting arboviruses, in T.E. Walton and B.I Osburn (eds.), *Bluetongue, African Horse Sickness and Related Orbiviruses*, CRC Press Inc., Boca Raton. pp.16-20.

Standfast, H.A., Dyce, A.L. and Muller, M.J. (1985a) Vectors of bluetongue virus in Australia. *Progress in Clinical and Biological Research* **178**, 177-186.

Standfast, H.A., Dyce, A.L. and Muller, M.J. (1985b) Vectors of bluetongue in Australia, in T.C Barber, M.S. Jochim and B.I Osburn, (eds.), *Bluetongue and Related Orbiviruses*, Alan R. Liss Inc., New York. pp.177-186.

Standfast, H.A., Dyce, A.L., St George, T.D., Cybinski, D.H. and Muller, M.J. (1979) Vectors of bluetongue virus in Australia. *Arbovirus Research in Australia* **2**, 20-28

Standfast, H.A., St George T.D. and Cybinski, D.H. (1986) Economics and epidemiology of Simbu group viruses in Australia. *Arbovirus Research in Australia* **4**, 227-228.

Sutherst, R.W. (2000) Climate variability, seasonal forecasting and invertebrate pests – The need for a synoptic view, in G.L. Hammer, N. Nicholls, and C. Mitchell (eds.), *Applications of Seasonal Climate Forecasting in Agricultural and Natural Ecosystems – The Australian Experience*. Kluwer Academic, The Netherlands. (this volume).

Turell, M.J. (1993) Effect of environmental temperature on the vector competence of *Aedes taeniorhynchus* for Rift Valley fever and Venezuelan encephalitis viruses. *American Journal of Tropical Medicine and Hygiene* **49**, 672-676.

Uren, M.F. (1989) Bovine ephemeral fever. *Australian Veterinary Journal* **66**, 233-236.

Voice, M., Brewster, A. and Skinner, C. (1990) El Nino, climate change and some health implications. Australian Meteorological and Oceanographic Society Newsletter, **3**, 102-110.

Voice, M., Brewster, A. and Skinner, C. (1993) El-Nino-Southern Oscillation: climatic variability and climate change, in C.E Ewan, E.A. Bryant, G.D. Calvert, and J.A. Garrick, (eds.), *Health in the Greenhouse*, Australian Government Publishing Service, Canberra. pp. 223-228.

Walker, A.R. and Davies, F.G. (1971) A preliminary survey of the epidemiology of bluetongue in Kenya. *Journal of Hygiene* **69**, 47-60.

Ward, M.P. (1994) Climatic factors associated with the prevalence of bluetongue virus infection of cattle herd in Queensland, Australia. *Veterinary Record* **134**, 407-410.

Ward, M.P. (1995) Bluetongue and Douglas virus activity in New South Wales in 1989: further evidence for long-distance dispersal of the biting midge *Culicoides brevitarsis*. *Australian Veterinary Journal* **72**, 197-198.

Ward, M.P. (1996) Climatic factors associated with the infection of herds of cattle with bluetongue viruses. *Veterinary Research Communications* **20**, 273-283.

Woodring, J.L., Higgs, S. and Beaty, B.J. (1996) Natural cycles of mosquito-borne pathogens, in B.J. Beaty and W.C. Marquardt (eds.) *The Biology of Disease Vectors*, University Press of Colorado. pp.51-72.

APPLYING SEASONAL CLIMATE FORECASTS IN AGRICULTURAL AND NATURAL ECOSYSTEMS – A SYNTHESIS

GRAEME HAMMER

Agricultural Production Systems Research Unit (APSRU)
Queensland Depts. Primary Industries and Natural Resources and
CSIRO Tropical Agriculture
PO Box 102, Toowoomba, Qld 4350, Australia

Abstract

The chapters contributed to this book express a wealth of knowledge and experience on generating and applying seasonal climate forecasts in agricultural and natural ecosystems. In the introductory section the impact of climate variability and the potential for using seasonal forecasts in managing climate risks were outlined. It was made clear that climate forecasts must be phrased as probabilistic statements. The concept of a general integrated systems approach to applying forecasts across a range of scale from farm to region and nation was introduced. This concept focussed on deriving the targeted information needs of decision-makers in a manner that captured the value of the climate forecast. It was suggested that, as the climate forecast only acquired value when decisions were modified in response to it, then a focus on decision-making in the target systems was an appropriate basis to structure the contributions to this book. While systems across the range of scale involved diverse decision-makers and issues ranging from resource management to policy, the general approach was relevant in all cases. The aim of this paper is to synthesise key issues from those contributions and to discuss them in a way that might guide future activities.

The contributions reviewed developments in seasonal climate forecasting and presented details of on-going work aimed at applying climate forecasts in relation to farm, regional, and national scale agricultural decisions and possibilities in natural systems. Numerous issues were raised from the insights presented. Five key issues are presented and discussed here as they represent the most common threads. These issues are –

- Understanding and predicting responses of the target system is critical
- Applications are about managing risks
- Information must be relevant to decision-makers
- Communication of probabilistic information can be problematic
- Connecting agricultural and climate models needs to be considered

In discussing these key issues the major general need was a focus on implications of decision options in the target system in a way that the unexplained variability of the

G.L. Hammer et al. (eds.), The Australian Experience, 453–462.

imprecise forecast was retained so that risk implications for modified decisions could be explored transparently and communicated simply and effectively. It was suggested that coupling of the key players in the entire applications milieu - practitioners, analysts, modellers, researchers - was essential for future progress in this regard. The various players each have their own models and behaviours, which contribute some unique elements. While these models have a degree of utility in their own right, it is argued that the dialogue they generate among the disparate players is equally important to effective applications of climate forecasting.

The degree of connectivity required among key players requires an environment where collaboration is encouraged and facilitated. This environment has tended to prevail in Australia and complements the integrated systems approach to effective applications of seasonal climate forecasting that has developed. It is suggested, however, that nurturing this emerging culture of connectivity is paramount for future progress.

1. Introduction

The chapters contributed to this book document a wealth of knowledge and experience relevant to the subject of generating and applying seasonal climate forecasts in agricultural and natural ecosystems. They also identify a range of critical issues and present some views about these issues. The intent of this chapter is to synthesise the key issues arising from the body of work presented in this book and discuss them in a way that might assist in deriving critical lessons for the future. In particular, I will be seeking unifying principles and concepts that might guide future activities.

The introductory section of the book was designed to set the scene for the sections that followed. White (Chapter 1) and Plant (Chapter 2) both described the large impact of climate variability on Australian systems at national and farm scales. They both argued that seasonal climate forecasts have considerable potential to assist in managing risks associated with this climate variability. Both authors noted, however, that the process of application was complex and there were significant issues of communication to overcome, in particular in relation to probabilistic information. In reviewing the scientific basis underpinning climate forecasts based on ENSO, Cane (Chapter 3) reinforced the notion that to be correct a forecast **must** be phrased as a probabilistic statement. He also added that on-going research on dynamic climate models and climate forcing factors other than ENSO would rapidly add to current knowledge and forecasting capacity. To conclude the introductory section, Hammer (Chapter 4) discussed the need for a decision focus as part of a general systems approach to applying seasonal climate forecasting. The distinction between climate impact and application of a forecast was clarified. Impact measures the extent of effect, whereas application is the management response to change the anticipated impact. The point was made that forecasts only have value if they change decisions in a way that improves outcomes. In the general systems approach, the climate forecast is placed in the context of all the factors influencing the system of interest. The focus becomes decision-makers, their decision options, and the information they require for informed decisions, including the potential use of seasonal climate forecasts. The use of models of target systems in simulation studies provides the means to generate discussions with the decision-maker about decision options for effective use of a climate forecast.

The essential components of an effective systems approach are the knowledge and understanding from research, integration and options analysis from modelling, and the insights and needs from practitioners (Figure 1). The interactions of these components and the connecting of research and application approaches, rather than their separation, is the quintessence of the systems approach. The implementation of this conceptual approach can occur at a range of scale and with a range of decision-makers (see Figure 6 (colour plate B), Chapter 4). The concept embodied in this figure was used to organise the subsequent sections of the book. The implementation of the general systems approach was evident to varying degrees in many of the studies presented in those sections and provides a basis for framing the discussion of key issues identified.

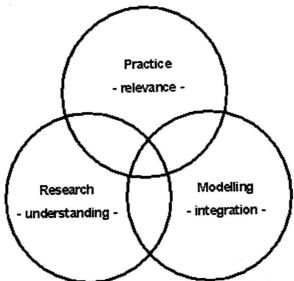

Figure 1. The conceptual basis of a systems approach showing the essential interactions among researchers, modellers, and practitioners.

In pursuing the generation and application of seasonal climate forecasts in agricultural and natural ecosystems throughout this book, five significant issues emerged –
• Understanding and predicting responses of the target system is critical
• Applications are about managing risks
• Information must be relevant to decision-makers
• Communication of probabilistic information can be problematic
• Connecting agricultural and climate models needs to be considered
I will discuss each of these issues as a means to lead to some thoughts that might help to guide approaches and activities in the future.

2. Understanding and Predicting the Target System

Effective applications of climate forecasts depend on responses to climate of relevant features in the target system and whether these responses are of sufficient importance to affect the behaviour of decision-makers. Many of the studies in Sections III, IV, and V (eg. Carberry *et al.*, Chapter 12; McKeon *et al.*, Chapter 15; Stafford Smith *et al.*, Chapter 17; Stephens *et al.*, Chapter 21; Chiew *et al.*, Chapter 25) presented details about biophysical and economic models as the critical tools to capture understanding of the target system dynamics in a manner that enabled prediction of consequences of change in climatic input and management.

Simulation studies with system models were critical for examining possible scenarios for responding to seasonal climate forecasts. But, apart from the need to predict system responses, the models have another important integrating role – they provide the basis for the dialectic among scientists, analysts, and practitioners (see Meinke and Hochman, Chapter 11). Models require scientists to structure and quantify their thoughts about system dynamics. This provides an avenue for testing hypotheses about system processes and hence, for debate among scientists, about ideas and concepts. It also assists in identifying areas where knowledge is lacking. If this discussion were left to the scientists alone though, it would miss some critical synergies. By introducing the views of analysts and practitioners, the debate becomes more aligned to purpose and parsimony in modelling. Some of the subtleties of the scientific debate about process may not be relevant. In turn, some of the practitioners may not really know what they want or what is possible. The model of the target system provides the focus for connecting the players in an iterative process of learning and discovery. The interaction forces consideration of different views and potential trade-offs.

Not all systems are sufficiently well understood to allow development of credible models. This is perhaps noted most often in Section V in chapters relating to natural systems (eg. Sutherst, Chapter 23; Mackenzie *et al.*, Chapter 26). The models, however, do not need to be comprehensive or complex to be useful. They can be based on sound observation and description (eg. Limpus and Nicholls, Chapter 24; Chiew *et al.*, Chapter 25; Stephens *et al.*, Chapter 21) combined with a good general understanding of system dynamics. As noted above, the discussion generated across players around a modelling process is likely to be as valuable as specific improvements in a particular model might be. For example, starting with a broad overview of system dynamics (eg. Sutherst, Chapter 23; Johnston *et al.*, Chapter 14) provides context for any modelling activity.

The main point is that modelling of the target system provides a means to predict responses to climate forecasts and management behaviours at the same time as providing the basis for discursive relationships among key players in the applications effort. Investment in target system modelling is a critical component of a program concerned with effective applications of seasonal climate forecasting.

3. Applications and Managing Risks

Applying seasonal forecasts in agricultural and natural ecosystems is concerned with risk management. Cane (Chapter 3) made it clear that seasonal climate forecasts must be phrased as probabilistic statements and chapters in Section II reinforce this (eg. Stone and de Hoedt, Chapter 5). White (Chapter 1) discusses the need to view application of seasonal forecasting in a risk management context, describes some economic tools that are commonly used in risk analysis, and reviews their use in relation to applying climate forecasting. Forecasts provide opportunities to adjust management to increase profits or reduce financial and environmental risks. The major consequence of the probabilistic nature of a forecast, however, is that modifying management in response to the forecast can, depending on the subsequent season that eventuates, generate either a gain or a loss when compared to not modifying. That is, there is a clear risk associated with modifying management on any single occasion even though there might be an advantage in the long run. Hammer (Chapter 4, see Figure 5) gives a good example of this point in describing a simple application in field crop management. Carberry *et al.* (Chapter 12) and Ash *et al.* (Chapter 16) present comprehensive case studies using a risk analysis approach that includes issues of resource condition as well as profitability. In some instances, reduced risks of financial loss or resource degradation may be as important as chance of increased profit when valuing the potential for applying seasonal climate forecasts.

White (Chapter 1) suggests that simple cost/loss approaches assuming risk indifference of decision-makers (see Table 5, Chapter 1), or simpler derivations from them, were appropriate in presenting risk analysis concepts to practitioners as they minimised complexity. Such approaches align a season type with an average outcome on the system variables of direct interest to decision-makers (eg. profit) and then consider the probability of a specific season type for a given forecast. Whilst superficially attractive, this approach can be misleading as it aligns a specific system outcome with a forecast season type. As a number of simulation studies in subsequent chapters show, this is rarely the case. In fact, by taking this approach, the information on risk that has been most relevant in discussion of applications of forecasting with practitioners is removed (see Meinke and Hochman, Chapter 11). This information relates to the distribution of outcomes that are likely within a forecast season type due to the unexplained variation arising from the imprecise nature of the forecast. This arises because factors such as the distribution of rainfall in a season can have greater consequences in some circumstances on the system variables of interest than the seasonal rainfall total. There are many other confounding interactions in the target systems that make simple generalisations from forecast season type to system outcome a matter of concern. The more comprehensive approach that considers the entire set of possible consequences enables improved analysis of risk and greater transparency in communicating risks associated with decision options. It does not necessarily involve increased complexity in communication. I will return to this point later when discussing communication issues.

Adopting a risk management approach requires an understanding or definition of risk. The dictionary definition of risk is exposure to the chance of loss and this appears to reflect common usage and behaviour. Carberry *et al.* (Chapter 12) use this definition to quantify decision risks in their case study. Most economic methods use measures of variance or utility functions to quantify risk, and while these methods capture some of the essence, their connection with decision-makers perceptions of risk is less obvious – a point also made by White (Chapter 1).

Hence, while it is clear that applications require a risk management approach there remain issues about approaches and communication to resolve. Again the case studies and models provide a robust means to foster the dialectic across the relevant disciplines that is needed.

4. Relevance of Information to the Decision-maker

Effective applications require targeted information on likely consequences of a climate forecast on system variables of interest to the decision-maker. In most cases, a probabilistic climate forecast, on its own, is not that useful. Most of the research and development reported on the applications presented in this book is about dealing with this non-trivial issue.

While system understanding and modelling is needed to derive targeted information, it is the interaction with the decision-maker that is critical in ensuring we are aiming at the right target. A broad systems analysis (eg. Johnston *et al.*, Chapter 14; Stafford Smith *et al.*, Chapter 17; Sutherst, Chapter 23) provides an ideal starting point for this interaction but cannot replace it. The developing of models and exercising of them in pursuit of this information requires considerable time and effort, beyond the investment in the underpinning knowledge (eg. Carberry *et al.*, Chapter 12; McKeon *et al.*, Chapter 15; Carter *et al.*, Chapter 20; Chiew *et al.*, Chapter 25). It is the iterations of the systems and simulation analyses with the decision-maker, however, that enforces relevance of the information. The detailed policy example of Carter *et al.* (Chapter 20) sets out procedures to derive targeted information required by policy advisors. Meinke and Hochman (Chapter 11) and Stafford Smith et al. (Chapter 17) also address the issue of relevance and interaction with the decision-maker. The earlier this interaction is introduced the more efficient the overall application will become.

Although a seasonal climate forecast offers new opportunities, it is just another component to be considered in managing or developing policy for target systems. It is not a replacement for lack of knowledge or management skill. Skilled management enables integration of all factors impinging on the system of interest. Effective application requires high levels of management skill as its value may be best captured by complementing more sophisticated management approaches, such as responsive management. Responsive management requires monitoring of the status of key variables in the system, such as soil water, pasture condition, or population levels and adjusting decisions accordingly (see discussions in White, Chapter 1; Ash *et al.,* Chapter 16; Sutherst, Chapter 23).

While involvement of the decision-maker is important, it must also be realised that practitioners may not know what they want. In the same way that the scientist and analyst may not be aware of what is most important to the decision-maker, the practitioner may not be aware of what is possible. The interactions and iterations around a model or analysis provide a means for demand and supply to merge in creative ways. For example, the study presented by Chapman *et al.* (Chapter 22) to explore influences of seasonal climate forecasts on grain prices was developed following discussions with practitioners on other system management issues.

5. Communication of Probabilistic Information

A number of contributions to this book have highlighted issues associated with communicating probabilistic information (White, Chapter 1; Clewett *et al.,* Chapter 18; Nicholls, Chapter 19) and have presented some of the research in this area. They have also included detailed discussions around the issue. While all of these contributions mention interactions of the climate forecast with the intricacies of the target system, they tend to focus on communication issues about the climate forecast. Perhaps a focus on the target rather than the input might assist the communication process.

The main issue concerns communicating forecast and outcome uncertainty, while retaining credibility with decision-makers. While there are clearly difficulties with communicating probabilistic information associated with framing, biases, etc (see White, Chapter 1; Nicholls, Chapter 19), there is some experience to show that in cases where direct interaction with decision-makers is possible, these difficulties are not great. In communicating outcome uncertainty it is possible to compare, on a year-by-year basis, the difference in outcome between responding to a forecast or not responding. This can be presented as simple bar graphs or summarised into frequency tables, but the former is simpler and much more valuable as a tool for discussion about outcome uncertainty. White (Chapter 1) makes the point about needing to know about false positives and false negatives, but this point is implicit and obvious in the approach just outlined. Further it avoids the confounding language about a forecast being right or wrong, which is inappropriate as the forecast is a probabilistic statement not a prediction of a particular state. While avoiding most of the language about probability and still embodying all of the concepts, this simple comprehensive approach also uses the targeted information (ie. the relevant target system outcome) that is of most interest to the decision-maker. Meinke and Hochman (Chapter 11) present examples of this approach in describing their interactions with farmers in responding to seasonal climate forecasts.

It may be easier to resolve communication issues relating to specific applications than it is to deal with less targeted communication, such as that associated with the general seasonal climate forecast. The lessons from the former, however, may well inform the latter. Rather than using statements of probability, with all their inherent interpretation difficulties, simple bar graphs showing occurrences of seasonal rainfall in historical analogues and their relation to the long-term mean or median might be more appropriate. The main point is to move the message away from an attempt to get it "right" to one that communicates the forecast uncertainty in a way that is transparent and simple.

Much of the general communication issue arises from approaches adopted in the media, which often seek to either sensationalise a point or to magnify differences. Climate forecast information is available from many sources and the only way to deal with concerns about media reporting is to establish credibility via relationships with key players in the media and be proactive. This could involve educative approaches as a means to develop a better-informed media. But it is not all one-way. The media provide significant opportunities to explain to the community the value of applications of seasonal climate forecasting, so it is in the interest of all involved to establish collaborative relationships.

Many concerns about communication of new technologies can be relieved over time via education. Clewett *et al.* (Chapter 18) describe adult education processes that have been developed to build knowledge and skills in rural communities concerning seasonal climate forecasts. These activities are essential to build awareness and understanding as a basis for considering effective applications. Placing these learning processes in a property management planning context facilitates dialogue about relevance and potential use of forecasts in that context.

6. Connecting Agricultural and Climate Models

Seasonal climate forecasting capability is dependent on scientific understanding of the world's climate systems. A number of contributors (Cane, Chapter 4; Hunt and Hirst, Chapter 7; Nicholls *et al.,* Chapter 8) have made it clear that dynamic climate models will play an increasing role in climate forecasting in the future. Although it is likely to still be sometime before model-based forecasts show superior skill to statistical systems (Nicholls *et al.,* Chapter 8; Drosdowsky and Allen, Chapter 6; Stone and de Hoedt, Chapter 5), issues such as interaction with climate trends (see McKeon et al., Chapter 15) will hasten the transition to climate models.

Irrespective of the origin of the climate forecast, the main issues of relevance in applications are the skill of the forecast in relation to impact on key decisions and the ability to utilise the forecast in analysing its application to target systems. The skill of most forecasting systems can be measured against climatological variables using hindcast procedures as shown by Drosdowsky and Allan (Chapter 6) for the forecasting system based on analysis of global SST's. They showed significant improvement in skill over simple correlation with the SOI alone. Such analyses provide a useful initial screening of various systems, but it is important to compare new systems against benchmarks of advanced existing systems. The SOI phase system used in many contributions in this book (see Stone and de Hoedt, Chapter 5) is known to be superior to simple correlation with the SOI alone, so a comparison using it as the benchmark would be considerably more informative and useful.

Although forecast skill in relation to raw climate variables is important, the relevance of this skill to decisions in the target system is arguably more so, as it includes target system sensitivities to the forecast. The contributions by Hammer *et al.* (Chapter 13) and Ash *et al.* (Chapter 16) compare the value of a range of forecasting systems in cropping and pasture systems. In addition to providing a measure of extent of relevance of forecast skill, these studies can also assist in identifying attributes of forecasts that are most useful. Such feedback is invaluable in targeting and design of improved forecasting systems as it goes considerably beyond the general notions of improved skill captured in general surveys of decision-makers.

The ability to connect a forecasting system to an analysis of decision-making in the target system is critical to the utility of the forecast. Many of the example applications in this book (Sections III, IV, and V) use the SOI-phase forecasting system because it generates historical seasonal analogues that can be used readily with target system models, which often require daily climate data as input. This enables analysis that retains the unexplained variability of the imprecise forecast in a way that risk implications for modified decisions can be explored transparently. The issue of connecting climate forecasting systems that have not been developed with an awareness of this need (such as global climate modelling systems) to target systems analyses is addressed by Bates *et al.* (Chapter 9) and Stone *et al.* (Chapter 10). These types of innovative approaches will be essential as reliance on dynamic models increases. Again, the dialogue among the players is as important as the models in this process and may ultimately generate changes in approach to modelling (of climate and target systems) rather than just clever interfaces.

7. The Way Ahead – The Emerging Culture of Connectivity

Throughout the discussion of issues arising from contributions to this book, it has become increasingly clear that coupling of the key players in the entire applications milieu **is** the way ahead. Given a shared goal of effectively developing and applying seasonal climate forecasts, the various models used by the key players provide the means to achieve this connectivity. The key groups represented in Figure 1 each have their own models and behaviours. The practitioners use simple heuristic models integrating the many factors in the system they need to deal with simultaneously – White's breadth not depth (Chapter 1). The target and climate system modellers seek explanatory models and systems analyses to capture understanding and predict consequences in ways that heuristic models cannot, but they sacrifice some realism and breadth for the increase in insight. The researchers look for ways to increase understanding of system dynamics and thus add to the knowledge base in areas they see as either relevant or interesting. While each of these descriptions would likely be accepted within each group, they clearly differ. So how is coupling going to help?

Coupling provides the means to generate the dialogue across these disparate models and behaviours that is needed for effective applications of climate forecasting. Each player has something useful to contribute but doesn't know it all – viz. Nicholls (Chapter 19) - ensuring that a climate forecast is useful is considerably more complicated and multi-faceted than is generally believed by climate scientists. One could add - it is likely that the attributes of climate forecasts are not as simple as is generally believed by practitioners. By confronting the disparity in a collaborative manner, a better appreciation of the entire issue and the roles of the various players can be gained by all. The practitioners provide the coupling to the real world; the modellers provide access to enhanced and possibly counterintuitive insight; and the researchers may open up new possibilities – but each needs to consider the view of the other in relation to their common goal. This can generate valuable interaction, leading to introspection about priorities of all involved, and hopefully to efficient synergistic action.

The simulation models developed and used by the players have a critical dual purpose. They can provide useful predictions of the biophysical and economic components of the system, while creating a need for dialogue among the players. The issues of developing methods to connect climate and agricultural models and of seeking improved communication processes between analysts and practitioners are two examples arising from this dialogue. In some instances the simulation models are less important than the discussions they generate. Once back into the breadth of the real world, with various sociopolitical dimensions, the models play a subservient role in the discussions about feasible decision options. But the point is that they are useful in generating and informing some aspects of the discussion.

This notion of connectivity among key players fits comfortably with the integrated systems approach as outlined in the introduction (Hammer, Chapter 4) and reinforced by other contributors (eg. Stafford Smith *et al.,* Chapter 17; Nicholls, Chapter 19). The integrated approach requires interdisciplinary action among specialists as well as interaction with practitioners. Given the complexities of the climate and target systems, the individuals involved reside in numerous organisations with a broad range of cultures and policies. This has the potential to generate tension among organisations and individuals unless an openness to collaborate is encouraged and facilitated. In Australia, this openness has tended to prevail and the consequent discussions have generated significant advance. However, although much remains to be done technically, as we move forward in the effective application of climate forecasts in agricultural and natural ecosystems, nurturing this emerging culture of connectivity is paramount.

COLOUR PLATE SECTION

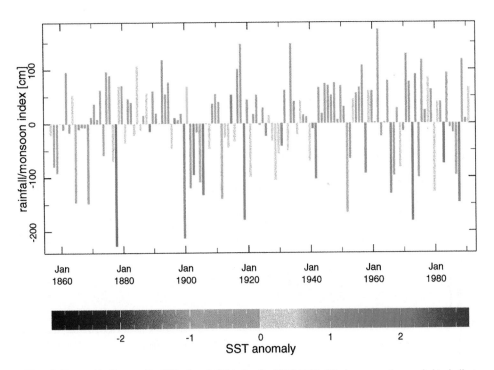

Plate A (Figure 10, Chapter 3). All India rainfall index for 1856-1988. The bars are colour coded to indicate the strength of the SST anomaly in the NINO3 area.

466

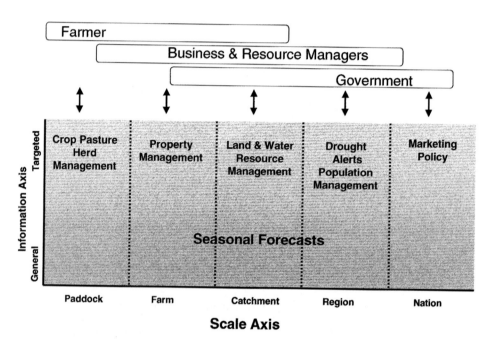

Plate B (Figure 6, Chapter 4). The relationships between scale, information content, and decision-makers in defining relevant systems and the systems approach to applying seasonal climate forecasts in agricultural and natural ecosystems.

Probability of Exceeding Median Growth
QLD - February to April 1998

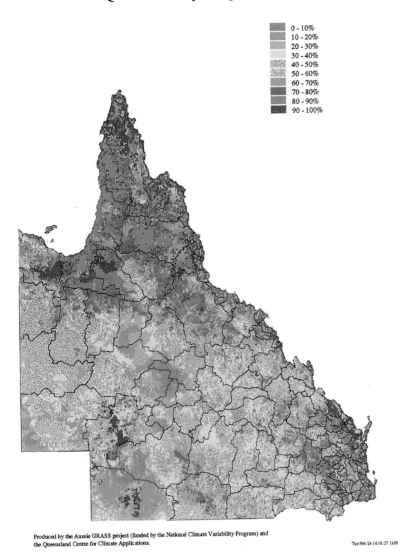

▮	0 - 10%
▮	10 - 20%
▮	20 - 30%
▮	30 - 40%
▮	40 - 50%
▮	50 - 60%
▮	60 - 70%
▮	70 - 80%
▮	80 - 90%
▮	90 - 100%

Produced by the Aussie GRASS project (funded by the National Climate Variability Program) and
the Queensland Centre for Climate Applications.
Tue Feb 24 14:01:27 1998

Plate C (Figure 4, Chapter 20). Probability of exceeding median pasture growth for the period February-
April 1998. Calculations were performed at the beginning of February 1998 using analogue climate years
from the historical climate record selected according to the Southern Oscillation Index phase system of Stone
and Auliciems (1992).

COMBINING SATELLITE DATA AND MODEL OUTPUTS
" SPATIAL VIEW "

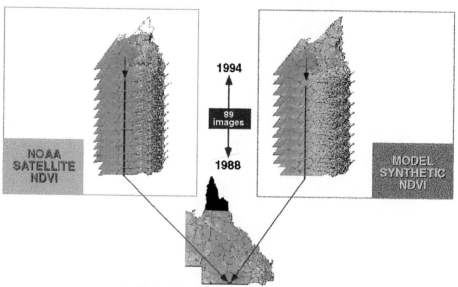

CORRELATION MAP, SATELLITE & MODEL

Plate D (Figure 6, Chapter 20). Spatial view of Aussie GRASS validation using NOAA NDVI imagery (1988-1994). Model calculated NDVI values for each pixel have been compared against the time series of 89 NOAA images. The black area on Cape York represents an area of no coverage.

Wool Gross Margins - 1995
(cents / ha, Base 1992-93)

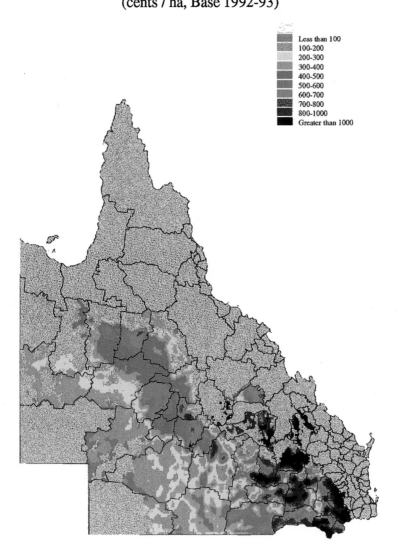

Less than 100
100-200
200-300
300-400
400-500
500-600
600-700
700-800
800-1000
Greater than 1000

Plate E (Figure 8, Chapter 20). Gross margin for Queensland wool production for 1995 calculated using the Aussie GRASS spatial analysis system.

ATMOSPHERIC AND OCEANOGRAPHIC SCIENCES LIBRARY

1. F.T.M. Nieuwstadt and H. van Dop (eds.): *Atmospheric Turbulence and Air Pollution Modelling.* 1982; rev. ed. 1984
ISBN 90-277-1365-6; Pb (1984) 90-277-1807-5
2. L.T. Matveev: *Cloud Dynamics.* Translated from Russian. 1984
ISBN 90-277-1737-0
3. H. Flohn and R. Fantechi (eds.): *The Climate of Europe: Past, Present and Future.* Natural and Man-Induced Climate Changes: A European Perspective. 1984
ISBN 90-277-1745-1
4. V.E. Zuev, A.A. Zemlyanov, Yu.D. Kopytin, and A.V. Kuzikovskii: *High-Power Laser Radiation in Atmospheric Aerosols.* Nonlinear Optics of Aerodispersed Media. Translated from Russian. 1985
ISBN 90-277-1736-2
5. G. Brasseur and S. Solomon: *Aeronomy of the Middle Atmosphere.* Chemistry and Physics of the Stratosphere and Mesosphere. 1984; rev. ed. 1986
ISBN (1986) 90-277-2343-5; Pb 90-277-2344-3
6. E.M. Feigelson (ed.): *Radiation in a Cloudy Atmosphere.* Translated from Russian. 1984
ISBN 90-277-1803-2
7. A.S. Monin: *An Introduction to the Theory of Climate.* Translated from Russian. 1986
ISBN 90-277-1935-7
8. S. Hastenrath: *Climate Dynamics of the Tropics*, Updated Edition from *Climate and Circulation of the Tropics.* 1985; rev. ed. 1991
ISBN 0-7923-1213-9; Pb 0-7923-1346-1
9. M.I. Budyko: *The Evolution of the Biosphere.* Translated from Russian.
1986
ISBN 90-277-2140-8
10. R.S. Bortkovskii: *Air-Sea Exchange of Heat and Moisture During Storms.* Translated from Russian, rev. ed. 1987
ISBN 90-277-2346-X
11. V.E. Zuev and V.S. Komarov: *Statistical Models of the Temperature and Gaseous Components of the Atmosphere.* Translated from Russian. 1987
ISBN 90-277-2466-0
12. H. Volland: *Atmospheric Tidal and Planetary Waves.* 1988   ISBN 90-277-2630-2
13. R.B. Stull: *An Introduction to Boundary Layer Meteorology.* 1988
ISBN 90-277-2768-6; Pb 90-277-2769-4
14. M.E. Berlyand: *Prediction and Regulation of Air Pollution.* Translated from Russian, rev. ed. 1991
ISBN 0-7923-1000-4
15. F. Baer, N.L. Canfield and J.M. Mitchell (eds.): *Climate in Human Perspective.* A tribute to Helmut E. Landsberg (1906-1985). 1991
ISBN 0-7923-1072-1
16. Ding Yihui: *Monsoons over China.* 1994
ISBN 0-7923-1757-2
17. A. Henderson-Sellers and A.-M. Hansen: *Climate Change Atlas.* Greenhouse Simulations from the Model Evaluation Consortium for Climate Assessment. 1995
ISBN 0-7923-3465-5
18. H.R. Pruppacher and J.D. Klett: *Microphysics of Clouds and Precipitation,* 2nd rev. ed.
1997
ISBN 0-7923-4211-9; Pb 0-7923-4409-X
19. R.L. Kagan: *Averaging of Meteorological Fields.* 1997   ISBN 0-7923-4801-X
20. G.L. Geernaert (ed.): *Air-Sea Exchange: Physics, Chemistry and Dynamics.* 1999
ISBN 0-7923-5937-2
21. G.L. Hammer, N. Nicholls and C. Mitchell (eds.): *Applications of Seasonal Climate Forecasting in Agricultural and Natural Ecosystems.* 2000   ISBN 0-7923-6270-5

KLUWER ACADEMIC PUBLISHERS – DORDRECHT / BOSTON / LONDON